MEASURES, INTEGRALS AND MARTINGALES

D0165862

This is a concise and elementary introduction to measure and integration theory as it is nowadays needed in many parts of analysis and probability theory. The basic theory – measures, integrals, convergence theorems, L^p-spaces and multiple integrals – is explored in the first part of the book. The second part then uses the notion of martingales to develop the theory further, covering topics such as Jacobi's general transformation theorem, the Radon–Nikodým theorem, differentiation of measures, Hardy–Littlewood maximal functions or general Fourier series.

Undergraduate calculus and an introductory course on rigorous analysis in $\mathbb{R}$ are the only essential prerequisites, making this text suitable for both lecture courses and for self-study. Numerous illustrations and exercises are included, and these are not merely drill problems but are there to consolidate what has already been learnt and to discover variants, sideways and extensions to the main material. Hints and solutions will be available on the internet.

RENÉ SCHILLING is Professor of Stochastics at the University of Marburg.

MEASURES, INTEGRALS AND MARTINGALES

RENÉ L. SCHILLING

CAMBRIDGE
UNIVERSITY PRESS

CAMBRIDGE UNIVERSITY PRESS
Cambridge, New York, Melbourne, Madrid, Cape Town, Singapore, São Paulo

CAMBRIDGE UNIVERSITY PRESS
The Edinburgh Building, Cambridge CB2 2RU, UK

Published in the United states of America by Cambridge University Press, New york

www.cambridge.org
Information on this title: www.cambridge.org/9780521850155

First published 2005

Printed in the United Kingdom at the University Press, Cambridge

A catalogue record for this publication is available from the British Library

Library of Congress Cataloguing in Publication data

ISBN-13 978-0-521-85015-5 hardback
ISBN-10 0-521-85015-0 hardback
ISBN-13 978-0-521-61525-9 paperback
ISBN-10 0-521-61525-9 paperback

Contents

v

Prelude

The purpose of this book is to give a straightforward and yet elementary introduction to measure and integration theory that is within the grasp of second or third year undergraduates. Indeed, apart from interest in the subject, the only prerequisites for Chapters 1–13 are a course on rigorous ϵ-δ-analysis on the real line and basic notions of linear algebra and calculus in $\mathbb{R}^n$. The first few chapters form a concise (not to say minimalist) introduction to Lebesgue's approach to measure and integration, based on a 10-week, 30-hour lecture course for Sussex University mathematics undergraduates. Chapters 14–24 are more advanced and contain a selection of results from measure theory, probability theory and analysis. This material can be read linearly but it is also possible to select certain topics; see the dependence chart on page xi. Although more challenging than the first part, the prerequisites stay essentially the same and a reader who has worked through and understood Chapters 1–13 will be well prepared for all that follows. At some points, one or another concept from point-set topology will be (mostly superficially) needed; those readers who are not familiar with the topic can look up the basic results in Appendix B whenever the need arises.

Each chapter is followed by a section of *Problems*. They are not just drill exercises but contain variants, excursions from and extensions of the material presented in the text. The proofs of the core material do not depend on any of the problems and it is an exception that I refer to a problem in one of the proofs. Nevertheless I do advise you to attempt as many problems as possible. The material in the *Appendices* – on upper and lower limits, basic topology and the Riemann integral – is primarily intended as back-up, for when you want to look something up.

Unlike many textbooks this is not an *introduction to integration for analysts* or a *probabilistic measure theory*. I want to reach both (future) analysts and (future) probabilists, and to provide a foundation which will be useful for both communities and for further, more specialized, studies. It goes without saying that I have to leave out many pet choices of each discipline. On the other hand, I try to intertwine the subjects as far as possible, resulting – mostly in the latter part of the book – in the consequent use of the martingale machinery which gives 'probabilistic' proofs of 'analytic' results.

Measure and integration theory is often seen as an abstract and dry subject, disliked by many students. There are several reasons for this. One of them is certainly the fact that measure theory has traditionally been based on a thorough knowledge of real analysis in one and several dimensions. Many excellent textbooks are written for such an audience but today's undergraduates find it increasingly hard to follow such tracts, which are often more aptly labelled *graduate* texts. Another reason lies within the subject: measure theory has come a long way and is, in its modern purist form, stripped of its motivating roots. If, for example, one starts out with the basic definition of measures, it takes unreasonably long until one arrives at interesting examples of measures – the proof of existence and uniqueness of something as basic as Lebesgue measure already needs the full abstract machinery – and it is not easy to entertain by constantly referring to examples made up of delta functions and artificial discrete measures. I try to alleviate this by postulating the existence and properties of Lebesgue measure early on, then justifying the claims as we proceed with the abstract theory.

Technically, measure and integration theory is no more difficult than, say, complex function theory or vector calculus. Most proofs are even shorter and have a very clear structure. The one big exception, Carathéodory's extension theorem, can be safely stated without proof since an understanding of the technique is not really needed at the beginning; we will refer to the details of it only in Chapter 14 in connection with regularity questions. The other exception is the (classical proof of the) Radon–Nikodým theorem, but we will follow a different route in this book and use martingales to prove this and other results.

I am grateful to all students who went to my classes, challenged me to write, rewrite and improve this text and who drew my attention – sometimes unbeknownst to them – to many weaknesses. I owe a great debt to the patience and interest of my colleagues, in particular to Niels Jacob, Nick Bingham, David Edmunds and Alexei Tyukov who read the whole text, and to Charles Goldie and Alex Sobolev who commented on large parts of the manuscript. Without their encouragement and help there would be more obscure passages, blunders and typos in the pages to follow. It is a pleasure to acknowledge the interest and skill of the Cambridge University Press and its editor, Roger Astley, in the preparation of this book.

A few words on notation before getting started. I tried to keep unusual and special notation to a minimum. However, a few remarks are in order: $\mathbb{N}$ means the natural numbers $\{1, 2, 3, \ldots\}$ and $\mathbb{N}_0 := \mathbb{N} \cup \{0\}$. *Positive* or *negative* is always understood in non-strict sense $\geqslant 0$ or $\leqslant 0$; to exclude 0, I say *strictly positive/negative*. A '$+$' as sub- or superscript refers to the positive part of a function or the positive members of a set. Finally, $a \vee b$ resp. $a \wedge b$ denote the maximum resp. minimum of the numbers $a, b \in \mathbb{R}$. For any other general notation there is a comprehensive index of notation at the end of the book.

In some statements I indicate alternatives using square brackets, i.e., '*if A [B] ... then P [Q]*' should be read as '*if A ... then P*' and '*if B ... then Q*'. The end of a proof is marked by Halmos' 'tombstone' symbol ∎, and Bourbaki's 'dangerous bend' symbol ⚡ in the margin identifies a passage which requires some attention.

As with every book, one cannot give all the details at every instance. On the other hand, the less experienced reader might glide over these places without even noticing that some extra effort is needed; for these readers – and, hopefully, not to the annoyance of all others – I use the symbol$^{[\checkmark]}$ to indicate where some little verification is appropriate.

Cross-referencing. Throughout the text chapters are numbered with arabic numerals and appendices with capital letters. Formulae are numbered $(n.k)$ referring to formula k from Chapter n. For theorems and the like I write $n.m$ for Theorem m from Chapter n. The abbreviation T$n.m$ is sometimes used for **Theorem** $n.m$ (with D standing for **D**efinition, L for **L**emma, P for **P**roposition and C for **C**orollary).

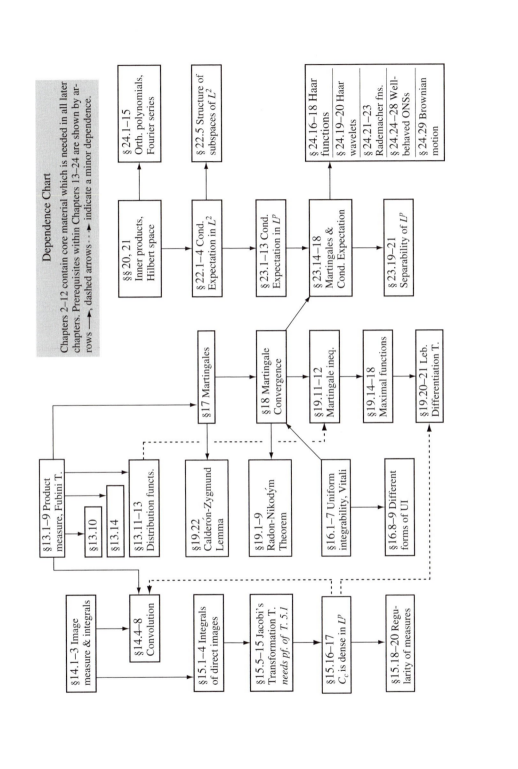

Dependence Chart

Chapters 2–12 contain core material which is needed in all later chapters. Prerequisites within Chapters 13–24 are shown by arrows ——▶, dashed arrows - - -▶ indicate a minor dependence.

§24.1–15 Orth. polynomials, Fourier series

§22.5 Structure of subspaces of L^2

§24.16–18 Haar functions

§24.19–20 Haar wavelets

§24.21–23 Rademacher fns.

§24.24–28 Well-behaved ONSs

§24.29 Brownian motion

§§20, 21 Inner products, Hilbert space

§22.1–4 Cond. Expectation in L^2

§23.1–13 Cond. Expectation in L^P

§23.14–18 Martingales & Cond. Expectation

§23.19–21 Separability of L^P

§17 Martingales

§18 Martingale Convergence

§19.11–12 Martingale ineq.

§19.14–18 Maximal functions

§19.20–21 Leb. Differentiation T.

§13.1–9 Product measure, Fubini T.

§13.10

§13.14

§13.11–13 Distribution functs.

§19.22 Calderón-Zygmund Lemma

§19.1–9 Radon-Nikodým Theorem

§16.1–7 Uniform integrability, Vitali

§16.8–9 Different forms of UI

§14.1–3 Image measure & integrals

§14.4–8 Convolution

§15.1–4 Integrals of direct images

§15.5–15 Jacobi's Transformation T. *needs pf. of T. 5.1*

§15.16–17 C_c is dense in L^P

§15.18–20 Regularity of measures

1

Prologue

The theme of this book is the problem of how to assign a size, a content, a probability, etc. to certain sets. In everyday life this is usually pretty straightforward; we

- count: $\{a, b, c, \dots, x, y, z\}$ has 26 letters;
- take measurements: length (in one dimension), area (in two dimensions), volume (in three dimensions) or time;
- calculate: rates of radioactive decay or the odds to win the lottery.

In each case we compare (and express the outcome) with respect to some base unit; most of the measurements just mentioned are intuitively clear. Nevertheless, let's have a closer look at areas:

$$\text{area} = \text{length}(\ell) \times \text{width}(w). \qquad (1.1)$$

An even more flexible shape than the rectangle is the triangle:

$$\text{area} = \frac{1}{2} \times \text{base}(b) \times \text{height}(h). \qquad (1.2)$$

1

Triangles are indeed more basic than rectangles since we can represent every rectangle, and actually any odd-shaped quadrangle, as the 'sum' of two non-overlapping triangles:

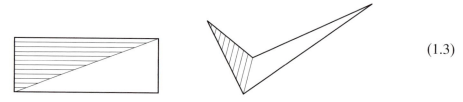

$$(1.3)$$

area = area of shaded triangle + area of white triangle.

In doing so we have *tacitly* assumed a few things. In (1.2) we have chosen *a particular* base line and the corresponding height arbitrarily. But the concept of *area* should not depend on such a choice and the calculation this choice entails. Independence of the area from the way we calculate it is called *well-definedness*. Plainly,

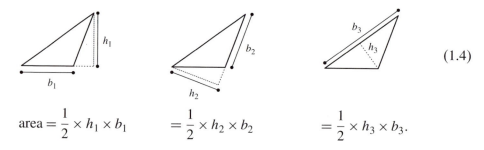

$$(1.4)$$

$$\text{area} = \frac{1}{2} \times h_1 \times b_1 \qquad = \frac{1}{2} \times h_2 \times b_2 \qquad = \frac{1}{2} \times h_3 \times b_3.$$

Notice that (1.4) allows us to pick the most convenient method to work out the area. In (1.3) we actually used two facts:

- the area of non-overlapping (disjoint) sets can be added, i.e.

$$\text{area}(A) = \alpha, \text{area}(B) = \beta, A \cap B = \emptyset \quad \Longrightarrow \quad \text{area}(A \cup B) = \alpha + \beta.$$

- congruent triangles have the same area, i.e. $\text{area}\left(\triangleright\right) = \text{area}\left(\triangle\right)$.

This shows that the least we should expect from a reasonable measure μ is that it is

$$\text{well-defined, takes values in } [0, \infty], \text{ and } \mu(\emptyset) = 0; \qquad (1.5)$$

$$\text{additive, i.e. } \mu(A \cup B) = \mu(A) + \mu(B) \text{ whenever } A \cap B = \emptyset. \qquad (1.6)$$

The additional property that the measure μ

$$\text{is invariant under congruences} \qquad (1.7)$$

turns out to be a very special property of length, area and volume, i.e. of *Lebesgue measure* on $\mathbb{R}^n$.

The above rules allow us to measure arbitrarily odd-looking *polygons* using the following recipe: dissect the polygon into non-overlapping triangles and add their areas. But what about *curved* or even more complicated shapes, say,

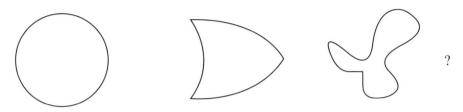

?

Here is *one* possibility for the circle: inscribe a regular 2^j-gon, $j \in \mathbb{N}$, into the circle, subdivide it into congruent triangles, find the area of each of these slices and then add all 2^j pieces.

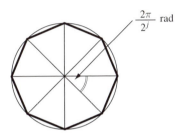

$\frac{2\pi}{2^j}$ rad

In the next step increase $j \rightsquigarrow j+1$ by doubling the number of points on the circumference and repeat the above procedure. Eventually,

$$\text{area of circle} = \lim_{j \to \infty} 2^j \times \text{area}\big(\text{triangle at step } j\big) \tag{1.8}$$

Again, there are a few problems: does the limit exist? Is it admissible to subdivide a set into arbitrarily many subsets? Is the procedure independent of the particular subdivision? In fact, nothing would have prevented us from paving the circle with ever smaller squares! For a reasonable notion of measure the answer to all of these questions should be *yes* and the way we pave the circle should not lead to different results, as long as our tiles are disjoint. However, finite additivity (1.6) is not enough for this and we have to use instead

$$(\sigma - \text{additivity}) \qquad \text{area}\Big(\biguplus_{j \in \mathbb{N}} A_j \Big) = \sum_{j \in \mathbb{N}} \text{area}(A_j) \tag{1.9}$$

where the notation $\biguplus_j A_j$ means the *disjoint union* of the sets A_j, i.e. the union where the sets A_j are pairwise disjoint: $A_j \cap A_k = \emptyset$ if $j \neq k$; a corresponding notation is used for unions of finitely many sets.

We will see that conditions (1.5) and (1.9) lead to the notion of measure which is powerful enough to cater for all our everyday measuring needs and for much more. We will also see that a good notion of measure allows us to introduce integrals, basically starting with the naïve idea that the integral of a positive function should stand for the area of the set between the graph of the function and the abscissa.

Problems

1.1. Consider the two figures below.

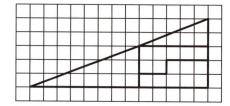 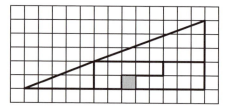

     They seem to indicate that there is no conclusive way to exhaust an area by squares (see the extra square in the second figure). Can that be?

1.2. Use (1.8) to find the area of a circle with radius r.

2

The pleasures of counting

Set algebra and countability play a major rôle in measure theory. In this chapter we review briefly notation and manipulations with sets and introduce then the notion of countability. If you are not already acquainted with set algebra, you should verify all statements in this chapter and work through the exercises.

Throughout this chapter X and Y denote two arbitrary sets. For any two sets A, B (which are not necessarily subsets of a common set) we write

$$A \cup B = \{x : x \in A \text{ or } x \in B \text{ or } x \in A \text{ and } B\}$$

$$A \cap B = \{x : x \in A \text{ and } x \in B\}$$

$$A \setminus B = \{x : x \in A \text{ and } x \notin B\};$$

in particular we write $A \uplus B$ for the *disjoint union*, i.e. for $A \cup B$ if $A \cap B = \emptyset$. $A \subset B$ means that A is contained in B including the possibility that $A = B$; to exclude the latter we write $A \subsetneq B$. If $A \subset X$, we set $A^c := X \setminus A$ for the *complement* of A (relative to X). Recall also the *distributive laws* for $A, B, C \subset X$

$$A \cap (B \cup C) = (A \cap B) \cup (A \cap C)$$
$$A \cup (B \cap C) = (A \cup B) \cap (A \cup C) \tag{2.1}$$

and *de Morgan's identities*

$$(A \cap B)^c = A^c \cup B^c$$
$$(A \cup B)^c = A^c \cap B^c \tag{2.2}$$

which also hold for *arbitrarily* many sets $A_i \subset X$, $i \in I$ (I stands for an arbitrary index set),

$$\left(\bigcap_{i \in I} A_i \right)^c = \bigcup_{i \in I} A_i^c$$
$$\left(\bigcup_{i \in I} A_i \right)^c = \bigcap_{i \in I} A_i^c. \tag{2.3}$$

5

A map $f: X \to Y$ is called

$$
\begin{aligned}
\textit{injective} \text{ (or one-one)} &\iff f(x) = f(x') \implies x = x' \\
\textit{surjective} \text{ (or onto)} &\iff f(X) := \{f(x) \in Y : x \in X\} = Y \\
\textit{bijective} &\iff f \text{ is injective and surjective.}
\end{aligned}
$$

Set operations and *direct images* under a map f are not necessarily compatible: indeed, we have, in general,

$$
f(A \cup B) = f(A) \cup f(B)
$$
$$
f(A \cap B) \neq f(A) \cap f(B) \tag{2.4}
$$
$$
f(A \setminus B) \neq f(A) \setminus f(B).
$$

Inverse images and set operations are, however, *always* compatible. For $C, C_i, D \subset Y$ one has

$$
f^{-1}\left(\bigcup_{i \in I} C_i\right) = \bigcup_{i \in I} f^{-1}(C_i)
$$
$$
f^{-1}\left(\bigcap_{i \in I} C_i\right) = \bigcap_{i \in I} f^{-1}(C_i) \tag{2.5}
$$
$$
f^{-1}(C \setminus D) = f^{-1}(C) \setminus f^{-1}(D).
$$

If we have more information about f we can, of course, say more.

2.1 Lemma $f: X \to Y$ *is injective if, and only if,* $f(A \cap B) = f(A) \cap f(B)$ *for all* $A, B \subset X$.

Proof '$\Rightarrow$': Since $f(A \cap B) \subset f(A)$ and $f(A \cap B) \subset f(B)$, we have always $f(A \cap B) \subset f(A) \cap f(B)$. Let us check the converse inclusion '$\supset$'. If $y \in f(A) \cap f(B)$, we have $y = f(a)$ and $y = f(b)$ for some $a \in A, b \in B$. So, $f(a) = y = f(b)$ and, by injectivity, $a = b$. This means that $a = b \in A \cap B$, hence $y \in f(A \cap B)$ and $f(A) \cap f(B) \subset f(A \cap B)$ follows.

'$\Leftarrow$': Take $x, x' \in X$ with $f(x) = f(x')$ and set $A := \{x\}$, $B := \{x'\}$. Then $\emptyset \neq f(\{x\}) \cap f(\{x'\}) = f(\{x\} \cap \{x'\})$ which is only possible if $\{x\} \cap \{x'\} \neq \emptyset$, i.e. if $x = x'$. This shows that f is injective. ∎

2.2 Lemma $f: X \to Y$ *is injective if, and only if,* $f(X \setminus A) = f(X) \setminus f(A)$ *for all* $A \subset X$.

Proof '$\Rightarrow$' Assume that f is injective. We show first that $f(x) \notin f(A)$ if, and only if, $x \notin A$. Indeed, if $f(x) \notin f(A)$, then $x \notin A$; if $x \notin A$ but $f(x) \in f(A)$, then we can find some $a \in A$ such that $f(a) = f(x) \in f(A)$. Since f is injective, $x = a \in A$ and we have found a contradiction. Thus

$$f(X) \setminus f(A) = \{y \in Y : y = f(x), \ f(x) \notin f(A)\}$$
$$= \{y \in Y : y = f(x), \ x \notin A\}$$
$$= f(X \setminus A).$$

'$\Leftarrow$': Let $f(x) = f(x')$ and assume that $x \neq x'$. Then

$$f(x) \in f(X \setminus \{x'\}) = f(X) \setminus f(\{x'\})$$

which cannot happen as $f(x) \in f(\{x'\})$. ∎

We can now start with the main topic of this chapter: counting.

2.3 Definition Two sets X, Y have the same *cardinality* if there exists a bijection $f : X \to Y$. In this case we write $\#X = \#Y$.

If there is an injection $g : X \to Y$, we say that the cardinality of X is less than or equal to the cardinality of Y and write $\#X \leqslant \#Y$. If $\#X \leqslant \#Y$ but $\#X \neq \#Y$, we say that X is of strictly smaller cardinality than Y and write $\#X < \#Y$ (in this case, no injection $g : X \to Y$ can be surjective).

That Definition 2.3 is indeed *counting* becomes clear if we choose $Y = \mathbb{N}$ since in this case $\#X = \#\mathbb{N}$ or $\#X \leqslant \#\mathbb{N}$ just means that we can label each $x \in X$ with a unique tag from the set $\{1, 2, 3, \ldots\}$, i.e. we are numbering X. This particular example is, in fact, of central importance.

2.4 Definition A set X is *countable* if $\#X \leqslant \#\mathbb{N}$. If $\#\mathbb{N} < \#X$, the set X is said to be *uncountable*. The cardinality of $\mathbb{N}$ is called $\aleph_0$, *aleph null*.

Plainly, Definition 2.4 requires that we can find for every countable set some *enumeration* $X = (x_1, x_2, x_3, \ldots)$ which may or may not be finite (and which may contain any x_j more than once).

Caution: Some authors reserve the word countable for the situation where $\#X = \#\mathbb{N}$ while sets where $\#X \leqslant \#\mathbb{N}$ are called *at most countable* or *finite or countable*. This has the effect that a countable set is always infinite. We do not adopt this convention.

The following examples show that (countable) sets with infinitely many elements can behave strangely.

2.5 Examples (i) Finite sets are countable: $\{a, b, \ldots, z\} \to \{1, 2, \ldots, 26\}$ where $a \leftrightarrow 1, \ldots, z \leftrightarrow 26$, is bijective and $\{1, 2, 3, \ldots, 26\} \to \mathbb{N}$ is clearly an injection. Thus

$$\#\{a, b, c, \ldots, z\} = \#\{1, 2, 3, \ldots, 26\} \leqslant \#\mathbb{N}.$$

(ii) The even numbers are countable. This follows from the fact that the map

$$f : \{2, 4, 6, \ldots, 2j, \ldots\} \to \mathbb{N}, \quad k \mapsto \frac{k}{2}$$

is an injection and even a bijection.[✓] This means that there are 'as many' even numbers as there are natural numbers.

(iii) The set of integers $\mathbb{Z} = \{0, \pm 1, \pm 2, \ldots\}$ is countable. The counting scheme is shown on the right (run through in clockwise orientation starting from 0) or, more formally,

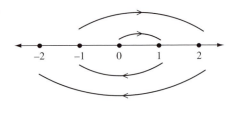

$$g : k \in \mathbb{Z} \mapsto \begin{cases} 2k & \text{if } k > 0, \\ 2|k| + 1 & \text{if } k \leqslant 0, \end{cases}$$

hence $\#\mathbb{Z} \leqslant \#\mathbb{N}$.[✓]

(iv) The Cartesian product $\mathbb{N} \times \mathbb{N} := \{(j, k) : j, k \in \mathbb{N}\}$ is countable. To see this, arrange the pairs (j, k) in an array and count along the diagonals:

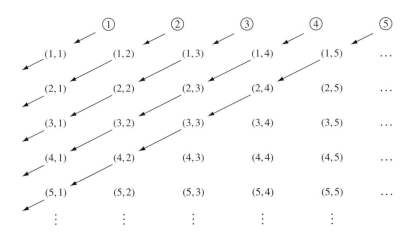

Notice that each line contains only finitely many elements, so that each diagonal can be dealt with in finitely many steps. The map for the above counting scheme is given by

$$h : (j, k) \mapsto \frac{(j+k)(j+k-1)}{2} - k + 1 \in \mathbb{N}, \qquad (j, k) \in \mathbb{N} \times \mathbb{N}. \qquad (2.6)$$

(v) The rational numbers $\mathbb{Q}$ are countable. To see this, set $Q_\pm := \{q \in \mathbb{Q} : \pm q > 0\}$. Every element $\frac{m}{n} \in Q_+$ can be identified with at least one pair $(m, n) \in \mathbb{N} \times \mathbb{N}$, so that

$$Q_+ \subset \{ \underbrace{\tfrac{1}{1}}_{\textcircled{1}}, \; \underbrace{\tfrac{1}{2}, \tfrac{2}{1}}_{\textcircled{2}}, \; \underbrace{\tfrac{1}{3}, \tfrac{2}{2}, \tfrac{3}{1}}_{\textcircled{3}}, \; \underbrace{\tfrac{1}{4}, \tfrac{2}{3}, \tfrac{3}{2}, \tfrac{4}{1}}_{\textcircled{4}}, \dots \};$$

in the set $\{\dots\}$ on the right we distinguish between cancelled and uncancelled forms of a rational, i.e. $\frac{6}{18}, \frac{1}{3}, \frac{2}{6}, \frac{3}{9}, \dots$ etc. are counted whenever they appear. The numbers $\textcircled{k}$ refer to the corresponding diagonals in the counting scheme in part (iv). This shows that we can find injections $Q_+ \xrightarrow{i} \{\dots\} \xrightarrow{j} \mathbb{N} \times \mathbb{N}$; the set $\mathbb{N} \times \mathbb{N}$ is countable, thus Q_+ is countable$^{[\checkmark]}$ and so is Q_-. Finally,

$$\mathbb{Q} = Q_- \cup \{0\} \cup Q_+ = \{r_1, r_2, r_3, \dots\} \cup \{0\} \cup \{q_1, q_2, q_3, \dots\}$$

and $p_1 := 0$, $p_{2k} := q_k$, $p_{2k+1} := r_k$ gives an enumeration $(p_1, p_2, p_3, \dots)$ of $\mathbb{Q}$.

2.6 Theorem *Let $A_1, A_2, A_3, \dots$ be countably many countable sets. Then $A = \bigcup_{j \in \mathbb{N}} A_j$ is countable, i.e. countable unions of countable sets are countable.*

Proof Since each A_j is countable we can find an enumeration

$$A_j = (a_{j,1}, a_{j,2}, \dots, a_{j,k}, \dots)$$

(if A_j is a finite set, we repeat the last element of the list infinitely often), so that

$$A = \bigcup_{j \in \mathbb{N}} A_j = (a_{j,k} : (j, k) \in \mathbb{N} \times \mathbb{N}).$$

Using Example 2.5(iv) we can relabel $\mathbb{N} \times \mathbb{N}$ by $\mathbb{N}$ and (after deleting all duplicates) we have found an enumeration. ∎

It is not hard to see that for cardinalities '$\leqslant$' is *reflexive* ($\#A \leqslant \#A$) and *transitive* ($\#A \leqslant \#B$, $\#B \leqslant \#C \implies \#A \leqslant \#C$). *Antisymmetry*, which makes '$\leqslant$' into a partial order relation, is less obvious. The proof of the following important result is somewhat technical and can be left out at first reading.

2.7 Theorem (Cantor–Bernstein) *Let X, Y be two sets. If $\#X \leqslant \#Y$ and $\#Y \leqslant \#X$, then $\#X = \#Y$.*

Proof By assumption,

$$\#X \leqslant \#Y \iff \text{there exists an injection} \quad f : X \to Y;$$
$$\#Y \leqslant \#X \iff \text{there exists an injection} \quad g : Y \to X.$$

In order to prove $\#X = \#Y$ we have to construct a bijection $h : X \to Y$.

Step 1. Without loss of generality we may assume that $Y \subset X$. Indeed, since $g : Y \to g(Y)$ is a bijection, we know that $\#Y = \#g(Y)$ and it is enough to show $\#g(Y) = \#X$. As $g(Y) \subset X$ we can simplify things and identify $g(Y)$ with Y, i.e. assume that $g = \mathrm{id}$ or, equivalently, $Y \subset X$.

Step 2. Let $Y \subset X$ and $g = \mathrm{id}$. Recursively we define

$$X_0 := X, \ldots, X_{j+1} := f(X_j); \qquad Y_0 := Y, \ldots, Y_{j+1} := f(Y_j).$$

As usual, we write $f^j := \underbrace{f \circ f \circ \cdots \circ f}_{j \text{ times}}$ and $f^0 := \mathrm{id}$. Then

$$f^{j+1}(X) = f^j(f(X)) \overset{f(X) \subset Y}{\subset} f^j(Y) \overset{Y \subset X}{\subset} f^j(X)$$

$$\| \qquad\qquad\qquad\qquad \| \qquad\qquad \|$$

$$X_{j+1} \qquad\qquad \subset \qquad\qquad Y_j \quad \subset \quad X_j$$

and we can define a map $h : X \to Y$ by

$$h(x) := \begin{cases} f(x) & \text{if} \quad x \in X_j \setminus Y_j \quad \text{for some } j \in \mathbb{N}_0, \\ x & \text{if} \quad x \notin \bigcup_{j \in \mathbb{N}_0} X_j \setminus Y_j. \end{cases}$$

Step 3. The map h is surjective: $h(X) = Y$. Indeed, we have by definition

$$h(X) = \bigcup_{j \in \mathbb{N}_0} f(X_j \setminus Y_j) \cup \Big(\bigcup_{j \in \mathbb{N}_0} X_j \setminus Y_j \Big)^c$$

$$\overset{1.2}{=} \bigcup_{j \in \mathbb{N}_0} (f(X_j) \setminus f(Y_j)) \cup \Big(X \setminus Y \cup \bigcup_{j \in \mathbb{N}} X_j \setminus Y_j \Big)^c$$

$$= \underbrace{\bigcup_{j \in \mathbb{N}_0} (X_{j+1} \setminus Y_{j+1})}_{=: A} \cup \Big[(X \setminus Y)^c \cap \Big(\bigcup_{j \in \mathbb{N}_0} X_{j+1} \setminus Y_{j+1} \Big)^c \Big]$$

$$= A \cup [(X^c \cup Y) \cap A^c] = A \cup [Y \cap A^c] = Y \cap X = Y$$

where we used that $A = \bigcup_{j \in \mathbb{N}_0} X_{j+1} \setminus Y_{j+1} \subset \bigcup_{j \in \mathbb{N}_0} X_{j+1} \subset X_1 = f(X) \subset Y$.

Step 4. The map h is injective. To see this, let $x, x' \in X$ and $h(x) = h(x')$. We have four possibilities

(a) $x, x' \in X_j \setminus Y_j$ for some $j \in \mathbb{N}_0$. Then $f(x) = h(x) = h(x') = f(x')$ so that $x = x'$ since f is injective.

(b) $x, x' \notin \bigcup_{j \in \mathbb{N}_0} X_j \setminus Y_j$. Then $x = h(x) = h(x') = x'$.

(c) $x \in X_j \setminus Y_j$ for some $j \in \mathbb{N}_0$ and $x' \notin X_k \setminus Y_k$ for all $k \in \mathbb{N}_0$. As $f(x) = h(x) = h(x') = x'$ we see

$$x' = f(x) \in f(X_j \setminus Y_j) \overset{1.2}{=} f(X_j) \setminus f(Y_j) = X_{j+1} \setminus Y_{j+1}$$

which is impossible, i.e. (c) cannot occur.

(d) $x' \in X_j \setminus Y_j$ for some $j \in \mathbb{N}_0$ and $x \notin X_k \setminus Y_k$ for all $k \in \mathbb{N}_0$. This is analogous to (c). ∎

Theorem 2.7 says that $\#X < \#Y$ and $\#Y < \#X$ cannot occur at the same time; it does not claim that we can compare the cardinality of any two sets X and Y, i.e. that '$\leqslant$' is a linear ordering. This is indeed true but its proof requires the *axiom of choice*, see Hewitt and Stromberg [20, p. 19].

Not all sets are countable. The following proof goes back to G. Cantor and is called *Cantor's diagonal method*.

2.8 Theorem *The interval* $(0, 1)$ *is uncountable; its cardinality* $\mathfrak{c} := \#(0, 1)$ *is called the* continuum.

Proof Recall that we can write each $x \in (0, 1)$ as a decimal fraction, i.e. $x = 0.y_1 y_2 y_3 \ldots$ with $y_j \in \{0, 1, \ldots, 9\}$. If x has a finite decimal representation, say $x = 0.y_1 y_2 y_3 \ldots y_n$, $y_n \neq 0$, we replace the last digit y_n by $y_n - 1$ and fill it up with trailing 9s. For example, $0.24 = 0.23\overline{99}\ldots$. This yields a *unique* representation of x by an *infinite* decimal expansion.

Assume that $(0, 1)$ were countable and let $\{x_1, x_2, \ldots\}$ be an enumeration (containing no element more than once!). Then we can write

$$x_1 = 0.\boldsymbol{a_{1,1}} a_{1,2} a_{1,3} a_{1,4} \cdots$$
$$x_2 = 0.a_{2,1} \boldsymbol{a_{2,2}} a_{2,3} a_{2,4} \cdots$$
$$x_3 = 0.a_{3,1} a_{3,2} \boldsymbol{a_{3,3}} a_{3,4} \cdots \tag{2.7}$$
$$x_4 = 0.a_{4,1} a_{4,2} a_{4,3} \boldsymbol{a_{4,4}} \cdots$$

$$\vdots \quad \vdots \qquad \qquad \vdots$$

and construct a new number $x := 0.y_1 y_2 y_3 \ldots \in (0, 1)$ with digits

$$y_j := \begin{cases} 1 & \text{if } a_{j,j} = 5 \\ 5 & \text{if } a_{j,j} \neq 5. \end{cases} \tag{2.8}$$

By construction, $x \neq x_j$ for any x_j from the list (2.7): x and x_j differ at the jth decimal. But then we have found a number $x \in (0, 1)$ which is not contained in our supposedly complete enumeration of $(0, 1)$ and we have arrived at a contradiction. ∎

By $(0, 1)^{\mathbb{N}}$ we denote the set of all sequences $(x_j)_{j \in \mathbb{N}}$ where $x_j \in (0, 1)$.

2.9 Theorem *We have* $\#(0, 1)^{\mathbb{N}} = \mathfrak{c}$.

Proof We have to assign to every sequence $(x_j)_{j \in \mathbb{N}} \subset (0, 1)$ a unique number $x \in (0, 1)$ – and vice versa. For this we write, as in the proof of Theorem 2.8, each x_j as a unique infinite decimal fraction

$$x_j = 0.a_{j,1}a_{j,2}a_{j,3}a_{j,4}\ldots, \quad j \in \mathbb{N},$$

and we organize the array $(a_{j,k})_{j,k \in \mathbb{N}}$ into one sequence with the help of the counting scheme of Example 2.5(iv):

$$x := 0. \underbrace{a_{1,1}}_{①} \underbrace{a_{1,2}\,a_{2,1}}_{②} \underbrace{a_{1,3}\,a_{2,2}\,a_{3,1}}_{③} \underbrace{a_{1,4}\,a_{2,3}\,a_{3,2}\,a_{4,1}}_{④} \ldots.$$

(The numbers $ⓚ$ refer to the corresponding diagonals in the counting scheme of Example 2.5(iv).) Since the counting scheme was bijective, this procedure is reversible, i.e. we can start with the decimal expansion of $x \in (0, 1)$ and get a unique sequence of x_js. We have thus found a bijection between $(0, 1)^{\mathbb{N}}$ and $(0, 1)$. ∎

We write $\mathcal{P}(X)$ for the *power set* $\{A : A \subset X\}$ which is the family of all subsets of a given set X. For finite sets it is clear that the power set is of strictly larger cardinality than X. This is still true for infinite sets.

2.10 Theorem *For any set X we have* $\#X < \#\mathcal{P}(X)$.

Proof We have to show that no injection $\Phi : X \to \mathcal{P}(X)$ can be surjective. Fix such an injection and define

$$B := \{x \in X : x \notin \Phi(x)\}$$

(mind: $\Phi(x)$ is a set!). Clearly $B \in \mathcal{P}(X)$. If Φ were surjective, $B = \Phi(z)$ for some element $z \in X$. Then, however,

$$z \in B \quad \overset{\text{def}}{\Longleftrightarrow} \quad z \notin \Phi(z) \quad \Longleftrightarrow \quad z \notin B \quad (\text{since } \Phi(z) = B)$$

which is impossible. Thus Φ cannot be surjective. ∎

Problems

2.1. Let $A, B, C \subset X$ be sets. Show that

    (i) $A \setminus B = A \cap B^c$;

    (ii) $(A \setminus B) \setminus C = A \setminus (B \cup C)$;

    (iii) $A \setminus (B \setminus C) = (A \setminus B) \cup (A \cap C)$;

    (iv) $A \setminus (B \cap C) = (A \setminus B) \cup (A \setminus C)$;

    (v) $A \setminus (B \cup C) = (A \setminus B) \cap (A \setminus C)$.

2.2. Let $A, B, C \subset X$. The *symmetric difference* of A and B is defined as $A \triangle B := (A \setminus B) \cup (B \setminus A)$. Verify that

$$(A \cup B \cup C) \setminus (A \cap B \cap C) = (A \triangle B) \cup (B \triangle C).$$

2.3. Prove de Morgan's identities (2.2) and (2.3).

2.4. (i) Find examples which illustrate that $f(A \cap B) \neq f(A) \cap f(B)$ and $f(A \setminus B) \neq f(A) \setminus f(B)$. In both relations one inclusion '$\subset$' or '$\supset$' is always true. Which one?

    (ii) Prove (2.5).

2.5. The *indicator function* of a set $A \subset X$ is defined by $\mathbf{1}_A(x) := \begin{cases} 1, & \text{if } x \in A, \\ 0, & \text{if } x \notin A. \end{cases}$

Check that

    (i) $\mathbf{1}_{A \cap B} = \mathbf{1}_A \mathbf{1}_B$;

    (ii) $\mathbf{1}_{A \cup B} = \min\{\mathbf{1}_A + \mathbf{1}_B, 1\}$;

    (iii) $\mathbf{1}_{A \setminus B} = \mathbf{1}_A - \mathbf{1}_{A \cap B}$;

    (iv) $\mathbf{1}_{A \cup B} = \mathbf{1}_A + \mathbf{1}_B - \mathbf{1}_{A \cap B}$;

    (v) $\mathbf{1}_{A \cup B} = \max\{\mathbf{1}_A, \mathbf{1}_B\}$;

    (vi) $\mathbf{1}_{A \cap B} = \min\{\mathbf{1}_A, \mathbf{1}_B\}$.

2.6. Let $A, B, C \subset X$ and denote by $A \triangle B$ the symmetric difference as in Problem 2.2. Show that

    (i) $\mathbf{1}_{A \triangle B} = \mathbf{1}_A + \mathbf{1}_B - 2 \mathbf{1}_A \mathbf{1}_B = \mathbf{1}_A + \mathbf{1}_B \mod 2$;

    (ii) $A \triangle (B \triangle C) = (A \triangle B) \triangle C$;

    (iii) $\mathcal{P}(X)$ is a commutative ring (in the usual algebraists' sense) with 'addition' $\triangle$ and 'multiplication' $\cap$.

[Hint: use indicator functions for (ii) and (iii).]

2.7. Let $f : X \to Y$ be a map, $A \subset X$ and $B \subset Y$. Show that, in general,

$$f \circ f^{-1}(B) \subsetneq B \qquad \text{and} \qquad f^{-1} \circ f(A) \supsetneq A.$$

When does '$=$' hold in these relations? Provide an example showing that the above inclusions are strict.

2.8. Let f and g be two injective maps. Show that $f \circ g$, if it exists, is injective.

2.9. Show that the following sets have the same cardinality as $\mathbb{N}$: $\{m \in \mathbb{N} : m \text{ is odd}\}$, $\mathbb{N} \times \mathbb{Z}$, $\mathbb{Q}^m$ ($m \in \mathbb{N}$), $\bigcup_{m \in \mathbb{N}} \mathbb{Q}^m$.

2.10. Use Theorem 2.7 to show that $\#\mathbb{N} \times \mathbb{N} = \#\mathbb{N}$.
[Hint: $\#\mathbb{N} = \#\mathbb{N} \times \{1\}$ and $\mathbb{N} \times \{1\} \subset \mathbb{N} \times \mathbb{N}$.]

2.11. Show that if $E \subset F$ we have $\#E \leqslant \#F$. In particular, subsets of countable sets are again countable.

2.12. Show that $\{0, 1\}^{\mathbb{N}} = \{$all infinite sequences consisting of 0 and 1$\}$ is uncountable.
[Hint: diagonal method.]

2.13. Show that the set $\mathbb{R}$ is uncountable and that $\#(0, 1) = \#\mathbb{R}$.
[Hint: find a bijection $f : (0, 1) \to \mathbb{R}$.]

2.14. Let $(A_j)_{j \in \mathbb{N}}$ be a sequence of sets of cardinality $\mathfrak{c}$. Show that $\#\bigcup_{j \in \mathbb{N}} A_j = \mathfrak{c}$.
[Hint: map A_j bijectively onto $(j - 1, j)$ and use that $(0, 1) \subset \bigcup_{j=1}^{\infty} (j - 1, j) \subset \mathbb{R}$.]

2.15. Adapt the proof of Theorem 2.8 to show that $\#\{1, 2\}^{\mathbb{N}} \leqslant \#(0, 1) \leqslant \#\{0, 1\}^{\mathbb{N}}$ and conclude that $\#(0, 1) = \#\{0, 1\}^{\mathbb{N}}$.
Remark. This is the reason for writing $\mathfrak{c} = 2^{\aleph_0}$.
[Hint: interpret $\{0, 1\}^{\mathbb{N}}$ as base-2 expansions of all numbers in $(0, 1)$ while $\{1, 2\}^{\mathbb{N}}$ are all infinite base-3 expansions lacking the digit 0.]

2.16. Extend Problem 2.15 to deduce $\#\{0, 1, 2, \ldots, n\}^{\mathbb{N}} = \#(0, 1)$ for all $n \in \mathbb{N}$.

2.17. Mimic the proof of Theorem 2.9 to show that $\#(0, 1)^2 = \mathfrak{c}$. Use the fact that $\#\mathbb{R} = \#(0, 1)$ to conclude that $\#\mathbb{R}^2 = \mathfrak{c}$.

2.18. Show that the set of all infinite sequences of natural numbers $\mathbb{N}^{\mathbb{N}}$ has cardinality $\mathfrak{c}$.
[Hint: use that $\#\{0, 1\}^{\mathbb{N}} = \#\{1, 2\}^{\mathbb{N}}$, $\{1, 2\}^{\mathbb{N}} \subset \mathbb{N}^{\mathbb{N}} \subset \mathbb{R}^{\mathbb{N}}$ and $\#\mathbb{R}^{\mathbb{N}} = \#(0, 1)^{\mathbb{N}}$.]

2.19. Let $\mathcal{F} := \{F \subset \mathbb{N} : \#F < \infty\}$. Show that $\#\mathcal{F} = \#\mathbb{N}$.
[Hint: embed $\mathcal{F}$ into $\bigcup_{k \in \mathbb{N}} \mathbb{N}^k$ or show that $F \mapsto \sum_{j \in F} 2^j$ is a bijection between $\mathcal{F}$ and $\mathbb{N}$.]

2.20. Show – not using Theorem 2.10 – that $\#\mathcal{P}(\mathbb{N}) > \#\mathbb{N}$. Conclude that there are more than countably many maps $f : \mathbb{N} \to \mathbb{N}$.
[Hint: diagonal method.]

2.21. If $A \subset \mathbb{N}$ we can identify the indicator function $\mathbf{1}_A : \mathbb{N} \to \{0, 1\}$ with the 0-1-sequence $(\mathbf{1}_A(j))_{j \in \mathbb{N}}$, i.e., $\mathbf{1}_A \in \{0, 1\}^{\mathbb{N}}$. Show that the map $\mathcal{P}(\mathbb{N}) \ni A \mapsto \mathbf{1}_A \in \{0, 1\}^{\mathbb{N}}$ is a bijection and conclude that $\#\mathcal{P}(\mathbb{N}) = \mathfrak{c}$.

3

σ-algebras

We have seen in the prologue that a reasonable measure should be able to deal with *countable* partitions. Therefore, a measure function should be defined on a system of sets which is stable whenever we repeat any of the basic set operations – $\cup, \cap, \,^c$ – countably many times.

3.1 Definition A σ-*algebra* $\mathcal{A}$ on a set X is a family of subsets of X with the following properties:

$$X \in \mathcal{A}, \tag{Σ_1}$$

$$A \in \mathcal{A} \implies A^c \in \mathcal{A}, \tag{Σ_2}$$

$$(A_j)_{j\in\mathbb{N}} \subset \mathcal{A} \implies \bigcup_{j\in\mathbb{N}} A_j \in \mathcal{A}. \tag{Σ_3}$$

A set $A \in \mathcal{A}$ is said to be ($\mathcal{A}$-)*measurable*.

3.2 Properties (of a σ-algebra)   **(i)** $\emptyset \in \mathcal{A}$.
  Indeed: $\emptyset = X^c \in \mathcal{A}$ by (Σ_1, Σ_2).
 (ii) $A, B \in \mathcal{A} \implies A \cup B \in \mathcal{A}$.
  Indeed: set $A_1 = A$, $A_2 = B$, $A_3 = A_4 = \ldots = \emptyset$. Then $A \cup B = \bigcup_{j\in\mathbb{N}} A_j \in \mathcal{A}$
  by (Σ_3).
(iii) $(A_j)_{j\in\mathbb{N}} \subset \mathcal{A} \implies \bigcap_{j\in\mathbb{N}} A_j \in \mathcal{A}$.
  Indeed: if $A_j \in \mathcal{A}$, then $A_j^c \in \mathcal{A}$ by (Σ_2), hence $\bigcup_{j\in\mathbb{N}} A_j^c \in \mathcal{A}$ by (Σ_3) and,
  again by (Σ_2), $\bigcap_{j\in\mathbb{N}} A_j = \left(\bigcup_{j\in\mathbb{N}} A_j^c\right)^c \in \mathcal{A}$.

3.3 Examples (i) $\mathcal{P}(X)$ is a σ-algebra (the maximal σ-algebra in X).
 (ii) $\{\emptyset, X\}$ is a σ-algebra (the minimal σ-algebra in X).
(iii) $\{\emptyset, B, B^c, X\}$, $B \subset X$, is a σ-algebra.
(iv) $\{\emptyset, B, X\}$ is no σ-algebra (unless $B = \emptyset$ or $B = X$).
 (v) $\mathcal{A} := \{A \subset X : \#A \leqslant \#\mathbb{N} \quad \text{or} \quad \#A^c \leqslant \#\mathbb{N}\}$ is a σ-algebra.

Proof: Let us verify (Σ_1)–(Σ_3).

(Σ_1): $X^c = \emptyset$ which is certainly countable.

(Σ_2): if $A \in \mathcal{A}$, either A or A^c is by definition countable, so $A^c \in \mathcal{A}$.

(Σ_3): if $(A_j)_{j \in \mathbb{N}} \subset \mathcal{A}$, then two cases can occur:

- All A_j are countable. Then $A = \bigcup_{j \in \mathbb{N}} A_j$ is a countable union of countable sets which is, by T2.6, itself countable.
- At least one A_{j_0} is uncountable. Then $A_{j_0}^c$ must be countable, so that

$$\left(\bigcup_{j \in \mathbb{N}} A_j \right)^c = \bigcap_{j \in \mathbb{N}} A_j^c \subset A_{j_0}^c.$$

Hence $\left(\bigcup_{j \in \mathbb{N}} A_j \right)^c$ is countable (Problem 2.11) and so $\bigcup_{j \in \mathbb{N}} A_j \in \mathcal{A}$.

(vi) (trace σ-algebra) Let $E \subset X$ be any set and let $\mathcal{A}$ be some σ-algebra in X. Then

$$\mathcal{A}_E := E \cap \mathcal{A} := \{ E \cap A : A \in \mathcal{A} \} \tag{3.1}$$

is a σ-algebra in E.

(vii) (pre-image σ-algebra) Let $f : X \to X'$ be a map and let $\mathcal{A}'$ be a σ-algebra in X'. Then

$$\mathcal{A} := f^{-1}(\mathcal{A}') := \left\{ f^{-1}(A') : A' \in \mathcal{A}' \right\}$$

is a σ-algebra in X.

3.4 Theorem (and Definition) (i) *The intersection $\bigcap_{i \in I} \mathcal{A}_i$ of arbitrarily many σ-algebras $\mathcal{A}_i$ in X is again a σ-algebra in X.*

(ii) *For every system of sets $\mathcal{G} \subset \mathcal{P}(X)$ there exists a smallest (also: minimal, coarsest) σ-algebra containing $\mathcal{G}$. This is the σ-algebra generated by $\mathcal{G}$, denoted by $\sigma(\mathcal{G})$, and $\mathcal{G}$ is called its generator.*

Proof (i) We check (Σ_1)–(Σ_3). (Σ_1): since $X \in \mathcal{A}_i$ for all $i \in I$, $X \in \bigcap_i \mathcal{A}_i$. (Σ_2): if $A \in \bigcap_i \mathcal{A}_i$, then $A^c \in \mathcal{A}_i$ for all $i \in I$, so $A^c \in \bigcap_i \mathcal{A}_i$. (Σ_3): let $(A_k)_{k \in \mathbb{N}} \subset \bigcap_i \mathcal{A}_i$. Then $A_k \in \mathcal{A}_i$ for all $k \in \mathbb{N}$ and all $i \in I$, hence $\bigcup_{k \in \mathbb{N}} A_k \in \mathcal{A}_i$ for each $i \in I$ and so $\bigcup_{k \in \mathbb{N}} A_k \in \bigcap_{i \in I} \mathcal{A}_i$.

(ii) Consider the family

$$\mathcal{A} := \bigcap_{\substack{\mathcal{C} \ \sigma\text{-alg.} \\ \mathcal{C} \supset \mathcal{G}}} \mathcal{C}.$$

Since $\mathcal{G} \subset \mathcal{P}(X)$ and since $\mathcal{P}(X)$ is a σ-algebra, the above intersection is non-void. This means that the definition of $\mathcal{A}$ makes sense and yields, by part (i), a σ-algebra containing $\mathcal{G}$. If $\mathcal{A}'$ is a further σ-algebra with $\mathcal{A}' \supset \mathcal{G}$, then $\mathcal{A}'$ would be included in the intersection used for the definition of $\mathcal{A}$, hence $\mathcal{A} \subset \mathcal{A}'$. In this sense, $\mathcal{A}$ is the smallest σ-algebra containing $\mathcal{G}$. ∎

3.5 Remarks **(i)** If $\mathcal{G}$ is a σ-algebra, then $\mathcal{G} = \sigma(\mathcal{G})$.

(ii) For $A \subset X$ we have $\sigma(\{A\}) = \{\emptyset, A, A^c, X\}$.

(iii) If $\mathcal{F} \subset \mathcal{G} \subset \mathcal{A}$, then $\sigma(\mathcal{F}) \subset \sigma(\mathcal{G}) \subset \sigma(\mathcal{A}) \overset{3.5(i)}{=} \mathcal{A}$.

On the Euclidean space $\mathbb{R}^n$ there is a canonical σ-algebra, which is generated by the open sets. Recall that

$$U \subset \mathbb{R}^n \quad \text{is } open \iff \forall x \in U \; \exists \epsilon > 0 : B_\epsilon(x) \subset U$$

where $B_\epsilon(x) := \{y \in \mathbb{R}^n : |x - y| < \epsilon\}$ is the open ball with centre x and radius ϵ. A set is *closed* if its complement is open. The system of open sets in $X = \mathbb{R}^n$, $\mathcal{O}^n$, has the following properties:

$$\emptyset, X \in \mathcal{O}^n, \tag{$\mathcal{O}_1$}$$

$$U, V \in \mathcal{O}^n \implies U \cap V \in \mathcal{O}^n, \tag{$\mathcal{O}_2$}$$

$$U_i \in \mathcal{O}^n, \; i \in I \text{ (arbitrary)} \implies \bigcup_{i \in I} U_i \in \mathcal{O}^n. \tag{$\mathcal{O}_3$}$$

Note, however, that countable or arbitrary intersections of open sets need not be open[✓]. A family of subsets $\mathcal{O}$ of a general space X satisfying the conditions $(\mathcal{O}_1)$–$(\mathcal{O}_3)$ is called a *topology*, and the pair $(X, \mathcal{O})$ is called a *topological space*; in analogy to $\mathbb{R}^n$, $U \in \mathcal{O}$ is said to be *open* while *closed* sets are exactly the complements of open sets; see Appendix B.

3.6 Definition The σ-algebra $\sigma(\mathcal{O}^n)$ generated by the open sets $\mathcal{O}^n$ of $\mathbb{R}^n$ is called *Borel σ-algebra*, and its members are the *Borel sets* or *Borel measurable sets*. We write $\mathcal{B}(\mathbb{R}^n)$ or $\mathcal{B}^n$ for the Borel sets in $\mathbb{R}^n$.

The Borel sets are fundamental for the study of measures on $\mathbb{R}^n$. Since the Borel σ-algebra depends on the topology of $\mathbb{R}^n$, $\mathcal{B}(\mathbb{R}^n)$ is often also called the *topological σ-algebra*.

3.7 Theorem *Denote by* $\mathcal{O}^n$, $\mathcal{C}^n$ *and* $\mathcal{K}^n$ *the families of open, closed and compact*[1] *sets in* $\mathbb{R}^n$. *Then*

$$\mathcal{B}(\mathbb{R}^n) = \sigma(\mathcal{O}^n) = \sigma(\mathcal{C}^n) = \sigma(\mathcal{K}^n).$$

Proof Since compact sets are closed, we have $\mathcal{K}^n \subset \mathcal{C}^n$ and by Remark 3.5(iii), $\sigma(\mathcal{K}^n) \subset \sigma(\mathcal{C}^n)$. On the other hand, if $C \in \mathcal{C}^n$, then $C_k := C \cap \overline{B_k(0)}$ is[2] closed and bounded, hence $\in \mathcal{K}^n$. By construction $C = \bigcup_{k \in \mathbb{N}} C_k$, thus $\mathcal{C}^n \subset \sigma(\mathcal{K}^n)$ and also $\sigma(\mathcal{C}^n) \subset \sigma(\mathcal{K}^n)$.

---

[1] i.e. closed and bounded.

[2] $B_k(0)$, $\overline{B_k(0)}$ denote the open, resp., closed balls with centre 0 and radius k.

Since $(\mathcal{O}^n)^c := \{U^c : U \in \mathcal{O}^n\} = \mathcal{C}^n$ (and $(\mathcal{C}^n)^c = \mathcal{O}^n$) we have $\mathcal{C}^n = (\mathcal{O}^n)^c \subset \sigma(\mathcal{O}^n)$, hence $\sigma(\mathcal{C}^n) \subset \sigma(\mathcal{O}^n)$ and the converse inclusion is similar.   ∎

The Borel σ-algebra $\mathcal{B}(\mathbb{R}^n)$ is generated by many different systems of sets. For our purposes the most interesting generators are the families of open rectangles

$$\mathcal{J}^o = \mathcal{J}^{o,n} = \mathcal{J}^o(\mathbb{R}^n) = \{(a_1, b_1) \times \cdots \times (a_n, b_n) : a_j, b_j \in \mathbb{R}\}$$

and (from the right) half-open rectangles

$$\mathcal{J} = \mathcal{J}^n = \mathcal{J}(\mathbb{R}^n) = \{[a_1, b_1) \times \cdots \times [a_n, b_n) : a_j, b_j \in \mathbb{R}\}.$$

We use the convention that $[a_j, b_j) = (a_j, b_j) = \emptyset$ if $b_j \geqslant a_j$ and, of course, that $[a_1, b_1) \times \cdots \times \emptyset \times \cdots \times [a_n, b_n) = \emptyset$. Sometimes we use the shorthand $[\![a, b)\!) = [a_1, b_1) \times \cdots \times [a_n, b_n)$ for vectors $a = (a_1, \ldots, a_n), b = (b_1, \ldots, b_n)$ from $\mathbb{R}^n$. Finally, we write $\mathcal{J}_{\mathrm{rat}}, \mathcal{J}^o_{\mathrm{rat}}$ for the (half-)open rectangles with only rational endpoints. Notice that the half-open rectangles are ...

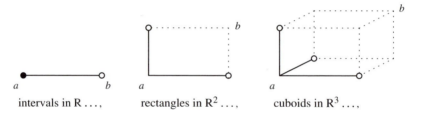

intervals in R ...,   rectangles in R^2 ...,   cuboids in R^3 ...,

and hypercubes in dimensions $n > 3$.

3.8 Theorem *We have* $\mathcal{B}(\mathbb{R}^n) = \sigma(\mathcal{J}^n_{\mathrm{rat}}) = \sigma(\mathcal{J}^{o,n}_{\mathrm{rat}}) = \sigma(\mathcal{J}^n) = \sigma(\mathcal{J}^{o,n})$.

Proof We begin with open rectangles having rational endpoints. Since the open rectangle $(\![a, b)\!)$ is an open set[✓], we find $\sigma(\mathcal{O}^n) \supset \sigma(\mathcal{J}^o) \supset \sigma(\mathcal{J}^o_{\mathrm{rat}})$.

Conversely, if $U \in \mathcal{O}^n$, we have

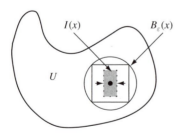

$$U = \bigcup_{I \in \mathcal{J}^o_{\mathrm{rat}}, I \subset U} I. \tag{3.2}$$

Here '$\supset$' is clear from the definition and for the other direction '$\subset$' we fix $x \in U$. Since U is open, there is some ball $B_\epsilon(x) \subset U$ – see the picture – and we can inscribe a square into $B_\epsilon(x)$ and then shrink this square to get a rectangle $I = I(x) \in \mathcal{J}^o_{\mathrm{rat}}$ containing x. Since every rectangle I is uniquely determined by its main diagonal, there are at most $\#(\mathbb{Q}^n \times \mathbb{Q}^n) = \#\mathbb{N}$ many I in the union (3.2). Thus

$$U \in \mathcal{O}^n \subset \sigma(\mathcal{J}_{\text{rat}}^o),$$

proving the other inclusion $\sigma(\mathcal{O}^n) \subset \sigma(\mathcal{J}_{\text{rat}}^o)$, and so $\sigma(\mathcal{O}^n) = \sigma(\mathcal{J}^o) = \sigma(\mathcal{J}_{\text{rat}}^o)$.

Every half-open rectangle (with rational endpoints) can be written as

$$[a_1, b_1) \times \cdots \times [a_n, b_n) = \bigcap_{j \in \mathbb{N}} (a_1 - \tfrac{1}{j}, b_1) \times \cdots \times (a_n - \tfrac{1}{j}, b_n),$$

while every open rectangle (with rational endpoints) can be represented as

$$(c_1, d_1) \times \cdots \times (c_n, d_n) = \bigcup_{j \in \mathbb{N}} [c_1 + \tfrac{1}{j}, d_1) \times \cdots \times [c_n + \tfrac{1}{j}, d_n).$$

These formulae imply that $\mathcal{J} \subset \sigma(\mathcal{J}^o)$ and $\mathcal{J}^o \subset \sigma(\mathcal{J})$ [resp. $\mathcal{J}_{\text{rat}} \subset \sigma(\mathcal{J}_{\text{rat}}^o)$ and $\mathcal{J}_{\text{rat}}^o \subset \sigma(\mathcal{J}_{\text{rat}})$], hence by Remark 3.5(iii), $\sigma(\mathcal{J}^o) = \sigma(\mathcal{J})$ [resp. $\sigma(\mathcal{J}_{\text{rat}}^o) = \sigma(\mathcal{J}_{\text{rat}})$] and the proof follows since we know $\sigma(\mathcal{J}_{\text{rat}}^o) = \sigma(\mathcal{J}^o) = \sigma(\mathcal{O}^n)$ from the first part. ∎

3.9 Remark The Borel sets of the real line $\mathbb{R}$ are also generated by any of the following systems

$$\{(-\infty, a) : a \in \mathbb{Q}\}, \qquad \{(-\infty, a) : a \in \mathbb{R}\},$$
$$\{(-\infty, b] : b \in \mathbb{Q}\}, \qquad \{(-\infty, b] : b \in \mathbb{R}\},$$
$$\{(c, \infty) : c \in \mathbb{Q}\}, \qquad \{(c, \infty) : c \in \mathbb{R}\},$$
$$\{[d, \infty) : d \in \mathbb{Q}\}, \qquad \{[d, \infty) : d \in \mathbb{R}\}.$$

3.10 Remark One might think that $\sigma(\mathcal{G})$ can be explicitly constructed for any given $\mathcal{G}$ by adding to the family $\mathcal{G}$ all possible countable unions of its members and their complements:

$$\mathcal{G}_{\sigma c} := \left\{ \bigcup_{j \in \mathbb{N}} G_j, \left(\bigcup_{j \in \mathbb{N}} G_j \right)^c : G_j \in \mathcal{G} \right\}.$$

But $\mathcal{G}_{\sigma c}$ is not necessarily a σ-algebra.[✓] Even if we repeat this procedure *countably* often, i.e.

$$\mathcal{G}_n := (\ldots (\mathcal{G} \underbrace{_{\sigma c})_{\sigma c} \cdots)_{\sigma c}}_{n \text{ times}}, \qquad \widehat{\mathcal{G}} := \bigcup_{n \in \mathbb{N}} \mathcal{G}_n,$$

we end up, in general, with a set that is too small: $\widehat{\mathcal{G}} \subsetneq \sigma(\mathcal{G})$.[3]

---

[3] A 'constructive' approach along these lines is nevertheless possible if we use transfinite induction, see Hewitt and Stromberg [20, Theorem 10.23] or Appendix D.

This shows that the σ-operation produces a pretty big family; so big, in fact, that no approach using countably many countable set operations will give the whole of $\sigma(\mathcal{G})$. On the other hand, it is rather typical that a σ-algebra is given through its generator. In order to deal with these cases, we need the notion of *Dynkin systems* which will be introduced in Chapter 5.

Problems

3.1. Let $\mathcal{A}$ be a σ-algebra. Show that

(i) if $A_1, A_2, \ldots, A_N \in \mathcal{A}$, then $A_1 \cap A_2 \cap \ldots \cap A_N \in \mathcal{A}$;

(ii) $A \in \mathcal{A}$ if, and only if, $A^c \in \mathcal{A}$;

(iii) if $A, B \in \mathcal{A}$, then $A \setminus B \in \mathcal{A}$ and $A \triangle B \in \mathcal{A}$.

3.2. Prove the assertions made in Example 3.3 (iv), (vi) and (vii).
[Hint: use (2.5) for (vii).]

3.3. Verify the assertions made in Remark 3.5.

3.4. Let $X = [0, 1]$. Find the σ-algebra generated by the sets

(i) $\left(0, \frac{1}{2}\right)$;

(ii) $\left[0, \frac{1}{4}\right), \left(\frac{3}{4}, 1\right]$;

(iii) $\left[0, \frac{3}{4}\right], \left[\frac{1}{4}, 1\right]$.

3.5. Let $A_1, A_2, \ldots, A_N$ be subsets of X.

(i) If the A_j are disjoint and $\bigcup A_j = X$, then $\#\sigma(A_1, A_2, \ldots, A_N) = 2^N$.
Remark. A set A in a σ-algebra $\mathcal{A}$ is called an *atom*, if there is no proper subset $B \subsetneq A$ such that $B \in \mathcal{A}$. In this sense all A_j are atoms.

(ii) Show that $\sigma(A_1, A_2, \ldots, A_N)$ consists of finitely many sets.
[Hint: show that $\sigma(A_1, A_2, \ldots, A_N)$ has only finitely many atoms.]

3.6. Verify the properties $(\mathcal{O}_1)$–$(\mathcal{O}_3)$ for open sets in $\mathbb{R}^n$. Is $\mathcal{O}^n$ a σ-algebra?

3.7. Find an example (e.g. in $\mathbb{R}$) showing that $\bigcap_{j \in \mathbb{N}} U_j$ need not be open even if all U_j are open sets.

3.8. Prove any one of the assertions made in Remark 3.9.

3.9. Is this still true for the family $\mathbb{B}' := \{B_r(x) : x \in \mathbb{Q}^n, \ r \in \mathbb{Q}^+\}$?
[Hint: mimic the Proof of T3.8.]

3.10. Let $\mathcal{O}^n$ be the collection of open sets (topology) in $\mathbb{R}^n$ and let $A \subset \mathbb{R}^n$ be an arbitrary subset. We can introduce a topology $\mathcal{O}_A$ on A as follows: a set $V \subset A$ is called open (relative to A) if $V = U \cap A$ for some $U \in \mathcal{O}^n$. We write $\mathcal{O}_A$ for the open sets relative to A.

(i) Show that $\mathcal{O}_A$ is a topology on A, i.e. a family satisfying $(\mathcal{O}_1)$–$(\mathcal{O}_3)$.

(ii) If $A \in \mathcal{B}(\mathbb{R}^n)$, show that the trace σ-algebra $A \cap \mathcal{B}(\mathbb{R}^n)$ coincides with $\sigma(\mathcal{O}_A)$ (the latter is usually denoted by $\mathcal{B}(A)$: the Borel sets relative to A).

3.11. **Monotone classes.** A family $\mathcal{M} \subset \mathcal{P}(X)$ is called a *monotone class* if it is stable under countable unions and countable intersections, i.e.

$$(A_j)_{j \in \mathbb{N}} \subset \mathcal{M} \implies \bigcup_{j \in \mathbb{N}} A_j, \ \bigcap_{j \in \mathbb{N}} A_j \in \mathcal{M}.$$

(i) Mimic the proof of T3.4 to show that for every $\mathcal{E} \subset \mathcal{P}(X)$ there is a smallest monotone class $\mathrm{m}(\mathcal{E})$ containing $\mathcal{E}$.

(ii) Assume that $\emptyset \in \mathcal{E}$ and that $E \in \mathcal{E} \implies E^c \in \mathcal{E}$. Show that the system $\Sigma := \{B \in \mathrm{m}(\mathcal{E}) : B^c \in \mathrm{m}(\mathcal{E})\}$ is a σ-algebra.

(iii) Show that in (ii) $\mathcal{E} \subset \Sigma \subset \mathrm{m}(\mathcal{E}) \subset \sigma(\mathcal{E})$ holds and conclude that $\mathrm{m}(\mathcal{E}) = \sigma(\mathcal{E})$.

3.12. **Alternative characterization of $\mathcal{B}(\mathbb{R}^n)$.** In older books the Borel sets are often introduced as the smallest family $\mathcal{M}$ of sets which is stable under countable intersections and countable unions and which contains all open sets $\mathcal{O}^n$. The purpose of this exercise is to verify that $\mathcal{M} = \mathcal{B}(\mathbb{R}^n)$. Show that

(i) $\mathcal{M}$ is well-defined and $\mathcal{M} \subset \sigma(\mathcal{O}^n)$;

(ii) $U \in \mathcal{O}^n \implies U^c \in \mathcal{M}$, i.e. $\mathcal{M}$ contains all closed sets;

(iii) $\{B \in \mathcal{M} : B^c \in \mathcal{M}\}$ is a σ-algebra;

(iv) $\sigma(\mathcal{O}^n) \subset \{B \in \mathcal{M} : B^c \in \mathcal{M}\} \subset \mathcal{M}$.

[Hints: (i) – mimic T3.4(ii); (ii) – every closed set F is the intersection of the open sets $U_n := F + B_{1/n}(0) := \{y : x \in F, \ |x - y| < 1/n\}, n \in \mathbb{N}.$]

4

Measures

We are now ready to introduce one of the central concepts of measure and integration theory: measures. As before, X is some set and $\mathcal{A}$ is a σ-algebra on X.

4.1 Definition A (positive) *measure* μ on X is a mapping $\mu : \mathcal{A} \to [0, \infty]$ defined on a σ-algebra $\mathcal{A}$ satisfying

$$\mu(\emptyset) = 0, \tag{M_1}$$

and, for any countable family of pairwise disjoint sets $(A_j)_{j \in \mathbb{N}} \subset \mathcal{A}$,

$$(\sigma\text{-}additivity) \qquad \mu\left(\biguplus_{j \in \mathbb{N}} A_j\right) = \sum_{j \in \mathbb{N}} \mu(A_j). \tag{M_2}$$

If (M_1), (M_2) hold, but $\mathcal{A}$ is not a σ-algebra, μ is said to be a *pre-measure*.

Ƨ **Caution:** (M_2) requires implicitly that $\biguplus_j A_j$ is again in $\mathcal{A}$ – this is clearly the case for σ-algebras, but needs special attention if one deals with pre-measures.

4.2 Definition Let X be a set and $\mathcal{A}$ be a σ-algebra on X. The pair $(X, \mathcal{A})$ is called *measurable space*. If μ is a measure on X, $(X, \mathcal{A}, \mu)$ is called *measure space*.

A *finite measure* is a measure with $\mu(X) < \infty$ and a *probability measure* is a measure with $\mu(X) = 1$. The corresponding measure spaces are called *finite measure space* resp. *probability space*.

An *exhausting sequence* $(A_j)_{j \in \mathbb{N}} \subset \mathcal{A}$ is an increasing sequence of sets $A_1 \subset A_2 \subset A_3 \subset \ldots$ such that $\bigcup_{j \in \mathbb{N}} A_j = X$. A measure μ is said to be σ-*finite* and $(X, \mathcal{A}, \mu)$ is called a σ-*finite measure space*, if $\mathcal{A}$ contains an exhausting sequence $(A_j)_{j \in \mathbb{N}}$ such that $\mu(A_j) < \infty$ for all $j \in \mathbb{N}$.

Let us derive some immediate properties of (pre-)measures.

22

4.3 Proposition *Let $(X, \mathcal{A}, \mu)$ be a measure space and $A, B \in \mathcal{A}$. Then*

(i) $A \cap B = \emptyset \implies \mu(A \uplus B) = \mu(A) + \mu(B)$ (*finitely additive*)

(ii) $A \subset B \implies \mu(A) \leqslant \mu(B)$ (*monotone*)

(iii) $A \subset B, \mu(A) < \infty \implies \mu(B \setminus A) = \mu(B) - \mu(A)$

(iv) $\mu(A \cup B) + \mu(A \cap B) = \mu(A) + \mu(B)$ (*strongly additive*)

(v) $\mu(A \cup B) \leqslant \mu(A) + \mu(B)$. (*subadditive*)

Proof (i) Set $A_1 := A$, $A_2 := B$, $A_3 = A_4 = \ldots = \emptyset$. Then $(A_j)_{j \in \mathbb{N}}$ is a family of pairwise disjoint sets from $\mathcal{A}$. Moreover $A \uplus B = \biguplus_j A_j$ and by (M_2)

$$\mu(A \uplus B) = \mu\left(\biguplus_{j \in \mathbb{N}} A_j\right) = \sum_{j \in \mathbb{N}} \mu(A_j) = \mu(A) + \mu(B) + \mu(\emptyset) + \ldots$$

$$= \mu(A) + \mu(B).$$

(ii) If $A \subset B$, we have $B = A \uplus (B \setminus A)$, and by (i)

$$\mu(B) = \mu(A \uplus (B \setminus A)) = \mu(A) + \mu(B \setminus A) \tag{4.1}$$

$$\geqslant \mu(A). \tag{4.2}$$

(iii) If $A \subset B$, we can subtract the finite number $\mu(A)$ from both sides of (4.1) to get $\mu(B) - \mu(A) = \mu(B \setminus A)$.

(iv) For all $A, B \in \mathcal{A}$ we have

$$A \cup B = \big(A \setminus (A \cap B)\big) \uplus \big(A \cap B\big) \uplus \big(B \setminus (A \cap B)\big)$$

and using (i) twice we get

$$\mu(A \cup B) = \mu(A \setminus (A \cap B)) + \mu(A \cap B) + \mu(B \setminus (A \cap B)).$$

Adding $\mu(A \cap B)$ (which may assume the value $+\infty$) on both sides and using again (4.1) yields

$$\mu(A \cup B) + \mu(A \cap B)$$

$$= \mu(A \setminus (A \cap B)) + \mu(A \cap B) + \mu(B \setminus (A \cap B)) + \mu(A \cap B)$$

$$= \mu(A) + \mu(B).$$

(v) From (iv) we get $\mu(A) + \mu(B) = \mu(A \cup B) + \mu(A \cap B) \geqslant \mu(A \cup B)$ for all $A, B \in \mathcal{A}$. ∎

So far we have not really used the σ-additivity of μ in its full strength. The next theorem shows that σ-additivity is, in fact, some kind of continuity condition for (pre-)measures.

We call a sequence of sets $(A_j)_{j\in\mathbb{N}}$ *increasing*, if $A_1 \subset A_2 \subset A_3 \subset \ldots$ and we write in this case $A_j \uparrow A$ with *limit* $A = \bigcup_j A_j$. *Decreasing* sequences of sets are defined accordingly and we write $A_j \downarrow A$ with *limit* $A = \bigcap_j A_j$. All σ-algebras are stable under increasing or decreasing limits of their members.

4.4 Theorem *Let $(X, \mathcal{A})$ be a measurable space. A map $\mu : \mathcal{A} \to [0, \infty]$ is a measure if, and only if,*

  (i) $\mu(\emptyset) = 0$,

 (ii) $\mu(A \cup B) = \mu(A) + \mu(B)$ *for all $A, B \in \mathcal{A}$ with $A \cap B = \emptyset$,*

(iii) *(continuity of measures from below)*

     for any increasing sequence $(A_j)_{j\in\mathbb{N}} \subset \mathcal{A}$ with $A_j \uparrow A \in \mathcal{A}$ we have

$$\mu(A) = \lim_{j\to\infty} \mu(A_j) \quad \left(= \sup_{j\in\mathbb{N}} \mu(A_j) \right).$$

If $\mu(A) < \infty$ for all $A \in \mathcal{A}$, (iii) can be replaced by either of the following equivalent conditions

(iii′) *(continuity of measures from above)*

     for any decreasing sequence $(A_j)_{j\in\mathbb{N}} \subset \mathcal{A}$ with $A_j \downarrow A \in \mathcal{A}$ we have

$$\mu(A) = \lim_{j\to\infty} \mu(A_j) \quad \left(= \inf_{j\in\mathbb{N}} \mu(A_j) \right);$$

(iii″) *(continuity of measures at $\emptyset$)*

     for any decreasing sequence $(A_j)_{j\in\mathbb{N}} \subset \mathcal{A}$ with $A_j \downarrow \emptyset$ we have

$$\lim_{j\to\infty} \mu(A_j) = 0.$$

4.5 Remark With some obvious rewordings, P4.3 and T4.4 are still valid for pre-measures, i.e. for families $\mathcal{A}$ which are not σ-algebras. Of course, one has to make sure that $\emptyset \in \mathcal{A}$ and that $\mathcal{A}$ is stable under finite unions, intersections and differences of sets[1] (for P4.3) and, for T4.4, that increasing and decreasing sequences of the sets under consideration have their limits in $\mathcal{A}$. The proofs are literally the same.

Proof (of Theorem 4.4) Let us, first of all, check that every measure μ enjoys all the properties (i)–(iii) and (iii′), (iii″). Property (i) is clear from the definition of a measure and (ii) follows from P4.3(i). Let $(A_j)_{j\in\mathbb{N}} \subset \mathcal{A}$ be an increasing sequence of sets $A_j \uparrow A$ and set

$$B_1 := A_1, \ldots, B_{j+1} := A_{j+1} \setminus A_j.$$

---

[1] Such a family is called a *ring* of sets.

Obviously, $B_j \in \mathcal{A}$, the B_j are pairwise disjoint, $A_k = \bigcup_{j=1}^{k} B_j$ and $\bigcup_{k \in \mathbb{N}} A_k = \biguplus_{j \in \mathbb{N}} B_j = A$. Thus

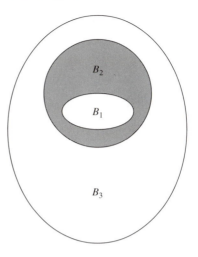

$$\mu(A) = \mu\left(\biguplus_{j \in \mathbb{N}} B_j\right) = \sum_{j \in \mathbb{N}} \mu(B_j)$$

$$= \lim_{k \to \infty} \sum_{j=1}^{k} \mu(B_j)$$

$$= \lim_{k \to \infty} \mu(B_1 \cup \ldots \cup B_k)$$

$$= \lim_{k \to \infty} \mu(A_k).$$

If $A_j \downarrow A$ we see easily that $(A_1 \setminus A_j) \uparrow (A_1 \setminus A)$ as $j \to \infty$. Since $\mu(A_1) < \infty$, the previous argument shows that

$$\mu(A_1 \setminus A) = \lim_{j \to \infty} \mu(A_1 \setminus A_j) = \lim_{j \to \infty} \big(\mu(A_1) - \mu(A_j)\big) = \mu(A_1) - \lim_{j \to \infty} \mu(A_j).$$

This means that $\mu(A_1) - \mu(A) = \mu(A_1) - \lim_{j \to \infty} \mu(A_j)$ and (iii′) follows. If we take, in particular, $A = \emptyset$, the above calculation also proves (iii″).

Let us now assume that (i)–(iii) hold for the set-function $\mu : \mathcal{A} \to [0, \infty]$. In order to see that μ is a measure, we have to check (M_2). For this take a sequence $(B_j)_{j \in \mathbb{N}} \subset \mathcal{A}$ of pairwise disjoint sets and define

$$A_k := B_1 \uplus \ldots \uplus B_k \in \mathcal{A}, \qquad A := \bigcup_{k \in \mathbb{N}} A_k = \biguplus_{j \in \mathbb{N}} B_j. \tag{4.3}$$

Clearly $A_k \uparrow A$, and using repeatedly property (ii) we get $\mu(A_k) = \mu(B_1) + \cdots + \mu(B_k)$. From (iii) we conclude

$$\mu(A) = \lim_{k \to \infty} \mu(A_k) = \lim_{k \to \infty} \sum_{j=1}^{k} \mu(B_j) = \sum_{j \in \mathbb{N}} \mu(B_j).$$

$$* \quad * \quad *$$

Finally assume that $\mu(A) < \infty$ for all $A \in \mathcal{A}$ and that (i), (ii) and (iii′) or (iii″) hold. We will show that under the finiteness assumption (iii′)$\Rightarrow$(iii″)$\Rightarrow$(iii); the assertion follows then from the considerations of the first part of the proof.

For (iii′)⇒(iii″) there is nothing to show. For the remaining implication take a sequence $(B_j)_{j\in\mathbb{N}} \subset \mathcal{A}$ of pairwise disjoint sets and define sets A_k and A as in (4.3). Then $A \setminus A_k \downarrow \emptyset$ and from (iii″) we conclude that $\lim_{k\to\infty} \mu(A \setminus A_k) = 0$. Since $\mu(A_k) < \infty$ we get $\mu(A) = \lim_{k\to\infty} \mu(A_k)$ and (iii) follows. ∎

4.6 Corollary *Every measure [pre-measure] is σ-subadditive, i.e.*

$$\mu\left(\bigcup_{j\in\mathbb{N}} A_j\right) \leqslant \sum_{j\in\mathbb{N}} \mu(A_j) \tag{4.4}$$

holds for all sequences $(A_j)_{j\in\mathbb{N}} \subset \mathcal{A}$ of not necessarily disjoint sets [such that $\bigcup_{j\in\mathbb{N}} A_j \in \mathcal{A}$].[2]

Proof Since the arguments are virtually the same, we may assume that $\mathcal{A}$ is a σ-algebra, so that μ becomes a measure. Set $B_k := A_1 \cup \ldots \cup A_k \uparrow \bigcup_{j\in\mathbb{N}} A_j$ as $k \to \infty$. By T4.4(iii) and repeated applications of P4.3(v),

$$\mu\left(\bigcup_{j\in\mathbb{N}} A_j\right) = \lim_{k\to\infty} \mu(A_1 \cup \ldots \cup A_k)$$

$$\leqslant \lim_{k\to\infty} \left(\mu(A_1) + \cdots + \mu(A_k)\right) = \sum_{j\in\mathbb{N}} \mu(A_j). \quad ∎$$

It is about time to give some examples of measures. At this stage this is, unfortunately, a somewhat difficult task! The main problem is that we have to explain for *every* set of the σ-algebra $\mathcal{A}$ what its measure $\mu(A)$ shall be. Since $\mathcal{A}$ can be very large – see Remark 3.10 – this is, in general, only (explicitly!) possible if either μ or $\mathcal{A}$ is very simple. Nevertheless ...

4.7 Examples   (i) (Dirac measure, unit mass) Let $(X, \mathcal{A})$ be any measurable space and let $x \in X$ be some point. Then $\delta_x : \mathcal{A} \to \{0, 1\}$, defined for $A \in \mathcal{A}$ by

$$\delta_x(A) := \begin{cases} 0 & \text{if } x \notin A, \\ 1 & \text{if } x \in A, \end{cases}$$

is a measure. It is called *Dirac's delta measure* or *unit mass* at the point x.

(ii) Consider $(\mathbb{R}, \mathcal{A})$ with $\mathcal{A}$ from Example 2.3(v) (i.e. $A \in \mathcal{A}$ if A or A^c is countable). Then $\gamma : \mathcal{A} \to \{0, 1\}$, defined for $A \in \mathcal{A}$ by

$$\gamma(A) := \begin{cases} 0 & \text{if } A \text{ is countable}, \\ 1 & \text{if } A^c \text{ is countable}, \end{cases}$$

is a measure.

---

[2] This is automatically fulfilled for a measure on a σ-algebra.

(iii) (Counting measure) Let $(X, \mathcal{A})$ be a measurable space. Then

$$|A| := \begin{cases} \#A & \text{if } A \text{ is finite,} \\ +\infty & \text{if } A \text{ is infinite,} \end{cases}$$

defines a measure. It is called *counting measure*.

(iv) (Discrete probability measure) Let $\Omega = \{\omega_1, \omega_2, \ldots\}$ be a countable set and $(p_j)_{j \in \mathbb{N}}$ be a sequence of real numbers $p_j \in [0, 1]$ such that $\sum_{j \in \mathbb{N}} p_j = 1$. On $(\Omega, \mathcal{P}(\Omega))$ the set-function

$$P(A) = \sum_{j : \omega_j \in A} p_j = \sum_{j \in \mathbb{N}} p_j \, \delta_{\omega_j}(A), \qquad A \subset \Omega,$$

defines a probability measure. The triplet $(\Omega, \mathcal{P}(\Omega), P)$ is called *discrete probability space*.

(v) (Trivial measures) Let $(X, \mathcal{A})$ be a measurable space. Then

$$\mu(A) := \begin{cases} 0 & \text{if } A = \emptyset \\ +\infty & \text{if } A \neq \emptyset, \end{cases} \qquad \text{and} \qquad \nu(A) := 0, \ A \in \mathcal{A}$$

are measures.

Note that our list of examples does not include the most familiar of all measures: length, area and volume.

4.8 Definition The set-function λ^n on $(\mathbb{R}^n, \mathcal{B}(\mathbb{R}^n))$ that assigns every half-open rectangle $[\![a, b)\!) = [a_1, b_1) \times \cdots \times [a_n, b_n) \in \mathcal{J}$ the value

$$\lambda^n([\![a, b)\!)) := \prod_{j=1}^{n} (b_j - a_j)$$

is called *n-dimensional Lebesgue measure*.

The problem here is that we do not know whether λ^n is a measure in the sense of Definition 4.1: λ^n is only explicitly given on the half-open rectangles $\mathcal{J}$ and it is not obvious at all that λ^n is a pre-measure on $\mathcal{J}$; much less clear is the question if and how we can extend this pre-measure from $\mathcal{J}$ to a proper measure on $\sigma(\mathcal{J})$. Over the next few chapters we will see that such an extension is indeed possible. But this requires some extra work and a more abstract approach. One of the main obstacles is, of course, that $\sigma(\mathcal{J})$ cannot be obtained by a bare-hands construction from $\mathcal{J}$.

Let us, meanwhile, note the upshot of what will be proved in the next chapters.

4.9 Theorem *Lebesgue measure* λ^n *exists, is a measure on the Borel sets* $\mathcal{B}(\mathbb{R}^n)$ *and is unique. Moreover,* λ^n *enjoys the following additional properties for* $B \in \mathcal{B}(\mathbb{R}^n)$:

(i) λ^n *is invariant under translations:* $\lambda^n(x+B) = \lambda^n(B)$, $x \in \mathbb{R}^n$;

(ii) λ^n *is invariant under motions:* $\lambda^n(R^{-1}(B)) = \lambda^n(B)$ *where* R *is a motion, i.e. a combination of translations, rotations and reflections;*

(iii) $\lambda^n(M^{-1}(B)) = |\det M|^{-1} \lambda^n(B)$ *for any invertible matrix* $M \in \mathbb{R}^{n \times n}$.

The attentive reader will have noticed that the sets $x + B := \{x + y : y \in B\}$, $R^{-1}(B) := \{R^{-1}(y) : y \in B\}$ and $M^{-1}(B)$ must again be Borel sets, otherwise the statement of T4.9 would be senseless, cf. T5.8 and Chapter 7.

Problems

4.1. Extend Proposition 4.3(i), (iv) and (v) to finitely many sets $A_1, A_2, \ldots, A_N \in \mathcal{A}$.

4.2. Check that the set-functions defined in Example 4.7 are measures in the sense of Definition 4.1.

4.3. Is the set-function γ of 4.7 (ii) still a measure on the measurable space $(\mathbb{R}, \mathcal{B}(\mathbb{R}))$? And on the measurable space $(\mathbb{Q}, \mathbb{Q} \cap \mathcal{B}(\mathbb{R}))$?

4.4. Let $X = \mathbb{R}$. For which σ-algebras are the following set-functions measures:

(i) $\mu(A) = \begin{cases} 0, & \text{if } A = \emptyset \\ 1, & \text{if } A \neq \emptyset; \end{cases}$
(ii) $\nu(A) = \begin{cases} 0, & \text{if } A \text{ is finite} \\ 1, & \text{if } A^c \text{ is finite?} \end{cases}$

4.5. Find an example showing that the finiteness condition in Theorem 4.4 (iii′) or (iii″) is essential.

[Hint: use Lebesgue measure or the counting measure on infinite tails $[k, \infty) \downarrow \emptyset$.]

4.6. Let $(X, \mathcal{A})$ be a measurable space.

(i) Let μ, ν be two measures on $(X, \mathcal{A})$. Show that for all $a, b \geqslant 0$ the set-function $\rho(A) := a\mu(A) + b\nu(A)$, $A \in \mathcal{A}$, is again a measure.

(ii) Let $\mu_1, \mu_2, \ldots$ be countably many measures on $(X, \mathcal{A})$ and let $(\alpha_j)_{j \in \mathbb{N}}$ be a sequence of positive numbers. Show that $\mu(A) := \sum_{j=1}^{\infty} \alpha_j \mu_j(A)$, $A \in \mathcal{A}$, is again a measure.

[Hint: to show σ-additivity use (and prove) the following helpful lemma: *for any double sequence* β_{ij}, $i, j \in \mathbb{N}$, *of real numbers we have*

$$\sup_{i \in \mathbb{N}} \sup_{j \in \mathbb{N}} \beta_{ij} = \sup_{j \in \mathbb{N}} \sup_{i \in \mathbb{N}} \beta_{ij}.$$

Thus $\lim_{i \to \infty} \lim_{j \to \infty} \beta_{ij} = \lim_{j \to \infty} \lim_{i \to \infty} \beta_{ij}$ if $i \mapsto \beta_{ij}$, and $j \mapsto \beta_{ij}$ increases when the other index is fixed.]

4.7. Let $(X, \mathcal{A}, \mu)$ be a measure space and $F \in \mathcal{A}$. Show that $\mathcal{A} \ni A \mapsto \mu(A \cap F)$ defines a measure.

4.8. Let $(\Omega, \mathcal{A}, P)$ be a probability space and $(A_j)_{j \in \mathbb{N}} \subset \mathcal{A}$ a sequence of sets with $P(A_j) = 1$ for all $j \in \mathbb{N}$. Show that $P(\bigcap_{j \in \mathbb{N}} A_j) = 1$.

4.9. Let $(X, \mathcal{A}, \mu)$ be a finite measure space and $(A_j)_{j \in \mathbb{N}}, (B_j)_{j \in \mathbb{N}} \subset \mathcal{A}$ such that $A_j \supset B_j$ for all $j \in \mathbb{N}$. Show that

$$\mu\left(\bigcup_{j \in \mathbb{N}} A_j\right) - \mu\left(\bigcup_{j \in \mathbb{N}} B_j\right) \leqslant \sum_{j \in \mathbb{N}} (\mu(A_j) - \mu(B_j)).$$

[Hint: show first that $\bigcup_j A_j \setminus \bigcup_k B_k \subset \bigcup_j (A_j \setminus B_j)$ then use C4.6.]

4.10. **Null sets.** Let $(X, \mathcal{A}, \mu)$ be a measure space. A set $N \in \mathcal{A}$ is called a *null set* or *μ-null set* if $\mu(N) = 0$. We write $\mathcal{N}_\mu$ for the family of all μ-null sets. Check that $\mathcal{N}_\mu$ has the following properties:

  (i) $\emptyset \in \mathcal{N}_\mu$;
  (ii) if $N \in \mathcal{N}_\mu$, $M \in \mathcal{A}$ and $M \subset N$, then $M \in \mathcal{N}_\mu$;
  (iii) if $(N_j)_{j \in \mathbb{N}} \subset \mathcal{N}_\mu$, then $\bigcup_{j \in \mathbb{N}} N_j \in \mathcal{N}_\mu$.

4.11. Let λ be one-dimensional Lebesgue measure.

  (i) Show that for all $x \in \mathbb{R}$ the set $\{x\}$ is a Borel set with $\lambda(\{x\}) = 0$.
  [Hint: consider the intervals $[x - 1/k, x + 1/k)$, $k \in \mathbb{N}$ and use Theorem 4.4.]

  (ii) Prove that $\mathbb{Q}$ is a Borel set and that $\lambda(\mathbb{Q}) = 0$ in two ways:

    a) by using the first part of the problem;
    b) by considering the set $C(\epsilon) := \bigcup_{k \in \mathbb{N}} [q_k - \epsilon\, 2^{-k}, q_k + \epsilon\, 2^{-k})$, where $(q_k)_{k \in \mathbb{N}}$ is an enumeration of $\mathbb{Q}$, and letting $\epsilon \to 0$.

  (iii) Use the trivial fact that $[0, 1] = \bigcup_{0 \leqslant x \leqslant 1} \{x\}$ to show that a non-countable union of null sets (here: $\{x\}$) is not necessarily a null set.

4.12. Determine all null sets of the measure $\delta_a + \delta_b$, $a, b \in \mathbb{R}$, on $(\mathbb{R}, \mathcal{B}(\mathbb{R}))$.

4.13. **Completion (1).** We have seen in Problem 4.10 that *measurable* subsets of null sets are again null sets: $M \in \mathcal{A}, M \subset N \in \mathcal{A}, \mu(N) = 0$ then $\mu(M) = 0$; but there might be subsets of N which are not in $\mathcal{A}$. This motivates the following definition: *a measure space $(X, \mathcal{A}^*, \mu)$ (or a measure μ) is* **complete** *if all subsets of μ-null sets are again in $\mathcal{A}^*$*. In other words: if all subsets of a null set are null sets.

The following exercise shows that a measure space $(X, \mathcal{A}, \mu)$ which is not yet complete can be completed.

  (i) $\mathcal{A}^* := \{A \cup N : A \in \mathcal{A}, \quad N \text{ is a subset of some } \mathcal{A}\text{-measurable null set}\}$ is a σ-algebra satisfying $\mathcal{A} \subset \mathcal{A}^*$.

  (ii) $\bar{\mu}(A^*) := \mu(A)$ for $A^* = A \cup N \in \mathcal{A}^*$ is well-defined, i.e. it is independent of the way we can write A^*, say as $A^* = A \cup N = B \cup M$ where $A, B \in \mathcal{A}$ and M, N are subsets of null sets.

  (iii) $\bar{\mu}$ is a measure on $\mathcal{A}^*$ and $\bar{\mu}(A) = \mu(A)$ for all $A \in \mathcal{A}$.

    (iv) $(X, \mathcal{A}^*, \bar{\mu})$ is complete.

    (v) We have $\mathcal{A}^* = \{A^* \subset X : \exists A, B \in \mathcal{A}, \quad A \subset A^* \subset B, \quad \mu(B \setminus A) = 0\}$.

4.14. Restriction. Let $(X, \mathcal{A}, \mu)$ be a measure space and let $\mathcal{B} \subset \mathcal{A}$ be a sub-σ-algebra. Denote by $\nu := \mu|_{\mathcal{B}}$ the *restriction* of μ to $\mathcal{B}$.

    (i) Show that ν is again a measure.

    (ii) Assume that μ is a finite measure [a probability measure]. Is ν still a finite measure [a probability measure]?

    (iii) Does ν inherit σ-finiteness from μ?

4.15. Show that a measure space $(X, \mathcal{A}, \mu)$ is σ-finite if, and only if, there exists a sequence of measurable sets $(E_j)_{j \in \mathbb{N}} \subset \mathcal{A}$ such that $\bigcup_{j \in \mathbb{N}} E_j = X$ and $\mu(E_j) < \infty$ for all $j \in \mathbb{N}$.

5

Uniqueness of measures

Before we embark on the proof of the existence of measures in the following chapter, let us first check whether it is enough to consider measures on some *generator* $\mathcal{G}$ of a σ-algebra – otherwise our construction of Lebesgue measure would be flawed from the start.

As mentioned in Remark 3.10 a major problem is that, apart from trivial cases, $\sigma(\mathcal{G})$ cannot be constructively obtained from $\mathcal{G}$. To overcome this obstacle we need a new concept.

5.1 Definition A family $\mathcal{D} \subset \mathcal{P}(X)$ is a *Dynkin system* if

$$X \in \mathcal{D}, \tag{Δ_1}$$

$$D \in \mathcal{D} \implies D^c \in \mathcal{D}, \tag{Δ_2}$$

$$(D_j)_{j \in \mathbb{N}} \subset \mathcal{D} \text{ pairwise disjoint} \implies \biguplus_{j \in \mathbb{N}} D_j \in \mathcal{D}. \tag{Δ_3}$$

5.2 Remark As for σ-algebras, cf. Properties 3.2, one sees that $\emptyset \in \mathcal{D}$ and that finite disjoint unions are again in $\mathcal{D}$: $D, E \in \mathcal{E}, D \cap E = \emptyset \implies D \uplus E \in \mathcal{D}$. Of course, every σ-algebra is a Dynkin system, but the converse is, in general, wrong[✓], Problem 5.2.

5.3 Proposition *Let* $\mathcal{G} \subset \mathcal{P}(X)$*. Then there is a smallest (also minimal, coarsest) Dynkin system* $\delta(\mathcal{G})$ *containing* $\mathcal{G}$*.* $\delta(\mathcal{G})$ *is called the* Dynkin system generated *by* $\mathcal{G}$*. Moreover,* $\mathcal{G} \subset \delta(\mathcal{G}) \subset \sigma(\mathcal{G})$*.*

Proof The proof that $\delta(\mathcal{G})$ exists parallels the proof of T3.4(ii). As in the case of σ-algebras, $\delta(\mathcal{D}) = \mathcal{D}$ if $\mathcal{D}$ is a Dynkin system (by minimality) and so $\delta(\sigma(\mathcal{G})) = \sigma(\mathcal{G})$. Hence, $\mathcal{G} \subset \sigma(\mathcal{G})$ implies that $\delta(\mathcal{G}) \subset \delta(\sigma(\mathcal{G})) = \sigma(\mathcal{G})$. ∎

It is important to know when a Dynkin system is already a σ-algebra.

5.4 Lemma *A Dynkin system $\mathcal{D}$ is a σ-algebra if, and only if, it is stable under finite intersections:[1] $D, E \in \mathcal{D} \implies D \cap E \in \mathcal{D}$.*

Proof Since a σ-algebra is $\cap$-stable (cf. Properties 3.2, Problem 3.1) as well as a Dynkin system (Remark 5.2) it only remains to show that a $\cap$-stable Dynkin system $\mathcal{D}$ is a σ-algebra.

Let $(D_j)_{j \in \mathbb{N}}$ be a sequence of subsets in $\mathcal{D}$. We have to show that $D := \bigcup_{j \in \mathbb{N}} D_j \in \mathcal{D}$. Set $E_1 := D_1 \in \mathcal{D}$ and

$$E_{j+1} := (\ldots((D_{j+1} \setminus D_j) \setminus D_{j-1}) \setminus \ldots) \setminus D_1$$
$$= D_{j+1} \cap D_j^c \cap D_{j-1}^c \cap \ldots \cap D_1^c \in \mathcal{D}$$

where we used (Δ_2) and the assumed $\cap$-stability of $\mathcal{D}$. The E_j are obviously mutually disjoint and $D = \biguplus_{j \in \mathbb{N}} E_j \in \mathcal{D}$ by (Δ_3).   ∎

Lemma 5.4 is not applicable if $\mathcal{D}$ is given in terms of a generator $\mathcal{G}$, which is often the case. The next theorem is very important for applications as it extends Lemma 5.4 to the much more convenient setting of generators.

5.5 Theorem *If $\mathcal{G} \subset \mathcal{P}(X)$ is stable under finite intersections, then $\delta(\mathcal{G}) = \sigma(\mathcal{G})$.*

Proof We have already established $\delta(\mathcal{G}) \subset \sigma(\mathcal{G})$ in P5.3. If we knew that $\delta(\mathcal{G})$ were a σ-algebra, the minimality of $\sigma(\mathcal{G})$ and $\mathcal{G} \subset \delta(\mathcal{G})$ would immediately imply $\sigma(\mathcal{G}) \subset \delta(\mathcal{G})$, hence equality.

In view of L5.4 it is enough to show that $\delta(\mathcal{G})$ is $\cap$-stable. For this we fix some $D \in \delta(\mathcal{G})$ and introduce the family

$$\mathcal{D}_D := \{Q \subset X : Q \cap D \in \delta(\mathcal{G})\}.$$

Let us check that $\mathcal{D}_D$ is a Dynkin system: (Δ_1) is obviously true. (Δ_2): take $Q \in \mathcal{D}_D$. Then

$$Q^c \cap D = (Q^c \cup D^c) \cap D = (Q \cap D)^c \cap D = (\underbrace{(Q \cap D)}_{\in \delta(\mathcal{G})} \uplus \underbrace{D^c}_{\in \delta(\mathcal{G})})^c \qquad (5.1)$$

and disjoint unions of sets from $\delta(\mathcal{G})$ are still in $\delta(\mathcal{G})$. Thus $Q^c \in \mathcal{D}_D$. (Δ_3): let $(Q_j)_{j \in \mathbb{N}}$ be a sequence of pairwise disjoint sets from $\mathcal{D}_D$. By definition, $(Q_j \cap D)_{j \in \mathbb{N}}$ is a disjoint sequence in $\delta(\mathcal{G})$ and (Δ_3) for the Dynkin system $\delta(\mathcal{G})$ shows

$$\left(\biguplus_{j \in \mathbb{N}} Q_j \right) \cap D = \biguplus_{j \in \mathbb{N}} (Q_j \cap D) \in \delta(\mathcal{G}),$$

which means that $\biguplus_{j \in \mathbb{N}} Q_j \in \mathcal{D}_D$.

---

[1] $\cap$-stable, for short.

Since $\mathcal{G} \subset \delta(\mathcal{G})$ and since $\mathcal{G}$ is $\cap$-stable, we have $\mathcal{G} \subset \mathcal{D}_G$ for all $G \in \mathcal{G}$.[✓] But $\mathcal{D}_G$ is a Dynkin system and so $\delta(\mathcal{G}) \subset \mathcal{D}_G$ for *all* $G \in \mathcal{G}$ (use P5.3, Problem 5.4). Consequently, if $D \in \delta(\mathcal{G})$ and $G \in \mathcal{G}$ we find because of $\delta(\mathcal{G}) \subset \mathcal{D}_G$ and the very definition of $\mathcal{D}_G$ that

$$G \cap D \in \delta(\mathcal{G}) \qquad \forall\, G \in \mathcal{G}, \quad \forall D \in \delta(\mathcal{G})$$

so $\qquad \mathcal{G} \subset \mathcal{D}_D \qquad \forall D \in \delta(\mathcal{G})$

and $\qquad \delta(\mathcal{G}) \subset \mathcal{D}_D \qquad \forall D \in \delta(\mathcal{G}).$

The latter just says that $\delta(\mathcal{G})$ is stable under intersections with $D \in \delta(\mathcal{G})$. By Lemma 5.4 $\delta(\mathcal{G})$ is a σ-algebra and the theorem is proved. ∎

5.6 Remark The technique used in the proof of Theorem 5.5 is an extremely important and powerful tool. We will use it almost exclusively in this chapter to prove the uniqueness of measures theorem and some properties of Lebesgue measure λ^n.

5.7 Theorem (Uniqueness of measures). *Assume that $(X, \mathcal{A})$ is a measurable space and that $\mathcal{A} = \sigma(\mathcal{G})$ is generated by a family $\mathcal{G}$ such that*

- *$\mathcal{G}$ is stable under finite intersections: $G, H \in \mathcal{G} \implies G \cap H \in \mathcal{G}$;*
- *there exists an exhausting sequence $(G_j)_{j \in \mathbb{N}} \subset \mathcal{G}$ with $G_j \uparrow X$.*

Any two measures μ, ν that coincide on $\mathcal{G}$ and are finite for all members of the exhausting sequence $\mu(G_j) = \nu(G_j) < \infty$, are equal on $\mathcal{A}$, i.e. $\mu(A) = \nu(A)$ for all $A \in \mathcal{A}$.

Proof For $j \in \mathbb{N}$ we define

$$\mathcal{D}_j := \left\{ A \in \mathcal{A} : \mu(G_j \cap A) = \nu(G_j \cap A) \quad (< \infty!) \right\}$$

and we claim that every $\mathcal{D}_j$ is a Dynkin system. (Δ_1) is clear. (Δ_2): if $A \in \mathcal{D}_j$ we have

$$\mu(G_j \cap A^c) = \mu(G_j \setminus A) = \mu(G_j) - \mu(G_j \cap A)$$
$$= \nu(G_j) - \nu(G_j \cap A)$$
$$= \nu(G_j \setminus A) = \nu(G_j \cap A^c),$$

so that $A^c \in \mathcal{D}_j$. (Δ_3): if $(A_k)_{k \in \mathbb{N}} \subset \mathcal{D}_j$ are mutually disjoint sets, we get

$$\mu\Big(G_j \cap \biguplus_{k \in \mathbb{N}} A_k\Big) = \mu\Big(\biguplus_{k \in \mathbb{N}}(G_j \cap A_k)\Big) = \sum_{k \in \mathbb{N}} \mu(G_j \cap A_k)$$

$$= \sum_{k \in \mathbb{N}} \nu(G_j \cap A_k) = \nu\Big(\biguplus_{k \in \mathbb{N}}(G_j \cap A_k)\Big) = \nu\Big(G_j \cap \biguplus_{k \in \mathbb{N}} A_k\Big),$$

and $\biguplus_{k \in \mathbb{N}} A_k \in \mathcal{D}_j$ follows.

Since $\mathcal{G}$ is $\cap$-stable, we know from T5.5 that $\delta(\mathcal{G}) = \sigma(\mathcal{G})$; therefore,

$$\mathcal{D}_j \supset \mathcal{G} \implies \mathcal{D}_j \supset \delta(\mathcal{G}) = \sigma(\mathcal{G}) \qquad \forall j \in \mathbb{N}.$$

On the other hand, $\mathcal{A} = \sigma(\mathcal{G}) \subset \mathcal{D}_j \subset \mathcal{A}$, which means that $\mathcal{A} = \mathcal{D}_j$ for all $j \in \mathbb{N}$, and so

$$\mu(G_j \cap A) = \nu(G_j \cap A) \qquad \forall j \in \mathbb{N}, \quad \forall A \in \mathcal{A}. \tag{5.2}$$

Using T4.4(iii) we can let $j \to \infty$ in (5.2) to get

$$\mu(A) = \lim_{j \to \infty} \mu(G_j \cap A) = \lim_{j \to \infty} \nu(G_j \cap A) = \nu(A) \qquad \forall A \in \mathcal{A}. \qquad \blacksquare$$

The following two theorems show why Lebesgue measure (if it exists) plays a very special rôle indeed.

5.8 Theorem (i) *n-dimensional Lebesgue measure λ^n is invariant under translations, i.e.*

$$\lambda^n(x + B) = \lambda^n(B) \qquad \forall x \in \mathbb{R}^n, \ \forall B \in \mathcal{B}(\mathbb{R}^n). \tag{5.3}$$

(ii) *Every measure μ on $(\mathbb{R}^n, \mathcal{B}(\mathbb{R}^n))$ which is invariant under translations and satisfies $\kappa = \mu([0, 1)^n) < \infty$ is a multiple of Lebesgue measure: $\mu = \kappa \lambda^n$.*

Proof First of all we should convince ourselves that

$$B \in \mathcal{B}(\mathbb{R}^n) \implies x + B \in \mathcal{B}(\mathbb{R}^n) \qquad \forall x \in \mathbb{R}^n \tag{5.4}$$

otherwise the statement of T5.8 would be senseless. For this set

$$\mathcal{A}_x := \big\{ B \in \mathcal{B}(\mathbb{R}^n) : x + B \in \mathcal{B}(\mathbb{R}^n) \big\} \subset \mathcal{B}(\mathbb{R}^n).$$

It is clear that $\mathcal{A}_x$ is a σ-algebra and that $\mathcal{J} \subset \mathcal{A}_x$.[✓] Hence, $\mathcal{B}(\mathbb{R}^n) = \sigma(\mathcal{J}) \subset \mathcal{A}_x \subset \mathcal{B}(\mathbb{R}^n)$ and (5.4) follows. We can now start the proof proper.

(i) Set $\nu(B) := \lambda^n(x+B)$ for some fixed $x = (x_1, \ldots, x_n) \in \mathbb{R}^n$. It is easy to check that ν is a measure on $(\mathbb{R}^n, \mathcal{B}(\mathbb{R}^n))^{[\checkmark]}$. Take $I = [a_1, b_1) \times \cdots \times [a_n, b_n) \in \mathcal{J}$ and observe that

$$x + I = [a_1 + x_1, b_1 + x_1) \times \cdots \times [a_n + x_n, b_n + x_n) \in \mathcal{J},$$

so that

$$\nu(I) = \lambda^n(x+I) = \prod_{j=1}^{n} \big((b_j + x_j) - (a_j + x_j)\big) = \prod_{j=1}^{n} (b_j - a_j) = \lambda^n(I).$$

This means that $\nu|_{\mathcal{J}} = \lambda^n|_{\mathcal{J}}$.[2] But $\mathcal{J}$ is $\cap$-stable,[3] generates $\mathcal{B}(\mathbb{R}^n)$ and admits the exhausting sequence

$$[-k, k)^n \uparrow \mathbb{R}^n, \qquad \lambda^n\big([-k, k)^n\big) = (2k)^n < \infty.$$

We can now invoke T5.5 to see that $\lambda^n = \nu$ on the whole of $\mathcal{B}(\mathbb{R}^n)$.

(ii) Take $I \in \mathcal{J}$ as in part (I) but with rational endpoints $a_j, b_j \in \mathbb{Q}$. Thus there is some $M \in \mathbb{N}$ and $k(I) \in \mathbb{N}$ and points $x^{(j)} \in \mathbb{R}^n$, such that

$$I = \overset{k(I)}{\underset{j=1}{\biguplus}} \big(x^{(j)} + [0, \tfrac{1}{M})^n\big)$$

i.e. we pave the rectangle I by little squares $[0, \tfrac{1}{M})^n$ of side-length $1/M$, where M is, say, the common denominator of all a_j and b_j. Using the translation invariance of μ and λ^n, we see

$$\mu(I) = k(I)\,\mu\big([0, \tfrac{1}{M})^n\big), \qquad\qquad \mu\big([0, 1)^n\big) = M^n\,\mu\big([0, \tfrac{1}{M})^n\big),$$

$$\lambda^n(I) = k(I)\,\lambda^n\big([0, \tfrac{1}{M})^n\big), \qquad\qquad \underbrace{\lambda^n\big([0, 1)^n\big)}_{=1} = M^n\,\lambda^n\big([0, \tfrac{1}{M})^n\big),$$

and dividing the top two and bottom two equalities gives

$$\mu(I) = \frac{k(I)}{M^n}\,\mu\big([0, 1)^n\big), \qquad\qquad \lambda^n(I) = \frac{k(I)}{M^n}\,\lambda^n\big([0, 1)^n\big) = \frac{k(I)}{M^n}.$$

Thus $\mu(I) = \mu\big([0, 1)^n\big)\,\lambda^n(I) = \kappa\,\lambda^n(I)$ for all $I \in \mathcal{J}$ and, as in part (I), an application of T5.5 finishes the proof. ∎

Incidentally, Theorem 5.8 proves Theorem 4.9(I). Further properties of Lebesgue measure will be studied in the following chapters, but first we concentrate on its existence.

---

[2] This is short for $\nu(I) = \lambda^n(I) \quad \forall I \in \mathcal{J}$.

[3] Use $\overset{n}{\underset{j=1}{\times}} [a_j, b_j) \cap \overset{n}{\underset{j=1}{\times}} [a_j', b_j') = \overset{n}{\underset{j=1}{\times}} [a_j \vee a_j', b_j \wedge b_j');.^{[\checkmark]}$

Problems

5.1. Verify the claims made in Remark 5.2.

5.2. The following exercise shows that Dynkin systems and σ-algebras are, in general, different: Let $X = \{1, 2, 3, \ldots, 2k-1, 2k\}$ for some fixed $k \in \mathbb{N}$. Then the family $\mathcal{D} = \{A \subset X : \#A \quad \text{is even}\}$ is a Dynkin system, but not a σ-algebra.

5.3. Let $\mathcal{D}$ be a Dynkin system. Show that for all $A, B \in \mathcal{D}$ the difference $B \setminus A \in \mathcal{D}$. [Hint: use $R \setminus Q = ((R \cap Q) \cup R^c)^c$ where $R, Q \subset X$.]

5.4. Let $\mathcal{A}$ be a σ-algebra, $\mathcal{D}$ be a Dynkin system and $\mathcal{G} \subset \mathcal{H} \subset \mathcal{P}(X)$ two collections of subsets of X. Show that

   (i) $\delta(\mathcal{A}) = \mathcal{A}$ and $\delta(\mathcal{D}) = \mathcal{D}$;
   (ii) $\delta(\mathcal{G}) \subset \delta(\mathcal{H})$;
   (iii) $\delta(\mathcal{G}) \subset \sigma(\mathcal{G})$.

5.5. Let $A, B \subset X$. Compare $\delta(\{A, B\})$ and $\sigma(\{A, B\})$. When are they equal?

5.6. Show that Theorem 5.7 is still valid, if $(G_j)_{j \in \mathbb{N}} \subset \mathcal{G}$ is not an increasing sequence but *any* countable family of sets such that

$$(1) \quad \bigcup_{j \in \mathbb{N}} G_j = X \qquad \text{and} \qquad (2) \quad \nu(G_j) = \mu(G_j) < \infty.$$

[Hint: set $F_N := G_1 \cup \ldots \cup G_N = F_{N-1} \cup G_N$ and check by induction that $\mu(F_N) = \nu(F_N)$; use then T5.7.]

5.7. Show that the half-open intervals $\mathcal{J}^n$ in $\mathbb{R}^n$ are stable under finite intersections. [Hint: check that $I = \underset{j=1}{\overset{n}{\times}} [a_j, b_j)$, $I' = \underset{j=1}{\overset{n}{\times}} [a'_j, b'_j)$ satisfy $I \cap I' = \underset{j=1}{\overset{n}{\times}} [a_j \vee a'_j, b_j \wedge b'_j)$.]

5.8. **Dilations.** Mimic the proof of Theorem 5.8(I) and show that $t \cdot B := \{tb : b \in B\}$ is a Borel set for all $B \in \mathcal{B}(\mathbb{R}^n)$ and $t > 0$. Moreover,

$$\lambda^n(t \cdot B) = t^n \lambda^n(B) \qquad \forall B \in \mathcal{B}(\mathbb{R}^n), \ \forall t > 0. \tag{5.5}$$

5.9. **Invariant measures.** Let $(X, \mathcal{A}, \mu)$ be a finite measure space where $\mathcal{A} = \sigma(\mathcal{G})$ for some $\cap$-stable generator $\mathcal{G}$. Assume that $\theta : X \to X$ is a map such that $\theta^{-1}(A) \in \mathcal{A}$ for all $A \in \mathcal{A}$. Prove that

$$\mu(G) = \mu(\theta^{-1}(G)) \quad \forall G \in \mathcal{G} \quad \Longrightarrow \quad \mu(A) = \mu(\theta^{-1}(A)) \quad \forall A \in \mathcal{A}.$$

(A measure μ with this property is called *invariant* w.r.t. the map θ.)

5.10. **Independence (1).** Let $(\Omega, \mathcal{A}, P)$ be a probability space and let $\mathcal{B}, \mathcal{C} \subset \mathcal{A}$ be two sub-σ-algebras of $\mathcal{A}$. We call $\mathcal{B}$ and $\mathcal{C}$ *independent*, if

$$P(B \cap C) = P(B) P(C) \qquad \forall B \in \mathcal{B}, \ C \in \mathcal{C}.$$

Assume now that $\mathcal{B} = \sigma(\mathcal{G})$ and $\mathcal{C} = \sigma(\mathcal{H})$ where $\mathcal{G}, \mathcal{H}$ are $\cap$-stable collections of sets. Prove that $\mathcal{B}$ and $\mathcal{C}$ are independent if, and only if,

$$P(G \cap H) = P(G) P(H) \qquad \forall G \in \mathcal{G}, \ H \in \mathcal{H}.$$

6

Existence of measures

In Chapter 4 we saw that it is not a trivial task to assign *explicitly* a μ-value to *every* set A from a σ-algebra $\mathcal{A}$. Rather than doing this it is often more natural to assign μ-values to, say, rectangles (in the case of the Borel σ-algebra) or, in general, to sets from some generator $\mathcal{G}$ of $\mathcal{A}$. Because of Theorem 4.4 (and Remark 4.5) $\mu|_{\mathcal{G}}$ should be a pre-measure. If $\mathcal{G}$ and μ satisfy the conditions of the uniqueness theorem 5.7, this approach will lead to a unique measure on $\mathcal{A}$, provided we can extend μ from $\mathcal{G}$ onto $\sigma(\mathcal{G}) = \mathcal{A}$.

To get such an automatic extension the following (technically motivated) class of generators is useful. A *semi-ring* is a family $\mathcal{S} \subset \mathcal{P}(X)$ with the following properties:

$$\emptyset \in \mathcal{S}, \qquad\qquad (S_1)$$

$$S, T \in \mathcal{S} \implies S \cap T \in \mathcal{S}, \qquad\qquad (S_2)$$

$$\text{for } S, T \in \mathcal{S} \text{ there exist finitely many disjoint}$$
$$S_1, S_2, \ldots, S_M \in \mathcal{S} \text{ such that } S \setminus T = \bigcup_{j=1}^{M} S_j. \qquad (S_3)$$

The solution to our problems is the following deep extension theorem for measures which goes back to Carathéodory [9].

6.1 Theorem (Carathéodory) *Let $\mathcal{S}$ be a semi-ring of subsets of X and $\mu : \mathcal{S} \to [0, \infty]$ be a pre-measure, i.e. a set-function with*

(i) $\mu(\emptyset) = 0$;

(ii) $(S_j)_{j \in \mathbb{N}} \subset \mathcal{S}$, *disjoint and* $S = \bigcup_{j \in \mathbb{N}} S_j \in \mathcal{S} \implies \mu(S) = \sum_{j \in \mathbb{N}} \mu(S_j)$.

Then μ has an extension to a measure μ on $\sigma(\mathcal{S})$. If, moreover, $\mathcal{S}$ contains an exhausting sequence $(S_j)_{j\in\mathbb{N}}$, $S_j \uparrow X$ such that $\mu(S_j) < \infty$ for all $j \in \mathbb{N}$, then the extension is unique.

6.2 Remark From the Definition 4.1 of a measure it is clear that the conditions 6.1(i) and (ii) are necessary for μ to become a measure. Theorem 6.1 says that they are even sufficient. Remarkable is the fact that (ii) is only needed relative to $\mathcal{S}$ – its extension to $\sigma(\mathcal{S})$ is then automatic.

The proof of Carathéodory's theorem is a bit involved and not particularly rewarding when read superficially. Therefore we recommend skipping the proof on first reading and resuming on p. 44.

Proof (of Theorem 6.1) We begin with the construction of an auxiliary set-function $\mu^*: \mathcal{P}(X) \to [0, \infty]$ which will, eventually, extend μ. Define for each $A \subset X$ the family of countable $\mathcal{S}$-coverings of A

$$\mathcal{C}(A) := \left\{ (S_j)_{j\in\mathbb{N}} \subset \mathcal{S} : \bigcup_{j\in\mathbb{N}} S_j \supset A \right\}$$

$(\mathcal{C}(A) = \emptyset$ is possible since we do not require $X \in \mathcal{S})$, and set

$$\mu^*(A) := \inf \left\{ \sum_{j\in\mathbb{N}} \mu(S_j) : (S_j)_{j\in\mathbb{N}} \in \mathcal{C}(A) \right\} \tag{6.1}$$

where, as usual, $\inf \emptyset := +\infty$.

Step 1: Claim: μ^* has the following three properties:[1]

$$\mu^*(\emptyset) = 0, \tag{OM_1}$$

(monotone) $$A \subset B \implies \mu^*(A) \leqslant \mu^*(B), \tag{OM_2}$$

$$(A_j)_{j\in\mathbb{N}} \subset \mathcal{P}(X) \implies$$

(σ-subadditive) $\qquad\qquad\qquad\qquad\qquad\qquad\qquad\qquad\qquad$ (OM_3)

$$\mu^*\left(\bigcup_{j\in\mathbb{N}} A_j \right) \leqslant \sum_{j\in\mathbb{N}} \mu^*(A_j).$$

(OM_1) is obvious since we can take in (6.1) the constant sequence $S_1 = S_2 = \ldots = \emptyset$ which is clearly in $\mathcal{C}(\emptyset)$.

(OM_2): if $B \supset A$, then each $\mathcal{S}$-cover of B also covers A, i.e. $\mathcal{C}(B) \subset \mathcal{C}(A)$. Therefore,

$$\mu^*(A) = \inf \left\{ \sum_{j\in\mathbb{N}} \mu(S_j) : (S_j)_{j\in\mathbb{N}} \in \mathcal{C}(A) \right\}$$

$$\leqslant \inf \left\{ \sum_{k\in\mathbb{N}} \mu(T_k) : (T_k)_{k\in\mathbb{N}} \in \mathcal{C}(B) \right\} = \mu^*(B).$$

---

[1] A set-function $\mu^*: \mathcal{P}(X) \to [0, \infty]$ satisfying (OM_1)–(OM_3) is called *outer measure*.

(OM_3): without loss of generality we can assume that $\mu^*(A_j) < \infty$ for all $j \in \mathbb{N}$ and so $\mathcal{C}(A_j) \neq \emptyset$. Fix $\epsilon > 0$ and observe that by the very nature of the infimum we find for each A_j a cover $(S_k^j)_{k \in \mathbb{N}} \in \mathcal{C}(A_j)$ with

$$\sum_{k \in \mathbb{N}} \mu(S_k^j) \leqslant \mu^*(A_j) + \frac{\epsilon}{2^j}, \qquad j \in \mathbb{N}. \tag{6.2}$$

The double sequence $(S_k^j)_{j,k \in \mathbb{N}}$ is an $\mathcal{S}$-cover of $A := \bigcup_{j \in \mathbb{N}} A_j$, and so

$$\mu^*(A) \leqslant \sum_{(j,k) \in \mathbb{N} \times \mathbb{N}} \mu(S_k^j) = \sum_{j \in \mathbb{N}} \sum_{k \in \mathbb{N}} \mu(S_k^j)$$

$$\overset{(6.2)}{\leqslant} \sum_{j \in \mathbb{N}} \left(\mu^*(A_j) + \frac{\epsilon}{2^j} \right)$$

$$= \sum_{j \in \mathbb{N}} \mu^*(A_j) + \epsilon,$$

where the second '$\leqslant$' follows from (6.2). Letting $\epsilon \to 0$ proves (OM_3).

Step 2. Claim: μ^* extends μ, i.e. $\mu^*(S) = \mu(S) \quad \forall S \in \mathcal{S}$.

Observe that μ can be uniquely extended to the set $\mathcal{S}_{\cup} := \{S_1 \uplus \ldots \uplus S_M : M \in \mathbb{N}, \ S_j \in \mathcal{S}\}$ of all finite unions of disjoint $\mathcal{S}$-sets by

$$\bar{\mu}(S_1 \uplus \ldots \uplus S_M) := \sum_{j=1}^{M} \mu(S_j). \tag{6.3}$$

Since (6.3) is *necessary* for an additive set-function on $\mathcal{S}_{\cup}$, (6.3) implies the uniqueness of the extension[✓] once we know that $\bar{\mu}$ is well-defined, that is, independent of the particular representation of sets in $\mathcal{S}_{\cup}$. To see this assume that

$$S_1 \uplus \ldots \uplus S_M = T_1 \uplus \ldots \uplus T_N, \qquad M, N \in \mathbb{N}, \ S_j, T_k \in \mathcal{S}.$$

Then

$$S_j = S_j \cap (T_1 \uplus \ldots \uplus T_N) = \biguplus_{k=1}^{N} (S_j \cap T_k),$$

and the additivity of μ on $\mathcal{S}$ shows

$$\mu(S_j) = \sum_{k=1}^{N} \mu(S_j \cap T_k).$$

Summing over $j = 1, 2, \ldots, M$ and swapping the rôles of S_j and T_k gives

$$\sum_{j=1}^{M} \mu(S_j) = \sum_{j=1}^{M} \sum_{k=1}^{N} \mu(S_j \cap T_k) = \sum_{k=1}^{N} \mu(T_k),$$

which proves that (6.3) does not depend on the representation of $\mathcal{S}_\cup$-sets.

The family $\mathcal{S}_\cup$ is clearly stable under finite *disjoint* unions. If $S, T \in \mathcal{S}_\cup$ we find (notation as before)

$$S \cap T = (S_1 \uplus \ldots \uplus S_M) \cap (T_1 \uplus \ldots \uplus T_N) = \biguplus_{j,k=1}^{M,N} \underbrace{(S_j \cap T_k)}_{\in \mathcal{S}} \in \mathcal{S}_\cup,$$

and, since by (S_3) $S_j \setminus T_k \in \mathcal{S}_\cup$, also

$$S \setminus T = (S_1 \uplus \ldots \uplus S_M) \setminus (T_1 \uplus \ldots \uplus T_N)$$

$$= \biguplus_{j=1}^{M} \bigcap_{k=1}^{N} (S_j \cap T_k^c) = \biguplus_{j=1}^{M} \underbrace{\bigcap_{k=1}^{N} \underbrace{S_j \setminus T_k}_{\in \mathcal{S}_\cup}}_{\in \mathcal{S}_\cup} \in \mathcal{S}_\cup,$$

where we used the $\cap$- and $\uplus$-stability of $\mathcal{S}_\cup$. Finally,

$$S \cup T = (S \setminus T) \uplus (S \cap T) \uplus (T \setminus S) \in \mathcal{S}_\cup,^2$$

and the prescription (6.3) can be used to extend μ to finite unions of $\mathcal{S}$-sets.

Let us show that $\bar{\mu}$ is σ-additive on $\mathcal{S}_\cup$, i.e. a pre-measure. For this take $(T_k)_{k \in \mathbb{N}} \subset \mathcal{S}_\cup$ such that $T := \bigcup_{k \in \mathbb{N}} T_k \in \mathcal{S}_\cup$. By the definition of the family $\mathcal{S}_\cup$ we find a sequence of disjoint sets $(S_j)_{j \in \mathbb{N}} \subset \mathcal{S}$ and a sequence of integers $0 = n(0) \leqslant n(1) \leqslant n(2) \leqslant \ldots$ such that

$$T_k = S_{n(k-1)+1} \uplus \ldots \uplus S_{n(k)}, \quad k \in \mathbb{N}$$

and $T = U_1 \uplus \ldots \uplus U_N$, where $U_\ell = \biguplus_{j \in J_\ell} S_j \in \mathcal{S}^{[\checkmark]}$ with disjoint index sets $J_1 \uplus J_2 \uplus \ldots \uplus J_N = \mathbb{N}$ partitioning $\mathbb{N}$. Thus

$$\bar{\mu}(T) \stackrel{def}{=} \sum_{\ell=1}^{N} \mu(U_\ell) \stackrel{6.1(ii)}{=} \sum_{\ell=1}^{N} \sum_{j \in J_\ell} \mu(S_j) = \sum_{k \in \mathbb{N}} \sum_{j=n(k-1)+1}^{n(k)} \mu(S_j) \stackrel{def}{=} \sum_{k \in \mathbb{N}} \bar{\mu}(T_k),$$

which proves σ-additivity of $\bar{\mu}$.

---

^{2} This shows that $\mathcal{S}_\cup$ is the ring generated by $\mathcal{S}$, i.e. the smallest ring containing $\mathcal{S}$.

Using the pre-measure $\bar{\mu}$ we get from Corollary 4.6 for any cover $(S_j)_{j\in\mathbb{N}} \in \mathcal{C}(S)$, $S \in \mathcal{S}$, that

$$\mu(S) = \bar{\mu}(S) = \bar{\mu}\Big(\bigcup_{j\in\mathbb{N}} S_j \cap S\Big) \leqslant \sum_{j\in\mathbb{N}} \bar{\mu}(S_j \cap S)$$

$$= \sum_{j\in\mathbb{N}} \mu(S_j \cap S) \leqslant \sum_{j\in\mathbb{N}} \mu(S_j),$$

and passing to the infimum over $\mathcal{C}(S)$ shows $\mu(S) \leqslant \mu^*(S)$. The special cover $(S, \emptyset, \emptyset, \ldots) \in \mathcal{C}(S)$, on the other hand, yields $\mu^*(S) \leqslant \mu(S)$ and this shows that $\mu|_{\mathcal{S}} = \mu^*|_{\mathcal{S}}$.

Step 3. Claim: $\mathcal{S} \subset \mathcal{A}^*$, where $\mathcal{A}^*$ is given by

$$\mathcal{A}^* := \big\{A \subset X : \mu^*(Q) = \mu^*(Q \cap A) + \mu^*(Q \setminus A) \quad \forall Q \subset X\big\}. \tag{6.4}$$

Let $S, T \in \mathcal{S}$. From (S_3) we get

$$T = (S \cap T) \uplus (T \setminus S) = (S \cap T) \uplus \biguplus_{j=1}^{M} S_j$$

for some mutually disjoint sets $S_j \in \mathcal{S}$, $j = 1, 2, \ldots, M$. Since μ is additive on $\mathcal{S}$ and μ^* is (σ-)subadditive by (OM_3), we find

$$\mu^*(S \cap T) + \mu^*(T \setminus S) \leqslant \mu(S \cap T) + \sum_{j=1}^{M} \mu(S_j) = \mu(T). \tag{6.5}$$

Take any $B \subset X$ and some $\mathcal{S}$-cover $(T_j)_{j\in\mathbb{N}} \in \mathcal{C}(B)$. Using $\mu^*(T_j) = \mu(T_j)$ and summing the inequality (6.5) for $T = T_j$ over $j \in \mathbb{N}$ yields

$$\sum_{j\in\mathbb{N}} \mu^*(T_j \setminus S) + \sum_{j\in\mathbb{N}} \mu^*(T_j \cap S) \leqslant \sum_{j\in\mathbb{N}} \mu^*(T_j),$$

and the σ-subadditivity (OM_3) and monotonicity (OM_2) of μ^* give (recall that $B \subset \bigcup_{j\in\mathbb{N}} T_j$)

$$\mu^*(B \setminus S) + \mu^*(B \cap S) \leqslant \mu^*\Big(\bigcup_{j\in\mathbb{N}} T_j \setminus S\Big) + \mu^*\Big(\bigcup_{j\in\mathbb{N}} T_j \cap S\Big)$$

$$\leqslant \sum_{j\in\mathbb{N}} \mu^*(T_j) = \sum_{j\in\mathbb{N}} \mu(T_j).$$

We can now pass to the inf over $\mathcal{C}(B)$ and find $\mu^*(B \setminus S) + \mu^*(B \cap S) \leqslant \mu^*(B)$. Since the reverse inequality follows easily from the (σ-)subadditivity (OM_3) of μ^*, $S \in \mathcal{A}^*$ holds for all $S \in \mathcal{S}$.

Step 4. Claim: $\mathcal{A}^*$ is a σ-algebra and μ^* is a measure on $(X, \mathcal{A}^*)$.

Clearly, $\emptyset \in \mathcal{A}^*$ and by the symmetry (w.r.t. A and A^c) of definition (6.4) of $\mathcal{A}^*$ we have $A \in \mathcal{A}^*$ if, and only if, $A^c \in \mathcal{A}^*$. Let us show that $\mathcal{A}^*$ is $\cup$-stable. Using the (σ-)subadditivity (OM_3) of μ^* we find for $A, A' \in \mathcal{A}^*$ and any $P \subset X$

$$\mu^*(P \cap (A \cup A')) + \mu^*(P \setminus (A \cup A'))$$
$$= \mu^*(P \cap (A \cup [A' \setminus A])) + \mu^*(P \setminus (A \cup A'))$$
$$\leqslant \mu^*(P \cap A) + \mu^*(P \cap (A' \setminus A)) + \mu^*(P \setminus (A \cup A'))$$
$$= \mu^*(P \cap A) + \mu^*((P \setminus A) \cap A') + \mu^*((P \setminus A) \setminus A')$$
$$\overset{(6.4)}{=} \mu^*(P \cap A) + \mu^*(P \setminus A) \tag{6.6}$$
$$\overset{(6.4)}{=} \mu^*(P), \tag{6.6'}$$

where we used in the last two steps the definition (6.4) of $\mathcal{A}^*$ with $Q \mathrel{\hat{=}} P \setminus A$ and $Q \mathrel{\hat{=}} P$, respectively. The reverse inequality follows from (OM_3), hence equality, and we conclude that $A \cup A' \in \mathcal{A}^*$.

If A, A' are disjoint, the equality (6.6)=(6.6') becomes, for $P := (A \uplus A') \cap Q$, $Q \subset X$,

$$\mu^*(Q \cap (A \uplus A')) = \mu^*(Q \cap A) + \mu^*(Q \cap A') \qquad \forall Q \subset X,$$

and a simple induction argument yields

$$\mu^*(Q \cap (A_1 \uplus \ldots \uplus A_M)) = \sum_{j=1}^{M} \mu^*(Q \cap A_j) \qquad \forall Q \subset X$$

for all mutually disjoint $A_1, A_2, \ldots, A_M \in \mathcal{A}^*$.

In particular, if $(A_j)_{j \in \mathbb{N}} \subset \mathcal{A}^*$ is a sequence of pairwise disjoint sets, we find for their union $A := \biguplus_{j \in \mathbb{N}} A_j$ that

$$\mu^*(Q \cap A) \geqslant \mu^*(Q \cap (A_1 \uplus \ldots \uplus A_M)) = \sum_{j=1}^{M} \mu^*(Q \cap A_j). \tag{6.7}$$

Since $A_1 \cup \ldots \cup A_M \in \mathcal{A}^*$, we can use (OM_3) and (6.7) to deduce

$$\mu^*(Q) = \mu^*(Q \cap (A_1 \cup \ldots \cup A_M)) + \mu^*(Q \setminus (A_1 \cup \ldots \cup A_M))$$
$$\geqslant \mu^*(Q \cap (A_1 \cup \ldots \cup A_M)) + \mu^*(Q \setminus A) \tag{6.8}$$
$$= \sum_{j=1}^{M} \mu^*(Q \cap A_j) + \mu^*(Q \setminus A).$$

The left-hand side is independent of M; therefore, we can let $M \to \infty$ and get

$$\mu^*(Q) \geqslant \sum_{j=1}^{\infty} \mu^*(Q \cap A_j) + \mu^*(Q \setminus A) \geqslant \mu^*(Q \cap A) + \mu^*(Q \setminus A). \qquad (6.9)$$

The reverse inequality $\mu^*(Q) \leqslant \mu^*(Q \cap A) + \mu^*(Q \setminus A)$ follows at once from the subadditivity of μ^*. This means that equality holds throughout (6.9) and we get $A \in \mathcal{A}^*$. If we take $Q := A$ in (6.9) we even see the σ-additivity of μ^* on $\mathcal{A}^*$.

So far we have seen that $\mathcal{A}^*$ is a $\cup$-stable Dynkin system. Because of $A \cap B = (A^c \cup B^c)^c$ we see that $\mathcal{A}^*$ is also $\cap$-stable and, by L5.4, a σ-algebra.

Step 5. Claim: μ^* is a measure on $\sigma(\mathcal{S})$ which extends μ.

By step 3, $\mathcal{S} \subset \mathcal{A}^*$ and thus $\sigma(\mathcal{S}) \subset \sigma(\mathcal{A}^*) = \mathcal{A}^*$ since $\mathcal{A}^*$ is itself a σ-algebra (step 4). Again by step 4, $\mu^*|_{\sigma(\mathcal{S})}$ is a measure which, by step 2, extends μ.

Step 6. Uniqueness of $\mu^*|_{\sigma(\mathcal{S})}$. If there is an exhausting sequence $(S_j)_{j \in \mathbb{N}} \subset \mathcal{S}$, $S_j \uparrow X$ such that $\mu(S_j) < \infty$ for all $j \in \mathbb{N}$, it follows from T5.7 that any two extensions of μ to $\sigma(\mathcal{S})$ coincide. ∎

6.3 Remark The core of Carathéodory's theorem 6.1 is the definition (6.4) of μ^*-measurable sets, i.e. of the σ-algebra $\mathcal{A}^*$. The proof shows that, in general, we cannot expect μ^* to be (σ-)additive outside $\mathcal{A}^*$. In many situations the σ-algebra $\mathcal{P}(X)$ is simply too big to support a non-trivial measure. Notable exceptions are countable sets X or Dirac measures[✓]. For n-dimensional Lebesgue measure, this was first remarked by Hausdorff [19, pp. 401–402]. The general case depends on the cardinality of X and the behaviour of μ on one-point sets; see the discussion in Oxtoby [33, Chapter 5].

Put in other words this says that even a household measure like Lebesgue measure cannot assign a content to *every* set! In $\mathbb{R}^3$ (and higher dimensions) we even have the *Banach–Tarski paradox*: the open balls $B_1(0)$ and $B_2(0)$ with centre 0 and radii 1 resp. 2 have finite disjoint decompositions $B_1(0) = \bigcup_{j=1}^{M} E_j$ and $B_2(0) = \bigcup_{j=1}^{M} F_j$ such that for every $j = 1, 2, \ldots, M$ the sets E_j and F_j are geometrically congruent (hence, should have the same Lebesgue measure); see Stromberg [49] or Wagon [52]. Of course, not all of the sets E_j and F_j can be Borel sets.

This brings us to the question if and how we can construct a *non-Borel measurable set*, i.e. a set $A \in \mathcal{P}(\mathbb{R}^n) \setminus \mathcal{B}(\mathbb{R}^n)$. Such constructions are possible but they are based on the axiom of choice, see for example Hewitt and Stromberg [20, pp. 136–7], Oxtoby [33, pp. 22–3] or Appendix D.

$$* \qquad * \qquad *$$

Let us now apply Theorem 6.1 to prove the existence of n-dimensional Lebesgue measure λ^n which was defined for half-open rectangles $\mathcal{J}^n = \mathcal{J}(\mathbb{R}^n)$ in D4.8:

$$\lambda^n([\![a, b)\!)) = \prod_{j=1}^{n} (b_j - a_j), \qquad [\![a, b)\!) = \underset{j=1}{\overset{n}{\bigtimes}} [a_j, b_j) \in \mathcal{J}^n.$$

6.4 Proposition *The family of n-dimensional rectangles $\mathcal{J}^n$ is a semi-ring.*

Proof (By induction) It is obvious that $\mathcal{J}^1$ satisfies the properties (S_1)–(S_3) from page 37. Assume that $\mathcal{J}^n$ is a semi-ring for some $n \geqslant 1$. From the definition of rectangles it is clear that

$$\mathcal{J}^{n+1} = \mathcal{J}^n \times \mathcal{J}^1 := \{ I_n \times I_1 : I_n \in \mathcal{J}^n,\ I_1 \in \mathcal{J}^1 \}.$$

(S_1) is obviously true and (S_2) follows from the identity

$$(I_n \times I_1) \cap (J_n \times J_1) = (I_n \cap J_n) \times (I_1 \cap J_1) \tag{6.10}$$

where $I_n, J_n \in \mathcal{J}^n$ and $I_1, J_1 \in \mathcal{J}^1$. Since

$$(J_n \times J_1)^c$$
$$= \{ (x, y) : x \notin J_n,\ y \notin J_1 \quad \text{or} \quad x \in J_n,\ y \notin J_1 \quad \text{or} \quad x \notin J_n,\ y \in J_1 \}$$
$$= (J_n^c \times J_1^c) \uplus (J_n \times J_1^c) \uplus (J_n^c \times J_1)$$

we see, using (6.10),

$$(I_n \times I_1) \setminus (J_n \times J_1)$$
$$= (I_n \times I_1) \cap (J_n \times J_1)^c$$
$$= \big[(I_n \setminus J_n) \times (I_1 \setminus J_1) \big] \uplus \big[(I_n \cap J_n) \times (I_1 \setminus J_1) \big] \uplus \big[(I_n \setminus J_n) \times (I_1 \cap J_1) \big].$$

Both $I_n \setminus J_n$ and $I_1 \setminus J_1$ are made up of finitely many disjoint rectangles from $\mathcal{J}^n$ and $\mathcal{J}^1$, and therefore $(I_n \times I_1) \setminus (J_n \times J_1)$ is a finite union of disjoint rectangles from $\mathcal{J}^n \times \mathcal{J}^1$; thus (S_3) holds. ∎

In $\mathbb{R}^2$ it is easy to depict the two typical situations that occur in the proof of (S_3) in Proposition 6.4:

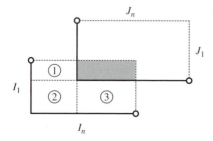
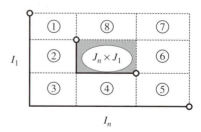

The proof of Proposition 6.4 reveals a bit more: the Cartesian product of any two semi-rings is again a semi-ring.[✓]

6.5 Proposition λ^n *is a pre-measure on $\mathcal{J}^n$*

Proof It is enough to verify (i), (ii) and (iii″) of Theorem 4.4 since λ^n assigns finite measure to every rectangle in $\mathcal{J}^n$.

We consider only the case $n = 2$, since $n = 1$ is similar but easier and $n \geqslant 3$ adds only notational complications. Obviously, $\lambda^2(\emptyset) = 0$. To see additivity on $\mathcal{J}^2$, we may as well cut $I = [a_1, b_1) \times [a_2, b_2)$ along one direction (say, along $j = 2$) to get $I_1 = [a_1, b_1) \times [a_2, \gamma)$, $I_2 = [a_1, b_1) \times [\gamma, b_2)$ and reassemble it $I = I_1 \uplus I_2$ (if $n \geqslant 3$ this is accomplished by a hyperplane). Thus

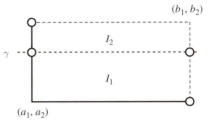

$$\lambda^2(I_1) + \lambda^2(I_2) = (b_1 - a_1)(\gamma - a_2) + (b_1 - a_1)(b_2 - \gamma)$$
$$= (b_1 - a_1)(b_2 - \gamma + \gamma - a_2)$$
$$= (b_1 - a_1)(b_2 - a_2)$$
$$= \lambda^2(I).$$

Now let $(I_j)_{j\in\mathbb{N}} \subset \mathcal{J}^2$, $I_j = [\![a^{(j)}, b^{(j)})\!)$, be a decreasing sequence of rectangles $I_j \downarrow \emptyset$. We have to show that $\lim_{j\to\infty} \lambda^2(I_j) = 0$. Since $I_j \downarrow \emptyset$, it is clear that at least in one coordinate direction, say $k = 2$, we have $\lim_{j\to\infty}\big(b_2^{(j)} - a_2^{(j)}\big) = 0$, otherwise $\bigcap_{j\in\mathbb{N}} I_j$ would contain a rectangle with side-lengths $\lim_{j\to\infty}\big(b_k^{(j)} - b_k^{(j)}\big) > 0$ for $k = 1, 2$. But then

$$\lambda^2(I_j) = \prod_{k=1}^{2}\big(b_k^{(j)} - a_k^{(j)}\big) \leqslant \max_{k\neq 2}\big(b_k^{(j)} - a_k^{(j)}\big)^{2-1}\big(b_2^{(j)} - a_2^{(j)}\big) \xrightarrow{j\to\infty} 0. \quad\blacksquare$$

6.6 Corollary (Existence of Lebesgue measure) *There is a unique extension of n-dimensional Lebesgue pre-measure λ^n from $\mathcal{J}^n$ (Definition 4.8) to a measure on the Borel sets $\mathcal{B}(\mathbb{R}^n)$. This extension is again denoted by λ^n and is called Lebesgue measure.*

Proof We know from Theorem 3.8 that $\mathcal{B}(\mathbb{R}^n) = \sigma(\mathcal{J}^n)$. Since $[-k, k)^n \uparrow \mathbb{R}^n$ is an exhausting sequence of cubes and since $\lambda^n\big([-k, k)^n\big) = (2k)^n < \infty$, all conditions of Carathéodory's theorem 6.1 are fulfilled, and λ^n extends to a measure on $\mathcal{B}(\mathbb{R}^n)$. $\quad\blacksquare$

6.7 Remark The uniqueness of Lebesgue measure and its properties (cf. Theorem 4.9) show that it is necessarily the familiar elementary-geometric volume (length, area ...)-function $\text{vol}^{(n)}(\bullet)$ in the sense that $\text{vol}^{(n)}$ can in only one way be extended to a measure on the Borel σ-algebra.

Problems

6.1. Consider on $\mathbb{R}$ the family Σ of all Borel sets which are symmetric w.r.t. the origin. Show that Σ is a σ-algebra. Is it possible to extend a pre-measure μ on Σ to a measure on $\mathcal{B}(\mathbb{R})$? If so, is this extension unique? Continues in Problem 9.12.

6.2. **Completion (2).** Recall from Problem 4.13 that a measure space $(X, \mathcal{A}, \mu)$ is complete, if every subset of a μ-null set is a μ-null set (thus, in particular, measurable). Let $(X, \mathcal{A}, \mu)$ be a σ-finite measure space – i.e. there is an exhausting sequence $(A_j)_{j\in\mathbb{N}} \subset \mathcal{A}$ such that $\mu(A_j) < \infty$. As in the proof of Theorem 6.1 we write μ^* for the outer measure (1) – now defined using $\mathcal{A}$-coverings – and $\mathcal{A}^*$ for the σ-algebra defined by (6.4).

    (i) Show that for every $Q \subset X$ there is some $A \in \mathcal{A}$ such that $\mu^*(Q) = \mu(A)$ and that $\mu(N) = 0$ for all $N \subset Q \setminus A$ with $N \in \mathcal{A}$.
    [Hint: since μ^* is defined as an infimum, every Q with $\mu^*(Q) < \infty$ admits a sequence $B_k \in \mathcal{A}$ with $B_k \supset Q$ and $\mu(B) - \mu^*(Q) \leqslant 1/k$. If $\mu^*(Q) = \infty$, consider for each $j \in \mathbb{N}$ the set $Q \cap A_j$.]
    (ii) Show that $(X, \mathcal{A}^*, \mu^*|_{\mathcal{A}^*})$ is a complete measure space.
    (iii) Show that $(X, \mathcal{A}^*, \mu^*|_{\mathcal{A}^*})$ is the completion of $(X, \mathcal{A}, \mu)$ in the sense of Problem 4.13.

6.3.   (i) Show that non-void open sets in $\mathbb{R}$ (resp. $\mathbb{R}^n$) have always strictly positive Lebesgue measure.
    [Hint: let U be open. Find a small ball in U and inscribe a cube.]
    (ii) Is (i) still true for closed sets?

6.4.   (i) Show that $\lambda^1((a, b)) = b - a$ for all $a, b \in \mathbb{R}, a \leqslant b$.
    [Hint: approximate (a, b) by half-open intervals and use Theorem 4.4.]
    (ii) Let $H \subset \mathbb{R}^2$ be a hyperplane which is perpendicular to the x_1-direction (that is to say: H is a translate of the x_2-axis). Show that $H \in \mathcal{B}(\mathbb{R}^2)$ and $\lambda^2(H) = 0$.
    [Hint: consider the sets $A_k = [-\epsilon 2^{-k}, \epsilon 2^{-k}) \times [-k, k)$ and note that $H \subset y + \cup_{k\in\mathbb{N}} A_k$ for some y.]
    (iii) State and prove the $\mathbb{R}^n$-analogues of (i) and (ii).

6.5. Let $(X, \mathcal{A}, \mu)$ be a measure space such that all *singletons* $\{x\} \in \mathcal{A}$. A point x is called an *atom*, if $\mu(\{x\}) > 0$. A measure is called *non-atomic* or *diffuse*, if there are no atoms.

    (i) Show that one-dimensional Lebesgue measure λ^1 is diffuse.
    (ii) Give an example of a non-diffuse measure on $(\mathbb{R}, \mathcal{B}(\mathbb{R}))$.
    (iii) Show that for a diffuse measure μ on $(X, \mathcal{A})$ all countable sets are null sets.

(iv) Show that every probability measure P on $(\mathbb{R}, \mathcal{B}(\mathbb{R}))$ can be decomposed into a sum of two measures $\mu + \nu$, where μ is diffuse and ν is a measure of the form $\nu = \sum_{j \in \mathbb{N}} \epsilon_j \delta_{x_j}$, $\epsilon_j > 0$, $x_j \in \mathbb{R}$.

[Hint: since $P(\mathbb{R}) = 1$, there are at most k points $y_1^{(k)}, y_2^{(k)}, \dots, y_k^{(k)}$ such that $\frac{1}{k-1} > P(\{y_j^{(k)}\}) \geqslant \frac{1}{k}$. Find by recursion (in k) all points satisfying such a relation. There are at most countably many of these $y_j^{(k)}$. Relabel them as $x_1, x_2, \dots$. These are the atoms of P. Now take $\epsilon_j = P(\{y_j\})$, define ν as stated and prove that ν and $P - \nu$ are measures.]

6.6. A set $A \subset \mathbb{R}^n$ is called *bounded*, if it can be contained in a ball $B_r(0) \supset A$ of finite radius r. A set $A \subset \mathbb{R}^n$ is called *connected*, if we can go along a curve from any point $a \in A$ to any other point $a' \in A$ without ever leaving A, cf. Appendix B.

  (i) Construct an open and unbounded set in $\mathbb{R}$ with finite, strictly positive Lebesgue measure.

  [Hint: try unions of ever smaller open intervals centred around $n \in \mathbb{N}$.]

 (ii) Construct an open, unbounded and connected set in $\mathbb{R}^2$ with finite, strictly positive Lebesgue measure.

  [Hint: try a union of adjacent, ever longer, ever thinner rectangles.]

(iii) Is there a connected, open and unbounded set in $\mathbb{R}$ with finite, strictly positive Lebesgue measure?

6.7. Let $\lambda := \lambda^1|_{[0,1]}$ be Lebesgue measure on $([0, 1], \mathcal{B}[0, 1])$. Show that for every $\epsilon > 0$ there is a dense open set $U \subset [0, 1]$ with $\lambda(U) \leqslant \epsilon$.

[Hint: take an enumeration $(q_j)_{j \in \mathbb{N}}$ of $\mathbb{Q} \cap (0, 1)$ and make each q_j the centre of a small open interval.]

6.8. Let $\lambda = \lambda^1$ be Lebesgue measure on $(\mathbb{R}, \mathcal{B}(\mathbb{R}))$. Show that $N \in \mathcal{B}(\mathbb{R})$ is a null set if, and only if, for every $\epsilon > 0$ there is an open set $U = U_\epsilon \supset N$ such that $\lambda(U) < \epsilon$.

[Hint: sufficiency is trivial, for necessity use λ^* constructed in Theorem 6.1 (6.1) from $\lambda|_{\mathcal{O}}$ and observe that by Theorem 6.1 $\lambda|_{\mathcal{B}(\mathbb{R}^n)} = \lambda^*|_{\mathcal{B}(\mathbb{R}^n)}$. This gives the required open cover.]

6.9. **Borel–Cantelli lemma (1) – the direct half.** Prove the following theorem.

Theorem (Borel–Cantelli lemma). *Let* $(\Omega, \mathcal{A}, P)$ *be a probability space. For every sequence* $(A_j)_{j \in \mathbb{N}} \subset \mathcal{A}$ *we have*

$$\sum_{j=1}^{\infty} P(A_j) < \infty \quad \Longrightarrow \quad P\Big(\bigcap_{n=1}^{\infty} \bigcup_{j=n}^{\infty} A_j \Big) = 0. \tag{6.11}$$

[Hint: use Theorem 4.4 and the fact that $P\big(\bigcup_{j \geqslant n} A_j \big) \leqslant \sum_{j \geqslant n} P(A_j)$.]

Remark. This is the 'easy' or direct half of the so-called Borel–Cantelli lemma; the more difficult part see T18.9. The condition $\omega \in \bigcap_{n=1}^{\infty} \bigcup_{j=n}^{\infty} A_j$ means that ω happens to be in *infinitely many* of the A_j and the lemma gives a simple sufficient condition when certain events happen almost surely *not* infinitely often, i.e. only finitely often with probability one.

6.10. **Non-measurable sets (1).** Let μ be a measure on $\mathcal{A} = \{\emptyset, [0, 1), [1, 2), [0, 2)\}$, $X = [0, 2)$, such that $\mu([0, 1)) = \mu([1, 2)) = \frac{1}{2}$ and $\mu([0, 2)) = 1$. Denote by μ^* and $\mathcal{A}^*$ the outer measure and σ-algebra which appear in the proof of Theorem 6.1.

(i) Find $\mu^*((a, b))$ and $\mu^*(\{a\})$ for all $0 \leqslant a < b < 2$ if we use $\mathcal{S} = \mathcal{A}$ in T6.1;

(ii) Show that $(0, 1), \{0\} \notin \mathcal{A}^*$.

6.11. **Non-measurable sets (2).** Consider on $X = \mathbb{R}$ the σ-algebra $\mathcal{A} := \{A \subset \mathbb{R} : A \text{ or } A^c \text{ is countable}\}$ from Example 3.3(v) and the measure $\gamma(A)$ from 4.7(ii) which is 0 or 1 according to A or A^c being countable. Denote by γ^* and $\mathcal{A}^*$ the outer measure and σ-algebra which appear in the proof of Theorem 6.1.

(i) Find γ^* if we use $\mathcal{S} = \mathcal{A}$ in T6.1;

(ii) Show that no set $B \subset \mathbb{R}$, such that both B and B^c are uncountable, is in $\mathcal{A}$ or in $\mathcal{A}^*$.

7

Measurable mappings

In this chapter we consider maps $T: X \to X'$ between two measurable spaces $(X, \mathcal{A})$ and $(X', \mathcal{A}')$ which respect the measurable structures, that is σ-algebras, on X and X'. Such maps can be used to transport a given measure μ, defined on $(X, \mathcal{A})$, onto $(X', \mathcal{A}')$. We have already used this technique in Theorem 5.8, where we considered shifts of sets: $A \rightsquigarrow x + A$, but it is in probability theory where this concept is truly fundamental: you use it whenever you speak of the 'distribution' of a 'random variable'.

7.1 Definition Let $(X, \mathcal{A}), (X', \mathcal{A}')$ be two measurable spaces. A map $T: X \to X'$ is called $\mathcal{A}/\mathcal{A}'$-*measurable* (or *measurable* unless this is too ambiguous) if the pre-image of every measurable set is a measurable set:

$$T^{-1}(A') \in \mathcal{A} \qquad \forall A' \in \mathcal{A}'. \tag{7.1}$$

A *random variable* is a measurable map from a probability space to any measurable space.

Note that $T^{-1}(\mathcal{A}') \subset \mathcal{A}$ is a common shorthand for (7.1).

In the language of Definition 7.1 the translation (and its inverse) which we used in Theorem 5.8 is a $\mathcal{B}(\mathbb{R}^n)/\mathcal{B}(\mathbb{R}^n)$-measurable map:

$$\begin{array}{ccc} \tau_x : \mathbb{R}^n \to \mathbb{R}^n & & \tau_x^{-1} : \mathbb{R}^n \to \mathbb{R}^n \\ y \mapsto y - x & \text{and} & y \mapsto y + x. \end{array} \tag{7.2}$$

In fact, Theorem 5.8 states that

$$\lambda^n(B) = \lambda^n(x + B) \qquad \forall B \in \mathcal{B}(\mathbb{R}^n) \tag{7.3}$$

and this requires $x + B = \tau_{-x}(B) = \tau_x^{-1}(B)$ to be a Borel set! Our proof of T5.8 needed (and proved) this for rectangles $B \in \mathcal{J}^n$ and not for all Borel sets – but this is good enough even in the most general case. The following lemma shows that measurability needs only to be checked for the sets of a generator.

7.2 Lemma *Let* $(X, \mathcal{A})$, $(X', \mathcal{A}')$ *be measurable spaces and let* $\mathcal{A}' = \sigma(\mathcal{G}')$. *Then* $T : X \to X'$ *is* $\mathcal{A}/\mathcal{A}'$-*measurable if, and only if,* $T^{-1}(\mathcal{G}') \subset \mathcal{A}$, *i.e. if*

$$T^{-1}(G') \in \mathcal{A} \qquad \forall\, G' \in \mathcal{G}'. \tag{7.4}$$

Proof If T is $\mathcal{A}/\mathcal{A}'$-measurable, we have $T^{-1}(\mathcal{G}') \subset T^{-1}(\mathcal{A}') \subset \mathcal{A}$, and (7.4) is obviously satisfied.

Conversely, consider the system $\Sigma' := \{A' \subset X' : T^{-1}(A') \in \mathcal{A}\}$. By (7.4), $\mathcal{G}' \subset \Sigma'$ and it is not difficult to see that Σ' is itself a σ-algebra since T^{-1} commutes with all set-operations.[✓] Therefore,

$$\mathcal{A}' = \sigma(\mathcal{G}') \subset \sigma(\Sigma') = \Sigma' \implies T^{-1}(A') \in \mathcal{A} \qquad \forall\, A' \in \mathcal{A}'. \qquad \blacksquare$$

On a topological space $(X, \mathcal{O})$ – see Appendix B – we consider usually the (topological) Borel σ-algebra $\mathcal{B}(X) := \sigma(\mathcal{O})$. The interplay between measurability and topology is often quite intricate. One of the simple and extremely useful aspects is the fact that *continuous maps are measurable*; let us check this for $\mathbb{R}^n$.

7.3 Example Every continuous map $T : \mathbb{R}^n \to \mathbb{R}^m$ is $\mathcal{B}^n/\mathcal{B}^m$-measurable.

From calculus[1] we know that T is continuous if, and only if,

$$T^{-1}(U) \subset \mathbb{R}^n \quad \text{is open} \qquad \forall \text{ open } U \subset \mathbb{R}^m. \tag{7.5}$$

Since the open sets $\mathcal{O}^m$ in $\mathbb{R}^m$ generate the Borel σ-algebra $\mathcal{B}^m$, we can use (7.5) to deduce

$$T^{-1}(\mathcal{O}^m) \subset \mathcal{O}^n \subset \sigma(\mathcal{O}^n) = \mathcal{B}^n.$$

By Lemma 7.2, $T^{-1}(\mathcal{B}^m) \subset \mathcal{B}^n$ which means that T is measurable.

Caution: Not every measurable map is continuous, e.g. $x \mapsto \mathbf{1}_{[-1,1]}(x)$.

7.4 Theorem *Let* $(X_j, \mathcal{A}_j)$, $j = 1, 2, 3$, *be measurable spaces and* $T : X_1 \to X_2$, $S : X_2 \to X_3$ *be* $\mathcal{A}_1/\mathcal{A}_2$- *resp.* $\mathcal{A}_2/\mathcal{A}_3$-*measurable maps. Then* $S \circ T : X_1 \to X_3$ *is* $\mathcal{A}_1/\mathcal{A}_3$-*measurable.*

Proof For $A_3 \in \mathcal{A}_3$ we have

$$(S \circ T)^{-1}(A_3) = T^{-1}\bigl(\underbrace{S^{-1}(A_3)}_{\in \mathcal{A}_2}\bigr) \in T^{-1}(\mathcal{A}_2) \subset \mathcal{A}_1. \qquad \blacksquare$$

---

[1] See also Appendix B, Theorem B.12 and B.19.

Often we find ourselves in a situation where $T: X \to X'$ is given and where X' is equipped with a natural σ-algebra $\mathcal{A}'$ – e.g. if $X' = \mathbb{R}$ and $\mathcal{A}' = \mathcal{B}(\mathbb{R})$ – but no σ-algebra is specified in X. Then the question arises: *is there a (smallest) σ-algebra on X which makes T measurable?* An obvious, but nevertheless useless, candidate is $\mathcal{P}(X)$, which renders every map measurable.[✓] From Example 3.3(vii) we know that $T^{-1}(\mathcal{A}')$ is a σ-algebra in X but we cannot remove a single set from it without endangering the measurability of T.[✓] Let us formalize this observation.

7.5 Definition (and Lemma) Let $(T_i)_{i \in I}$ be arbitrarily many mappings $T_i : X \to X_i$ from the same space X into measurable spaces $(X_i, \mathcal{A}_i)$. The smallest σ-algebra on X that makes all T_i simultaneously measurable[✓] is

$$\sigma(T_i : i \in I) := \sigma\Big(\bigcup_{i \in I} T_i^{-1}(\mathcal{A}_i)\Big). \tag{7.6}$$

We say that $\sigma(T_i : i \in I)$ is *generated by the family* $(T_i)_{i \in I}$.

Although $T_i^{-1}(\mathcal{A}_i)$ is a σ-algebra this is, in general, no longer true for $\bigcup_{i \in I} T_i^{-1}(\mathcal{A}_i)$ if $\#I > 1$; this explains why we have to use the σ-hull in (7.6).

7.6 Theorem *Let $(X, \mathcal{A}), (X', \mathcal{A}')$ be measurable spaces and $T: X \to X'$ be an $\mathcal{A}/\mathcal{A}'$-measurable map. For every measure μ on $(X, \mathcal{A})$,*

$$\mu'(A') := \mu(T^{-1}(A')), \qquad A' \in \mathcal{A}', \tag{7.7}$$

defines a measure on $(X', \mathcal{A}')$.

Proof If $A' = \emptyset$, then $T^{-1}(\emptyset) = \emptyset$ and $\mu'(\emptyset) = \mu(\emptyset) = 0$. If $(A'_j)_{j \in \mathbb{N}} \subset \mathcal{A}'$ is a sequence of mutually disjoint sets, then

$$\mu'\Big(\bigcup_{j \in \mathbb{N}} A'_j\Big) = \mu\Big(T^{-1}\Big(\bigcup_{j \in \mathbb{N}} A'_j\Big)\Big) \overset{[✓]}{=} \mu\Big(\bigcup_{j \in \mathbb{N}} T^{-1}(A'_j)\Big)$$

$$= \sum_{j \in \mathbb{N}} \mu\big(T^{-1}(A'_j)\big) = \sum_{j \in \mathbb{N}} \mu'(A'_j). \qquad \blacksquare$$

Notice that we have seen a special case of Theorem 7.6 in the proof of Theorem 5.8 when considering translates of Lebesgue measure: $\lambda^n(x + B) = \lambda(\tau_x^{-1}(B))$.

7.7 Definition The measure $\mu'(\bullet)$ of Theorem 7.6 is called the *image measure* of μ under T and is denoted by $T(\mu)(\bullet)$ or $\mu \circ T^{-1}(\bullet)$.

7.8 Example Let $(\Omega, \mathcal{A}, P)$ be a probability space and $\xi : \Omega \to \mathbb{R}$ be a random variable, i.e. an $\mathcal{A}/\mathcal{B}$-measurable map. Then[2]

$$\xi(P)(A) = P(\xi^{-1}(A)) = P(\{\omega : \xi(\omega) \in A\}) = P(\xi \in A)$$

is again a probability measure, called the *law* or *distribution* of the random variable ξ.

More concretely, if $(\Omega, \mathcal{A}, P)$ describes throwing two fair dice, i.e. $\Omega := \{(j, k) : 1 \leqslant j, k \leqslant 6\}$, $\mathcal{A} = \mathcal{P}(\Omega)$ and $P(\{(j, k)\}) = 1/36$, we could ask for the total number of points thrown: $\xi : \Omega \to \{2, 3, \ldots, 12\}$, $\xi((j, k)) := j + k$, which is a measurable map.[✓] The law of ξ is then given in the table below:

j	2	3	4	5	6	7	8	9	10	11	12
$P(\xi = j)$	$\frac{1}{36}$	$\frac{1}{18}$	$\frac{1}{12}$	$\frac{1}{9}$	$\frac{5}{36}$	$\frac{1}{6}$	$\frac{5}{36}$	$\frac{1}{9}$	$\frac{1}{12}$	$\frac{1}{18}$	$\frac{1}{36}$

We close this section with some transformation formulae for Lebesgue measure.

Recall that $O(n)$ is the set of all *orthogonal $n \times n$ matrices*: $T \in O(n)$ if, and only if, $\,^t T \cdot T = \mathrm{id}$. Orthogonal matrices preserve lengths and angles, i.e. we have for all $x, y \in \mathbb{R}^n$

$$\langle x, y \rangle = \langle Tx, Ty \rangle \iff \|x\| = \|Tx\| \tag{7.8}$$

where $\langle x, y \rangle = \sum_{j=1}^{n} x_j y_j$ and $\|x\|^2 = \langle x, x \rangle$ denote the usual Euclidean scalar product, resp. norm.

7.9 Theorem *If $T \in O(n)$, then $\lambda^n = T(\lambda^n)$.*

Proof Since T is a linear orthogonal map it is continuous and by (7.8) even an isometry,

$$\|Tx - Ty\| = \|T(x - y)\| = \|x - y\|,$$

hence measurable by Example 7.3. Therefore, the image measure $\mu(B) := \lambda^n(T^{-1}(B))$ is well-defined (by T7.6) and satisfies for all $x \in \mathbb{R}^n$

$$\mu(x + B) = \lambda^n(T^{-1}(x + B)) = \lambda^n(T^{-1}x + T^{-1}B)$$

$$\overset{5.8}{=} \lambda^n(T^{-1}B)$$

$$= \mu(B)$$

---

[2] We use the shorthand $\{\xi \in A\}$ for $\xi^{-1}(A)$ and $P(\xi \in A)$ for $P(\{\xi \in A\})$.

and, again by Theorem 5.8, $\mu(B) = \kappa \lambda^n(B)$ for all $B \in \mathcal{B}(\mathbb{R}^n)$. To determine the constant κ we choose $B = B_1(0)$. Since $T \in O(n)$, (7.8) implies $B_1(0) = \{x : \|x\| < 1\} = \{x : \|Tx\| < 1\} = T^{-1}B_1(0)$ and thus

$$\lambda^n(B_1(0)) = \lambda^n(T^{-1}B_1(0)) = \mu(B_1(0)) = \kappa \lambda^n(B_1(0)).$$

As $0 < \lambda^n(B_1(0)) < \infty$, we have $\kappa = 1$, and the theorem follows. ∎

Theorem 7.9 is a particular case of the following general *change-of-variable* formula for Lebesgue measure. Recall that $GL(n, \mathbb{R})$ is the set of all invertible $n \times n$ matrices, i.e. $S \in GL(n, \mathbb{R}) \iff \det S \neq 0$.

7.10 Theorem *Let* $S \in GL(n, \mathbb{R})$. *Then*

$$S(\lambda^n) = |\det S^{-1}| \lambda^n = \frac{1}{|\det S|} \lambda^n. \tag{7.9}$$

Proof Since S is invertible, both S and S^{-1} are linear maps on $\mathbb{R}^n$, and as such continuous and measurable (Example 7.3). Set $\mu(B) := \lambda^n(S^{-1}(B))$ for $B \in \mathcal{B}(\mathbb{R}^n)$. Then we have for all $x \in \mathbb{R}^n$

$$\mu(x + B) = \lambda^n(S^{-1}(x + B)) = \lambda^n(S^{-1}x + S^{-1}B)$$
$$\overset{5.8}{=} \lambda^n(S^{-1}B)$$
$$= \mu(B),$$

and from Theorem 5.8 we conclude that

$$\mu(B) = \mu([0, 1)^n) \lambda^n(B) = \lambda^n(S^{-1}([0, 1)^n)) \lambda^n(B).$$

From elementary geometry we know that $S^{-1}([0, 1)^n)$ is a parallelepiped spanned by the vectors $S^{-1}e_j, j = 1, 2, \ldots n$, $e_j = (\underbrace{0, \ldots, 0, 1, 0, \ldots}_{j})$. Its geometric volume is

$$\text{vol}^{(n)}(S^{-1}([0, 1)^n)) = |\det S^{-1}| = \frac{1}{|\det S|},$$

see also Appendix C. By Remark 6.7, $\text{vol}^{(n)} = \lambda^n$ (at least on the Borel sets) and the proof is finished. ∎

Theorem 7.9 or 7.10 allow us to complete the characterization of Lebesgue measure announced earlier in Theorem 4.9. A *motion* is a linear transformation of the form

$$M_x = \tau_x \circ T$$

where $\tau_x(y) = y - x$ is a translation and $T \in O(n)$ is an orthogonal map $({}^t T \cdot T = \mathrm{id})$. In particular, congruent sets are connected by motions.

7.11 Corollary *Lebesgue measure is invariant under motions:* $\lambda^n = M(\lambda^n)$ *for all motions M in $\mathbb{R}^n$. In particular, congruent sets have the same measure.*

Proof We know that M is of the form $\tau_x \circ T$. Since $\det T = \pm 1$, we get

$$M(\lambda^n) = \tau_x(T(\lambda^n)) \overset{7.10}{=} \tau_x(\lambda^n) \overset{5.8}{=} \lambda^n. \qquad \blacksquare$$

Problems

7.1. Use Lemma 7.2 to show that τ_x of (7.2), i.e. $\tau_x(y) = y - x$, $x, y \in \mathbb{R}^n$, is $\mathcal{B}^n / \mathcal{B}^n$-measurable.

7.2. Show that Σ' defined in the proof of Lemma 7.2 is a σ-algebra.

7.3. Let X be a set, $(X_i, \mathcal{A}_i)$, $i \in I$, be arbitrarily many measurable spaces, and $T_i : X \to X_i$ be a family of maps.

   (i) Show that for every $i \in I$ the smallest σ-algebra in X that makes T_i measurable is given by $T_i^{-1}(\mathcal{A}_i)$.

   (ii) Show that $\sigma\left(\bigcup_{i \in I} T_i^{-1}(\mathcal{A}_i)\right)$ is the smallest σ-algebra in X that makes all T_i, $i \in I$, *simultaneously* measurable.

7.4. Let X be a set, $(X_i, \mathcal{A}_i)$, $i \in I$, be arbitrarily many measurable spaces, and $T_i : X \to X_i$ be a family of maps. Show that a map f from a measurable space $(F, \mathcal{F})$ to $(X, \sigma(T_i : i \in I))$ is measurable if, and only if, all maps $T_i \circ f$ are $\mathcal{F} / \mathcal{A}_i$-measurable.

7.5. Use Problem 7.4 to show that a function $f : \mathbb{R}^n \to \mathbb{R}^m$, $x \mapsto (f_1(x), \dots, f_m(x))$ is 'take out' measurable if, and only if, all coordinate maps $f_j : \mathbb{R}^n \to \mathbb{R}$, $j = 1, 2, \dots, m$, are measurable.
   [Hint: show that the coordinate projections $x = (x_1, \dots, x_n) \mapsto x_j$ are measurable.]

7.6. Let $T : (X, \mathcal{A}) \to (X', \mathcal{A}')$ be a measurable map. Under which circumstances is the family of sets $T(\mathcal{A})$ a σ-algebra?

7.7. Use image measures to give a new proof of Problem 5.8, i.e. show that

$$\lambda^n(t \cdot B) = t^n \lambda^n(B) \qquad \forall B \in \mathcal{B}(\mathbb{R}^n), \ \forall t > 0.$$

7.8. Let $T : X \to Y$ be any map. Show that $T^{-1}(\sigma(\mathcal{G})) = \sigma(T^{-1}(\mathcal{G}))$ holds for arbitrary families $\mathcal{G}$ of subsets of Y.

7.9. **Stieltjes measure (1).** Throughout this exercise $(X, \mathcal{A}) = (\mathbb{R}, \mathcal{B}^1)$ and $\lambda = \lambda^1$ is one-dimensional Lebesgue measure.

   (i) Let μ be a measure on $(\mathbb{R}, \mathcal{B}^1)$. Show that $F_\mu(x) := \begin{cases} \mu([0, x)), & \text{if } x > 0 \\ 0, & \text{if } x = 0 \\ -\mu([x, 0)), & \text{if } x < 0 \end{cases}$

   is a monotonically increasing and left-continuous function $F_\mu : \mathbb{R} \to \mathbb{R}$.

Remark. Increasing and left-continuous functions are called *Stieltjes functions*.

(ii) Let $F: \mathbb{R} \to \mathbb{R}$ be a Stieltjes function (cf. part (i)). Show that

$$\nu_F([a, b)) := F(b) - F(a), \qquad \forall a, b \in \mathbb{R}, \ a < b,$$

has a unique extension to a measure on $\mathcal{B}^1$.

[Hint: check the assumptions of Theorem 6.1 with $\mathcal{S} = \{[a, b) : a \leqslant b\}$.]

(iii) Use part (i) to show that every measure μ on $(\mathbb{R}, \mathcal{B}^1)$ with $\mu([-n, n)) < \infty$, $n \in \mathbb{N}$, can be written in the form ν_F as in (ii) with some Stieltjes function $F = F_\mu$ as in (i).

(iv) Which Stieltjes function F corresponds to λ?

(v) Which Stieltjes function F corresponds to δ_0?

(vi) Show that F_μ as in (i) is continuous at $x \in \mathbb{R}$ if, and only if, $\mu(\{x\}) = 0$.

(vii) Show that every measure μ on $(\mathbb{R}, \mathcal{B}^1)$ which has no atoms (see Problem 6.5) can be written as image measure of λ.

[Hint: μ has no atoms implies that F_μ is continuous. So $G = F^{-1}$ exists and can be made left-continuous. Finally $\mu([a, b)) = F_\mu(b) - F_\mu(a) = \lambda(G^{-1}\{[a, b)\})$]

(viii) Is (vii) true for measures with atoms, say, $\mu = \delta_0$?

[Hint: determine $F_{\delta_0}^{-1}$. Is it measurable?]

7.10. **Cantor's ternary set.** Let $(X, \mathcal{A}) = ([0, 1], [0, 1] \cap \mathcal{B}^1)$, $\lambda = \lambda^1|_{[0,1]}$, and set $E_0 = [0, 1]$. Remove the open middle third of E_0 to get $E_1 = I_1^1 \uplus I_1^2$. Remove the open middle thirds of I_1^j, $j = 1, 2$, to get $E_2 = I_2^1 \uplus I_2^2 \uplus I_2^3 \uplus I_2^4$ and so forth.

(i) Make a sketch of E_0, E_1, E_2, E_3.

(ii) Prove that each E_n is compact. Conclude that $C := \bigcap_{n \in \mathbb{N}_0} E_n$ is non-void and compact.

(iii) The set C is called the *Cantor set* or *Cantor's discontinuum*. It satisfies $C \cap \bigcup_{n \in \mathbb{N}} \bigcup_{k \in \mathbb{N}_0} \left(\frac{3k+1}{3^n}, \frac{3k+2}{3^n} \right) = \emptyset$.

(iv) Find the value of $\lambda(E_n)$ and show that $\lambda(C) = 0$.

(v) Show that C does not contain any open interval. Conclude that the interior (of the closure) of C is empty.

Remark. Sets with empty interior are called *nowhere dense*.

(vi) We can write $x \in [0, 1)$ as a base-3 ternary fraction, i.e. $x = 0.x_1 x_2 x_3 \ldots$ where $x_j \in \{0, 1, 2\}$, which is short for $x = \sum_{j=1}^{\infty} x_j 3^{-j}$. (E.g. $\frac{1}{3} = 0.1 = 0.02222\ldots$; note that this representation is not unique[✓], which is important for this exercise.)

Show that $x \in C$ if, and only if, x has a ternary representation involving only 0s and 2s.

[Hint: the numbers in $(\frac{1}{3}, \frac{2}{3})$, the first interval to be removed, are all of the form $0.1*** \ldots$, i.e. they contain at least one '1', while in $[0, \frac{1}{3}]$ and $[\frac{2}{3}, 1]$ we have numbers of the form $0.0*** \ldots - 0.022222 \ldots$ and $0.2*** \ldots - 0.2222 \ldots$, respectively. The next step eliminates the $0.01***\ldots s$ and $0.21***\ldots s -$ etc.]

(vii) Use (vi) to show that C is not countable and has even the same cardinality as $[0, 1]$. Nevertheless, $\lambda(C) = 0 \neq 1 = \lambda([0, 1])$.

7.11. **Factorization lemma.** Let X be a set, $(Y, \mathcal{A})$ be a measurable space and $T : X \to Y$ be a surjective map. Show that a function $f : X \to \mathbb{R}$ is $\sigma(T)/\mathcal{B}^1$-measurable if, and only if, there exists some $\mathcal{A}/\mathcal{B}^1$-measurable function $g : Y \to \mathbb{R}$ such that $f = g \circ T$. [Hint: show first that $T(x) = T(x')$ implies $f(x) = f(x')$.]

Remark. The result is actually true for any map $T : X \to Y$, but the proof is quite difficult if $T(X) \notin \mathcal{A}$. The problem is that one has to extend the $T(X) \cap \mathcal{A}/\mathcal{B}^1$ measurable function $g : T(X) \to \mathbb{R}$ to an $\mathcal{A}/\mathcal{B}^1$-measurable function $g : Y \to \mathbb{R}$.

8

Measurable functions

A *measurable function* is a measurable map $u : X \to \mathbb{R}$ from some measurable space $(X, \mathcal{A})$ to $(\mathbb{R}, \mathcal{B}(\mathbb{R}))$. Measurable functions will play a central rôle in the theory of integration. Recall that $u : X \to \mathbb{R}$ is $\mathcal{A}/\mathcal{B}$-measurable[1] $(\mathcal{B} = \mathcal{B}(\mathbb{R}))$ if

$$u^{-1}(B) \in \mathcal{A} \qquad \forall B \in \mathcal{B} \tag{8.1}$$

which is, due to Lemma 7.2, equivalent to

$$u^{-1}(G) \in \mathcal{A} \qquad \forall G \text{ from a generator } \mathcal{G} \text{ of } \mathcal{B}. \tag{8.2}$$

As we have seen in Remark 3.9, $\mathcal{B}$ is generated by all sets of the form $[a, \infty)$ (or (b, ∞) or $(-\infty, c)$ or $(-\infty, d]$) with $a, b, c, d \in \mathbb{R}$ or $\mathbb{Q}$, and we need

$$u^{-1}([a, \infty)) = \{x \in X : u(x) \in [a, \infty)\} = \{x \in X : u(x) \geqslant a\} \in \mathcal{A}, \tag{8.3}$$

with similar expressions for the other types of intervals. Let us introduce the following useful shorthand notation:

$$\{u \geqslant v\} := \{x \in X : u(x) \geqslant v(x)\} \tag{8.4}$$

and $\{u > v\}, \{u \leqslant v\}, \{u < v\}, \{u = v\}, \{u \neq v\}, \{u \in B\}$, etc. which are defined in a similar fashion.

In this new notation measurability of functions reads as

8.1 Lemma *Let $(X, \mathcal{A})$ be a measurable space. The function $u : X \to \mathbb{R}$ is $\mathcal{A}/\mathcal{B}$-measurable if, and only if, one, hence all, of the following conditions hold*

(i) $\{u \geqslant a\} \in \mathcal{A}$ *for all $a \in \mathbb{R}$ (or all $a \in \mathbb{Q}$),*

(ii) $\{u > a\} \in \mathcal{A}$ *for all $a \in \mathbb{R}$ (or all $a \in \mathbb{Q}$),*

---

[1] We will frequently drop the $\mathcal{B}$ since $\mathbb{R}$ is naturally equipped with the Borel σ-algebra and just say that u is $\mathcal{A}$-measurable.

(iii) $\{u \leqslant a\} \in \mathcal{A}$ for all $a \in \mathbb{R}$ (or all $a \in \mathbb{Q}$),
(iv) $\{u < a\} \in \mathcal{A}$ for all $a \in \mathbb{R}$ (or all $a \in \mathbb{Q}$).

Proof Combine Remark 3.9 and Lemma 7.2.   ∎

It is sometimes practical to admit the values $+\infty$ and $-\infty$ in some calculations. To do this properly, consider the *extended real line* $\bar{\mathbb{R}} := [-\infty, +\infty]$. If we agree that $-\infty < x$ and $y < +\infty$ for all $x, y \in \mathbb{R}$, then $\bar{\mathbb{R}}$ inherits the ordering from $\mathbb{R}$ as well as the usual rules of addition and multiplication of elements from $\mathbb{R}$. The latter need to be augmented as follows: for all $x \in \mathbb{R}$ we have

$$x + (+\infty) = (+\infty) + x = +\infty, \qquad x + (-\infty) = (-\infty) + x = -\infty,$$

$$(+\infty) + (+\infty) = +\infty, \qquad (-\infty) + (-\infty) = -\infty,$$

and, if $x \in (0, \infty]$,

$$(\pm x)(+\infty) = (+\infty)(\pm x) = \pm\infty,$$

$$(\pm x)(-\infty) = (-\infty)(\pm x) = \mp\infty,$$

$$0 \cdot (\pm\infty) = (\pm\infty) \cdot 0 = 0,^2 \quad \tfrac{1}{\pm\infty} = 0.$$

Caution: $\bar{\mathbb{R}}$ is not a field. Expressions of the form

$$\infty - \infty \quad \text{and} \quad \tfrac{\pm\infty}{\pm\infty} \qquad \textbf{must be avoided.}^2$$

Functions which take values in $\bar{\mathbb{R}}$ are called *numerical functions*. The Borel σ-algebra $\bar{\mathcal{B}} = \mathcal{B}(\bar{\mathbb{R}})$ is defined by

$$B^* \in \bar{\mathcal{B}} \iff \begin{array}{l} B^* = B \cup S \quad \text{for some } B \in \mathcal{B} \text{ and} \\ S \in \{\emptyset, \{-\infty\}, \{+\infty\}, \{-\infty, +\infty\}\} \end{array} \tag{8.5}$$

and it is not hard to see that $\bar{\mathcal{B}}$ is again a σ-algebra whose trace w.r.t. $\mathbb{R}$ is $\mathcal{B}(\mathbb{R})$.[✓]

8.2 Lemma $\mathcal{B}(\mathbb{R}) = \mathbb{R} \cap \mathcal{B}(\bar{\mathbb{R}})$.

Moreover,

8.3 Lemma $\bar{\mathcal{B}}$ is generated by all sets of the form $[a, \infty]$ (or $(b, \infty]$ or $[-\infty, c)$ or $[-\infty, d]$) where a (or b, c, d) is from $\mathbb{R}$ or $\mathbb{Q}$.

---

[2] Conventions are tricky. The rationale behind our definitions is to understand '$\pm\infty$' in every instance as the limit of some (possibly each time different) sequence, and '0' as a bona fide zero. Then $0 \cdot (\pm\infty) = 0 \cdot \lim_n a_n = \lim_n (0 \cdot a_n) = \lim_n 0 = 0$ while expressions of the type $\infty - \infty$ or $\tfrac{\pm\infty}{\pm\infty}$ become $\lim_n a_n - \lim_n b_n$ or $\tfrac{\lim_n a_n}{\lim_n b_n}$ where two sequences compete and do not lead to unique results.

Proof Set $\Sigma := \sigma(\{[a, \infty] : a \in \mathbb{R}\})$. Since

$$[a, \infty] = [a, \infty) \cup \{+\infty\} \quad \text{and} \quad [a, \infty) \in \mathcal{B},$$

we see that $[a, \infty] \in \bar{\mathcal{B}}$ and $\Sigma \subset \bar{\mathcal{B}}$. On the other hand,

$$[a, b) = [a, \infty] \setminus [b, \infty] \in \Sigma \qquad \forall -\infty < a \leqslant b < \infty$$

which means that $\mathcal{B} \subset \Sigma \subset \bar{\mathcal{B}}$. Since also

$$\{+\infty\} = \bigcap_{j \in \mathbb{N}} [j, \infty], \qquad \{-\infty\} = \bigcap_{j \in \mathbb{N}} [-\infty, -j) = \bigcap_{j \in \mathbb{N}} [-j, \infty]^c$$

we have $\{-\infty\}, \{+\infty\} \in \Sigma$ which entails that all sets of the form

$$B, \; B \cup \{+\infty\}, \, B \cup \{-\infty\}, \, B \cup \{-\infty, +\infty\} \; \in \Sigma, \qquad \forall B \in \mathcal{B},$$

therefore, $\bar{\mathcal{B}} \subset \Sigma$.

The proofs for $a \in \mathbb{Q}$ and the other generating systems are similar. ∎

8.4 Definition Let $(X, \mathcal{A})$ be a measurable space. We write $\mathcal{M} := \mathcal{M}(\mathcal{A})$ and $\mathcal{M}_{\bar{\mathbb{R}}} := \mathcal{M}_{\bar{\mathbb{R}}}(\mathcal{A})$ for the families of real-valued $\mathcal{A}/\mathcal{B}$-measurable and numerical $\mathcal{A}/\bar{\mathcal{B}}$-measurable functions on X.

8.5 Examples Let $(X, \mathcal{A})$ be a measurable space.

(i) The indicator function $f(x) := \mathbf{1}_A(x)$ is measurable if, and only if, $A \in \mathcal{A}$. This follows easily from Lemma 8.1 and

$$\{\mathbf{1}_A > \lambda\} = \begin{cases} \emptyset, & \text{if } \lambda \geqslant 1 \\ A, & \text{if } 1 > \lambda \geqslant 0 \\ X, & \text{if } \lambda < 0. \end{cases}$$

(ii) Let $A_1, A_2, \ldots, A_M \in \mathcal{A}$ be mutually disjoint sets and $y_1, \ldots, y_M \in \mathbb{R}$. Then the function

$$g(x) := \sum_{j=1}^{M} y_j \mathbf{1}_{A_j}(x) \tag{8.6}$$

is measurable.

This follows from Lemma 8.1 and the fact (compare with the picture!) that

$$\{g > \lambda\} = \biguplus_{j : y_j > \lambda} A_j \in \mathcal{A}.$$

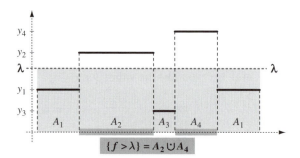

$$\{f > \lambda\} = A_2 \cup A_4$$

Functions of the form (8.6) are the building blocks for all measurable functions as well as for the definition of the integral.

8.6 Definition A *simple function* $g : X \to \mathbb{R}$ on a measurable space $(X, \mathcal{A})$ is a function of the form (8.6) with finitely many sets $A_1, \ldots, A_M \in \mathcal{A}$ and $y_1, \ldots, y_M \in \mathbb{R}$. The set of simple functions is denoted by $\mathcal{E}$ or $\mathcal{E}(\mathcal{A})$.

If the sets $A_j, 1 \leqslant j \leqslant M$, are mutually disjoint we call

$$\sum_{j=0}^{M} y_j \mathbf{1}_{A_j}(x) \tag{8.7}$$

with $y_0 := 0$ and $A_0 := (A_1 \cup \ldots \cup A_M)^c$ a *standard representation* of g.

Caution: The representations (8.7) are *not unique*.

8.7 Examples (continued)

(iii) If a measurable function $h : X \to \mathbb{R}$ attains only finitely many values $y_1, y_2, \ldots, y_M \in \mathbb{R}$, then it is a simple function.
Indeed: set $B_j := \{h = y_j\} = \{h \leqslant y_j\} \setminus \{h < y_j\} \in \mathcal{A}, j = 1, 2, \ldots, M$, and note that the B_j are disjoint. Thus

$$h(x) = \sum_{j=1}^{M} y_j \mathbf{1}_{B_j}(x) = \sum_{j=1}^{M} y_j \mathbf{1}_{\{h=y_j\}}(x).$$

Since every simple function attains only finitely many values, this shows that every simple function has at least one standard representation. In particular, $\mathcal{E}(\mathcal{A}) \subset \mathcal{M}(\mathcal{A})$ consists of measurable functions.

(iv) $f, g \in \mathcal{E}(\mathcal{A}) \implies f \pm g, \ fg \in \mathcal{E}(\mathcal{A})$.
Indeed: let $f = \sum_{j=0}^{M} y_j \mathbf{1}_{A_j}$ and $g = \sum_{k=0}^{N} z_k \mathbf{1}_{B_k}$ be standard representations of f and g.

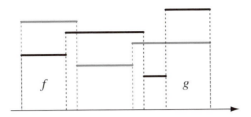

It is not hard to see (use the picture!) that

$$f \pm g = \sum_{j=0}^{M} \sum_{k=0}^{N} (y_j \pm z_k) \mathbf{1}_{A_j \cap B_k}$$

$$fg = \sum_{j=0}^{M} \sum_{k=0}^{N} y_j z_k \mathbf{1}_{A_j \cap B_k}$$

and that $(A_j \cap B_k) \cap (A_{j'} \cap B_{k'}) = \emptyset$ whenever $(j, k) \neq (j', k')$. After relabelling and merging the double indexation into a single index, this shows that $f \pm g$, $fg \in \mathcal{E}(\mathcal{A})$. Notice that $(A_j \cap B_k)_{j,k}$ is the common refinement of the partitions $(A_j)_j$ and $(B_k)_k$ and that inside each of the sets $A_j \cap B_k$ the functions f and g do not change their respective values.

(v) $f \in \mathcal{E}(\mathcal{A}) \implies f^+, f^- \in \mathcal{E}(\mathcal{A})$.

Here we use the following notation: for a function $u : X \to \mathbb{R}$ we write for the

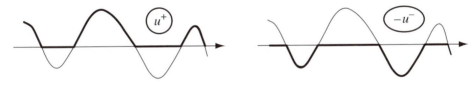

$$u^+(x) := \max\{u(x), 0\} \qquad u^-(x) := -\min\{u(x), 0\} \qquad (8.8)$$

for the *positive* (u^+) and *negative* (u^-) *parts* of u. Obviously,

$$u = u^+ - u^- \qquad \text{and} \qquad |u| = u^+ + u^-. \qquad (8.9)$$

(vi) $f \in \mathcal{E}(\mathcal{A}) \implies |f| \in \mathcal{E}(\mathcal{A})$.

Our next theorem reveals the fundamental rôle of simple functions.

8.8 Theorem *Let $(X, \mathcal{A})$ be a measurable space. Every $\mathcal{A}/\bar{\mathcal{B}}$-measurable numerical function $u : X \to \bar{\mathbb{R}}$ is the pointwise limit of simple functions:* $u(x) = \lim_{j \to \infty} f_j(x)$, $f_j \in \mathcal{E}(\mathcal{A})$ *and* $|f_j| \leqslant |u|$.

If $u \geqslant 0$, all f_j can be chosen to be positive and increasing towards u so that $u = \sup_{j \in \mathbb{N}} f_j$.

Proof Assume first that $u \geqslant 0$. Fix $j \in \mathbb{N}$ and define level sets

$$A_k^j := \begin{cases} \{k2^{-j} \leqslant u < (k+1)2^{-j}\} & k = 0, 1, 2, \dots j2^j - 1 \\ \{u \geqslant j\} & k = j2^j \end{cases}$$

which slice up the graph of u horizontally as shown in the picture. The approximating simple functions are

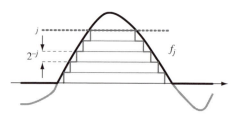

$$f_j(x) := \sum_{k=0}^{j2^j} k2^{-j} \mathbf{1}_{A_k^j}(x)$$

and from the picture it is easy to see that

- $|f_j(x) - u(x)| \leqslant 2^{-j}$ if $x \in \{u < j\}$;
- $A_k^j = \{k2^{-j} \leqslant u\} \cap \{u < (k+1)2^{-j}\}$, $\{u \geqslant j\} \in \mathcal{A}$;
- $0 \leqslant f_j \leqslant u$ and $f_j \uparrow u$.

For a general u, we consider its positive and negative parts $u^{\pm}$. Since

$$\{u^+ > \lambda\} = \begin{cases} \{u > \lambda\}, & \text{if } \lambda \geqslant 0, \\ \varnothing, & \text{if } \lambda < 0, \end{cases}$$

and since $u^- = (-u)^+$, we have $\{u^{\pm} > \lambda\} \in \mathcal{A}$ for all $\lambda \in \bar{\mathbb{R}}$. Thus $u^{\pm}$ are positive measurable functions, and we can construct, as above, simple functions $g_j \uparrow u^+$ and $h_j \uparrow u^-$. Clearly, $f_j := g_j - h_j \xrightarrow{j \to \infty} u^+ - u^- = u$ as well as $|f_j| = g_j + h_j \leqslant u^+ + u^- = |u|$, and we are done. $\blacksquare$

8.9 Corollary *Let $(X, \mathcal{A})$ be a measurable space. If $u_j : X \to \bar{\mathbb{R}}$, $j \in \mathbb{N}$, are measurable functions, then so are*

$$\sup_{j \in \mathbb{N}} u_j, \quad \inf_{j \in \mathbb{N}} u_j, \quad \limsup_{j \to \infty} u_j, \quad \liminf_{j \to \infty} u_j,$$

and, whenever it exists, $\lim_{j \to \infty} u_j$.

Before we prove Corollary 8.9 let us stress again that expressions of the type $\sup_{j \in \mathbb{N}} u_j$ or $u_j \xrightarrow{j \to \infty} u$, etc. are always understood in a *pointwise, x-by-x sense*, i.e. they are short for $\sup_{j \in \mathbb{N}} u_j(x) := \sup\{u_j(x) : j \in \mathbb{N}\}$ or $\lim_{j \to \infty} u_j(x) = u(x)$ at each x (or for a specified range).

The infimum 'inf' and supremum 'sup' are familiar from calculus. Recall the following useful formula

$$\inf_{j \in \mathbb{N}} u_j(x) = -\sup_{j \in \mathbb{N}}(-u_j(x)) \tag{8.10}$$

which allows us to express an inf as a sup, and vice versa. Recall also the definition of the *lower* resp. *upper limits* lim inf and lim sup,

$$\liminf_{j\to\infty} u_j(x) := \sup_{k\in\mathbb{N}}\left(\inf_{j\geqslant k} u_j(x)\right) = \lim_{k\to\infty}\left(\inf_{j\geqslant k} u_j(x)\right), \tag{8.11}$$

$$\limsup_{j\to\infty} u_j(x) := \inf_{k\in\mathbb{N}}\left(\sup_{j\geqslant k} u_j(x)\right) = \lim_{k\to\infty}\left(\sup_{j\geqslant k} u_j(x)\right); \tag{8.12}$$

more details can be found in Appendix A.

In the extended real line $\bar{\mathbb{R}}$ lim inf and lim sup *always* exist – but they may attain the values $+\infty$ and $-\infty$ – and we have

$$\liminf_{j\to\infty} u_j(x) \leqslant \limsup_{j\to\infty} u_j(x). \tag{8.13}$$

Moreover, $\lim_{j\to\infty} u_j(x)$ exists [and is finite] if, and only if, upper and lower limits coincide $\liminf_{j\to\infty} u_j(x) = \limsup_{j\to\infty} u_j(x)$ [and are finite]; in this case all three limits have the same value.

Proof (of Corollary 8.9) We show that $\sup_j u_j$ and $(-1)u = -u$ (for a measurable function u) are again measurable. Observe that for all $a \in \mathbb{R}$

$$\left\{\sup_{j\in\mathbb{N}} u_j > a\right\} = \bigcup_{j\in\mathbb{N}} \underbrace{\{u_j > a\}}_{\in\mathcal{A}} \in \mathcal{A}.$$

The inclusion '$\supset$' is trivial since $a < u_j(x) \leqslant \sup_{j\in\mathbb{N}} u_j(x)$ always holds; the direction '$\subset$' follows by contradiction: if $u_j(x) \leqslant a$ for all $j \in \mathbb{N}$, then also $\sup_{j\in\mathbb{N}} u_j(x) \leqslant a$. This proves the measurability of $\sup_{j\in\mathbb{N}} u_j$.

If u is measurable, we have for all $a \in \mathbb{R}$

$$\{-u > a\} = \{u < -a\} \in \mathcal{A},$$

which shows that $-u$ is also measurable.

The measurability of $\inf_{j\in\mathbb{N}} u_j$, $\liminf_{j\to\infty} u_j$ and $\limsup_{j\to\infty} u_j$ follows now from formulae (8.10)–(8.12), which can be written down in terms of $\sup_j$s and several multiplications by (-1). If $\lim_{j\to\infty} u_j$ exists, it coincides with $\liminf_{j\to\infty} u_j = \limsup_{j\to\infty} u_j$ and inherits their measurability. ■

8.10 Corollary *Let u, v be $\mathcal{A}/\bar{\mathcal{B}}$-measurable numerical functions. Then the functions*

$$u \pm v, \quad uv, \quad u \vee v := \max\{u, v\}, \quad u \wedge v := \min\{u, v\} \tag{8.14}$$

are $\mathcal{A}/\bar{\mathcal{B}}$-measurable (whenever they are defined).

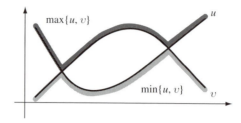

The maximum $u \vee v$ and minimum $u \wedge v$ of two functions is always meant pointwise, i.e.

$$(u \vee v)(x) = \max\{u(x), v(x)\} \overset{[\checkmark]}{=} \frac{1}{2}\big(u(x) + v(x) + |u(x) - v(x)|\big),$$

$$(u \wedge v)(x) = \min\{u(x), v(x)\} \overset{[\checkmark]}{=} \frac{1}{2}\big(u(x) + v(x) - |u(x) - v(x)|\big).$$

Proof (of Corollary 8.10) If $u, v \in \mathcal{E}$ are simple functions, all functions in (8.14) are again simple functions[$\checkmark$] and, therefore, measurable. For general $u, v \in \mathcal{M}_{\bar{\mathbb{R}}}$ choose sequences $(f_j)_{j \in \mathbb{N}}, (g_j)_{j \in \mathbb{N}} \subset \mathcal{E}$ of simple functions such that $f_j \xrightarrow{j \to \infty} u$ and $g_j \xrightarrow{j \to \infty} v$. The claim now follows from the usual rules for limits.   ∎

8.11 Corollary *A function u is $\mathcal{A}/\bar{\mathcal{B}}$-measurable if, and only if, $u^{\pm}$ are $\mathcal{A}/\bar{\mathcal{B}}$-measurable.*

8.12 Corollary *If u, v are $\mathcal{A}/\bar{\mathcal{B}}$-measurable numerical functions, then*

$$\{u < v\}, \quad \{u \leqslant v\}, \quad \{u = v\}, \quad \{u \neq v\} \in \mathcal{A}.$$

Let us finally show an interesting result on the structure of $\sigma(T)$-measurable functions; see also Problem 7.11 of the previous chapter.

8.13 Lemma (Factorization lemma) *Let $T : (X, \mathcal{A}) \to (X', \mathcal{A}')$ be an $\mathcal{A}/\mathcal{A}'$-measurable map and let $\sigma(T) \subset \mathcal{A}$ be the σ-algebra generated by T. Then $u = w(T)$ for some $\mathcal{A}'/\bar{\mathcal{B}}$-measurable function $w : X' \to \bar{\mathbb{R}}$ if, and only if, $u : X \to \bar{\mathbb{R}}$ is $\sigma(T)/\bar{\mathcal{B}}$-measurable.*

Proof Suppose that u is $\sigma(T)$-measurable. If u is an indicator function, $u = \mathbf{1}_A$ with $A \in \sigma(T)$, we know from the definition of $\sigma(T)$ that $A = T^{-1}(A')$ for some $A' \in \mathcal{A}'$. Thus

$$u = \mathbf{1}_A = \mathbf{1}_{T^{-1}(A')} = \mathbf{1}_{A'} \circ T,$$

and $w := \mathbf{1}_{A'}$ will do. This consideration remains true for simple functions $u \in \mathcal{E}(\sigma(T))$ since they are just sums of scalar multiples of indicator functions;[$\checkmark$] hence $u = w \circ T$ for a suitable $w \in \mathcal{E}(\mathcal{A}')$.

We can now use Theorem 8.8 and approximate the $\sigma(T)$-measurable function u by a sequence $(u_j)_{j\in\mathbb{N}} \subset \mathcal{E}(\sigma(T))$. By what was said above, $u_j = w_j \circ T$ for suitable $w_j \in \mathcal{E}(\mathcal{A}')$. Then $w := \limsup_{j\to\infty} w_j$ is measurable by C8.9 and satisfies

$$w(T) = \limsup_{j\to\infty} w_j(T) \overset{(*)}{=} \lim_{j\to\infty} w_j(T) = \lim_{j\to\infty} u_j = u$$

where we used for the equality marked $(*)$ the fact that the limit $\lim_{j\to\infty} u_j$ exists. The converse, that $u := w \circ T$ is $\sigma(T)$-measurable, is obvious. ∎

Problems

8.1. Show *directly* that condition (i) of Lemma 8.1 is equivalent to either of (ii), (iii), (iv).

8.2. Verify that $\bar{\mathcal{B}} = \mathcal{B}(\bar{\mathbb{R}})$ defined in (8.5) is a σ-algebra. Moreover, prove that $\mathcal{B}(\mathbb{R}) = \mathbb{R} \cap \mathcal{B}(\bar{\mathbb{R}})$.

8.3. Let $(X, \mathcal{A})$ be a measurable space.

   (i) Let $f, g : X \to \mathbb{R}$ be measurable functions. Show that for every $A \in \mathcal{A}$ the function $h(x) := f(x)$, if $x \in A$, and $h(x) := g(x)$, if $x \notin A$, is measurable.

   (ii) Let $(f_j)_{j\in\mathbb{N}}$ be a sequence of measurable functions and let $(A_j)_{j\in\mathbb{N}} \subset \mathcal{A}$ such that $\bigcup_{j\in\mathbb{N}} A_j = X$. Suppose that $f_j|_{A_j \cap A_k} = f_k|_{A_j \cap A_k}$ for all $j, k \in \mathbb{N}$ and set $f(x) := f_j(x)$ if $x \in A_j$. Show that $f : X \to \mathbb{R}$ is measurable.

8.4. Let $(X, \mathcal{A})$ be a measurable space and let $\mathcal{B} \subsetneq \mathcal{A}$ be a sub-σ-algebra. Show that $\mathcal{M}(\mathcal{B}) \subsetneq \mathcal{M}(\mathcal{A})$.

8.5. Show that $f \in \mathcal{E}$ implies that $f^\pm \in \mathcal{E}$. Is the converse valid?

8.6. Show that for every real-valued function $u = u^+ - u^-$ and $|u| = u^+ + u^-$.

8.7. Scrutinize the proof of Theorem 8.8 and check that bounded [positive] measurable functions $u \in \mathcal{M}(\mathcal{A})$ can be approximated *uniformly* by an [increasing] sequence $(f_j)_{j\in\mathbb{N}} \subset \mathcal{E}(\mathcal{A})$ of [positive] simple functions.

8.8. Show that every continuous function $u : \mathbb{R} \to \mathbb{R}$ is $\mathcal{B}/\mathcal{B}$ measurable.

   [Hint: check that for continuous functions $\{f > \alpha\}$ is an open set.]

8.9. Show that $x \mapsto \max\{x, 0\}$ and $x \mapsto \min\{x, 0\}$ are continuous, and by Problem 8.8 or Example 7.3, measurable functions from $\mathbb{R} \to \mathbb{R}$. Conclude that on any measurable space $(X, \mathcal{A})$ positive and negative parts $u^\pm$ of a measurable function $u : X \to \mathbb{R}$ are measurable.

8.10. Check that the approximating sequence $(f_j)_{j\in\mathbb{N}}$ for u in Theorem 8.8 consists of $\sigma(u)$-measurable functions.

8.11. Complete the proofs of Corollaries 8.11 and 8.12.

8.12. Let $u : \mathbb{R} \to \mathbb{R}$ be differentiable. Explain why u and $u' = du/dx$ are measurable.

8.13. Find $\sigma(u)$, i.e. the σ-algebra generated by u, for the following functions:

$$f, g, h : \mathbb{R} \to \mathbb{R}, \quad \text{(i)} \quad f(x) = x; \quad \text{(ii)} \quad g(x) = x^2; \quad \text{(iii)} \quad h(x) = |x|;$$

$$F, G : \mathbb{R}^2 \to \mathbb{R}, \quad \text{(iv)} \quad F(x, y) = x + y; \quad \text{(v)} \quad G(x, y) = x^2 + y^2.$$

[Hint: under f, g, h the pre-images of intervals are (unions of) intervals, under F we get strips in the plane, under G annuli and discs.]

8.14. Consider $(\mathbb{R}, \mathcal{B})$ and $u : \mathbb{R} \to \mathbb{R}$. Show that $\{x\} \in \sigma(u)$ for all $x \in \mathbb{R}$ if, and only if, u is injective.

8.15. Let λ be one-dimensional Lebesgue measure. Find $\lambda \circ u^{-1}$, if $u(x) = |x|$.

8.16. Let $u : \mathbb{R} \to \mathbb{R}$ be measurable. Which of the following functions are measurable:

$$u(x-2) \qquad e^{u(x)} \qquad \sin(u(x)+8) \qquad u''(x) \qquad \operatorname{sgn} u(x-7)?$$

8.17. One can show that there are non-Borel measurable sets $A \subset \mathbb{R}$, cf. Appendix D. Taking this fact for granted, show that measurability of $|u|$ does not, in general, imply the measurability of u. (The converse is, of course, true: measurability of u always guarantees that of $|u|$.)

8.18. Show that every increasing function $u : \mathbb{R} \to \mathbb{R}$ is $\mathcal{B}/\mathcal{B}$ measurable. Under which additional condition(s) do we have $\sigma(u) = \mathcal{B}$?

[Hint: show that $\{u < \lambda\}$ is an interval by distinguishing three cases: u is continuous and strictly increasing when passing the level λ, u jumps over the level λ, u is 'flat' at level λ. Make a picture of these situations.]

9

Integration of positive functions

Throughout this chapter $(X, \mathcal{A}, \mu)$ will be some measurable space.

Recall that $\mathcal{M}^+$ $[\mathcal{M}_{\overline{\mathbb{R}}}^+]$ are the $\mathcal{A}$-measurable positive real [numerical] functions and $\mathcal{E}$ $[\mathcal{E}^+]$ are the [positive] simple functions.

The fundamental idea of *integration* is to measure the area between the graph of a function and the abscissa. For a positive simple function $f \in \mathcal{E}^+$ in standard representation[1] this is easily done:

$$\text{if} \qquad f = \sum_{j=0}^{M} y_j \mathbf{1}_{A_j} \in \mathcal{E}^+ \qquad \text{then} \qquad \sum_{j=0}^{M} y_j \mu(A_j) \tag{9.1}$$

should be the μ-area enclosed by the graph and the abscissa.

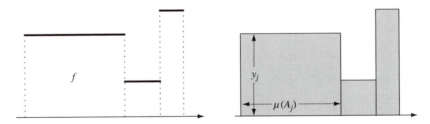

There is only the problem that (9.1) might depend on the particular (standard) representation of f – and this should not happen.

9.1 Lemma *Let* $\sum_{j=0}^{M} y_j \mathbf{1}_{A_j} = \sum_{k=0}^{N} z_k \mathbf{1}_{B_k}$ *be two standard representations of the same function* $f \in \mathcal{E}^+$. *Then*

$$\sum_{j=0}^{M} y_j \mu(A_j) = \sum_{k=0}^{N} z_k \mu(B_k).$$

---

[1] In the sense of Definition 8.6. By 8.5(iii) every $f \in \mathcal{E}$ has a standard representation.

Proof Since $A_0 \uplus A_1 \uplus \ldots \uplus A_M = X = B_0 \uplus B_1 \uplus \ldots \uplus B_N$ we get

$$A_j = \bigcup_{k=0}^{N} (A_j \cap B_k) \qquad \text{and} \qquad B_k = \bigcup_{j=0}^{M} (B_k \cap A_j).$$

Using the (finite) additivity of μ we see that

$$\sum_{j=0}^{M} y_j \mu(A_j) = \sum_{j=0}^{M} y_j \sum_{k=0}^{N} \mu(A_j \cap B_k) = \sum_{j=0}^{M} \sum_{k=0}^{N} y_j \mu(A_j \cap B_k) \qquad (9.2)$$

(since all y_j are positive, the above sums always exist in $[0, \infty]$). Similarly,

$$\sum_{k=0}^{N} z_k \mu(B_k) = \sum_{k=0}^{N} z_k \sum_{j=0}^{M} \mu(A_j \cap B_k) = \sum_{k=0}^{N} \sum_{j=0}^{M} z_k \mu(A_j \cap B_k). \qquad (9.3)$$

But $y_j = z_k$ whenever $A_j \cap B_k \neq \emptyset$, while for $A_j \cap B_k = \emptyset$ we have $\mu(A_j \cap B_k) = \mu(\emptyset) = 0$. Thus

$$y_j \mu(A_j \cap B_k) = z_k \mu(A_j \cap B_k) \qquad \forall (j, k),$$

and (9.2) and (9.3) have the same value.   ∎

Lemma 9.1 justifies the following definition based on (9.1).

9.2 Definition Let $f = \sum_{j=0}^{M} y_j \mathbf{1}_{A_j} \in \mathcal{E}^+$ be a simple function in standard representation. Then the number

$$I_\mu(f) := \sum_{j=0}^{M} y_j \mu(A_j) \in [0, \infty]$$

(which is independent of the representation of f) is called the $(\mu\text{-})$*integral* of f.

9.3 Properties (of $I_\mu : \mathcal{E}^+ \to [0, \infty]$). Let $f, g \in \mathcal{E}^+$. Then

(i) $I_\mu(\mathbf{1}_A) = \mu(A) \qquad \forall A \in \mathcal{A}$;

(ii) $I_\mu(\lambda f) = \lambda I_\mu(f) \qquad \forall \lambda \geq 0$;   (*positive homogeneous*)

(iii) $I_\mu(f + g) = I_\mu(f) + I_\mu(g)$;   (*additive*)

(iv) $f \leq g \implies I_\mu(f) \leq I_\mu(g)$.   (*monotone*)

Proof (i) and (ii) are obvious from the definition of I_μ. (iii): take standard representations

$$f = \sum_{j=0}^{M} y_j \mathbf{1}_{A_j} \qquad \text{and} \qquad g = \sum_{k=0}^{N} z_k \mathbf{1}_{B_k}$$

and observe that, as in Example 8.5(iv),

$$f + g = \sum_{j=0}^{M} \sum_{k=0}^{N} (y_j + z_k) \mathbf{1}_{A_j \cap B_k} \in \mathcal{E}^+$$

is a standard representation of $f + g$. Thus

$$\begin{aligned}
I_\mu(f + g) &= \sum_{j=0}^{M} \sum_{k=0}^{N} (y_j + z_k) \mu(A_j \cap B_k) \\
&= \sum_{j=0}^{M} y_j \sum_{k=0}^{N} \mu(A_j \cap B_k) + \sum_{k=0}^{N} z_k \sum_{j=0}^{M} \mu(A_j \cap B_k) \\
&\overset{(9.2),(9.3)}{=} \sum_{j=0}^{M} y_j \mu(A_j) + \sum_{k=0}^{N} z_k \mu(B_k) \\
&= I_\mu(f) + I_\mu(g).
\end{aligned}$$

(iv): If $f \leqslant g$, then $g = f + (g - f)$ where $g - f \in \mathcal{E}^+$, see examples 8.7(iv). By part (iii) of this proof,

$$I_\mu(g) = I_\mu(f) + I_\mu(g - f) \geqslant I_\mu(f)$$

since $I_\mu(\cdot)$ is positive. ∎

In Theorem 8.8 we have seen that every $u \in \mathcal{M}^+$ can be written as an increasing limit of simple functions; by Corollary 8.9, suprema of simple functions are again measurable, so that

$$u \in \mathcal{M}^+ \iff u = \sup_{j \in \mathbb{N}} f_j, \quad f_j \in \mathcal{E}^+, \quad f_j \leqslant f_{j+1} \leqslant \ldots$$

We will use this to 'inscribe' simple functions (which we know how to integrate) below the graph of a positive measurable function u and exhaust the μ-area below u.

9.4 Definition Let $(X, \mathcal{A}, \mu)$ be a measure space. The (μ-)*integral* of a positive numerical function $u \in \mathcal{M}_{\overline{\mathbb{R}}}^+$ is given by

$$\int u \, d\mu := \sup \left\{ I_\mu(g) : g \leqslant u, \ g \in \mathcal{E}^+ \right\} \in [0, \infty]. \tag{9.4}$$

If we need to emphasize the *integration variable*, we also write $\int u(x) \, \mu(dx)$ or $\int u(x) \, d\mu(x)$.

The key observation is that the integral $\int \ldots d\mu$ extends I_μ, i.e.

9.5 Lemma *For all $f \in \mathcal{E}^+$ we have $\int f \, d\mu = I_\mu(f)$.*

Proof Let $f \in \mathcal{E}^+$. Since $f \leqslant f$, f is an admissible function in the supremum appearing in (9.4), hence

$$I_\mu(f) \leqslant \sup \left\{ I_\mu(g) : g \leqslant f, \ g \in \mathcal{E}^+ \right\} \overset{\text{def}}{=} \int f \, d\mu.$$

On the other hand, $\mathcal{E}^+ \ni g \leqslant f$ implies that $I_\mu(g) \leqslant I_\mu(f)$ by Properties 9.3(iv), and

$$\int f \, d\mu \overset{\text{def}}{=} \sup \left\{ I_\mu(g) : g \leqslant f, \ g \in \mathcal{E}^+ \right\} \leqslant I_\mu(f). \qquad \blacksquare$$

The next result is the first of several *convergence theorems*. It shows, in particular, that we could have defined (9.4) using *any* increasing sequence $f_j \uparrow u$ of simple functions $f_j \in \mathcal{E}^+$.

9.6 Theorem (Beppo Levi) *Let $(X, \mathcal{A}, \mu)$ be a measure space. For an increasing sequence of numerical functions $(u_j)_{j \in \mathbb{N}} \subset \mathcal{M}_{\overline{\mathbb{R}}}^+$, $0 \leqslant u_j \leqslant u_{j+1} \leqslant \ldots$, we have $u := \sup_{j \in \mathbb{N}} u_j \in \mathcal{M}_{\overline{\mathbb{R}}}^+$ and*

$$\int \sup_{j \in \mathbb{N}} u_j \, d\mu = \sup_{j \in \mathbb{N}} \int u_j \, d\mu. \qquad (9.5)$$

Note that we can write $\lim_{j \to \infty}$ instead of $\sup_{j \in \mathbb{N}}$ in (9.5) since the supremum of an increasing sequence is its limit. Moreover, (9.5) holds in $[0, +\infty]$, i.e. the case '$+\infty = +\infty$' is possible.

Proof (of Theorem 9.6) That $u \in \mathcal{M}_{\overline{\mathbb{R}}}^+$ follows from Corollary 8.9.

Step 1. Claim: $u, v \in \mathcal{M}_{\overline{\mathbb{R}}}^+$, $u \leqslant v \implies \int u \, d\mu \leqslant \int v \, d\mu$.

This follows from the monotonicity of the supremum since every simple $f \in \mathcal{E}^+$ with $f \leqslant u$ also satisfies $f \leqslant v$, and so

$$\int u \, d\mu = \sup \left\{ I_\mu(f) : f \leqslant u, \ f \in \mathcal{E}^+ \right\}$$

$$\leqslant \sup \left\{ I_\mu(f) : f \leqslant v, \ f \in \mathcal{E}^+ \right\} = \int v \, d\mu.$$

Step 2. Claim: $\sup_{j \in \mathbb{N}} \int u_j \, d\mu \leqslant \int \sup_{j \in \mathbb{N}} u_j \, d\mu$; this shows '$\geqslant$' in (9.5). Because of step 1 and $u_j \leqslant u = \sup_{j \in \mathbb{N}} u_j$ we see

$$\int u_j \, d\mu \leqslant \int u \, d\mu \qquad \forall j \in \mathbb{N}.$$

The right-hand side is independent of j, so that we may take the supremum over all $j \in \mathbb{N}$ on the left.

Step 3. Claim: $f \leqslant u$, $f \in \mathcal{E}^+ \implies I_\mu(f) \leqslant \sup_{j \in \mathbb{N}} \int u_j \, d\mu$.

This will prove '$\leqslant$' in (9.5) since the right-hand side does not depend on f and so we may take the supremum over all $f \in \mathcal{E}^+$ with $f \leqslant u$ on the left (which is, by definition, the integral $\int u \, d\mu$).

To prove the claim we fix some $f \in \mathcal{E}^+$, $f \leqslant u$. Since $u = \sup_{j \in \mathbb{N}} u_j$ we can find[✓] for every $\alpha \in (0, 1)$ and every $x \in X$ some $N(x, \alpha) \in \mathbb{N}$ with

$$\alpha f(x) \leqslant u_j(x) \qquad \forall j \geqslant N(x, \alpha),$$

which means that the sets $B_j := \{\alpha f \leqslant u_j\}$ increase as $j \uparrow \infty$ towards X and are, by Corollary 8.12, measurable as $f, u_j \in \mathcal{M}_{\overline{\mathbb{R}}}^+$. By the very definition of the B_j

$$\alpha \mathbf{1}_{B_j} f \leqslant \mathbf{1}_{B_j} u_j \leqslant u_j$$

and, if $f = \sum_{k=0}^M y_k \mathbf{1}_{A_k}$, we get from Lemma 9.5 and step 1

$$\alpha \sum_{k=0}^M y_k \, \mu(A_k \cap B_j) = I_\mu(\alpha \mathbf{1}_{B_j} f) \leqslant \int u_j \, d\mu \leqslant \sup_{j \in \mathbb{N}} \int u_j \, d\mu. \qquad (9.6)$$

At this point we use the σ-additivity of μ (in the guise of T4.4(iii)) to get

$$\mu(A_k \cap B_j) \uparrow \mu(A_k \cap X) = \mu(A_k) \qquad \text{as} \quad B_j \uparrow X, \ j \uparrow \infty,$$

which implies (the far right of (9.6) no longer depends on j)

$$\alpha I_\mu(f) = \alpha \sum_{k=0}^M y_k \, \mu(A_k) \leqslant \sup_{j \in \mathbb{N}} \int u_j \, d\mu.$$

Since we were free in our choice of $\alpha \in (0, 1)$, we can make $\alpha \to 1$, and the claim and the theorem follow. ∎

One can see the next corollary just as a special case of Theorem 9.6. Its true meaning, however, is that it allows us to calculate the integral of a measurable function using *any* approximating *sequence* of elementary functions—and this is a considerable simplification of the original definition (9.4).

9.7 Corollary *Let $u \in \mathcal{M}_{\overline{\mathbb{R}}}^+$. Then*

$$\int u \, d\mu = \lim_{j \to \infty} \int f_j \, d\mu$$

holds for every increasing sequence $(f_j)_{j \in \mathbb{N}} \subset \mathcal{E}^+$ with $\lim_{j \to \infty} f_j = u$.

9.8 Properties (of the integral) Let $u, v \in \mathcal{M}_{\overline{\mathbb{R}}}^{+}$. Then

(i) $\displaystyle\int \mathbf{1}_A \, d\mu = \mu(A) \qquad \forall A \in \mathcal{A};$

(ii) $\displaystyle\int \alpha u \, d\mu = \alpha \int u \, d\mu \qquad \forall \alpha \geqslant 0;$ *(positive homogeneous)*

(iii) $\displaystyle\int (u + v) \, d\mu = \int u \, d\mu + \int v \, d\mu;$ *(additive)*

(iv) $u \leqslant v \implies \displaystyle\int u \, d\mu \leqslant \int v \, d\mu.$ *(monotone)*

Proof **(i)** follows from Properties 9.3(i) and Lemma 9.5 and **(ii)**, **(iii)** follow from the corresponding properties of I_μ, Corollary 9.7 and the usual rules for limits. **(iv)** has been proved in step 1 of the proof of Theorem 9.6. ∎

9.9 Corollary *Let* $(u_j)_{j \in \mathbb{N}} \subset \mathcal{M}_{\overline{\mathbb{R}}}^{+}$. *Then* $\sum_{j=1}^{\infty} u_j$ *is measurable and we have*

$$\int \sum_{j=1}^{\infty} u_j \, d\mu = \sum_{j=1}^{\infty} \int u_j \, d\mu \tag{9.7}$$

(including the possibility $+\infty = +\infty$*).*

Proof Set $s_M := u_1 + u_2 + \cdots + u_M$ and apply Properties 9.8(iii) and T9.6. ∎

9.10 Examples Let $(X, \mathcal{A})$ be a measurable space.

(i) Let $\mu = \delta_y$ be the Dirac measure for fixed $y \in X$. Then

$$\int u \, d\delta_y = \int u(x) \, \delta_y(dx) = u(y) \qquad \forall u \in \mathcal{M}_{\overline{\mathbb{R}}}^{+}.$$

Indeed: for any $f \in \mathcal{E}^{+}$ with standard representation $f = \sum_{j=0}^{M} \phi_j \mathbf{1}_{A_j}$, we know that $y \in X$ lies in exactly one of the A_j, say $y \in A_{j_0}$. Then

$$\int f(x) \, \delta_y(dx) = \int \sum_{j=0}^{M} \phi_j \mathbf{1}_{A_j}(x) \, \delta_y(dx) = \sum_{j=0}^{M} \phi_j \, \delta_y(A_j) = \phi_{j_0} = f(y).$$

Now take any sequence of simple functions $f_k \uparrow u$. By Corollary 9.7

$$\int u(x) \, \delta_y(dx) = \lim_{k \to \infty} \int f_k(x) \, \delta_y(dx) = \lim_{k \to \infty} f_k(y) = u(y).$$

(ii) Let $(X, \mathcal{A}, \mu) = \big(\mathbb{N}, \mathcal{P}(\mathbb{N}), \sum_{j=1}^{\infty} \alpha_j \delta_j\big)$. As we have seen in Problem 4.6(ii), μ is indeed a measure and $\mu(\{k\}) = \alpha_k$. On the other hand, all $\mathcal{P}(\mathbb{N})$-measurable functions $u \in \mathcal{M}_{\overline{\mathbb{R}}}^{+}$ are of the form

$$u(k) = \sum_{j=1}^{\infty} u_j \mathbf{1}_{\{j\}}(k) \qquad \forall k \in \mathbb{N}$$

for a suitable sequence $(u_j)_{j \in \mathbb{N}} \subset [0, \infty]$.[2] Thus by Corollary 9.9,

$$\int u \, d\mu = \int \sum_{j=1}^{\infty} u_j \mathbf{1}_{\{j\}} \, d\mu = \sum_{j=1}^{\infty} \int u_j \mathbf{1}_{\{j\}} \, d\mu$$

$$= \sum_{j=1}^{\infty} u_j \mu(\{j\}) = \sum_{j=1}^{\infty} u_j \alpha_j.$$

We close this chapter with another *convergence theorem* due to P. Fatou and which is often called *Fatou's lemma*.

9.11 Theorem (Fatou) *Let* $(u_j)_{j \in \mathbb{N}} \subset \mathcal{M}_{\overline{\mathbb{R}}}^{+}$ *be a sequence of positive measurable numerical functions. Then* $u := \liminf_{j \to \infty} u_j$ *is measurable and*

$$\int \liminf_{j \to \infty} u_j \, d\mu \leqslant \liminf_{j \to \infty} \int u_j \, d\mu.$$

Proof Recall that $\liminf_{j \to \infty} u_j = \sup_{k \in \mathbb{N}} \inf_{j \geqslant k} u_j$ always exists in $\overline{\mathbb{R}}$; the measurability of $\liminf$ was shown in C8.9. Applying T9.6 to the increasing sequence $\left(\inf_{j \geqslant k} u_j \right)_{k \in \mathbb{N}}$ – which is in $\mathcal{M}_{\overline{\mathbb{R}}}^{+}$ by C8.9 – we find

$$\int \liminf_{j \to \infty} u_j \, d\mu \overset{9.6}{=} \sup_{k \in \mathbb{N}} \left(\int \inf_{j \geqslant k} u_j \, d\mu \right)$$

$$\overset{9.8(iv)}{\leqslant} \sup_{k \in \mathbb{N}} \left(\inf_{\ell \geqslant k} \int u_\ell \, d\mu \right)$$

$$= \liminf_{\ell \to \infty} \int u_\ell \, d\mu$$

where we used that $\inf_{j \geqslant k} u_j \leqslant u_\ell$ for all $\ell \geqslant k$ and the monotonicity of the integral, cf. Properties 9.8(iv). ∎

Problems

9.1. Let $f : X \to \mathbb{R}$ be a positive simple function of the form $f(x) = \sum_{j=1}^{m} \xi_j \mathbf{1}_{A_j}(x)$, $\xi_j \geqslant 0$, $A_j \in \mathcal{A}$—but not necessarily disjoint. Show that $I_\mu(f) = \sum_{j=1}^{m} \xi_j \mu(A_j)$. [Hint: use additivity and positive homogeneity of I_μ.]

9.2. Complete the proof of Properties 9.8 (of the integral).

9.3. Find an example showing that an '*increasing sequence of functions*' is, in general, different from a '*sequence of increasing functions*'.

---

[2] This means that we can identify $\mathcal{P}(\mathbb{N})$-measurable functions $f : \mathbb{N} \to [0, \infty]$ and arbitrary sequences $(u_j)_{j \in \mathbb{N}} \subset [0, \infty]$ by $u(k) = u_k$.

9.4. Complete the proof of Corollary 9.9 and show that (9.7) is actually *equivalent* to (9.5) in Beppo Levi's theorem 9.6.

9.5. Let $(X, \mathcal{A}, \mu)$ be a measure space and $u \in \mathcal{M}^+(\mathcal{A})$. Show that the set-function $A \mapsto \int \mathbf{1}_A u \, d\mu$, $A \in \mathcal{A}$, is a measure.

9.6. Prove: *Every* function $u : \mathbb{N} \to \mathbb{R}$ on $(\mathbb{N}, \mathcal{P}(\mathbb{N}))$ is measurable.

9.7. Let $(X, \mathcal{A})$ be a measurable space and $(\mu_j)_{j \in \mathbb{N}}$ be a sequence of measures thereon. Set, as in 9.10(ii), $\mu = \sum_{j \in \mathbb{N}} \mu_j$. By Problem 4.6(ii) this is again a measure. Show that

$$\int u \, d\mu = \sum_{j \in \mathbb{N}} \int u \, d\mu_j \qquad \forall u \in \mathcal{M}^+(\mathcal{A}).$$

[Instructions: (1) consider $u = \mathbf{1}_A$. (2) consider $u = f \in \mathcal{E}^+$. (3) approximate $u \in \mathcal{M}^+$ by an increasing sequence of simple functions and use Theorem 9.6. To interchange increasing limits/suprema use the hint to Problem 4.6(ii).]

9.8. **Reverse Fatou lemma.** Let $(X, \mathcal{A}, \mu)$ be a measure space and $(u_j)_{j \in \mathbb{N}} \subset \mathcal{M}^+(\mathcal{A})$. If $u_j \leqslant u$ for all $j \in \mathbb{N}$ and some $u \in \mathcal{M}^+(\mathcal{A})$ with $\int u \, d\mu < \infty$, then

$$\limsup_{j \to \infty} \int u_j \, d\mu \leqslant \int \limsup_{j \to \infty} u_j \, d\mu.$$

9.9. **Fatou's lemma for measures.** Let $(X, \mathcal{A}, \mu)$ be a measure space and let $(A_j)_{j \in \mathbb{N}}$, $A_j \in \mathcal{A}$, be a sequence of measurable sets. We set

$$\liminf_{j \to \infty} A_j := \bigcup_{k \in \mathbb{N}} \bigcap_{j \geqslant k} A_j \qquad \text{and} \qquad \limsup_{j \to \infty} A_j := \bigcap_{k \in \mathbb{N}} \bigcup_{j \geqslant k} A_j. \qquad (9.8)$$

(i) Prove that $\mathbf{1}_{\liminf_{j \to \infty} A_j} = \liminf_{j \to \infty} \mathbf{1}_{A_j}$ and $\mathbf{1}_{\limsup_{j \to \infty} A_j} = \limsup_{j \to \infty} \mathbf{1}_{A_j}$.

[Hint: check first that $\mathbf{1}_{\bigcap_{j \in \mathbb{N}} A_j} = \inf_{j \in \mathbb{N}} \mathbf{1}_{A_j}$ and $\mathbf{1}_{\bigcup_{j \in \mathbb{N}} A_j} = \sup_{j \in \mathbb{N}} \mathbf{1}_{A_j}$.]

(ii) Prove that $\mu\left(\liminf_{j \to \infty} A_j\right) \leqslant \liminf_{j \to \infty} \mu(A_j)$.

(iii) Prove that $\limsup_{j \to \infty} \mu(A_j) \leqslant \mu\left(\limsup_{j \to \infty} A_j\right)$ if μ is a finite measure.

(iv) Provide an example showing that (iii) fails if μ is not finite.

9.10. Let $(A_j)_{j \in \mathbb{N}} \subset \mathcal{A}$ be a sequence of disjoint sets such that $\biguplus_{j \in \mathbb{N}} A_j = X$. Show that for every $u \in \mathcal{M}^+(\mathcal{A})$

$$\int u \, d\mu = \sum_{j=1}^{\infty} \int \mathbf{1}_{A_j} u \, d\mu.$$

Use this to construct on a σ-finite measure space $(X, \mathcal{A}, \mu)$ a function w which satisfies $w(x) > 0$ for all $x \in X$ and $\int w \, d\mu < \infty$.

9.11. **Kernels.** Let $(X, \mathcal{A}, \mu)$ be a measure space. A map $N : X \times \mathcal{A} \to [0, \infty]$ is called *kernel* if

$$A \mapsto N(x, A) \qquad \text{is a measure for every } x \in X$$

$$x \mapsto N(x, A) \qquad \text{is a measurable function for every } A \in \mathcal{A}.$$

(i) Show that $\mathcal{A} \ni A \mapsto \mu N(A) := \int N(x, A)\,\mu(dx)$ is a measure on $(X, \mathcal{A})$.

(ii) For $u \in \mathcal{M}^+(\mathcal{A})$ define $Nu(x) := \int u(y)\,N(x, dy)$. Show that $u \mapsto Nu$ is additive, positive homogeneous and $Nu(\cdot) \in \mathcal{M}^+(\mathcal{A})$.

(iii) Let μN be the measure introduced in (i). Show that $\int u\,d(\mu N) = \int Nu\,d\mu$ for all $u \in \mathcal{M}^+(\mathcal{A})$.

[Hint: consider in each part of this problem first indicator functions $u = \mathbf{1}_A$, then simple functions $u \in \mathcal{E}^+(\mathcal{A})$ and then approximate $u \in \mathcal{M}^+(\mathcal{A})$ by simple functions using 8.8 and 9.6]

9.12. (Continuation of Problem 6.1) Consider on $\mathbb{R}$ the σ-algebra Σ of all Borel sets which are symmetric w.r.t. the origin. Set $A^+ := A \cap [0, \infty)$, $A^- := (-\infty, 0] \cap A$ and consider their symmetrizations $A^\pm_\sigma := A^\pm \cup (-A^\pm) \in \Sigma$. Show that for every $u \in \mathcal{M}^+(\Sigma)$ with $0 \leqslant u \leqslant 1$ and for every measure μ on $(\mathbb{R}, \Sigma)$ the set-function

$$\mathcal{B}(\mathbb{R}) \ni A \mapsto \int \mathbf{1}_{A^+_\sigma} u\,d\mu + \int \mathbf{1}_{A^-_\sigma} (1 - u)\,d\mu$$

is a measure on $\mathcal{B}(\mathbb{R})$ that extends μ.

Why does this not contradict the uniqueness theorem 5.7 for measures?

10

Integrals of measurable functions and null sets

Throughout this chapter $(X, \mathcal{A}, \mu)$ will be a measure space.

Let us briefly review how we constructed the integral for positive measurable functions $u \in \mathcal{M}_{\bar{\mathbb{R}}}^+$. Guided by the idea that the integral should be the area between the graph of a function and the x-axis, we defined for indicator functions $\int \mathbf{1}_A \, d\mu = \mu(A)$ and extended this definition by linearity to all positive simple functions $\mathcal{E}^+$ which are just linear combinations of indicator functions (there was an issue about well-definedness which was addressed in L9.1). Since all positive measurable functions can be obtained as increasing limits of simple functions (Theorem 8.8), we could then define the integral of $u \in \mathcal{M}_{\bar{\mathbb{R}}}^+$ by exhausting the area below u with elementary functions $f \leqslant u$, see Definition 9.4. Beppo Levi's theorem (in the form of C9.7) finally allowed us to replace the sup by an increasing limit. The integral turned out to be positive homogeneous, additive and monotone.

We want to extend this integral now to not necessarily positive measurable functions $u \in \mathcal{M}_{\bar{\mathbb{R}}}$ by linearity. The fundamental observation here is that

$$u \in \mathcal{M}_{\bar{\mathbb{R}}} \iff u = u^+ - u^-, \quad u^+, u^- \in \mathcal{M}_{\bar{\mathbb{R}}}^+$$

(cf. Corollary 8.11). This remark suggests the following definition.

10.1 Definition A function $u : X \to \bar{\mathbb{R}}$ on a measure space $(X, \mathcal{A}, \mu)$ is said to be $(\mu\text{-})integrable$, if it is $\mathcal{A}/\bar{\mathcal{B}}$-measurable and if the integrals $\int u^+ \, d\mu, \int u^- \, d\mu < \infty$ are finite. In this case we call

$$\int u \, d\mu := \int u^+ \, d\mu - \int u^- \, d\mu \in (-\infty, \infty) \tag{10.1}$$

the $(\mu\text{-})integral$ of u. We write $\mathcal{L}^1(\mu)$ $[\mathcal{L}_{\bar{\mathbb{R}}}^1(\mu)]$ for the set of all real-valued [numerical] μ-integrable functions.

76

In case we need to exhibit the integration variable, we write

$$\int u \, d\mu = \int u(x) \, \mu(dx) = \int u(x) \, d\mu(x).$$

If $\mu = \lambda^n$, we call $\int u \, d\lambda^n$ the (n-dimensional) *Lebesgue integral* and $u \in \mathcal{L}^1_{\mathbb{R}}(\lambda^n)$ is said to be *Lebesgue integrable*.[1] Traditionally one writes $\int u(x) \, dx$ or $\int u \, dx$ for the formally more correct $\int u \, d\lambda^n$. If we want to stress X or $\mathcal{A}$, etc., we will also write $\mathcal{L}^1_{\mathbb{R}}(X)$ or $\mathcal{L}^1_{\mathbb{R}}(\mathcal{A})$, etc.

10.2 Remark In the definition of the integral for positive $u \in \mathcal{M}^+_{\overline{\mathbb{R}}}$ we did allow that $\int u \, d\mu = \infty$. Since we want to avoid the case '$\infty - \infty$' in (10.1), we impose the finiteness condition $\int u^{\pm} \, d\mu < \infty$. In particular, a positive function is said to be *integrable* only if the integral is finite:

$$u \in \mathcal{L}^1_{\mathbb{R}}(\mu), \ u \geqslant 0 \iff u \in \mathcal{M}^+_{\overline{\mathbb{R}}} \quad \text{and} \quad \int u \, d\mu < \infty$$

(which is clear since for positive functions $u^+ = u$ and $u^- = 0$).

Caution: Some authors call u μ-integrable (in the wide sense) whenever $\int u^+ \, d\mu - \int u^- \, d\mu$ makes sense in $\overline{\mathbb{R}}$, i.e. whenever it is not of the form '$\infty - \infty$'. We will not use this convention.

Let us briefly summarize the most important integrability criteria.

10.3 Theorem *Let $u \in \mathcal{M}_{\overline{\mathbb{R}}}$. Then the following conditions are equivalent:*

 (i) $u \in \mathcal{L}^1_{\mathbb{R}}(\mu)$;
 (ii) $u^+, u^- \in \mathcal{L}^1_{\mathbb{R}}(\mu)$;
 (iii) $|u| \in \mathcal{L}^1_{\mathbb{R}}(\mu)$;
 (iv) $\exists \, w \in \mathcal{L}^1_{\mathbb{R}}(\mu), \ w \geqslant 0$ *such that* $|u| \leqslant w$.

Proof (i)$\Leftrightarrow$(ii): this is just the definition of integrability.

   (ii)$\Rightarrow$(iii): since $|u| = u^+ + u^-$, we can use the additivity of the integral on $\mathcal{M}^+_{\overline{\mathbb{R}}}$, see 9.8(iii), to get $\int |u| \, d\mu = \int u^+ \, d\mu + \int u^- \, d\mu < \infty$.

   (iii)$\Rightarrow$(iv): take $w := |u|$.

   (iv)$\Rightarrow$(i): we have to show that $u^{\pm} \in \mathcal{L}^1_{\mathbb{R}}(\mu)$. Since $u^{\pm} \leqslant |u| \leqslant w$ we find by the monotonicity of the integral 9.8(iv) that $\int u^{\pm} \, d\mu \leqslant \int w \, d\mu < \infty$. ∎

---

[1] The letter $\mathcal{L}$ is in honour of H. Lebesgue who was one of the pioneers of modern integration theory. If μ is other than λ^n, $\int \ldots d\mu$ is sometimes called the *abstract Lebesgue integral*.

It is now easy to see that the properties 9.8 of the integral on $\mathcal{M}_{\overline{\mathbb{R}}}^+$ extend to the set $\mathcal{L}_{\overline{\mathbb{R}}}^1(\mu)$:

10.4 Theorem *Let* $(X, \mathcal{A}, \mu)$ *be a measure space and* $u, v \in \mathcal{L}_{\overline{\mathbb{R}}}^1(\mu)$, $\alpha \in \mathbb{R}$. *Then*

(i) $\alpha u \in \mathcal{L}_{\overline{\mathbb{R}}}^1(\mu)$ *and* $\displaystyle\int \alpha u \, d\mu = \alpha \int u \, d\mu;$ (*homogeneous*)

(ii) $u + v \in \mathcal{L}_{\overline{\mathbb{R}}}^1(\mu)$ *and* $\displaystyle\int (u + v) \, d\mu = \int u \, d\mu + \int v \, d\mu$ (*additive*)
(whenever $u + v$ *is defined);*

(iii) $\min\{u, v\}, \max\{u, v\} \in \mathcal{L}_{\overline{\mathbb{R}}}^1(\mu);$ (*lattice property*)

(iv) $u \leqslant v \implies \displaystyle\int u \, d\mu \leqslant \int v \, d\mu;$ (*monotone*)

(v) $\left| \displaystyle\int u \, d\mu \right| \leqslant \displaystyle\int |u| \, d\mu.$ (*triangle inequality*)

Proof There are principally two ways to prove this theorem: either we consider positive and negative parts for (i)–(v) and show that their integrals are finite, or we use T10.3(iii), (iv). Doing this we find

(i) $|\alpha u| = |\alpha| \cdot |u| \in \mathcal{L}_{\overline{\mathbb{R}}}^1(\mu)$ by 9.8(ii).

(ii) $|u + v| \leqslant |u| + |v| \in \mathcal{L}_{\overline{\mathbb{R}}}^1(\mu)$ by 9.8(iii).

(iii) $|\max\{u, v\}| \leqslant |u| + |v| \in \mathcal{L}_{\overline{\mathbb{R}}}^1(\mu)$ and
$|\min\{u, v\}| \leqslant |u| + |v| \in \mathcal{L}_{\overline{\mathbb{R}}}^1(\mu)$ by 9.8(iii).

(iv) If $u \leqslant v$, we find that $u^+ \leqslant v^+$ and $v^- \leqslant u^-$. Thus

$$\int u \, d\mu = \int u^+ \, d\mu - \int u^- \, d\mu \overset{9.8(iv)}{\leqslant} \int v^+ \, d\mu - \int v^- \, d\mu = \int v \, d\mu.$$

(v) Using $\pm u \leqslant |u|$ we deduce from (iv) that

$$\left| \int u \, d\mu \right| = \max \left\{ \int u \, d\mu, \ -\int u \, d\mu \right\}$$

$$\leqslant \max \left\{ \int |u| \, d\mu, \int |-u| \, d\mu \right\} = \int |u| \, d\mu. \qquad \blacksquare$$

10.5 Remark If $u(x) \pm v(x)$ is defined in $\overline{\mathbb{R}}$ for all $x \in X$ – i.e. if we can exclude '$\infty - \infty$' – then T10.4(i),(ii) just say that the integral is *linear*:

$$\int (\alpha u + \beta v) \, d\mu = \alpha \int u \, d\mu + \beta \int v \, d\mu, \qquad \alpha, \beta \in \mathbb{R}. \qquad (10.2)$$

This is always true for real-valued $u, v \in \mathcal{L}^1(\mu)$, i.e. $\mathcal{L}^1(\mu)$ is a *vector space* with addition and scalar ($\mathbb{R}$) multiplication defined by

$$(u + v)(x) := u(x) + v(x), \qquad (\alpha \cdot u)(x) := \alpha \cdot u(x),$$

and

$$\int \ldots d\mu : \mathcal{L}^1(\mu) \to \mathbb{R}, \quad u \mapsto \int u \, d\mu,$$

is a *positive linear functional*.

10.6 Examples Let us reconsider the examples from 9.10:

(i) On $(X, \mathcal{A}, \delta_y)$, $y \in X$ fixed, we have $\int u(x) \, \delta_y(dx) = u(y)$ and

$$u \in \mathcal{L}^1_{\bar{\mathbb{R}}}(\delta_y) \iff u \in \mathcal{M}_{\bar{\mathbb{R}}} \text{ and } |u(y)| < \infty.$$

(ii) On $\left(\mathbb{N}, \mathcal{P}(\mathbb{N}), \mu := \sum_{j=1}^{\infty} \alpha_j \delta_j\right)$ every $u : \mathbb{N} \to \mathbb{R}$ is measurable, cf. Problem 9.6. From 9.10(ii) we know that $\int |u| \, d\mu = \sum_{j=1}^{\infty} \alpha_j |u(j)|$, so that

$$u \in \mathcal{L}^1(\mu) \iff \sum_{j=1}^{\infty} \alpha_j |u(j)| < \infty.$$

If $\alpha_1 = \alpha_2 = \ldots = 1$, $\mathcal{L}^1(\mu)$ is called the set of *summable sequences* and customarily denoted by $\ell^1(\mathbb{N}) = \left\{ (x_j)_{j \in \mathbb{N}} \subset \mathbb{R} : \sum_{j=1}^{\infty} |x_j| < \infty \right\}$. This space is important in functional analysis.

(iii) Let $(\Omega, \mathcal{A}, P)$ be a probability space. Then every bounded measurable function ('random variable') $\xi \in \mathcal{M}(\mathcal{A})$, $C := \sup_{\omega \in \Omega} |\xi(\omega)| < \infty$, is integrable. This follows immediately from

$$\int |\xi| \, dP \leqslant \int \sup_{\omega \in \Omega} |\xi(\omega)| \, P(d\omega) = C \int P(d\omega) = C < \infty.$$

Caution: Not every P-integrable function is bounded.[✓]

For $A \in \mathcal{A}$ and $u \in \mathcal{M}_{\bar{\mathbb{R}}}^+$ [or $\mathcal{L}^1_{\bar{\mathbb{R}}}(\mu)$] we know from 8.5(i) and C8.10 [and 10.3(iv) using $|\mathbf{1}_A u| \leqslant |u|$] that $\mathbf{1}_A u$ is again measurable [or integrable].

10.7 Definition Let $(X, \mathcal{A}, \mu)$ be a measure space and $u \in \mathcal{L}^1_{\bar{\mathbb{R}}}(\mu)$ or $u \in \mathcal{M}_{\bar{\mathbb{R}}}^+(\mathcal{A})$. Then

$$\int_A u \, d\mu := \int \mathbf{1}_A u \, d\mu = \int \mathbf{1}_A(x) u(x) \, \mu(dx), \qquad \forall A \in \mathcal{A}.$$

Of course, $\int_X u \, d\mu = \int u \, d\mu$.

10.8 Lemma *On the measure space* $(X, \mathcal{A}, \mu)$ *let* $u \in \mathcal{M}^+$. *The set-function*

$$\nu : A \mapsto \int_A u \, d\mu = \int \mathbf{1}_A u \, d\mu, \qquad A \in \mathcal{A},$$

is a measure on $(X, \mathcal{A})$. It is called the measure with density (function) u with respect to μ and denoted by $\nu = u\,\mu$.

Proof Exercise.   ∎

If ν has a density w.r.t. μ, one writes traditionally $d\nu/d\mu$ for the density function. This notation is to be understood in a purely symbolical way; it is motivated by the well-known *fundamental theorem of integral and differential calculus* (for Riemann integrals)

$$u(b) - u(a) = \int_a^b u'(x)\,dx$$

where $u' = d(u'\lambda^1)/d\lambda^1$ (in our notation) $= du/dx$. At least if $u'(x) \geqslant 0$ one can show that $\nu[a, b) := u(b) - u(a)$ defines a measure and that, taking the fundamental theorem of integral calculus for granted, $\nu = u'\,\lambda^1 = u'\,dx$, compare with Problem 7.9. A more advanced discussion of derivatives can be found in Chapter 19, Theorem 19.20 and Appendix E.16–E.19.

Null sets and the 'a.e.'

We will now discuss the behaviour of integrable functions on null sets which we have already encountered in Problem 4.10.

Let $(X, \mathcal{A}, \mu)$ be a measure space. A $(\mu\text{-})$*null set* $N \in \mathcal{N}_\mu$ is a measurable set $N \in \mathcal{A}$ satisfying

$$N \in \mathcal{N}_\mu \iff N \in \mathcal{A} \quad \text{and} \quad \mu(N) = 0. \tag{10.3}$$

If a property $\Pi = \Pi(x)$ is true for all $x \in X$ apart from some x contained in a null set $N \in \mathcal{N}_\mu$, we say that $\Pi(x)$ *holds for* $(\mu\text{-})$ *almost all (a.a.)* $x \in X$ or that Π *holds* $(\mu\text{-})$ *almost everywhere (a.e.)*. In other words,

$$\Pi \quad \text{holds a.e.} \iff \{x : \Pi(x) \quad \text{fails}\} \subset N \in \mathcal{N}_\mu,$$

but we do not *a priori* require that the set $\{\Pi \text{ fails}\}$ is itself measurable. Typically we are interested in properties $\Pi(x)$ of the type: $u(x) = v(x)$, $u(x) \leqslant v(x)$, etc. and we say, for example,

$$u = v \quad \text{a.e.} \iff \{x : u(x) \neq v(x)\} \quad \text{is (contained in) a } \mu\text{-null set.}$$

Caution: The assertions 'u enjoys a property Π a.e.' and 'u is a.e. equal to v which satisfies Π everywhere' are, in general, *far apart*; see in this connection Problem 10.14.

10.9 Theorem *Let $u \in \mathcal{L}^1_{\overline{\mathbb{R}}}(\mu)$ be a numerical integrable function on a measure space $(X, \mathcal{A}, \mu)$. Then*

(i) $\int |u|\, d\mu = 0 \iff |u| = 0 \quad a.e. \iff \mu(\{u \neq 0\}) = 0;$

(ii) $\int_N u\, d\mu = 0 \quad \forall N \in \mathcal{N}_\mu.$

Proof Let us begin with **(ii)**. Obviously, $\min\{|u|, j\} \uparrow |u|$ as $j \uparrow \infty$. By Beppo Levi's theorem 9.6 we find

$$\left| \int_N u\, d\mu \right| = \left| \int \mathbf{1}_N u\, d\mu \right| \overset{10.4(\text{v})}{\leqslant} \int \mathbf{1}_N |u|\, d\mu$$

$$\overset{9.6}{=} \sup_{j \in \mathbb{N}} \int \mathbf{1}_N \min\{|u|, j\}\, d\mu \leqslant \sup_{j \in \mathbb{N}} \int j \mathbf{1}_N\, d\mu$$

$$= \sup_{j \in \mathbb{N}} \left(j \int \mathbf{1}_N\, d\mu \right) = \sup_{j \in \mathbb{N}} \big(j \underbrace{\mu(N)}_{=0} \big) = 0.$$

The second equivalence in **(i)** is clear since, due to the measurability of u, the set $\{u \neq 0\}$ is not just a subset of a null set, but measurable, hence a proper null set. In order to see '$\Leftarrow$' of the first equivalence, we use (ii) with $N = \{u \neq 0\}$:

$$\int |u|\, d\mu = \int_{\{|u| \neq 0\}} |u|\, d\mu + \int_{\{|u| = 0\}} |u|\, d\mu$$

$$= \int_{\{|u| \neq 0\}} |u|\, d\mu + \int_{\{|u| = 0\}} 0\, d\mu \overset{(\text{ii})}{=} 0.$$

For '$\Rightarrow$' we use the so-called *Markov inequality*: for $A \in \mathcal{A}$ and $c > 0$ we have

$$\mu(\{|u| \geqslant c\} \cap A) = \int \mathbf{1}_{\{|u| \geqslant c\} \cap A}(x)\, \mu(dx)$$

$$= \int_A \frac{c}{c} \mathbf{1}_{\{|u| \geqslant c\}}(x)\, \mu(dx)$$

$$\leqslant \frac{1}{c} \int_A |u(x)|\, \mathbf{1}_{\{|u| \geqslant c\}}(x)\, \mu(dx) \qquad (10.4)$$

$$\leqslant \frac{1}{c} \int_A |u(x)|\, \mu(dx),$$

and for $A = X$ this inequality implies that

$$\mu(\{|u| > 0\}) \overset{[\checkmark]}{=} \mu\Big(\bigcup_{j \in \mathbb{N}} \{|u| \geqslant \tfrac{1}{j}\}\Big) \overset{4.6}{\leqslant} \sum_{j \in \mathbb{N}} \mu(\{|u| \geqslant \tfrac{1}{j}\})$$

$$\leqslant \sum_{j \in \mathbb{N}} \Big(j \underbrace{\int |u| \, d\mu}_{=0}\Big) = 0. \qquad \blacksquare$$

10.10 Corollary *Let $u, v \in \mathcal{M}_{\overline{\mathbb{R}}}$ such that $u = v$ μ-almost everywhere. Then*

(i) $u, v \geqslant 0 \implies \int u \, d\mu = \int v \, d\mu;$[2]

(ii) $u \in \mathcal{L}^1_{\overline{\mathbb{R}}}(\mu) \implies v \in \mathcal{L}^1_{\overline{\mathbb{R}}}(\mu)$ *and* $\int u \, d\mu = \int v \, d\mu.$

Proof Since u, v are measurable, $N := \{u \neq v\} \in \mathcal{N}_\mu$. Therefore **(i)** follows from

$$\int u \, d\mu = \int_{N^c} u \, d\mu + \int_N u \, d\mu$$

$$\overset{10.9(\mathrm{i})}{=} \int_{N^c} v \, d\mu + 0 \qquad (\text{use that } u = v \text{ on } N^c)$$

$$\overset{10.9(\mathrm{i})}{=} \int_{N^c} v \, d\mu + \int_N v \, d\mu = \int v \, d\mu.$$

For **(ii)** we observe first that $u = v$ a.e. implies that $u^{\pm} = v^{\pm}$ a.e. and then apply (i) to positive and negative parts: $\int v^{\pm} \, d\mu = \int u^{\pm} \, d\mu < \infty$; the claim follows. $\blacksquare$

10.11 Corollary *If $u \in \mathcal{M}_{\overline{\mathbb{R}}}$ and $v \in \mathcal{L}^1_{\overline{\mathbb{R}}}(\mu)$, $v \geqslant 0$, then*

$$|u| \leqslant v \quad a.e. \implies u \in \mathcal{L}^1_{\overline{\mathbb{R}}}(\mu).$$

Proof We have $u^{\pm} \leqslant |u| \leqslant v$ a.e., and by C10.10 $\int u^{\pm} \, d\mu \leqslant \int v \, d\mu < \infty$. This shows that u is integrable. $\blacksquare$

10.12 Proposition (Markov inequality) *For all $u \in \mathcal{L}^1_{\overline{\mathbb{R}}}(\mu)$, $A \in \mathcal{A}$ and $c > 0$*

$$\mu(\{|u| \geqslant c\} \cap A) \leqslant \frac{1}{c} \int_A |u| \, d\mu, \qquad (10.5)$$

and if $A = X$, in particular,

$$\mu(\{|u| \geqslant c\}) \leqslant \frac{1}{c} \int |u| \, d\mu. \qquad (10.6)$$

Proof See (10.4) in the proof of Theorem 10.9(i). $\blacksquare$

---

[2] including, possibly, $+\infty = +\infty$.

10.13 Corollary *If $u \in \mathcal{L}^1_{\bar{\mathbb{R}}}(\mu)$, then u is almost everywhere $\mathbb{R}$-valued. In particular, we can find a version $\tilde{u} \in \mathcal{L}^1(\mu)$ such that $\tilde{u} = u$ a.e. and $\int \tilde{u} \, d\mu = \int u \, d\mu$.*

Proof Set $N := \{|u| = \infty\} = \{u = +\infty\} \uplus \{u = -\infty\} \in \mathcal{A}$. Now

$$N = \bigcap_{j \in \mathbb{N}} \{|u| \geqslant j\}$$

and by 3.4(iii′)[3] and the Markov inequality we get

$$\mu(N) = \lim_{j \to \infty} \mu(\{|u| \geqslant j\}) \leqslant \lim_{j \to \infty} \left(\frac{1}{j} \underbrace{\int |u| \, d\mu}_{< \infty} \right) = 0.$$

The function $\tilde{u} := \mathbf{1}_{N^c} u$ is real-valued, measurable and coincides outside N with u. From C10.10 we deduce that $\tilde{u}$ is integrable (and even $\in \mathcal{L}^1(\mu)$) with $\int \tilde{u} \, d\mu = \int u \, d\mu$. ∎

Corollary 10.13 allows us to identify (up to null sets) functions from $\mathcal{L}^1_{\bar{\mathbb{R}}}$ and $\mathcal{L}^1$. Since $\mathcal{L}^1$ is a much nicer space – it is a vector space and we need not take any precautions when adding functions, etc. – we will work from now on only with $\mathcal{L}^1$. The corresponding statements for $\mathcal{L}^1_{\bar{\mathbb{R}}}$ are then easily derived.

We close this section with a technique which will be useful in many applications later on.

10.14 Corollary *Let $\mathcal{G} \subset \mathcal{A}$ be a sub-σ-algebra.*

(i) *If $u, w \in \mathcal{L}^1(\mathcal{G})$ and if $\int_G u \, d\mu = \int_G w \, d\mu$ for all $G \in \mathcal{G}$, then $u = w$ μ-a.e.*

(ii) *If $u, w \in \mathcal{M}^+(\mathcal{G})$ and if $\int_G u \, d\mu = \int_G w \, d\mu$ for all $G \in \mathcal{G}$, then $u = w$ μ-a.e. under the additional assumption that $\mu|_{\mathcal{G}}$ is σ-finite.[4]*

Proof (i) Since u and w are $\mathcal{G}$-measurable functions, we have $G \cap \{u \geqslant w\}, G \cap \{u < w\} \in \mathcal{G}$ for all $G \in \mathcal{G}$. Thus

$$\int_G |u - w| \, d\mu = \int_{G \cap \{u \geqslant w\}} (u - w) \, d\mu + \int_{G \cap \{u < w\}} (w - u) \, d\mu, \qquad (10.7)$$

while

$$\int_{G \cap \{u \geqslant w\}} (u - w) \, d\mu = \left(\int_{G \cap \{u \geqslant w\}} u \, d\mu - \int_{G \cap \{u \geqslant w\}} w \, d\mu \right) = 0 \qquad (10.8)$$

---

[3] This is applicable since μ is finite on $\{|u| \geqslant 1\}$, say$^{[\checkmark]}$.

[4] i.e. there exists an exhausting sequence $(G_j)_j \subset \mathcal{G}$ with $G_j \uparrow X$ and $\mu(G_j) < \infty$.

by the linearity of the integral and because of our assumption. The other term on the right-hand side of (10.7) can be treated similarly, and the conclusion follows from Theorem 10.9(i).

(ii) If u, w are positive measurable functions, we cannot use the linearity of the integral in (10.8) as this may yield an undefined expression of the form '$\infty - \infty$'. We can avoid this by an approximation procedure which relies on the fact that $\mu|_{\mathcal{G}}$ is σ-finite. Pick an exhausting sequence $(G_j)_{j\in\mathbb{N}}$ with $G_j \uparrow X$ and $\mu(G_j) < \infty$. Then the sets $F_j := G \cap \{u \geq w\} \cap G_j \cap \{u \leq j\}$ are in $\mathcal{G}$ and have finite μ-measure. Moreover, the function $(u - w)\mathbf{1}_{F_j}$ is integrable[✓] and increases towards $(u - w)\mathbf{1}_{G\cap\{u\geq w\}}$, so that by Beppo Levi's Theorem 9.6,

$$\int_{G\cap\{u\geq w\}} (u-w)\,d\mu = \sup_{j\in\mathbb{N}} \int_{F_j} (u-w)\,d\mu = \sup_{j\in\mathbb{N}} \underbrace{\left(\int_{F_j} u\,d\mu - \int_{F_j} w\,d\mu \right)}_{=0 \quad \forall j\in\mathbb{N}} = 0.$$

A similar argument applies to the other term in (10.7) and the claim follows. ∎

Problem 10.16 below shows that the σ-finiteness of $\mathcal{G}$ in Corollary 10.14(ii) is really needed.

Problems

10.1. Prove Remark 10.5, i.e. prove the linearity of the integral.

10.2. Let $(\Omega, \mathcal{A}, P)$ be a probability space. Find a counterexample to the claim: *every P-integrable function $u \in \mathcal{L}^1(P)$ is bounded.*

   [Hint: you could try to take $\Omega = (0, 1)$, $P = \lambda^1$ and show that $1/\sqrt{x}$ is Lebesgue integrable on $(0, 1)$ by finding a sequence of suitable simple functions that is above $1/\sqrt{x}$ on, say, $(1/m, 1)$ and then let $m \to \infty$ using Beppo Levi's Theorem 9.6.]

10.3. True or false: if $f \in \mathcal{L}^1$ we can change f on a set N of measure zero, (e.g. by

$$\tilde{f}(x) := \begin{cases} f(x) & \text{if } x \notin N \\ \beta & \text{if } x \in N. \end{cases}$$

where $\beta \in \mathbb{R}$ is any number) and $\tilde{f}$ is still integrable, even $\int f\,d\mu = \int \tilde{f}\,d\mu$?

10.4. Every countable set is a λ^1-null set. Use the Cantor ternary set C (cf. Problem 7.10) to illustrate that the converse is not true. What happens if we change λ^1 to λ^2?

10.5. Prove the following variants of the Markov inequality P10.12: For all $c > 0$ and whenever the expressions involved make sense/are finite, then:

   (i) $\mu(\{|u| > c\}) \leq \dfrac{1}{c} \int |u|\,d\mu$;

   (ii) $\mu(\{|u| > c\}) \leq \dfrac{1}{c^p} \int |u|^p\,d\mu$ for all $0 < p < \infty$;

(iii) $\mu(\{|u| \geq c\}) \leq \dfrac{1}{\phi(c)} \displaystyle\int \phi(|u|)\, d\mu$ for an increasing function $\phi : \mathbb{R}_+ \to \mathbb{R}_+$;

(iv) $\mu(\{u \geq \alpha \int u\, d\mu\}) \leq \dfrac{1}{\alpha}$;

(v) $\mu(\{|u| < c\}) \leq \dfrac{1}{\psi(c)} \displaystyle\int \psi(|u|)\, d\mu$ for a decreasing function $\psi : \mathbb{R}_+ \to \mathbb{R}_+$;

(vi) $P(|X - EX| \geq \alpha \sqrt{VX}) \leq \dfrac{1}{\alpha^2}$, where $(\Omega, \mathcal{A}, P)$ is a probability space and, in probabilistic jargon, X is a random variable (i.e. a measurable function $X : \Omega \to \mathbb{R}$), $EX = \int X\, dP$ the expectation or mean value and $VX = \int (X - EX)^2\, dP$ the variance.

Remark. This is *Chebyshev's inequality*.

10.6. Show that $\int |u|^p\, d\mu < \infty$ implies that $|u|$ is a.e. real-valued (in the sense $(-\infty, \infty)$-valued!). Is this still true if we have $\int \arctan(u)\, d\mu < \infty$?.

10.7. Let $(A_j)_{j \in \mathbb{N}} \subset \mathcal{A}$ be a sequence of pairwise disjoint sets. Show that

$$u \mathbf{1}_{\bigcup_j A_j} \in \mathcal{L}^1(\mu) \iff u \mathbf{1}_{A_n} \in \mathcal{L}^1(\mu) \quad \text{and} \quad \sum_{j=1}^{\infty} \int_{A_j} |u|\, d\mu < \infty.$$

10.8. **Generalized Fatou lemma.** Assume that $(u_j)_{j \in \mathbb{N}} \subset \mathcal{L}^1(\mu)$. Prove:

(i) If $u_j \geq v$ for all $j \in \mathbb{N}$ and some $v \in \mathcal{L}^1(\mu)$, then

$$\int \liminf_{j \to \infty} u_j\, d\mu \leq \liminf_{j \to \infty} \int u_j\, d\mu.$$

(ii) If $u_j \leq w$ for all $j \in \mathbb{N}$ and some $w \in \mathcal{L}^1(\mu)$, then

$$\limsup_{j \to \infty} \int u_j\, d\mu \leq \int \limsup_{j \to \infty} u_j\, d\mu.$$

(iii) Find examples that show that the upper and lower bounds in (i) and (ii) are necessary.

[Hint: mimic and scrutinize the proof of Fatou's Lemma 9.11 especially when it comes to the application of Beppo Levi's theorem. What goes wrong if we do not have this upper/lower bound? Note that we have an 'invisible' $v = 0$ in T9.11.]

10.9. Let $(\Omega, \mathcal{A}, P)$ be a probability space. Show that for $u \in \mathcal{M}(\mathcal{A})$

$$u \in \mathcal{L}^1(P) \iff \sum_{j=0}^{\infty} P(\{u \geq j\}) < \infty.$$

10.10. **Independence (2).** Let $(\Omega, \mathcal{A}, P)$ be a probability space. Recall the notion of independence of two σ-algebras $\mathcal{B}, \mathcal{C} \subset \mathcal{A}$ introduced in Problem 5.10. Show that $u \in \mathcal{M}^+(\mathcal{B})$ and $w \in \mathcal{M}^+(\mathcal{C})$ satisfy

$$\int uw\, dP = \int u\, dP \cdot \int w\, dP$$

and that for $u \in \mathcal{M}(\mathcal{B})$ and $w \in \mathcal{M}(\mathcal{C})$

$$u \in \mathcal{L}^1(\mathcal{B}) \quad \text{and} \quad w \in \mathcal{L}^1(\mathcal{C}) \Rightarrow uw \in \mathcal{L}^1(\mathcal{A}).$$

Find an example proving that this fails if $\mathcal{B}$ and $\mathcal{C}$ are not independent.
[Hint: start with simple functions and use Beppo Levi's theorem 9.6.]

10.11. **Completion (3).** Let $(X, \mathcal{A}^*, \bar{\mu})$ be the completion of $(X, \mathcal{A}, \mu)$ – cf. Problems 4.13, 6.2.

(i) Show that for every $f^* \in \mathcal{E}^+(\mathcal{A}^*)$ there are $f, g \in \mathcal{E}^+(\mathcal{A})$ with $f \leqslant f^* \leqslant g$ and $\mu(f \neq g) = 0$ as well as $\int f \, d\mu = \int f^* \, d\bar{\mu} = \int g \, d\mu$.

(ii) $u^* : X \to \mathbb{R}$ is $\mathcal{A}^*$-measurable if, and only if, there exist $\mathcal{A}$-measurable functions $u, w : X \to \bar{\mathbb{R}}$ with $u \leqslant u^* \leqslant w$ and $u = w$ μ-a.e.

(iii) If $u^* \in \mathcal{L}^1(\bar{\mu})$, then u, w from (ii) can be chosen from $\mathcal{L}^1(\mu)$ such that $\int u \, d\mu = \int u^* \, d\bar{\mu} = \int w \, d\mu$.

[Hint: (i) use Problem 4.13(v). (ii) for '$\Rightarrow$' consider the sets $\{u^* > \alpha\}$ and use 4.13(v). The other direction is harder. For this consider first step functions using again 4.13(v) and then general functions by monotone convergence. (iii) by 4.13(iii), $\mu = \bar{\mu}$ on $\mathcal{A}$, and thus $\int f \, d\mu = \int f \, d\bar{\mu}$ for $\mathcal{A}$-measurable f.]

10.12. **Completion (4). Inner measure and outer measure.** Let $(X, \mathcal{A}, \mu)$ be a *finite* measure space. Define for every $E \subset X$ the outer resp. inner measure

$$\mu^*(E) := \inf\{\mu(A) : A \in \mathcal{A}, \ A \supset E\} \quad \text{and} \quad \mu_*(E) := \sup\{\mu(A) : A \in \mathcal{A}, \ A \subset E\}.$$

(i) Show that

$$\mu_*(E) \leqslant \mu^*(E), \qquad\qquad \mu_*(E) + \mu^*(E^c) = \mu(X),$$

$$\mu^*(E \cup F) \leqslant \mu^*(E) + \mu^*(F), \qquad \mu_*(E) + \mu_*(F) \leqslant \mu_*(E \cup F).$$

(ii) For every $E \subset X$ there exist sets $E_*, E^* \in \mathcal{A}$ such that $\mu(E_*) = \mu_*(E)$ and $\mu(E^*) = \mu^*(E)$.
[Hint: use the definition of the infimum to find sets $E^n \supset E$ such that $\mu(E^n) - \mu^*(E) \leqslant \frac{1}{n}$ and consider $\bigcap_n E^n \in \mathcal{A}$.]

(iii) Show that $\mathcal{A}^* := \{E \subset X : \mu_*(E) = \mu^*(E)\}$ is a σ-algebra and that it is the completion of $\mathcal{A}$ w.r.t. μ. Conclude, in particular, that $\mu^*|_{\mathcal{A}^*} = \mu_*|_{\mathcal{A}^*} = \bar{\mu}$ if $\bar{\mu}$ is the completion of μ.

10.13. Let $(X, \mathcal{A}, \mu)$ be a measure space and $u \in \mathcal{M}(\mathcal{A})$. Assume that $u \in \mathcal{M}(\mathcal{A})$ and $u = w$ almost everywhere w.r.t. μ. When can we say that $w \in \mathcal{M}(\mathcal{A})$?

10.14. **'a.e.' is a tricky business.** When working with 'a.e.' properties one has to be extremely careful. For example, the assertions 'u is continuous a.e.' and 'u is a.e. equal to an (everywhere) continuous function' are *far apart*! Illustrate this by considering the functions $u = \mathbf{1}_{\mathbb{Q}}$ and $u = \mathbf{1}_{[0,\infty)}$.

10.15. Let μ be a σ-finite measure on the measurable space $(X, \mathcal{A})$. Show that there exists a finite measure P on $(X, \mathcal{A})$ such that $\mathcal{N}_\mu = \mathcal{N}_P$, i.e. μ and P have the same null sets.

10.16. Construct an example showing that for $u, w \in \mathcal{M}^+(\mathcal{G})$ the equality $\int_G u \, d\mu = \int_G w \, d\mu$ for all $G \in \mathcal{G}$ does not necessarily imply that $u = w$ almost everywhere. [Hint: In view of 10.14 $\mu|_{\mathcal{G}}$ cannot be σ-finite. Consider on $(\mathbb{R}, \mathcal{B}(\mathbb{R}))$ the measure $\mu = m \, \lambda^1$ where $m = \mathbf{1}_{\{|x| \leqslant 1\}} + \infty \, \mathbf{1}_{\{|x| > 1\}}$, $u \equiv 1$ and $w = \mathbf{1}_{\{|x| \leqslant 1\}} + 2 \, \mathbf{1}_{\{|x| > 1\}}$. Then all Borel subsets of $\{|x| > 1\}$ have either μ-measure 0 or $+\infty$, thus $\int_B u \, d\mu = \int_B w \, d\mu$ for all $B \in \mathcal{B}(\mathbb{R})$ while $\mu(u \neq w) = \infty$.]

11

Convergence theorems and their applications

Throughout this chapter $(X, \mathcal{A}, \mu)$ will be some measure space.

One of the shortfalls of the Riemann integral is the fact that we do not have sufficiently general results that allow us to interchange limits and integrals – typically one has to assume uniform convergence for this. This has partly to do with the fact that the set of Riemann integrable functions is somewhat limited, see Theorem 11.8. The classical counterexample for this defect is *Dirichlet's jump function* $x \mapsto \mathbf{1}_{\mathbb{Q} \cap [0,1]}(x)$ which is not Riemann integrable since its upper function is $\mathbf{1}_{[0,1]}$ while the lower function is $0 \cdot \mathbf{1}_{[0,1]}$.[✓]

For the Lebesgue integral on $\mathcal{M}_{\overline{\mathbb{R}}}^{+}$ we have already seen more powerful convergence results in the form of Beppo Levi's theorem 9.6 or Fatou's lemma 9.11. They can deal with Dirichlet's jump function: for any enumeration of $\mathbb{Q} = \{q_j : j \in \mathbb{N}\}$ we get

$$\int \mathbf{1}_{\mathbb{Q} \cap [0,1]} \, d\lambda^1 = \int \sup_{N \in \mathbb{N}} \mathbf{1}_{\{q_1, \ldots, q_N\} \cap [0,1]} \, d\lambda^1$$

$$\overset{9.6}{=} \sup_{N \in \mathbb{N}} \int \mathbf{1}_{\{q_1, \ldots, q_N\} \cap [0,1]} \, d\lambda^1$$

$$= \sup_{N \in \mathbb{N}} \lambda^1 \big(\{q_j \in [0,1] : 1 \leqslant j \leqslant N\} \big) = 0.$$
$$\underbrace{\phantom{\sup_{N \in \mathbb{N}} \lambda^1 \big(\{q_j \in [0,1] : 1 \leqslant j \leqslant N\} \big)}}_{=0}$$

In this chapter we study systematically convergence theorems for $\mathcal{L}^1(\mu)$ and some of their most important applications. The first is a generalization of Beppo Levi's theorem 9.6.

11.1 Theorem (Monotone convergence). *Let $(X, \mathcal{A}, \mu)$ be a measure space.*

(i) *Let $(u_j)_{j \in \mathbb{N}} \subset \mathcal{L}^1(\mu)$ be an increasing sequence of integrable functions $u_1 \leqslant u_2 \leqslant \ldots$ with limit $u := \sup_{j \in \mathbb{N}} u_j$. Then $u \in \mathcal{L}^1(\mu)$ if, and only if,*

$\sup_{j \in \mathbb{N}} \int u_j \, d\mu < +\infty$, *in which case*

$$\sup_{j \in \mathbb{N}} \int u_j \, d\mu = \int \sup_{j \in \mathbb{N}} u_j \, d\mu.$$

(ii) *Let* $(v_k)_{k \in \mathbb{N}} \subset \mathcal{L}^1(\mu)$ *be a decreasing sequence of integrable functions* $v_1 \geqslant v_2 \geqslant \ldots$ *with limit* $v := \inf_{k \in \mathbb{N}} v_k$. *Then* $v \in \mathcal{L}^1(\mu)$ *if, and only if,* $\inf_{k \in \mathbb{N}} \int v_k \, d\mu > -\infty$, *in which case*

$$\inf_{k \in \mathbb{N}} \int v_k \, d\mu = \int \inf_{k \in \mathbb{N}} v_k \, d\mu.$$

Proof Obviously, (i) implies (ii) as $u_j := -v_j$ fulfils all the assumptions of (i). To see (i), we remark that $u_j - u_1 \in \mathcal{L}^1(\mu)$ defines an increasing sequence of *positive* functions

$$0 \leqslant u_j - u_1 \leqslant u_{j+1} - u_1 \leqslant \ldots,$$

for which we may use the Beppo Levi theorem 9.6:

$$0 \leqslant \sup_{j \in \mathbb{N}} \int (u_j - u_1) \, d\mu = \int \sup_{j \in \mathbb{N}} (u_j - u_1) \, d\mu. \tag{11.1}$$

Assume that $u \in \mathcal{L}^1(\mu)$. Since the 'sup' in (11.1) stands for an increasing limit, we find that

$$\sup_{j \in \mathbb{N}} \int u_j \, d\mu = \int (u - u_1) \, d\mu + \int u_1 \, d\mu$$

$$\overset{(10.2)}{=} \int u \, d\mu - \int u_1 \, d\mu + \int u_1 \, d\mu = \int u \, d\mu < \infty.$$

Conversely, if $\sup_{j \in \mathbb{N}} \int u_j \, d\mu < \infty$, we see from (11.1) that $u - u_1 \in \mathcal{L}^1(\mu)$ and, as $u_1 \in \mathcal{L}^1(\mu)$, $u = (u - u_1) + u_1 \in \mathcal{L}^1(\mu)$ by (10.2). Therefore, (11.1) implies

$$\int u \, d\mu = \int (u - u_1) \, d\mu + \int u_1 \, d\mu = \sup_{j \in \mathbb{N}} \int u_j \, d\mu < \infty. \qquad \blacksquare$$

One of the most useful and versatile convergence theorems is the following.

11.2 Theorem (Lebesgue. Dominated convergence). *Let* $(X, \mathcal{A}, \mu)$ *be a measure space and* $(u_j)_{j \in \mathbb{N}} \subset \mathcal{L}^1(\mu)$ *be a sequence of functions such that* $|u_j| \leqslant w$ *for all* $j \in \mathbb{N}$ *and some* $w \in \mathcal{L}_+^1(\mu)$. *If* $u(x) = \lim_{j \to \infty} u_j(x)$ *exists for almost every* $x \in X$, *then* $u \in \mathcal{L}^1(\mu)$ *and we have*

(i) $\lim_{j \to \infty} \int |u_j - u| \, d\mu = 0;$

(ii) $\lim_{j \to \infty} \int u_j \, d\mu = \int \lim_{j \to \infty} u_j \, d\mu = \int u \, d\mu.$

Proof Since all u_j are measurable, $N := \{x : \lim_j u_j(x) \text{ does not exist}\}$ is measurable, hence $N \in \mathcal{N}_\mu$, and we can assume that $N = \emptyset$ as the integral over the null set N gives no contribution, cf. Theorem 10.9(ii) – alternatively we could consider $\mathbf{1}_{N^c} u$ and $\mathbf{1}_{N^c} u_j$ instead of u and u_j.

From $|u_j| \leqslant w$ we get $|u| = \lim_{j \to \infty} |u_j| \leqslant w$, and $u \in \mathcal{L}^1(\mu)$ by C10.11(iv). Therefore,

$$\left| \int u_j \, d\mu - \int u \, d\mu \right| = \left| \int (u_j - u) \, d\mu \right| \overset{10.4(v)}{\leqslant} \int |u_j - u| \, d\mu$$

which means that (i) implies (ii). Since

$$|u_j - u| \leqslant |u_j| + |u| \leqslant 2w \qquad \forall j \in \mathbb{N}$$

we get $2w - |u_j - u| \geqslant 0$ and Fatou's lemma 9.11 tells us that

$$\int 2w \, d\mu = \int \liminf_{j \to \infty} (2w - |u_j - u|) \, d\mu$$

$$\leqslant \liminf_{j \to \infty} \int (2w - |u_j - u|) \, d\mu$$

$$= \int 2w \, d\mu - \limsup_{j \to \infty} \int |u_j - u| \, d\mu.^1$$

Thus $0 \leqslant \liminf_{j \to \infty} \int |u_j - u| \, d\mu \leqslant \limsup_{j \to \infty} \int |u_j - u| \, d\mu \leqslant 0$, and consequently $\lim_{j \to \infty} \int |u_j - u| \, d\mu = 0$. ■

11.3 Remark The uniform boundedness assumption

$$|u_j| \leqslant w \qquad \forall j \in \mathbb{N} \quad \text{and some} \quad w \in \mathcal{L}^1_+ \tag{11.2}$$

is very important for Theorem 11.2. To see this, consider $(\mathbb{R}, \mathcal{B}, \lambda^1)$ and set

$$u_j(x) := j \mathbf{1}_{[0, \frac{1}{j}]}(x) \xrightarrow{j \to \infty} \infty \mathbf{1}_{\{0\}}(x) \overset{\text{a.e.}}{=} 0$$

whereas $\int u_j \, d\lambda = j \frac{1}{j} = 1 \neq 0 = \int \infty \mathbf{1}_{\{0\}} \, d\lambda$.

The only obvious possibility to weaken (11.2) would be to require it to hold only almost everywhere.$^{[\checkmark]}$ Lebesgue's theorem gives merely sufficient – but easily verifiable – conditions for the interchange of limits and integrals; the ultimate version for such a result with necessary and sufficient conditions will be given in the form of Vitali's convergence theorem 16.6 in Chapter 16 below.

$$* \qquad * \qquad *$$

---

^{1} Recall that $\liminf_{j \to \infty} (-x_j) = -\limsup_{j \to \infty} x_j$.

Let us now have a look at two of the most important applications of the convergence theorems.

Parameter-dependent integrals

Again $(X, \mathcal{A}, \mu)$ is some measure space.

11.4 Theorem (Continuity lemma) *Let* $\emptyset \neq (a, b) \subset \mathbb{R}$ *be a non-degenerate open interval and* $u : (a, b) \times X \to \mathbb{R}$ *be a function satisfying*

(a) $x \mapsto u(t, x)$ *is in* $\mathcal{L}^1(\mu)$ *for every fixed* $t \in (a, b)$;
(b) $t \mapsto u(t, x)$ *is continuous for every fixed* $x \in X$;
(c) $|u(t, x)| \leqslant w(x)$ *for all* $(t, x) \in (a, b) \times X$ *and some* $w \in \mathcal{L}^1_+(\mu)$.

Then the function $v : (a, b) \to \mathbb{R}$ *given by*

$$t \mapsto v(t) := \int u(t, x)\, \mu(dx) \tag{11.3}$$

is continuous.

Proof Let us, first of all, remark that (11.3) is well-defined thanks to assumption (a). We are going to show that for any $t \in (a, b)$ and every sequence $(t_j)_{j \in \mathbb{N}} \subset (a, b)$ with $\lim_{j \to \infty} t_j = t$ we have $\lim_{j \to \infty} v(t_j) = v(t)$. This proves continuity of v at the point t.

Because of (b), $u(\cdot, x)$ is continuous and, therefore,

$$u_j(x) := u(t_j, x) \xrightarrow{j \to \infty} u(t, x) \quad \text{and} \quad |u_j(x)| \leqslant w(x) \qquad \forall x \in X.$$

Thus we can use Lebesgue's dominated convergence theorem, and conclude

$$\lim_{j \to \infty} v(t_j) = \lim_{j \to \infty} \int u(t_j, x)\, \mu(dx)$$

$$= \int \lim_{j \to \infty} u(t_j, x)\, \mu(dx)$$

$$= \int u(t, x)\, \mu(dx) = v(t). \qquad \blacksquare$$

A very similar consideration leads to

11.5 Theorem (Differentiability lemma) *Let* $\emptyset \neq (a, b) \subset \mathbb{R}$ *be a non-degenerate open interval and* $u : (a, b) \times X \to \mathbb{R}$ *be a function satisfying*

(a) $x \mapsto u(t, x)$ *is in* $\mathcal{L}^1(\mu)$ *for every fixed* $t \in (a, b)$;
(b) $t \mapsto u(t, x)$ *is differentiable for every fixed* $x \in X$;
(c) $|\partial_t u(t, x)| \leqslant w(x)$ *for all* $(t, x) \in (a, b) \times X$ *and some* $w \in \mathcal{L}^1_+(\mu)$.

Then the function $v : (a, b) \to \mathbb{R}$ given by

$$t \mapsto v(t) := \int u(t, x)\, \mu(dx) \tag{11.4}$$

is differentiable and its derivative is

$$\partial_t v(t) = \int \partial_t u(t, x)\, \mu(dx).^2 \tag{11.5}$$

Proof Let $t \in (a, b)$ and fix some sequence $(t_j)_{j \in \mathbb{N}} \subset (a, b)$ such that $t_j \neq t$ and $\lim_{j \to \infty} t_j = t$. Set

$$u_j(x) := \frac{u(t_j, x) - u(t, x)}{t_j - t} \xrightarrow{j \to \infty} \partial_t u(t, x).$$

which shows, in particular, that $x \mapsto \partial_t u(t, x)$ is measurable. By the mean value theorem of differential calculus and (c) we see for some intermediate value $\theta = \theta(j, x) \in (a, b)$

$$|u_j(x)| = \left| \partial_t u(t, x) \big|_{t=\theta} \right| \leqslant w(x) \qquad \forall j \in \mathbb{N}_0.$$

Thus $u_j \in \mathcal{L}^1$, and the sequence $(u_j)_{j \in \mathbb{N}}$ satisfies all conditions of the dominated convergence theorem 11.2. Finally,

$$
\begin{aligned}
\partial_t v(t) &= \lim_{j \to \infty} \frac{v(t_j) - v(t)}{t_j - t} \\
&= \lim_{j \to \infty} \int \frac{u(t_j, x) - u(t, x)}{t_j - t}\, \mu(dx) \\
&= \lim_{j \to \infty} \int u_j(x)\, \mu(dx) \\
&\stackrel{11.2}{=} \int \lim_{j \to \infty} u_j(x)\, \mu(dx) \\
&= \int \partial_t u(t, x)\, \mu(dx). \qquad \blacksquare
\end{aligned}
$$

Later in this chapter we will give examples of how to apply the continuity and differentiability lemmas.

Riemann vs. Lebesgue integration

From here to the end of this chapter we choose $(X, \mathcal{A}, \mu) = (\mathbb{R}, \mathcal{B}, \lambda)$.

---

[2] This formula is very effectively remembered as '$\partial_t \int \ldots = \int \partial_t \ldots$'

Let us briefly recall the definition of the *Riemann integral* (see Appendix E for a more detailed discussion). Consider on the finite interval $[a, b] \subset \mathbb{R}$ the partitions

$$\pi := \left\{ a = t_0 < t_1 < \ldots < t_{k(\pi)} = b \right\},$$

define for a given function $u : [a, b] \to \mathbb{R}$

$$m_j := \inf_{x \in [t_{j-1}, t_j]} u(x), \qquad M_j := \sup_{x \in [t_{j-1}, t_j]} u(x), \qquad j = 1, 2, \ldots, k(\pi),$$

and introduce the *lower* resp. *upper Darboux sums*

$$S_\pi[u] := \sum_{j=1}^{k(\pi)} m_j (t_j - t_{j-1}) \quad \text{resp.} \quad S^\pi[u] := \sum_{j=1}^{k(\pi)} M_j (t_j - t_{j-1}).$$

11.6 Definition A bounded function $u : [a, b] \to \mathbb{R}$ is said to be *Riemann integrable*, if the values

$$\int_* u := \sup_\pi S_\pi[u] = \inf_\pi S^\pi[u] =: \int^* u$$

(sup, inf range over all partitions π of $[a, b]$) coincide and are finite. Their common value is called the *Riemann integral* of u and denoted by $(R) \int_a^b u(x) \, dx$ or $\int_a^b u(x) \, dx$.

What is going on here? First of all, it is not difficult to see that lower [upper] Darboux sums increase [decrease] if we add points to the partition $\pi(N)$, i.e. the sup [inf] in Definition 11.6 makes sense.

Moreover, to $S_\pi[u]$ and $S^\pi[u]$ there correspond simple functions, namely $\sigma_\pi[u]$ and $\Sigma^\pi[u]$ given by

$$\sigma_\pi[u](x) = \sum_{j=1}^{k(\pi)} m_j \mathbf{1}_{[t_{j-1}, t_j)}(x) \quad \text{and} \quad \Sigma^\pi[u](x) = \sum_{j=1}^{k(\pi)} M_j \mathbf{1}_{[t_{j-1}, t_j)}(x)$$

which satisfy $\sigma_\pi[u](x) \leqslant u(x) \leqslant \Sigma^\pi[u](x)$ and which increase resp. decrease as π refines.

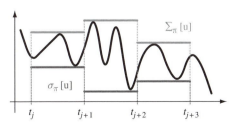

11.7 Remark The above construction gives the 'usual' integral which is often introduced as the anti-derivative. Unfortunately, this notion of integration is somewhat insufficient. Nice general convergence theorems (such as monotone or dominated convergence) hold only under unnatural restrictions or are not available at all. Moreover, it cannot deal with functions of the type $x \mapsto \mathbf{1}_{\mathbb{Q}\cap[0,1]}(x)$: the smallest upper function Σ^π is $\mathbf{1}_{[0,1]}$ while the largest lower function σ_π is identically 0.[✓] Thus the Riemann integral of $\mathbf{1}_{\mathbb{Q}\cap[0,1]}$ does not exist, whereas by T10.9(ii) the Lebesgue integral $\int \mathbf{1}_{\mathbb{Q}\cap[0,1)} \, d\lambda = 0$.

Roughly speaking, the reason for this is the fact that the Riemann sums partition the domain of the function without taking into account the shape of the function, thus slicing up the area under the function *vertically*. Lebesgue's approach is exactly the opposite: the domain is partitioned according to the values of the function at hand, leading to a *horizontal* decomposition of the area.

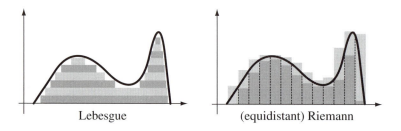

| Lebesgue | (equidistant) Riemann |

There is a beautifully simple connection with Lebesgue integrals which characterizes at the same time the class of Riemann integrable functions. It may come as a surprise that one needs the notion of Lebesgue null sets to understand Riemann's integral completely.

11.8 Theorem *Let $u : [a, b] \to \mathbb{R}$ be a measurable function.*

(i) *If u is Riemann integrable, then u is in $\mathcal{L}^1(\lambda)$ and the Lebesgue and Riemann integrals coincide: $\int_{[a,b]} u \, d\lambda = (R) \int_a^b u(x) \, dx$.*

(ii) *A bounded function $f : [a, b] \to \mathbb{R}$ is Riemann integrable if, and only if, the points in (a, b) where f is discontinuous are a Lebesgue null set.*

Caution: Theorem 11.8(ii) is often phrased in the following way: *f is Riemann integrable if, and only if, f is (Lebesgue) a.e. continuous.* Although correct, this is a dangerous way of putting things since one is led to read this statement (incorrectly) as 'if $f = \phi$ a.e. with $\phi \in C[a, b]$, then f is Riemann integrable'. That this is wrong is easily seen from $f = \mathbf{1}_{\mathbb{Q}\cap[a,b]}$ and $\phi \equiv 0$; see Problem 10.14 and 11.16.

Proof (of Theorem 11.8) (i) As u is Riemann integrable, we find a sequence of partitions $\pi(j)$ of $[a, b]$ such that

$$\lim_{j \to \infty} S_{\pi(j)}[u] = \int_* u = \int^* u = \lim_{j \to \infty} S^{\pi(j)}[u].$$

Without loss of generality we may assume that the partitions are nested $\pi(j) \subset \pi(j+1) \subset \ldots$ – otherwise we could switch to the increasing sequence $\pi(1) \cup \ldots \cup \pi(j)$ of partitions, where we also observe that the lower [upper] Riemann sums increase [decrease] as the partitions refine. The corresponding simple functions $\sigma_{\pi(j)}[u]$ and $\Sigma^{\pi(j)}[u]$ increase and decrease towards

$$\sigma[u] := \sup_{j \in \mathbb{N}} \sigma_{\pi(j)}[u] \leqslant u \leqslant \inf_{j \in \mathbb{N}} \Sigma^{\pi(j)}[u] =: \Sigma[u],$$

and from the monotone convergence theorem 11.1 we conclude

$$\int_* u = \lim_{j \to \infty} S_{\pi(j)}[u] = \lim_{j \to \infty} \int_{[a,b]} \sigma_{\pi(j)}[u] \, d\lambda = \int_{[a,b]} \sigma[u] \, d\lambda \qquad (11.6)$$

and also

$$\int^* u = \lim_{j \to \infty} S^{\pi(j)}[u] = \lim_{j \to \infty} \int_{[a,b]} \Sigma^{\pi(j)}[u] \, d\lambda = \int_{[a,b]} \Sigma[u] \, d\lambda. \qquad (11.7)$$

In other words $\sigma[u], \Sigma[u] \in \mathcal{L}^1(\lambda)$. Since u is Riemann integrable,

$$\int_{[a,b]} \underbrace{(\Sigma[u] - \sigma[u])}_{\geqslant 0} \, d\lambda = \int_{[a,b]} \Sigma[u] \, d\lambda - \int_{[a,b]} \sigma[u] \, d\lambda = \int^* u - \int_* u = 0,$$

which implies by Theorem 10.9(i) that $\Sigma[u] = \sigma[u]$ Lebesgue a.e. Thus

$$\{u \neq \Sigma[u]\} \cup \{u \neq \sigma[u]\} \subset \{\sigma[u] \neq \Sigma[u]\} \in \mathcal{N}_\lambda, \qquad (11.8)$$

and by Corollary 10.10(ii) we conclude that u is Lebesgue integrable.

    (ii) We continue to use the notation from part (i). The set $\Pi := \bigcup_{j \in \mathbb{N}} \pi(j)$ of all partition points is countable, and by Problem 6.5(i),(iii) a Lebesgue null set. If f is Riemann integrable, we can find for $\epsilon > 0$ and each $x \in [a, b]$ some $n_{\epsilon,x} \in \mathbb{N}$ such that for some suitable $t_{j_0-1}, t_{j_0} \in \pi(n_{\epsilon,x})$ we have $x \in (t_{j_0-1}, t_{j_0})$ and

$$\left| \sigma_{\pi(j)}[f](x) - \sigma[f](x) \right| + \left| \Sigma^{\pi(j)}[f](x) - \Sigma[f](x) \right| \leqslant \epsilon \qquad \forall j \geqslant n_{\epsilon,x}.$$

By construction of the Riemann integral, all $x, y \in (t_{j_0-1}, t_{j_0})$ satisfy

$$|f(x) - f(y)| \leqslant M_{j_0} - m_{j_0} = \Sigma^{\pi(n_{\epsilon,x})}[f](x) - \sigma_{\pi(n_{\epsilon,x})}[f](x)$$

$$\leqslant \epsilon + \left(\Sigma[f](x) - \sigma[f](x) \right).$$

This inequality shows on the one hand that[✓]

$$\{x : f(x) \text{ is not continuous}\} \subset \Pi \cup \underbrace{\{\Sigma[f] \neq \sigma[f]\}}_{\in \mathcal{N}_\lambda \text{ by (i)}} \in \mathcal{N}_\lambda$$

is a null set if f is Lebesgue integrable. On the other hand, the above inequality shows also[✓] that

$$\{\Sigma[f] = \sigma[f]\} \subset \{x : f(x) \text{ is continuous}\} \cup \Pi,$$

so that (11.6), (11.7) imply $\int^* f = \int_* f$, i.e. f is Riemann integrable.   ∎

*   *   *

Let us finally discuss *improper Riemann integrals* of the type

$$(R)\int_0^\infty u(x)\,dx := \lim_{a\to\infty} (R)\int_0^a u(x)\,dx, \tag{11.9}$$

provided the limit exists (cf. Appendix E for other types of improper integrals).

11.9 Corollary *Let $u : [0, \infty) \to \mathbb{R}$ be a measurable function which is Riemann integrable for every interval $[0, N]$, $N \in \mathbb{N}$. Then $u \in \mathcal{L}^1[0, \infty)$ if, and only if,*

$$\lim_{N\to\infty} (R)\int_0^N |u(x)|\,dx < \infty. \tag{11.10}$$

In this case, $(R)\int_0^\infty u(x)\,dx = \int_{[0,\infty)} u\,d\lambda$.

Proof Using Theorem 11.8 we see that Riemann integrability of u implies Riemann integrability of $u^\pm$.[✓] Moreover,

$$(R)\int_0^N u^\pm(x)\,dx = \int_{[0,N]} u^\pm(x)\,\lambda(dx) = \int u^\pm \mathbf{1}_{[0,N]}\,d\lambda. \tag{11.11}$$

If u is Riemann integrable and satisfies (11.9) and (11.10), the limit $N \to \infty$ of the left side of (11.11) exists and guarantees that the right-hand side has also a finite limit. The monotone convergence theorem 11.1 together with Theorem 10.3(ii) shows that $u \in \mathcal{L}^1[0, \infty)$.

Conversely, if u is Lebesgue integrable, then so are $u^\pm$, $u\mathbf{1}_{[0,a]}$ and $u^\pm \mathbf{1}_{[0,a]}$ for every $a > 0$. Since u is Riemann integrable over each interval $[0, N]$, we see from Theorem 11.8 that u and $u^\pm$ are Riemann integrable over each interval $[0, a]$. The monotone convergence theorem 11.1 shows that for *every* increasing sequence $a_j \uparrow \infty$

$$\lim_{j\to\infty} \int u^\pm \mathbf{1}_{[0,a_j]}\,d\lambda = \int u^\pm\,d\lambda < \infty,$$

which yields that the limits (11.9), (11.10) exist.   ∎

11.10 Remark We can avoid in T11.8(i) and C11.9 the assumption that u is Borel measurable. If we admit an arbitrary u, our proofs show that u is *outside a subset of a null set* equal to the Borel measurable function Σ – to wit: $\{u \neq \Sigma\} \subset \{\sigma \neq \Sigma\} \in \mathcal{N}_\lambda$, but $\{u \neq \Sigma\}$ is not necessarily measurable. In other words, u becomes automatically measurable w.r.t. the completed Borel σ-algebra. This entails, of course, that we have to replace λ and $\mathcal{L}^1(\lambda)$ with the completed versions $\bar{\lambda}$ and $\mathcal{L}^1(\bar{\lambda})$, see Problems 4.13, 6.2, 10.11 and 10.12.

11.11 Remark Lebesgue integration does not allow cancellations, but improper Riemann integrals do. More precisely: the limit (11.9) can make sense even if $\lim_{N \to \infty}(R) \int_0^N |u(x)|\, dx = \infty$. This is illustrated by the following example, which is typical in the theory of Fourier series:

The function $x \mapsto s(x) := \dfrac{\sin x}{x}$, $x \in (0, \infty)$, is improperly Riemann integrable but not Lebesgue integrable.

For $a > 0$ we can find $N = N(a) \in \mathbb{N}$ such that $N\pi \leqslant a < (N+1)\pi$. Thus

$$\int_0^\infty \frac{\sin x}{x}\, dx = \lim_{a \to \infty} \left(\int_0^{N\pi} \frac{\sin x}{x}\, dx + \int_{N\pi}^a \frac{\sin x}{x}\, dx \right)$$

$$= \lim_{N \to \infty} \sum_{j=0}^{N-1} \underbrace{\int_{j\pi}^{(j+1)\pi} \frac{\sin x}{x}\, dx}_{=: a_j},$$

where we used

$$\left| \lim_{a \to \infty} \int_{N\pi}^a \frac{\sin x}{x}\, dx \right| \leqslant \lim_{N \to \infty} \int_{N\pi}^{(N+1)\pi} \left| \frac{\sin x}{x} \right| dx \leqslant \lim_{N \to \infty} \frac{\pi}{N\pi} = 0.$$

Observe that the a_j have alternating signs since

$$a_j = \int_{j\pi}^{(j+1)\pi} \frac{\sin x}{x}\, dx = \int_0^\pi \frac{\sin(y + j\pi)}{y + j\pi}\, dy = (-1)^j \int_0^\pi \frac{\sin y}{y + j\pi}\, dy$$

both as Riemann and Lebesgue integrals, by Theorem 11.8. Further,

$$|a_j| = \int_0^\pi \frac{\sin y}{y + j\pi}\, dy \leqslant \int_0^\pi \frac{\sin y}{y + jy}\, dy = \frac{1}{j+1} \int_0^\pi \frac{\sin y}{y}\, dy,$$

and also

$$|a_j| = \int_0^\pi \frac{\sin y}{y + j\pi}\, dy \geqslant \overbrace{\int_0^\pi \frac{\sin y}{y + (j+1)\pi}\, dy}^{= |a_{j+1}|}$$

$$\geqslant \int_0^\pi \frac{\sin y}{\pi + (j+1)\pi}\, dy = \frac{2}{(j+2)\pi}.$$

Since the function $y \mapsto \frac{\sin y}{y}$ is continuous and has a finite limit as $y \downarrow 0^{[\checkmark]}$, we see that $C := \int_0^\pi \frac{\sin y}{y} \, dy < \infty$, so that

$$\frac{2/\pi}{j+2} \leqslant |a_{j+1}| \leqslant |a_j| \leqslant \frac{C}{j+1}.$$

This and Leibniz's convergence test prove that the alternating series $\sum_{j=0}^\infty a_j$ converges conditionally but not absolutely, i.e. we get a finite improper Riemann integral, but the Lebesgue integral does not exist.

Examples

As we have seen in this chapter, the Lebesgue integral provides very powerful tools that justify the interchange of limits and integrals. On the other hand, the Riemann theory is quite handy when it comes to calculating the primitive (anti-derivative) of some concrete integrand. Theorem 11.8 tells us when we can switch between these two notions.

11.12 Example *Let $f_\alpha(x) := x^\alpha$, $x > 0$ and $\alpha \in \mathbb{R}$. Then*

$$f_\alpha \in \mathcal{L}^1(0, 1) \iff \alpha > -1$$
$$f_\alpha \in \mathcal{L}^1[1, \infty) \iff \alpha < -1.$$

We show only the first assertion; the second follows similarly (or, indeed, from C11.9). Since f_α is continuous, it is Borel measurable, and since $f_\alpha \geqslant 0$ it is enough to show that $\int_{(0,1)} f_\alpha \, d\lambda < \infty$. We find

$$\int_{(0,1)} x^\alpha \, dx \overset{9.6}{=} \lim_{j\to\infty} \int x^\alpha \mathbf{1}_{[1/j,1)}(x) \, \lambda(dx)$$

$$\overset{11.8}{=} \lim_{j\to\infty} (R) \int_{1/j}^1 x^\alpha \, dx$$

$$= \lim_{j\to\infty} \left[\frac{x^{\alpha+1}}{\alpha+1} \right]_{1/j}^1$$

$$= \lim_{j\to\infty} \left(\frac{1}{\alpha+1} - \frac{1}{j^{\alpha+1}(\alpha+1)} \right),$$

and the last limit is finite if, and only if, $\alpha > -1$.

11.13 Example *The function $f(x) := x^\alpha e^{-\beta x}$, $x > 0$, is Lebesgue integrable over $(0, \infty)$ for all $\alpha > -1$ and $\beta \geqslant 0$.*

Measurability of f follows from its continuity. Using the exponential series, we find for all $N \in \mathbb{N}$ and $x > 0$

$$\frac{(\beta x)^N}{N!} \leqslant \sum_{j=0}^{\infty} \frac{(\beta x)^j}{j!} = e^{\beta x} \quad \Longrightarrow \quad e^{-\beta x} \leqslant \frac{N!}{\beta^N} x^{-N}.$$

As $e^{-\beta x} \leqslant 1$ for $x > 0$, we obtain the following majorization:

$$f(x) = x^{\alpha} e^{-\beta x} \leqslant \underbrace{x^{\alpha} \mathbf{1}_{(0,1)}(x)}_{\substack{\in \mathcal{L}^1(0,1),\ \text{if } \alpha > -1 \\ \text{by Example 11.12}}} + \underbrace{\frac{N!}{\beta^N} x^{\alpha - N} \mathbf{1}_{[1,\infty)}(x)}_{\substack{\in \mathcal{L}^1[1,\infty),\ \text{if } \alpha - N < -1 \\ \text{by Example 11.12}}} \in \mathcal{L}^1(0, \infty), \quad (11.12)$$

and $f \in \mathcal{L}^1(0, \infty)$ follows from T10.3(iv).

11.14 Example (Euler's Gamma function) *The parameter-dependent integral*

$$\Gamma(t) := \int_{(0,\infty)} x^{t-1} e^{-x} \lambda(dx), \qquad t > 0 \tag{11.13}$$

is called the Gamma function. It has the following properties:

  (i) Γ *is continuous;*
 (ii) Γ *is arbitrarily often differentiable;* (see Problem 11.13(i))
(iii) $t\Gamma(t) = \Gamma(t+1)$, *in particular* $\Gamma(n+1) = n!$; (see Problem 11.13(ii))
 (iv) $\ln \Gamma(t)$ *is convex.* (see Problem 11.13(iii))

Example 11.13 shows that the Gamma function is well-defined for all $t > 0$. We prove (i) and (ii) first for every interval (a, b) where $0 < a < b < \infty$. Since both continuity and differentiability are local properties, i.e. they need to be checked locally at each point, (i) and (ii) follow for the half-line if we let $a \to 0$ and $b \to \infty$.

  (i) We apply the continuity lemma T11.4. Set $u(t, x) := x^{t-1} e^{-x}$. We have already seen in Example 11.13 that $u(t, \cdot) \in \mathcal{L}^1(0, \infty)$ for all $t > 0$; the continuity of $u(\cdot, x)$ is clear and all that remains is to find a uniform (for $t \in (a, b)$) dominating function. An argument similar to (11.12) gives for $N > b + 1$

$$x^{t-1} e^{-x} \leqslant x^{t-1} \mathbf{1}_{(0,1)}(x) + N! \, x^{t-1-N} \mathbf{1}_{[1,\infty)}(x)$$

$$\leqslant x^{a-1} \mathbf{1}_{(0,1)}(x) + N! \, x^{b-1-N} \mathbf{1}_{[1,\infty)}(x)$$

$$\leqslant x^{a-1} \mathbf{1}_{(0,1)}(x) + N! \, x^{-2} \mathbf{1}_{[1,\infty)}(x).$$

The expression on the right no longer depends on t, and is integrable according to Example 11.12. (Note that $N = N(b)$ depends on the fixed interval (a, b), but not on t.) This shows that $\Gamma(t) = \int_{(0,\infty)} u(t, x) \lambda(dx)$ is continuous for all $t \in (a, b)$.

(ii) We apply the differentiability lemma T11.5. The integrand $u(t, \cdot)$ is integrable, and $u(\cdot, x)$ is differentiable for fixed $x > 0$. In fact,

$$\frac{\partial u(t, x)}{\partial t} = \frac{\partial}{\partial t} x^{t-1} e^{-x} = x^{t-1} e^{-x} \ln x.$$

We still have to show that $\frac{\partial}{\partial t} u(t, x)$ has an integrable majorant uniformly for all $t \in (a, b)$. First we observe that $\ln x \leqslant x$, thus

$$\left| \frac{\partial}{\partial t} u(t, x) \right| \leqslant x^t e^{-x} \leqslant x^b e^{-x} \qquad \forall\, a < t < b,\; x \geqslant 1.$$

For $0 < x < 1$ we use $|\ln x| = \ln \frac{1}{x}$, so that

$$\left| \frac{\partial}{\partial t} u(t, x) \right| = x^{t-1} e^{-x} \ln \frac{1}{x} \leqslant x^{a-1} e^{-x} \ln \frac{1}{x} \qquad \forall\, a < t < b,\; 0 < x < 1,$$

and since $a > 0$, we find some $\epsilon > 0$ with $a - \epsilon - 1 > -1$, so that

$$\left| \frac{\partial}{\partial t} u(t, x) \right| \leqslant x^{a-1-\epsilon} e^{-x} \underbrace{x^\epsilon \ln \frac{1}{x}}_{\to 0 \text{ as}^3 \; x \to 0} \leqslant C x^{a-1-\epsilon} e^{-x} \qquad \forall\, a < t < b,\; 0 < x < 1.$$

Combining these calculations, we arrive at

$$\left| \frac{\partial}{\partial t} u(t, x) \right| \leqslant C x^{a-1-\epsilon} e^{-x} \mathbf{1}_{(0,1)}(x) + x^b e^{-x} \mathbf{1}_{[1,\infty)}(x) \qquad \forall\, a < t < b,$$

which is an integrable majorant (by Examples 11.12, 11.13) independent of $t \in (a, b)$. This shows that $\Gamma(t)$ is differentiable on (a, b), with derivative

$$\Gamma'(t) = \int_{(0,\infty)} x^{t-1} e^{-x} \ln x\, \lambda(dx), \qquad t \in (a, b).$$

A similar calculation proves that $\Gamma^{(n)}$ exists for every $n \in \mathbb{N}$; see Problem 11.13.

Problems

11.1. Adapt the proof of Theorem 11.2 to show that any sequence $(u_j)_{j\in\mathbb{N}} \subset \mathcal{M}$ with $\lim_{j\to\infty} u_j(x) = u(x)$ and $|u_j| \leqslant g$ for some g with $g^p \in \mathcal{L}^1_+$ satisfies

$$\lim_{j\to\infty} \int |u_j - u|^p\, d\mu = 0.$$

  [Hint: mimic the proof of 11.2 using $|u_j - u|^p \leqslant (|u_j| + |u|)^p \leqslant 2^p g^p$.]

11.2. Give an alternative proof of Lebesgue's dominated convergence theorem 11.2(ii) using the generalized Fatou theorem from Problem 10.8.

---

^{3} To see this, use $\lim_{x\to 0} x^\epsilon \ln \frac{1}{x} \overset{x=\exp(-t)}{=} \lim_{t\to\infty} e^{-\epsilon t} t = 0$ if $\epsilon > 0$.

11.3. Prove the following result of W. H. Young [56]; among statisticians it is also known as *Pratt's lemma*, cf. J. W. Pratt [36].

Theorem (Young; Pratt): *Let $(f_k)_k$, $(g_k)_k$ and $(G_k)_k$ be sequences of integrable functions on a measure space $(X, \mathcal{A}, \mu)$. If*

(i) $f_k(x) \xrightarrow{k\to\infty} f(x)$, $g_k(x) \xrightarrow{k\to\infty} g(x)$, $G_k(x) \xrightarrow{k\to\infty} G(x)$ *for all $x \in X$,*

(ii) $g_k(x) \leqslant f_k(x) \leqslant G_k(x)$ *for all $k \in \mathbb{N}$ and all $x \in X$,*

(iii) $\int g_k \, d\mu \xrightarrow{k\to\infty} \int g \, d\mu$ *and* $\int G_k \, d\mu \xrightarrow{k\to\infty} \int G \, d\mu$ *with $\int g \, d\mu$ and $\int G \, d\mu$ finite,*

then $\lim_{k\to\infty} \int f_k \, d\mu = \int f \, d\mu$ and $\int f \, d\mu$ is finite.

Explain why this generalizes Lebesgue's dominated convergence theorem 11.2(ii).

11.4. Let $(u_j)_{j\in\mathbb{N}}$ be a sequence of integrable functions on $(X, \mathcal{A}, \mu)$. Show that, if $\sum_{j=1}^\infty \int |u_j| \, d\mu < \infty$, the series $\sum_{j=1}^\infty u_j$ converges a.e. to a real-valued function $u(x)$, and that in this case

$$\int \sum_{j=1}^\infty u_j \, d\mu = \sum_{j=1}^\infty \int u_j \, d\mu.$$

[Hint: use C9.9 to see that the series $\sum_j u_j$ converges absolutely for almost all $x \in X$. The rest is then dominated convergence.]

11.5. Let $(u_j)_{j\in\mathbb{N}}$ be a sequence of positive integrable functions on a measure space $(X, \mathcal{A}, \mu)$. Assume that the sequence decreases to 0: $u_1 \geqslant u_2 \geqslant u_3 \geqslant \dots$ and $u_j \downarrow 0$. Show that $\sum_{j=1}^\infty (-1)^j u_j$ converges, is integrable and that the integral is given by

$$\int \sum_{j=1}^\infty (-1)^j u_j \, d\mu = \sum_{j=1}^\infty (-1)^j \int u_j \, d\mu.$$

[Hint: mimic the proof of the Leibniz test for alternating series.]

11.6. Give an example of a sequence of integrable functions $(u_j)_{j\in\mathbb{N}}$ with $u_j(x) \xrightarrow{j\to\infty} u(x)$ for all x and an integrable function u but such that $\lim_{j\to\infty} \int u_j \, d\mu \neq \int u \, d\mu$. Does this contradict Lebesgue's dominated convergence theorem 11.2?

11.7. Let λ be one-dimensional Lebesgue measure. Show that for every integrable function u, the *integral function*

$$x \mapsto \int_{(0,x)} u(t) \, \lambda(dt), \quad x > 0.$$

is continuous. What happens if we exchange λ for a general measure μ?

11.8. Consider the functions

(i) $u(x) = \dfrac{1}{x}, \quad x \in [1, \infty);$ (ii) $v(x) = \dfrac{1}{x^2}, \quad x \in [1, \infty);$

(iii) $w(x) = \dfrac{1}{\sqrt{x}}, \quad x \in (0, 1];$ (iv) $y(x) = \dfrac{1}{x}, \quad x \in (0, 1];$

and check whether they are Lebesgue integrable in the regions given – what would happen if we consider $[\frac{1}{2}, 2]$ instead?

[Hint: consider first $u_k = u \mathbf{1}_{[1,k]}$, resp., $w_k = w \mathbf{1}_{[1/k,1]}$, etc. and use monotone convergence and the fact that Riemann and Lebesgue integrals coincide if both exist.]

11.9. Show that the function $\mathbb{R} \ni x \mapsto \exp(-x^\alpha)$ is $\lambda^1(dx)$-integrable over the set $[0, \infty)$ for every $\alpha > 0$.

[Hint: find dominating integrable functions u resp. w if $0 \leqslant x \leqslant 1$ resp. $1 < x < \infty$ and glue them together by $u \mathbf{1}_{[0,1]} + w \mathbf{1}_{(1,\infty)}$ to get an overall integrable upper bound.]

11.10. Show that for every parameter $\alpha > 0$ the function $x \mapsto \left(\frac{\sin x}{x}\right)^3 e^{-\alpha x}$ is integrable over $(0, \infty)$ and continuous as a function of the parameter.

[Hint: find piecewise dominating integrable functions like in Problem 11.9; use the continuity lemma 11.4.]

11.11. Show that the function

$$G : \mathbb{R} \to \mathbb{R}, \qquad G(x) := \int_{\mathbb{R} \setminus \{0\}} \frac{\sin(tx)}{t(1+t^2)} \, dt$$

is differentiable and find $G(0)$ and $G'(0)$. Use a limit argument, integration by parts for $\int_{(-n,n)} \ldots dt$ and the formula $t \, \partial_t \sin(tx) = x \, \partial_x \sin(tx)$ to show that

$$x \, G'(x) = \int_{\mathbb{R}} \frac{2t \sin(tx)}{(1+t^2)^2} \, dt.$$

11.12. Denote by λ one-dimensional Lebesgue measure. Prove that

(i) $\displaystyle \int_{(1,\infty)} e^{-x} \ln(x) \, \lambda(dx) = \lim_{k \to \infty} \int_{(1,k)} \left(1 - \frac{x}{k}\right)^k \ln(x) \, \lambda(dx)$.

(ii) $\displaystyle \int_{(0,1)} e^{-x} \ln(x) \, \lambda(dx) = \lim_{k \to \infty} \int_{(0,1)} \left(1 - \frac{x}{k}\right)^k \ln(x) \, \lambda(dx)$.

11.13. **Euler's Gamma function.** Show that the function

$$\Gamma(t) := \int_{(0,\infty)} e^{-x} x^{t-1} \, dx, \qquad t > 0,$$

(i) ... is m-times differentiable with $\Gamma^{(m)}(t) = \int_{(0,\infty)} e^{-x} x^{t-1} (\log x)^m \, dx$.
[Hint: take $t \in (a, b)$, use induction in m. Note that $|e^{-x} x^{t-1} (\log x)^m| \leqslant x^{m+t-1} e^{-x} \leqslant M x^{-2}$ for $x \geqslant 1$, and $\leqslant M' x^{\delta-1}$ for $x < 1$ and some $\delta > 0$ because $\lim_{x \to 0} x^{a-\delta} |\log x|^m = 0$ – use, e.g. the substitution $x = e^{-y}$.]

(ii) ... satisfies $\Gamma(t+1) = t \Gamma(t)$.
[Hint: use integration by parts for $\int_{1/n}^n \ldots dt$ and let $n \to \infty$.]

(iii) ... and is *logarithmically convex*, i.e. $t \mapsto \ln \Gamma(t)$ is convex.
[Hint: calculate $\left(\ln \Gamma(t)\right)''$ and show that this is positive.]

11.14. Show that $x \mapsto x^n f(u, x)$, $f(u, x) = e^{ux}/(e^x + 1)$, $0 < u < 1$, is integrable over $\mathbb{R}$ and that $g(u) := \int x^n f(u, x) \, dx$, $0 < u < 1$, is arbitrarily often differentiable.

11.15. **Moment generating function.** Let X be a random variable on the probability space $(\Omega, \mathcal{A}, P)$. The function $\phi_X(t) := \int e^{-tX} \, dP$ is called the *moment generating function*. Show that ϕ_X is m-times differentiable at $t = 0+$ if the absolute mth moment $\int |X|^m \, dP$ exists. If this is the case, the following formulae hold:

(i) $\int X^k \, dP = (-1)^k \dfrac{d^k}{dt^k} \, \phi_X(t) \Big|_{t=0+}$ for all $0 \leqslant k \leqslant m$.

(ii) $\phi_X(t) = \displaystyle\sum_{k=0}^{m} \dfrac{\int X^k \, dP}{k!} (-1)^k t^k + o(t^m).$ $(f(t) = o(t^m)$ means that $\lim_{t \to 0} f(t)/t^m = 0.)$

(iii) $\left| \phi_X(t) - \displaystyle\sum_{k=0}^{m-1} \dfrac{\int X^k \, dP}{k!} (-1)^k t^k \right| \leqslant \dfrac{|t|^m}{m!} \int |X|^m \, dP.$

(iv) If $\int |X|^k \, dP < \infty$ for all $k \in \mathbb{N}$, then

$$\phi_X(t) = \sum_{k=0}^{\infty} \frac{\int X^k \, dP}{k!} (-1)^k t^k$$

for all t within the convergence radius of the series.

11.16. Consider the functions $u(x) = \mathbf{1}_{\mathbb{Q} \cap [0,1]}$ and $v(x) = \mathbf{1}_{\{n^{-1} : n \in \mathbb{N}\}}(x)$. Prove or disprove:

  (i) The function u is 1 on the rationals and 0 otherwise. Thus u is continuous everywhere except the set $\mathbb{Q} \cap [0,1]$. Since this is a null set, u is a.e. continuous, hence Riemann integrable by Theorem 11.8.

 (ii) The function v is 0 everywhere but for the values $x = 1/n$, $n \in \mathbb{N}$. Thus v is continuous everywhere except a countable set, i.e. a null set, and v is a.e. continuous, hence Riemann integrable by Theorem 11.8.

(iii) The functions u and v are Lebesgue integrable and $\int u \, d\lambda = \int v \, d\lambda = 0$.

(iv) The function u is not Riemann integrable.

11.17. Construct a sequence of functions $(u_j)_{j \in \mathbb{N}}$ which are Riemann integrable but converge to a limit $u_j \xrightarrow{\ j \to \infty\ } u$ which is not Riemann integrable.
[Hint: consider, e.g. $u_j = \mathbf{1}_{\{q_1, q_2, \dots, q_j\}}$ where $(q_j)_j$ is an enumeration of $\mathbb{Q}$.]

11.18. Assume that $u : [0, \infty) \to \mathbb{R}$ is positive and improperly Riemann integrable. Show that u is also Lebesgue integrable.

11.19. **Fresnel integrals.** Show that the following improper Riemann integrals exist:

$$\int_0^\infty \sin x^2 \, dx \qquad \text{and} \qquad \int_0^\infty \cos x^2 \, dx.$$

Do they exist as Lebesgue integrals?
Remark. The above integrals have the value $\frac{1}{2}\sqrt{\frac{\pi}{2}}$. This can be proved by Cauchy's theorem or the residue theorem.

11.20. **Frullani's integral.** Let $f : (0, \infty) \to \mathbb{R}$ be a continuous function such that $\lim_{x \to 0} f(x) = m$ and $\lim_{x \to \infty} f(x) = M$. Show that the two-sided improper Riemann integral

$$\lim_{\substack{r \to 0 \\ s \to \infty}} \int_r^s \frac{f(bx) - f(ax)}{x} \, dx = (M - m) \ln \frac{b}{a}$$

exists for all $a, b > 0$. Does this integral have a meaning as Lebesgue integral?
[Hint: use the mean value theorem for integrals, E.12.]

11.21. Denote by λ one-dimensional Lebesgue measure on the interval $(0, 1)$.

  (i) Show that for all $k \in \mathbb{N}_0$ one has

$$\int_{(0,1)} (x \ln x)^k \, \lambda(dx) = (-1)^k \left(\frac{1}{k+1} \right)^{k+1} \Gamma(k+1).$$

  (ii) Use (i) to conclude that $\displaystyle \int_{(0,1)} x^{-x} \, \lambda(dx) = \sum_{k=1}^{\infty} k^{-k}.$

  [Hint: note that $x^{-x} = e^{-x \ln x}$ and use the exponential series.]

12

The function spaces $\mathcal{L}^p$, $1 \leqslant p \leqslant \infty$

Throughout this chapter $(X, \mathcal{A}, \mu)$ will be some measure space.

We will now discuss functions whose (absolute) pth power or pth (absolute) moment is integrable. More precisely, we are interested in the sets

$$\mathcal{L}^p(\mu) := \left\{ u : X \to \mathbb{R} : u \in \mathcal{M}, \int |u|^p \, d\mu < \infty \right\}, \quad p \in [1, \infty). \quad (12.1)$$

As usual, we suppress μ if the choice of measure is clear, and we write $\mathcal{L}^p(X)$ or $\mathcal{L}^p(\mathcal{A})$ if we want to stress the underlying space or σ-algebra. It is convenient to have the following notation:

$$\|u\|_p := \left(\int |u(x)|^p \, \mu(dx) \right)^{1/p}. \quad (12.2)$$

Clearly, $u \in \mathcal{L}^p(\mu) \iff u \in \mathcal{M}$ and $\|u\|_p < \infty$. It is no accident that the notation $\|\cdot\|_p$ resembles the symbol for a *norm*:[1] indeed, we have because of T10.9(i)

$$\|u\|_p = 0 \iff u = 0 \quad \text{a.e.}, \quad (12.3)$$

and for all $\alpha \in \mathbb{R}$

$$\|\alpha u\|_p = \left(\int |\alpha u|^p \, d\mu \right)^{1/p} = \left(|\alpha|^p \int |u|^p \, d\mu \right)^{1/p} = |\alpha| \|u\|_p. \quad (12.4)$$

The triangle inequality for $\|\cdot\|_p$ and deeper results on $\mathcal{L}^p$ depend much on the following elementary inequality.

12.1 Lemma (Young's inequality) *Let* $p, q \in (1, \infty)$ *be* conjugate *numbers, i.e.* $\frac{1}{p} + \frac{1}{q} = 1$ *or* $q = \frac{p}{p-1}$. *Then*

$$AB \leqslant \frac{A^p}{p} + \frac{B^q}{q} \quad (12.5)$$

holds for all $A, B \geqslant 0$; *equality occurs if, and only if,* $B = A^{p-1}$.

---

[1] See Appendix B.

Proof There are various different methods to prove (12.5) but probably the most intuitive one is through the following picture:

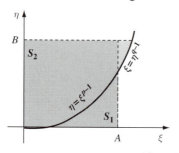

The shaded area representing the pieces S_1 and S_2 between the graph and the ξ- resp. η-axis is given by

$$\int_0^A \xi^{p-1}\, d\xi = \frac{A^p}{p} \quad \text{and} \quad \int_0^B \eta^{q-1}\, d\eta = \frac{B^q}{q}$$

respectively. The picture shows that their combined area is greater than the area of the darker rectangle, thus $\dfrac{A^p}{p} + \dfrac{B^q}{q} \geqslant AB$. Equality obtains if, and only if, the lighter shaded area vanishes, i.e. if $B = A^{p-1}$. ∎

We can now prove the following fundamental inequality.

12.2 Theorem (Hölder's inequality) *Assume that $u \in \mathcal{L}^p(\mu)$ and $v \in \mathcal{L}^q(\mu)$ where $p, q \in (1, \infty)$ are conjugate numbers: $\frac{1}{p} + \frac{1}{q} = 1$. Then $uv \in \mathcal{L}^1(\mu)$, and the following inequality holds:*

$$\left| \int uv\, d\mu \right| \leqslant \int |uv|\, d\mu \leqslant \|u\|_p \cdot \|v\|_q. \tag{12.6}$$

Equality occurs if, and only if, $|u(x)|^p / \|u\|_p^p = |v(x)|^q / \|v\|_q^q$ a.e.

Proof The first inequality of (12.6) follows directly from T10.4(v). To see the other inequality we use (12.5) with

$$A := \frac{|u(x)|}{\|u\|_p} \quad \text{and} \quad B := \frac{|v(x)|}{\|v\|_q}$$

to get

$$\frac{|u(x)v(x)|}{\|u\|_p \|v\|_q} \leqslant \frac{|u(x)|^p}{p\,\|u\|_p^p} + \frac{|v(x)|^q}{q\,\|v\|_q^q}.$$

Integrating both sides of this inequality over x yields

$$\frac{\int |u(x)v(x)|\, \mu(dx)}{\|u\|_p \|v\|_q} \leqslant \frac{\|u\|_p^p}{p\,\|u\|_p^p} + \frac{\|v\|_q^q}{q\,\|v\|_q^q} = \frac{1}{p} + \frac{1}{q} = 1.$$

Equality can only happen if we have equality in (12.5). Because of our choice of A and B, the condition for equality from L12.1 becomes $|v(x)|/\|v\|_q = \left(|u(x)|/\|u\|_p\right)^{p-1}$ a.e. Raising both sides to the qth power gives $|v(x)|^q/\|v\|_q^q = |u(x)|^p/\|u\|_p^p$ since $(p-1)q = p$. ∎

Hölder's inequality with $p = q = 2$ is usually called the *Cauchy–Schwarz inequality*.

12.3 Corollary (Cauchy–Schwarz inequality) *Let* $u, v \in \mathcal{L}^2(\mu)$. *Then* $uv \in \mathcal{L}^1(\mu)$ *and*

$$\int |uv| \, d\mu \leqslant \|u\|_2 \cdot \|v\|_2. \tag{12.7}$$

Equality occurs if, and only if, $|u(x)|^2/\|u\|_2^2 = |v(x)|^2/\|v\|_2^2$ *a.e.*

Another consequence of Hölder's inequality is the Minkowski or triangle inequality for $\|\cdot\|_p$.

12.4 Corollary (Minkowski's inequality) *Let* $u, v \in \mathcal{L}^p(\mu)$, $p \in [1, \infty)$. *Then* $u + v \in \mathcal{L}^p(\mu)$ *and*

$$\|u + v\|_p \leqslant \|u\|_p + \|v\|_p. \tag{12.8}$$

Proof Since

$$|u + v|^p \leqslant (|u| + |v|)^p \leqslant 2^p \max\{|u|^p, |v|^p\} \leqslant 2^p \left(|u|^p + |v|^p\right)$$

we get that $|u + v|^p \in \mathcal{L}^1(\mu)$ or $u + v \in \mathcal{L}^p(\mu)$. Now

$$\int |u+v|^p \, d\mu = \int |u+v| \cdot |u+v|^{p-1} \, d\mu$$

$$\leqslant \int |u| \cdot |u+v|^{p-1} \, d\mu + \int |v| \cdot |u+v|^{p-1} \, d\mu$$

(if $p = 1$ the proof stops here)

$$\overset{12.2}{\leqslant} \|u\|_p \cdot \left\||u+v|^{p-1}\right\|_q + \|v\|_p \cdot \left\||u+v|^{p-1}\right\|_q.$$

Dividing both sides by $\left\||u+v|^{p-1}\right\|_q$ proves our claim since

$$\left\||u+v|^{p-1}\right\|_q = \left(\int |u+v|^{(p-1)q} \, d\mu\right)^{1/q} = \left(\int |u+v|^p \, d\mu\right)^{1-1/p},$$

where we also used that $q = \frac{p}{p-1}$. ∎

12.5 Remarks (i) Formulae (12.4) and (12.8) imply

$$u, v \in \mathcal{L}^p(\mu) \quad \Longrightarrow \quad \alpha u + \beta v \in \mathcal{L}^p(\mu) \qquad \forall \alpha, \beta \in \mathbb{R},$$

which shows that $\mathcal{L}^p(\mu)$ is a vector space.

(ii) Formulae (12.3), (12.4) and (12.8) show that $\|\bullet\|_p$ is a *semi-norm* for $\mathcal{L}^p$: the definiteness of a norm is not fulfilled since

$$\|u\|_p = 0 \qquad \text{only implies that} \qquad u(x) = 0 \text{ for almost every } x$$

but not for all x. There is a standard recipe to fix this: since $\mathcal{L}^p$-functions can be altered on null sets without affecting their integration behaviour, we introduce the following *equivalence relation*: we call $u, v \in \mathcal{L}^p(\mu)$ *equivalent* if they differ on at most a μ-null set, i.e.

$$u \sim v \iff \{u \neq v\} \in \mathcal{N}_\mu.$$

The *quotient space* $L^p(\mu) := \mathcal{L}^p(\mu)/_\sim$ consists of all equivalence classes of $\mathcal{L}^p$-functions. If $[u]_p \in L^p(\mu)$ denotes the equivalence class induced by the function $u \in \mathcal{L}^p(\mu)$, it is not hard to see that

$$[\alpha u + \beta v]_p = \alpha [u]_p + \beta [v]_p \qquad \text{and} \qquad [uv]_1 = [u]_p [v]_q$$

hold, turning $L^p(\mu)$ into a bona fide vector space with the canonical *norm*

$$\|[u]_p\|_p := \inf \left\{ \|w\|_p : w \in \mathcal{L}^p, \ w \sim u \right\}$$

for quotient spaces. Fortunately, $\|[u]_p\|_p = \|u\|_p$ and later on we will often follow the usual abuse of notation and identify $[u]$ with u.

(iii) All results of this chapter are still valid for $\bar{\mathbb{R}}$-valued numerical functions. Indeed, if $f \in \mathcal{M}_{\bar{\mathbb{R}}}$ and $\int |f|^p \, d\mu < \infty$, then

$$\mu(\{|f| = \infty\}) = \mu(\{|f|^p = \infty\}) = \mu\left(\bigcap_{j \in \mathbb{N}} \{|f|^p > j\} \right)$$

$$\overset{4.4}{=} \lim_{j \to \infty} \mu(\{|f|^p > j\})$$

$$\overset{10.12}{\leqslant} \lim_{j \to \infty} \frac{1}{j} \int |f|^p \, d\mu = 0,$$

by the Markov inequality. This means, however, that f is a.e. $\mathbb{R}$-valued, so sums and products of such functions are always defined outside a μ-null set. In particular there is no need to distinguish between the classes $L^p(\mu) \ (= L^p_{\mathbb{R}}(\mu))$ and $L^p_{\bar{\mathbb{R}}}(\mu)$.

$$* \qquad * \qquad *$$

We will need the concept of *convergence* of a sequence in the space $\mathcal{L}^p(\mu)$. A sequence $(u_j)_{j\in\mathbb{N}} \subset \mathcal{L}^p(\mu)$ is said to be *convergent* in $\mathcal{L}^p(\mu)$ with *limit* $\mathcal{L}^p\text{-}\lim_{j\to\infty} u_j = u$ if, and only if,

$$\lim_{j\to\infty} \|u_j - u\|_p = 0.$$

Remember, however, that $\mathcal{L}^p$-limits are only almost everywhere unique. If u, w are both $\mathcal{L}^p$-limits of the same sequence $(u_j)_{j\in\mathbb{N}}$, we have

$$\|u - w\|_p \overset{12.4}{\leqslant} \lim_{j\to\infty} \left(\|u - u_j\|_p + \|u_j - w\|_p \right) = 0,$$

implying only $u = w$ almost everywhere.

We call $(u_j)_{j\in\mathbb{N}} \subset \mathcal{L}^p(\mu)$ a $(\mathcal{L}^p\text{-})$ *Cauchy sequence*, if

$$\forall \epsilon > 0 \quad \exists N_\epsilon \in \mathbb{N} \quad \forall j, k \geqslant N_\epsilon : \|u_j - u_k\|_p < \epsilon.$$

Note that these definitions reduce convergence in $\mathcal{L}^p$ to convergence questions of the semi-norm $\|\bullet\|_p$ in $\mathbb{R}^+$. This means that, apart from uniqueness, many formal properties of limits in $\mathbb{R}$ carry over to $\mathcal{L}^p$ – most of them even with the same proof!

Caution: Pointwise convergence of a sequence $u_j(x) \to u(x)$ of $\mathcal{L}^p$-functions $(u_j)_{j\in\mathbb{N}} \subset \mathcal{L}^p$ does not guarantee convergence in $\mathcal{L}^p$ – but in view of Lebesgue's dominated convergence theorem 11.2, the additional condition that

$$|u_j(x)| \leqslant g(x) \qquad \text{for some function} \quad g \in \mathcal{L}^p_+$$

is sufficient since $|u_j - u|^p \leqslant (|u_j| + |u|)^p \leqslant 2^p\, g^p$ and $|u_j(x) - u(x)| \to 0$.[✓]

Clearly, a convergent sequence $(u_j)_{j\in\mathbb{N}}$ is also a Cauchy sequence,

$$\|u_j - u_k\|_p \leqslant \|u_j - u\|_p + \|u - u_k\|_p < 2\epsilon \qquad \forall j, k \geqslant N_\epsilon;$$

the converse of this assertion is also true, but much more difficult to prove. We start with a simple observation:

12.6 Lemma *For any sequence* $(u_j)_{j\in\mathbb{N}} \subset \mathcal{L}^p(\mu)$, $p \in [1, \infty)$, *of positive functions* $u_j \geqslant 0$ *we have*

$$\left\| \sum_{j=1}^{\infty} u_j \right\|_p \leqslant \sum_{j=1}^{\infty} \|u_j\|_p. \tag{12.9}$$

Proof Repeated applications of Minkowski's inequality (12.8) show that

$$\left\| \sum_{j=1}^{N} u_j \right\|_p \leqslant \sum_{j=1}^{N} \|u_j\|_p \leqslant \sum_{j=1}^{\infty} \|u_j\|_p,$$

and since the right-hand side is independent of N, the inequality remains valid even if we pass to the sup on the left. By Beppo Levi's theorem 9.6, we find

$$\sup_{N \in \mathbb{N}} \left\| \sum_{j=1}^{N} u_j \right\|_p^p = \sup_{N \in \mathbb{N}} \int \left(\sum_{j=1}^{N} u_j \right)^p d\mu$$

$$= \int \left(\sup_{N \in \mathbb{N}} \sum_{j=1}^{N} u_j \right)^p d\mu = \int \left(\sum_{j=1}^{\infty} u_j \right)^p d\mu,$$

and the proof follows.   ∎

The completeness of $\mathcal{L}^p$ was proved by E. Fischer (for $p = 2$) and F. Riesz (for $1 \leqslant p < \infty$).

12.7 Theorem (Riesz–Fischer) *The spaces $\mathcal{L}^p(\mu)$, $p \in [1, \infty)$, are complete, i.e. every Cauchy sequence $(u_j)_{j \in \mathbb{N}} \subset \mathcal{L}^p(\mu)$ converges to some limit $u \in \mathcal{L}^p(\mu)$.*

Proof The main difficulty here is to identify the limit u. By the definition of a Cauchy sequence we find numbers

$$1 < n(1) < n(2) < \ldots < n(k) < \ldots$$

such that

$$\left\| u_{n(k+1)} - u_{n(k)} \right\|_p < 2^{-k} \qquad k \in \mathbb{N}.$$

To find u, we turn the sequence into a series by

$$u_{n(k+1)} = \sum_{j=0}^{k} \left(u_{n(j+1)} - u_{n(j)} \right), \qquad u_{n(0)} := 0, \qquad (12.10)$$

and the limit as $k \to \infty$ would formally be $u := \sum_{j=0}^{\infty} \left(u_{n(j+1)} - u_{n(j)} \right)$ – if we can make sense of this infinite sum. Since

$$\left\| \sum_{j=0}^{\infty} \left| u_{n(j+1)} - u_{n(j)} \right| \right\|_p \overset{(12.9)}{\leqslant} \sum_{j=0}^{\infty} \left\| u_{n(j+1)} - u_{n(j)} \right\|_p$$

$$(12.11)$$

$$\leqslant \left\| u_{n(1)} \right\|_p + \sum_{j=1}^{\infty} \frac{1}{2^j},$$

we conclude with C10.13 that $\left(\sum_{j=0}^{\infty} \left| u_{n(j+1)} - u_{n(j)} \right| \right)^p < \infty$ a.e., so that $u = \sum_{j=0}^{\infty} \left(u_{n(j+1)} - u_{n(j)} \right)$ is a.e. (absolutely) convergent.

Let us show that $u = \mathcal{L}^p\text{-}\lim_{k\to\infty} u_{n(k)}$. For this, observe that by the (ordinary) triangle inequality and (12.11),

$$\left\| u - u_{n(k)} \right\|_p = \left\| \sum_{j=k+1}^{\infty} \left(u_{n(j+1)} - u_{n(j)} \right) \right\|_p \overset{\text{def}}{=} \left\| \sum_{j=k+1}^{\infty} \left(u_{n(j+1)} - u_{n(j)} \right) \right\|_p$$

$$\leqslant \left\| \sum_{j=k+1}^{\infty} \left| u_{n(j+1)} - u_{n(j)} \right| \right\|_p$$

$$\overset{(12.9)}{\leqslant} \sum_{j=k+1}^{\infty} \left\| u_{n(j+1)} - u_{n(j)} \right\|_p \overset{k\to\infty}{\longrightarrow} 0.$$

Finally, using that $(u_j)_{j\in\mathbb{N}}$ is a Cauchy sequence, we get, for all $\epsilon > 0$ and suitable $N_\epsilon \in \mathbb{N}$,

$$\left\| u - u_j \right\|_p \leqslant \left\| u - u_{n(k)} \right\|_p + \left\| u_{n(k)} - u_j \right\|_p$$

$$\leqslant \left\| u - u_{n(k)} \right\|_p + \epsilon \qquad \forall j, n(k) \geqslant N_\epsilon.$$

Letting $k \to \infty$ shows $\left\| u - u_j \right\|_p \leqslant \epsilon$ if $j \geqslant N_\epsilon$. $\blacksquare$

The proof of Theorem 12.7 shows even a weak form of pointwise convergence:

12.8 Corollary *Let* $(u_j)_{j\in\mathbb{N}} \subset \mathcal{L}^p(\mu)$, $p \in [1, \infty)$ *with* $\mathcal{L}^p\text{-}\lim_{j\to\infty} u_j = u$. *Then there exists a subsequence* $(u_{n(k)})_{k\in\mathbb{N}}$ *such that* $\lim_{k\to\infty} u_{n(k)}(x) = u(x)$ *holds for almost every* $x \in X$.

Proof Since $(u_j)_{j\in\mathbb{N}}$ converges in $\mathcal{L}^p$, it is also an $\mathcal{L}^p$-Cauchy sequence and the claim follows from (12.11). $\blacksquare$

As we have already remarked, pointwise convergence alone does not guarantee convergence in $\mathcal{L}^p$, not even of a subsequence, see Problem 12.7. Let us repeat the following sufficient criterion, which we have already proved on page 109.

12.9 Theorem *Let* $(u_j)_{j\in\mathbb{N}} \subset L^p(\mu)$, $p \in [1, \infty)$, *be a sequence of functions such that* $|u_j| \leqslant w$ *for all* $j \in \mathbb{N}$ *and some* $w \in \mathcal{L}^p_+(\mu)$. *If* $u(x) = \lim_{j\to\infty} u_j(x)$ *exists for almost every* $x \in X$, *then*

$$u \in \mathcal{L}^p(\mu) \qquad and \qquad \lim_{j\to\infty} \left\| u - u_j \right\|_p = 0.$$

Of a different flavour is the next result which is sometimes called F. Riesz's convergence theorem.

12.10 Theorem (Riesz) *Let $(u_j)_{j\in\mathbb{N}} \subset \mathcal{L}^p(\mu)$, $p \in [1,\infty)$, be a sequence such that $\lim_{j\to\infty} u_j(x) = u(x)$ for almost every $x \in X$ and some $u \in \mathcal{L}^p(\mu)$. Then*

$$\lim_{j\to\infty} \|u_j - u\|_p = 0 \quad \Longleftrightarrow \quad \lim_{j\to\infty} \|u_j\|_p = \|u\|_p. \qquad (12.12)$$

Proof The direction '$\Rightarrow$' in (12.12) follows from the lower triangle inequality[2] $\big| \|u_j\|_p - \|u\|_p \big| \leqslant \|u_j - u\|_p$ for $\|\cdot\|_p$.

For '$\Leftarrow$' we observe that

$$|u_j - u|^p \leqslant (|u_j| + |u|)^p \leqslant 2^p \max\{|u_j|^p, |u|^p\} \leqslant 2^p (|u_j|^p + |u|^p),$$

and we can apply Fatou's lemma 9.11 to the sequence

$$2^p (|u_j|^p + |u|^p) - |u_j - u|^p \geqslant 0$$

to get

$$2^{p+1} \int |u|^p \, d\mu = \int \liminf_{j\to\infty} \left(2^p (|u_j|^p + |u|^p) - |u_j - u|^p \right) d\mu$$

$$\leqslant \liminf_{j\to\infty} \left(2^p \int |u_j|^p \, d\mu + 2^p \int |u|^p \, d\mu - \int |u_j - u|^p \, d\mu \right)$$

$$= 2^{p+1} \int |u|^p \, d\mu - \limsup_{j\to\infty} \int |u_j - u|^p \, d\mu,$$

where we used that $\lim_{j\to\infty} \int |u_j|^p \, d\mu = \int |u|^p \, d\mu$. This shows that

$$\limsup_{j\to\infty} \int |u_j - u|^p \, d\mu = 0, \quad \text{hence} \quad \lim_{j\to\infty} \int |u_j - u|^p \, d\mu = 0. \qquad \blacksquare$$

Let us note the following structural result on $\mathcal{L}^p$, which will become important later on.

12.11 Corollary *The simple p-integrable functions $\mathcal{E} \cap \mathcal{L}^p(\mu)$, $p \in [1,\infty)$, are a dense subset of $\mathcal{L}^p(\mu)$, i.e. for every $u \in \mathcal{L}^p(\mu)$ one can find a sequence $(f_j)_{j\in\mathbb{N}} \subset \mathcal{E}$ such that $\lim_{j\to\infty} \|f_j - u\|_p = 0$.*

Proof Assume first that $u \in \mathcal{L}^p_+(\mu)$ is positive. By Theorem 8.8 we find an increasing sequence $(f_j)_{j\in\mathbb{N}}$ of positive simple functions with $\sup_{j\in\mathbb{N}} f_j = u$. Since $0 \leqslant f_j \leqslant u$, we have $f_j \in \mathcal{L}^p(\mu)$ as well as $\sup_{j\in\mathbb{N}} \int |f_j|^p \, d\mu = \int |u|^p \, d\mu$.[✓] We can now apply Theorem 12.10 and deduce that $\lim_{j\to\infty} \|f_j - u\|_p = 0$.

---

[2] Follows exactly as $\big||a| - |b|\big| \leqslant |a - b|$ follows from $|a + b| \leqslant |a| + |b|$, $a, b \in \mathbb{R}$.

For a general $u \in \mathcal{L}^p(\mu)$, we consider its positive and negative parts $u^{\pm}$ and construct, as before, sequences $g_j, h_j \in \mathcal{E} \cap \mathcal{L}^p(\mu)$ with $g_j \to u^+$ and $h_j \to u^-$ in $\mathcal{L}^p(\mu)$. But then $g_j - h_j \in \mathcal{E} \cap \mathcal{L}^p(\mu)$, and

$$\|u - (g_j - h_j)\|_p \leqslant \|u^+ - g_j\|_p + \|u^- - h_j\|_p \xrightarrow{j \to \infty} 0$$

finishes the proof. ∎

With a special choice of $(X, \mathcal{A}, \mu)$ we can see that integrals generalize infinite series.

12.12 Example Consider the counting measure $\mu = \sum_{j=1}^{\infty} \delta_j$, cf. Example 4.7(iii), on the measurable space $(\mathbb{N}, \mathcal{P}(\mathbb{N}))$. As we have seen in Examples 9.10(ii) and 10.6(ii), a function $u : \mathbb{N} \to \mathbb{R}$ is μ-integrable if, and only if,

$$\sum_{j=1}^{\infty} |u(j)| < \infty, \qquad \text{in which case} \qquad \int_{\mathbb{N}} u \, d\mu = \sum_{j=1}^{\infty} u(j).$$

In a similar way one shows that $v \in \mathcal{L}^p(\mu)$ if, and only if, $\sum_{j=1}^{\infty} |v(j)|^p < \infty$. Functions $u : \mathbb{N} \to \mathbb{R}$ are determined by their values $(u(1), u(2), u(3), \ldots)$ and every sequence $(a_j)_{j \in \mathbb{N}} \subset \mathbb{R}$ defines a function u by $u(j) := a_j$. This means that we can identify the function u with the sequence $(u(j))_{j \in \mathbb{N}}$ of real numbers. Thus

$$\mathcal{L}^p(\mu) = \left\{ u : \mathbb{N} \to \mathbb{R} : \sum_{j=1}^{\infty} |u(j)|^p < \infty \right\}$$

$$= \left\{ (a_j)_{j \in \mathbb{N}} \subset \mathbb{R} : \sum_{j=1}^{\infty} |a_j|^p < \infty \right\} =: \ell^p(\mathbb{N}),$$

the latter being a so-called *sequence space*. Note that in this context Hölder's and Minkowski's inequalities become

$$\sum_{j=1}^{\infty} |a_j b_j| \leqslant \left(\sum_{j=1}^{\infty} |a_j|^p \right)^{1/p} \left(\sum_{j=1}^{\infty} |b_j|^q \right)^{1/q} \tag{12.13}$$

if $p, q \in (1, \infty)$ are conjugate numbers, and

$$\left(\sum_{j=1}^{\infty} |a_j \pm b_j|^p \right)^{1/p} \leqslant \left(\sum_{j=1}^{\infty} |a_j|^p \right)^{1/p} + \left(\sum_{j=1}^{\infty} |b_j|^p \right)^{1/p}. \tag{12.14}$$

We close this chapter with a useful convexity, resp. concavity, inequality. Recall that a function $\Phi : [a, b] \to \mathbb{R}$ on an interval $[a, b] \subset \bar{\mathbb{R}}$ is *convex* [*concave*] if

$$\Phi(tx + (1-t)y) \leqslant t\Phi(x) + (1-t)\Phi(y), \qquad 0 < t < 1,$$
$$\left[\Phi(tx + (1-t)y) \geqslant t\Phi(x) + (1-t)\Phi(y), \qquad 0 < t < 1, \right] \tag{12.15}$$

holds for all $x, y \in [a, b]$. Geometrically this means that the graph of a convex

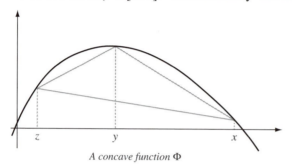

A concave function Φ

[concave] function Φ between the points $(x, \Phi(x))$ and $(y, \Phi(y))$ lies below [above] the chord linking $(x, \Phi(x))$ and $(y, \Phi(y))$. Convex [concave] functions have nice properties: they are continuous in (a, b) and if Φ' exists, it is increasing [decreasing].

If Φ is twice differentiable, convexity [concavity] is equivalent to $\Phi'' \geqslant 0$ [$\Phi'' \leqslant 0$]. Further details and proofs can be found in Boas [8]. For our purposes we need the following lemma.

12.13 Lemma *A convex [concave] function $\Phi : [a, b] \to \mathbb{R}$ has at every point in the open interval (a, b) a finite right-hand derivative Φ'_+ and satisfies*

$$\Phi(x) \geqslant \Phi'_+(y)\,(x-y) + \Phi(y), \qquad \forall\, x, y \in (a, b)$$
$$\left[\Phi(x) \leqslant \Phi'_+(y)\,(x-y) + \Phi(y), \qquad \forall\, x, y \in (a, b) \right]. \tag{12.16}$$

In particular, a convex [concave] function is the upper [lower] envelope of all linear functions below [above] its graph

$$\Phi(x) = \sup\{\ell(x) : \ell(z) = \alpha z + \beta \leqslant \Phi(z) \quad \forall\, z \in (a, b)\}$$
$$\left[\Phi(x) = \inf\{\ell(x) : \ell(z) = \alpha z + \beta \geqslant \Phi(z) \quad \forall\, z \in (a, b)\} \right]. \tag{12.17}$$

Proof Since the graph of a convex [concave] function looks like a smile [frown], the last statement of the lemma is intuitively clear. A rigorous argument uses (12.16) which says that Φ admits at every point a tangent below [above] its graph. We show (12.16) only for concave functions. Pick numbers

$$z < \eta < y < \xi < x$$

in (a, b) and choose $t = t(y, \xi, x) \in (0, 1)$ such that $\xi = ty + (1 - t)x$, $t' = t'(\eta, y, \xi)$ such that $y = t'\eta + (1 - t')\xi$ and $t'' = t''(z, \eta, \xi)$ such that $\eta = t''z + (1 - t'')\xi$. Using these values in (12.15) we see after some simple manipulations that

$$\frac{\Phi(x) - \Phi(y)}{x - y} \leqslant \frac{\Phi(\xi) - \Phi(y)}{\xi - y} \leqslant \frac{\Phi(\xi) - \Phi(\eta)}{\xi - \eta} \leqslant \frac{\Phi(z) - \Phi(\eta)}{z - \eta},$$

cf. the picture on page 114. This shows that $\frac{\Phi(\xi) - \Phi(y)}{\xi - y}$ is bounded and increasing as $\xi \downarrow y$. Therefore, the right-hand derivative $\Phi'_+(y) := \lim_{\xi \downarrow y} \frac{\Phi(\xi) - \Phi(y)}{\xi - y}$ exists and is finite. In particular, Φ is continuous from the right at the point y, so that $\lim_{\xi \downarrow y} \Phi(\xi) = \Phi(y)$. Letting $\xi \to y$ in the above chain of inequalities therefore yields

$$\frac{\Phi(x) - \Phi(y)}{x - y} \leqslant \Phi'_+(y) \leqslant \frac{\Phi(y) - \Phi(\eta)}{y - \eta} \qquad \forall \, \eta < y < x,$$

and rearranging the first of these inequalities gives (12.16). ∎

12.14 Theorem (Jensen's inequality) *Let* $\Lambda : [0, \infty) \to [0, \infty)$ *be a concave and* $V : [0, \infty) \to [0, \infty)$ *be a convex function. For any* $w \in \mathcal{L}^1_+(\mu)$ *we have*

$$\frac{\int \Lambda(u) \, w \, d\mu}{\int w \, d\mu} \leqslant \Lambda \left(\frac{\int uw \, d\mu}{\int w \, d\mu} \right), \qquad \forall u \in \mathcal{M}^+; \qquad (12.18)$$

$$V \left(\frac{\int uw \, d\mu}{\int w \, d\mu} \right) \leqslant \frac{\int V(u) \, w \, d\mu}{\int w \, d\mu}, \qquad \forall u \in \mathcal{M}^+. \qquad (12.19)$$

If $uw \in \mathcal{L}^1(\mu)$, *then* $\Lambda(u) \, w \in \mathcal{L}^1(\mu)$.

Proof We prove only (12.18) since (12.19) is similar. If the right-hand side of (12.18) is infinite, there is nothing to show. Therefore we may assume $\int uw \, d\mu < \infty$. Since $\Lambda(x)$ is concave, we find for any $\ell(x) := \alpha x + \beta \geqslant \Lambda(x)$ that

$$\frac{\int \Lambda(u) \, w \, d\mu}{\int w \, d\mu} \leqslant \frac{\int (\alpha u + \beta) \, w \, d\mu}{\int w \, d\mu} = \alpha \frac{\int uw \, d\mu}{\int w \, d\mu} + \beta = \ell \left(\frac{\int uw \, d\mu}{\int w \, d\mu} \right)$$

and the inequality follows from (12.17) if we pass to the inf over all linear functions satisfying $\ell \geqslant \Lambda$. ∎

12.15 The case $p = \infty$. In Theorem 12.2 and Corollary 12.4 we avoided the cases $p = 1$ or ∞. This can be overcome by introducing the space $\mathcal{L}^\infty(\mu)$:

$$\mathcal{L}^\infty(\mu) = \left\{ u \in \mathcal{M}(\mathcal{A}) : u \text{ is a.e. bounded} \right\}. \qquad (12.20)$$

Obviously, $\mathcal{L}^\infty(\mu)$ is a vector space, and we can introduce by

$$\|u\|_\infty := \inf\{C : \mu(\{|u| > C\}) = 0\} \tag{12.21}$$

a norm$^{[\checkmark]}$ which is, for continuous u, just $\|u\|_\infty = \sup|u|$. Interpreting $p = 1$ and $q = \infty$ as conjugate numbers, it is not hard to verify T12.2 and C12.4 for these values of p and q. The completeness of $\mathcal{L}^\infty(\mu)$ is much easier to prove than T12.7: if $(u_j)_{j\in\mathbb{N}}$ is a Cauchy sequence in $\mathcal{L}^\infty(\mu)$, we set

$$A_{k,\ell} := \{|u_k| > \|u_k\|_\infty\} \cup \{|u_k - u_\ell| > \|u_k - u_\ell\|_\infty\}, \quad A := \bigcup_{k,\ell\in\mathbb{N}} A_{k,\ell}.$$

By definition, $\mu(A_{k,\ell}) = 0$ and $\mu(A) = 0$, so that $\|u_j \mathbf{1}_A\|_\infty = 0$ for all $j \in \mathbb{N}$. On the set A^c, however, $(u_j)_{j\in\mathbb{N}}$ converges uniformly to a bounded function u, i.e. $u\mathbf{1}_{A^c} \in \mathcal{L}^\infty(\mu)$ as well as $\|u_j - u\mathbf{1}_{A^c}\|_\infty \to 0$.

As in Remark 12.5 we write $L^\infty(\mu)$ for $\mathcal{L}^\infty(\mu)/_\sim$, where $u \sim v$ means that $\{u \neq v\} \in \mathcal{N}_\mu$ is a μ-null set.

Note also that T12.10 and C12.11 are no longer true for $p = \infty$. This can be seen on $(\mathbb{R}, \mathcal{B}, \lambda)$ from $u_j(x) := e^{-|x/j|} \xrightarrow{j\to\infty} \mathbf{1}_{\mathbb{R}}(x)$ for the former and from $u(x) = \sum_{j=-\infty}^{\infty} \mathbf{1}_{[2j,2j+1]}(x)$ for the latter.$^{[\checkmark]}$

Problems

12.1. Let $(X, \mathcal{A}, \mu)$ be a finite measure space and let $1 \leqslant q < p < \infty$.

 (i) Show that $\|u\|_q \leqslant \mu(X)^{1/q-1/p} \|u\|_p$.
   [Hint: use Hölder's inequality for $u \cdot 1$.]
 (ii) Conclude that $\mathcal{L}^p(\mu) \subset \mathcal{L}^q(\mu)$ for all $p \geqslant q \geqslant 1$ and that a Cauchy sequence in $\mathcal{L}^p$ is also a Cauchy sequence in $\mathcal{L}^q$.
 (iii) Is this still true if the measure μ is not finite?

12.2. Let $(X, \mathcal{A}, \mu)$ be a general measure space and $1 \leqslant p \leqslant r \leqslant q \leqslant \infty$. Prove that $\mathcal{L}^p(\mu) \cap \mathcal{L}^q(\mu) \subset \mathcal{L}^r(\mu)$ by establishing the inequality

$$\|u\|_r \leqslant \|u\|_p^\lambda \cdot \|u\|_q^{1-\lambda} \qquad \forall u \in \mathcal{L}^p(\mu) \cap \mathcal{L}^q(\mu),$$

with $\lambda = (\frac{1}{r} - \frac{1}{q})/(\frac{1}{p} - \frac{1}{q})$.
[Hint: use Hölder's inequality.]

12.3. Extend the proof of Hölder's inequality 12.2 to $p = 1$ and $q = \infty$, i.e. show that

$$\int uv\,d\mu \leqslant \|u\|_1 \cdot \|v\|_\infty \tag{12.22}$$

holds for all $u \in \mathcal{L}^1(\mu)$ and $v \in \mathcal{L}^\infty(\mu)$.

12.4. Generalized Hölder inequality. Iterate Hölder's inequality to derive the following generalization:

$$\int |u_1 \cdot u_2 \cdot \ldots \cdot u_N| \, d\mu \leqslant \|u_1\|_{p_1} \cdot \|u_2\|_{p_2} \cdot \ldots \cdot \|u_N\|_{p_N} \qquad (12.23)$$

for all $p_j \in (1, \infty)$ such that $\sum_{j=1}^{N} p_j^{-1} = 1$ and all measurable $u_j \in \mathcal{M}$.

12.5. Young functions. Let $\phi : [0, \infty) \to [0, \infty)$ be a strictly increasing continuous function such that $\phi(0) = 0$ and $\lim_{\xi \to \infty} \phi(\xi) = \infty$. Denote by $\psi(\eta) := \phi^{-1}(\eta)$ the inverse function. The functions

$$\Phi(A) := \int_{[0, A)} \phi(\xi) \, \lambda^1(d\xi) \qquad \text{and} \qquad \Psi(B) := \int_{[0, B)} \psi(\eta) \, \lambda^1(d\eta) \qquad (12.24)$$

are called *conjugate Young functions*. Adapt the proof of L12.1 to show the following general *Young's inequality*:

$$AB \leqslant \Phi(A) + \Psi(B) \qquad \forall A, B \geqslant 0. \qquad (12.25)$$

[Hint: interpret $\Phi(A)$ and $\Psi(B)$ as areas below the graph of $\phi(\xi)$, resp. $\psi(\eta)$.]

12.6. Let $1 \leqslant p < \infty$ and $u, u_k \in \mathcal{L}^p(\mu)$ such that $\sum_{k=1}^{\infty} \|u - u_k\|_p < \infty$. Show that $\lim_{k \to \infty} u_k(x) = u(x)$ almost everywhere.

[Hint: mimic the proof of the Riesz–Fischer theorem using $\sum_j (u_{j+1} - u_j)$.]

12.7. Consider one-dimensional Lebesgue measure on $[0, 1]$. Verify that the sequence $u_n(x) := n \mathbf{1}_{(0, 1/n)}(x)$, $n \in \mathbb{N}$, converges pointwise to the function $u \equiv 0$, but that no subsequence of u_n converges in $\mathcal{L}^p$-sense for any $p \geqslant 1$.

12.8. Let $p, q \in [1, \infty]$ be conjugate indices, i.e. $p^{-1} + q^{-1} = 1$ and assume that $(u_k)_{k \in \mathbb{N}} \subset \mathcal{L}^p$ and $(w_k)_{k \in \mathbb{N}} \subset \mathcal{L}^q$ are sequences with limits u and w in $\mathcal{L}^p$, resp. $\mathcal{L}^q$-sense. Show that $u_k w_k$ converges in $\mathcal{L}^1$ to the function uw.

12.9. Prove that $(u_j)_{j \in \mathbb{N}} \subset \mathcal{L}^2$ converges in $\mathcal{L}^2$ if, and only if, $\lim_{n, m \to \infty} \int u_n u_m \, d\mu$ exists.

[Hint: verify and use the identity $\|u - w\|_2^2 = \|u\|_2^2 + \|w\|_2^2 - 2 \int uw \, d\mu$.]

12.10. Let $(X, \mathcal{A}, \mu)$ be a finite measure space. Show that every measurable $u \geqslant 0$ with $\int \exp(hu(x)) \, \mu(dx) < \infty$ for some $h > 0$ is in $\mathcal{L}^p$ for every $p \geqslant 1$.

[Hint: check that $|t|^N / N! \leqslant e^{|t|}$ implies $u \in \mathcal{L}^N$, $N \in \mathbb{N}$; then use Problem 12.1.]

12.11. Let λ be Lebesgue measure in $(0, \infty)$ and $p, q \geqslant 1$ arbitrary.

   (i) Show that $u_n(x) := n^\alpha (x + n)^{-\beta}$ ($\alpha \in \mathbb{R}$, $\beta > 1$) is for every $n \in \mathbb{N}$ in $\mathcal{L}^p(\lambda)$.

   (ii) Show that $v_n(x) := n^\gamma e^{-nx}$ ($\gamma \in \mathbb{R}$) is for every $n \in \mathbb{N}$ in $\mathcal{L}^q(\lambda)$.

12.12. Let $u(x) = (x^\alpha + x^\beta)^{-1}$, $x, \alpha, \beta > 0$. For which $p \geqslant 1$ is $u \in \mathcal{L}^p(\lambda^1, (0, \infty))$?

12.13. Consider the measure space $(\Omega = \{1, 2, \ldots, n\}, \mathcal{P}(\Omega), \mu)$, $n \geqslant 2$ where μ is the counting measure. Show that $\left(\sum_{j=1}^{n} |x_j|^p\right)^{1/p}$ is a norm if $p \in [1, \infty)$, but not for $p \in (0, 1)$.

[Hint: you can identify $\mathcal{L}^p(\mu)$ with $\mathbb{R}^n$.]

12.14. Let $(X, \mathcal{A}, \mu)$ be a measure space. The space $\mathcal{L}^p(\mu)$ is called *separable*, if there exists a countable dense subset $\mathcal{D}_p \subset \mathcal{L}^p(\mu)$. Show that $\mathcal{L}^p(\mu)$, $p \in (1, \infty)$, is separable if, and only if, $\mathcal{L}^1(\mu)$ is separable.

[Hint: use Riesz's convergence theorem 12.10.]

12.15. Let $u_n \in \mathcal{L}^p$, $p \geqslant 1$, for all $n \in \mathbb{N}$. What can you say about u and w if you know that $\lim_{n\to\infty} \int |u_n - u|^p \, d\mu = 0$ and $\lim_{n\to\infty} u_n(x) = w(x)$ for almost every x?

12.16. Let $(X, \mathcal{A}, \mu)$ be a finite measure space and let $u \in \mathcal{L}^1$ be strictly positive with $\int u \, d\mu = 1$. Show that

$$\int (\log u) \, d\mu \leqslant \mu(X) \log \frac{1}{\mu(X)}.$$

12.17. Let u be a positive measurable function on $[0, 1]$. Which of the following is larger:

$$\int_{(0,1)} u(x) \log u(x) \, \lambda(dx) \qquad \text{or} \qquad \int_{(0,1)} u(s) \, \lambda(ds) \cdot \int_{(0,1)} \log u(t) \, \lambda(dt)?$$

[Hint: show that $\log x \leqslant x \log x$, $x > 0$, and assume first that $\int u \, d\lambda = 1$, then consider $u/\int u \, d\lambda$.]

12.18. Let $(X, \mathcal{A}, \mu)$ be a measure space and $p \in (0, 1)$. The conjugate index is given by $q := 1/(p-1) < 0$. Prove for all measurable $u, v, w : X \to (0, \infty)$ with $\int u^p \, d\mu, \int v^p \, d\mu < \infty$ and $0 < \int w^q \, d\mu < \infty$ the inequalities

$$\int uw \, d\mu \geqslant \left(\int u^p \, d\mu \right)^{1/p} \left(\int w^q \, d\mu \right)^{1/q}$$

and

$$\left(\int (u+v)^p \, d\mu \right)^{1/p} \geqslant \left(\int u^p \, d\mu \right)^{1/p} + \left(\int v^p \, d\mu \right)^{1/p}.$$

[Hint: consider Hölder's inequality for u and $1/w$.]

12.19. Let $(X, \mathcal{A}, \mu)$ be a finite measure space and $u \in \mathcal{M}(\mathcal{A})$ be a bounded function with $\|u\|_\infty > 0$. Prove that for all $n \in \mathbb{N}$:

(i) $M_n := \int |u|^n \, d\mu \in (0, \infty)$;
(ii) $M_{n+1} M_{n-1} \geqslant M_n^2$;
(iii) $\mu(X)^{-1/n} \|u\|_n \leqslant M_{n+1}/M_n \leqslant \|u\|_\infty$;
(iv) $\lim_{n\to\infty} M_{n+1}/M_n = \|u\|_\infty$.

[Hint: (ii) – use Hölder's inequality; (iii) – use Jensen's inequality for the lower estimate, Hölder's inequality for the upper estimate; (iv) – observe that $\int u^n \, d\mu \geqslant \int_{\{u > \|u\|_\infty - \epsilon\}} (\|u\|_\infty - \epsilon)^n \, d\mu = \mu(\{u > \|u\|_\infty - \epsilon\}) (\|u\|_\infty - \epsilon)^n$, take the nth root and let $n \to \infty$.]

12.20. Let $(X, \mathcal{A}, \mu)$ be a general measure space and let $u \in \bigcap_{p \geqslant 1} \mathcal{L}^p(\mu)$. Then

$$\lim_{p\to\infty} \|u\|_p = \|u\|_\infty$$

where $\|u\|_\infty = \infty$ if u is unbounded.

[Hint: start with $\|u\|_\infty < \infty$. Show that for any sequence $q_n \to \infty$ one has $\|u\|_{p+q_n} \leqslant \|u\|_\infty^{q_n/(p+q_n)} \cdot \|u\|_p^{p/(p+q_n)}$ and conclude that $\limsup_{p\to\infty} \|u\|_p \leqslant \|u\|_\infty$. The other estimate follows from $\|u\|_p \geqslant \mu(\{|u| > (1-\epsilon)\|u\|_\infty\})^{1/p} (1-\epsilon)\|u\|_\infty$ and $p \to \infty, \epsilon \to 0$, see also the hint to Problem 12.19, where $\mu(\ldots)$ is finite in view of the Markov inequality.

If $\|u\|_\infty = \infty$, use part one of the hint and observe that

$$\liminf_{p\to\infty} \sup_{k\in\mathbb{N}} \||u| \wedge k\|_p \geqslant \sup_{k\in\mathbb{N}} \lim_{p\to\infty} \||u| \wedge k\|_p = \sup_{k\in\mathbb{N}} \||u| \wedge k\|_\infty$$

$$= \sup_{k\in\mathbb{N}} \sup_x (|u(x)| \wedge k) = \sup_x \sup_{k\in\mathbb{N}} (|u(x)| \wedge k)$$

$$= \|u\|_\infty = \infty.]$$

12.21. Let $(X, \mathcal{A}, \mu)$ be a measure space and $1 \leqslant p < \infty$. Show that $f \in \mathcal{E} \cap \mathcal{L}^p(\mu)$ if, and only if, $f \in \mathcal{E}$ and $\mu(\{f \neq 0\}) < \infty$. In particular, $\mathcal{E} \cap \mathcal{L}^p(\mu) = \mathcal{E} \cap \mathcal{L}^1(\mu)$.

12.22. Use Jensen's inequality (12.18) to derive Hölder's and Minkowski's inequalities. Instructions: use

$$\Lambda(x) = x^{1/q}, \ x \geqslant 0, \quad w = |f|^p \quad \text{and} \quad u = |g|^q |f|^{-p} \mathbf{1}_{\{f \neq 0\}}$$

for Hölder's inequality and

$$\Lambda(x) = (x^{1/p} + 1)^p, \ x \geqslant 0, \quad w = |f|^p \mathbf{1}_{\{f \neq 0\}} \quad \text{and} \quad u = |f|^{-p} |g|^p \mathbf{1}_{\{f \neq 0\}}$$

for Minkowski's inequality.

13

Product measures and Fubini's theorem

Lebesgue measure on $\mathbb{R}^n$ has, inherent in its definition, an interesting additional property: if $n > d \geqslant 1$

$$
\begin{aligned}
\lambda^n\big([a_1, b_1) \times \cdots \times [a_n, b_n)\big) & \\
= (b_1 - a_1) \cdot \ldots \cdot (b_d - a_d) \cdot (b_{d+1} - a_{d+1}) \cdot \ldots \cdot (b_n - a_n) & \quad (13.1) \\
= \lambda^d\big([a_1, b_1) \times \cdots \times [a_d, b_d)\big) \cdot \lambda^{n-d}\big([a_{d+1}, b_{d+1}) \times \cdots \times [a_n, b_n)\big), &
\end{aligned}
$$

i.e. it is – at least for rectangles – the product of Lebesgue measures in lower-dimensional spaces. In this chapter we will see that (13.1) remains true for any product $A \times B$ of sets $A \in \mathcal{B}(\mathbb{R}^d)$ and $B \in \mathcal{B}(\mathbb{R}^{n-d})$. More importantly, we will prove the following version of Cavalieri's principle

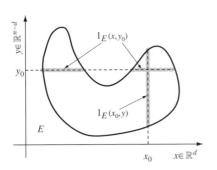

$$
\begin{aligned}
\lambda^n(E) &= \int \mathbf{1}_E \, d\lambda^n \\
&= \int \left[\int \mathbf{1}_E(x, y_0) \, \lambda^d(dx) \right] \lambda^{n-d}(dy_0) \\
&= \int \left[\int \mathbf{1}_E(x_0, y) \, \lambda^{n-d}(dy) \right] \lambda^d(dx_0)
\end{aligned}
$$

which just says that we carve up the set $E \subset \mathbb{R}^n$ horizontally or vertically, measure the volume of the slices and 'sum' them up along the other direction to get the volume of the whole set E.

Clearly, we should be careful about the measurability of products of sets. Recall the following simple rules for Cartesian products of sets $A, A', A_i \subset X, i \in I$, and $B, B' \subset Y$:

$$\left(\bigcup_{i \in I} A_i\right) \times B = \bigcup_{i \in I} (A_i \times B),$$

$$\left(\bigcap_{i \in I} A_i\right) \times B = \bigcap_{i \in I} (A_i \times B),$$

$$(A \times B) \cap (A' \times B') = (A \cap A') \times (B \cap B'), \qquad (13.2)$$

$$A^c \times B = (X \times B) \setminus (A \times B),$$

$$A \times B \subset A' \times B' \iff A \subset A' \text{ and } B \subset B',$$

which are easily derived from the formula

$$A \times B = (A \times Y) \cap (X \times B) = \pi_1^{-1}(A) \cap \pi_2^{-1}(B),$$

where $\pi_1 : X \times Y \to X$ and $\pi_2 : X \times Y \to Y$ are the coordinate projections, and the compatibility of inverse mappings and set operations. To treat measurability, we assume throughout this chapter that

$(X, \mathcal{A}, \mu)$ and $(Y, \mathcal{B}, \nu)$ are σ-finite measure spaces.

Following (13.1) we want to define a measure ρ on rectangles of the form $A \times B$ such that $\rho(A \times B) = \mu(A)\nu(B)$ for $A \in \mathcal{A}$ and $B \in \mathcal{B}$. The first problem which we encounter is that the family

$$\mathcal{A} \times \mathcal{B} := \{A \times B : A \in \mathcal{A}, B \in \mathcal{B}\} \qquad (13.3)$$

is, in general, no σ-algebra.

13.1 Lemma *Let $\mathcal{A}$ and $\mathcal{B}$ be two σ-algebras (or only semi-rings). Then $\mathcal{A} \times \mathcal{B}$ is a semi-ring.*[1]

Proof Literally the same as the induction step in the proof of P6.4. ∎

13.2 Definition Let $(X, \mathcal{A})$ and $(Y, \mathcal{B})$ be two measurable spaces. Then $\mathcal{A} \otimes \mathcal{B} := \sigma(\mathcal{A} \times \mathcal{B})$ is called a *product σ-algebra*, and $(X \times Y, \mathcal{A} \otimes \mathcal{B})$ is the *product of measurable spaces*.

The following lemma is quite useful since it allows us to reduce considerations for $\mathcal{A} \otimes \mathcal{B}$ to generators $\mathcal{F}$ and $\mathcal{G}$ of $\mathcal{A}$ and $\mathcal{B}$ – just as we did in (13.1).

---

[1] See (S_1)–(S_3) on p. 37 for the definition of a semi-ring.

13.3 Lemma *If* $\mathcal{A} = \sigma(\mathcal{F})$ *and* $\mathcal{B} = \sigma(\mathcal{G})$ *and if* $\mathcal{F}, \mathcal{G}$ *contain exhausting sequences* $(F_j)_{j \in \mathbb{N}} \subset \mathcal{F}$, $F_j \uparrow X$ *and* $(G_j)_{j \in \mathbb{N}} \subset \mathcal{G}$, $G_j \uparrow Y$, *then*

$$\sigma(\mathcal{F} \times \mathcal{G}) = \sigma(\mathcal{A} \times \mathcal{B}) \stackrel{def}{=} \mathcal{A} \otimes \mathcal{B}.$$

Proof Since $\mathcal{F} \times \mathcal{G} \subset \mathcal{A} \times \mathcal{B}$ we have $\sigma(\mathcal{F} \times \mathcal{G}) \subset \mathcal{A} \otimes \mathcal{B}$. On the other hand, the system

$$\Sigma := \left\{ A \in \mathcal{A} : A \times G \in \sigma(\mathcal{F} \times \mathcal{G}) \quad \forall G \in \mathcal{G} \right\}$$

is a σ-algebra: Let $A, A_j \in \Sigma$, $j \in \mathbb{N}$, and $G \in \mathcal{G}$; (Σ_1) follows from

$$X \times G = \bigcup_{j \in \mathbb{N}} \underbrace{(F_j \times G)}_{\in \mathcal{F} \times \mathcal{G}} \in \sigma(\mathcal{F} \times \mathcal{G}),$$

(Σ_2) from $A^c \times G = (X \times G) \setminus (A \times G) \in \sigma(\mathcal{F} \times \mathcal{G})$, and (Σ_3) from

$$\left(\bigcup_{j \in \mathbb{N}} A_j \right) \times G = \bigcup_{j \in \mathbb{N}} \underbrace{(A_j \times G)}_{\in \sigma(\mathcal{F} \times \mathcal{G})} \in \sigma(\mathcal{F} \times \mathcal{G}).$$

Obviously, $\mathcal{F} \subset \Sigma \subset \mathcal{A}$, and therefore $\Sigma = \mathcal{A}$; by the very definition of Σ we conclude that $\mathcal{A} \times \mathcal{G} \subset \sigma(\mathcal{F} \times \mathcal{G})$. A similar consideration shows $\mathcal{F} \times \mathcal{B} \subset \sigma(\mathcal{F} \times \mathcal{G})$. This means that for all $A \in \mathcal{A}$ and $B \in \mathcal{B}$

$$A \times B = (A \times X) \cap (Y \times B) = \bigcup_{j,k \in \mathbb{N}} \underbrace{(A \times G_k)}_{\in \sigma(\mathcal{F} \times \mathcal{G})} \cap \underbrace{(F_j \times B)}_{\in \sigma(\mathcal{F} \times \mathcal{G})},$$

so that $\mathcal{A} \times \mathcal{B} \subset \sigma(\mathcal{F} \times \mathcal{G})$ and thus $\mathcal{A} \otimes \mathcal{B} \subset \sigma(\mathcal{F} \times \mathcal{G})$. ∎

If the generators $\mathcal{F}$, $\mathcal{G}$ are rich enough, we have not too many choices of measures ρ with $\rho(F \times G) = \mu(F)\nu(G)$. In fact,

13.4 Theorem (Uniqueness of product measures) *Let* $(X, \mathcal{A}, \mu)$ *and* $(Y, \mathcal{B}, \nu)$ *be two measure spaces and assume that* $\mathcal{A} = \sigma(\mathcal{F})$ *and* $\mathcal{B} = \sigma(\mathcal{G})$. *If*

- $\mathcal{F}, \mathcal{G}$ *are* $\cap$-*stable*,
- $\mathcal{F}, \mathcal{G}$ *contain exhausting sequences* $F_j \uparrow X$ *and* $G_k \uparrow Y$ *with* $\mu(F_j) < \infty$ *and* $\nu(G_k) < \infty$ *for all* $j, k \in \mathbb{N}$,

then there is at most one measure ρ *on* $(X \times Y, \mathcal{A} \otimes \mathcal{B})$ *satisfying*

$$\rho(F \times G) = \mu(F)\nu(G) \qquad \forall F \in \mathcal{F}, \ G \in \mathcal{G}.$$

Proof By Lemma 13.3 $\mathcal{F} \times \mathcal{G}$ generates $\mathcal{A} \otimes \mathcal{B}$. Moreover, $\mathcal{F} \times \mathcal{G}$ inherits the $\cap$-stability of $\mathcal{F}$ and $\mathcal{G}^{[\checkmark]}$, the sequence $F_j \times G_j$ increases towards $X \times Y$ and $\rho(F_j \times G_j) = \mu(F_j)\nu(G_j) < \infty$. These were the assumptions of the uniqueness theorem 5.7, showing that there is at most one such product measure ρ.   ∎

As so often, it is the existence which is more difficult than uniqueness.

13.5 Theorem (Existence of product measures) *Let* $(X, \mathcal{A}, \mu)$ *and* $(Y, \mathcal{B}, \nu)$ *be* σ-*finite measure spaces. Then the set-function*

$$\rho : \mathcal{A} \times \mathcal{B} \to [0, \infty], \quad \rho(A \times B) := \mu(A)\nu(B),$$

extends uniquely to a σ-*finite measure on* $(X \times Y, \mathcal{A} \otimes \mathcal{B})$ *such that*

$$\rho(E) = \iint \mathbf{1}_E(x, y) \, \mu(dx) \, \nu(dy) = \iint \mathbf{1}_E(x, y) \, \nu(dy) \, \mu(dx) \qquad (13.4)$$

holds[2] *for all* $E \in \mathcal{A} \otimes \mathcal{B}$. *In particular, the functions*

$$x \mapsto \mathbf{1}_E(x, y), \ y \mapsto \mathbf{1}_E(x, y), \ x \mapsto \int \mathbf{1}_E(x, y) \, \nu(dy), \ y \mapsto \int \mathbf{1}_E(x, y) \, \mu(dx)$$

are $\mathcal{A}$, *resp.* $\mathcal{B}$-*measurable for every fixed* $y \in Y$, *resp.* $x \in X$.

Proof *Uniqueness* of ρ follows from T13.4. *Existence*: Let $(A_j)_{j \in \mathbb{N}}$, $(B_j)_{j \in \mathbb{N}}$ be sequences in $\mathcal{A}$ resp. $\mathcal{B}$ with $A_j \uparrow X$, $B_j \uparrow Y$ and $\mu(A_j), \nu(B_j) < \infty$. Clearly, $E_j := A_j \times B_j \uparrow X \times Y$.

For every $j \in \mathbb{N}$ we consider the family $\mathcal{D}_j$ of all subsets $D \subset X \times Y$ satisfying the following conditions:

- $x \mapsto \mathbf{1}_{D \cap E_j}(x, y)$ and $y \mapsto \mathbf{1}_{D \cap E_j}(x, y)$ are measurable,
- $x \mapsto \int \mathbf{1}_{D \cap E_j}(x, y) \, \nu(dy)$ and $y \mapsto \int \mathbf{1}_{D \cap E_j}(x, y) \, \mu(dx)$ are measurable,
- $\iint \mathbf{1}_{D \cap E_j}(x, y) \, \mu(dx) \, \nu(dy) = \iint \mathbf{1}_{D \cap E_j}(x, y) \, \nu(dy) \, \mu(dx)$.

That $\mathcal{A} \times \mathcal{B} \subset \mathcal{D}_j$ follows from

$$\iint \mathbf{1}_{(A \times B) \cap E_j}(x, y) \, \mu(dx) \, \nu(dy) = \iint \mathbf{1}_{A \cap A_j}(x) \mathbf{1}_{B \cap B_j}(y) \, \mu(dx) \, \nu(dy)$$

$$= \mu(A \cap A_j) \int \mathbf{1}_{B \cap B_j}(y) \, \nu(dy)$$

$$= \mu(A \cap A_j)\nu(B \cap B_j)$$

$$= \ldots = \iint \mathbf{1}_{(A \times B) \cap E_j}(x, y) \, \nu(dy) \, \mu(dx),$$

---

[2] We use the symbols $\int \ldots d\mu$ like brackets, i.e. $\iint \ldots d\mu \, d\nu = \int \left(\int \ldots d\mu \right) d\nu$.

where the ellipsis ... stands for the same calculations run through backwards. In each step the measurability conditions needed to perform the integrations are fulfilled because of the product structure.[✓] In particular, $X \times Y, \emptyset, E_k \in \mathcal{D}_j$. If $D \in \mathcal{D}_j$, then $\mathbf{1}_{D^c \cap E_j} = \mathbf{1}_{E_j} - \mathbf{1}_{E_j \cap D}$ and

$$\iint \mathbf{1}_{D^c \cap E_j}(x, y)\, \mu(dx)\, \nu(dy)$$

$$= \int \left(\int \mathbf{1}_{E_j}(x, y)\, \mu(dx) - \int \mathbf{1}_{E_j \cap D}(x, y)\, \mu(dx) \right) \nu(dy)$$

$$= \iint \mathbf{1}_{E_j}(x, y)\, \mu(dx)\, \nu(dy) - \iint \mathbf{1}_{E_j \cap D}(x, y)\, \mu(dx)\, \nu(dy)$$

$$= \iint \mathbf{1}_{E_j}(x, y)\, \nu(dy)\, \mu(dx) - \iint \mathbf{1}_{E_j \cap D}(x, y)\, \nu(dy)\, \mu(dx)$$

(by definition, since $E_j, D \in \mathcal{D}_j$)

$$= \ldots = \iint \mathbf{1}_{D^c \cap E_j}(x, y)\, \nu(dy)\, \mu(dx).$$

Again, in each step the measurability conditions hold since measurable functions form a vector space. If $(D_k)_{k \in \mathbb{N}} \subset \mathcal{D}_j$ are mutually disjoint sets, $D := \bigcup_{k \in \mathbb{N}} D_k$, the linearity of the integral and Beppo Levi's theorem in the form of C9.9 show that

$$\iint \mathbf{1}_{D \cap E_j}(x, y)\, \mu(dx)\, \nu(dy) = \int \left(\sum_{k=1}^{\infty} \int \mathbf{1}_{D_k \cap E_j}(x, y)\, \mu(dx) \right) \nu(dy)$$

$$= \sum_{k=1}^{\infty} \iint \mathbf{1}_{D_k \cap E_j}(x, y)\, \mu(dx)\, \nu(dy)$$

$$= \sum_{k=1}^{\infty} \iint \mathbf{1}_{D_k \cap E_j}(x, y)\, \nu(dy)\, \mu(dx)$$

(by definition, since $D_k \in \mathcal{D}_j$)

$$= \ldots = \iint \mathbf{1}_{D \cap E_j}(x, y)\, \nu(dy)\, \mu(dx)$$

and the measurability conditions hold since measurability is preserved under sums and increasing limits.

The last three calculations show that $\mathcal{D}_j$ is a Dynkin system containing the $\cap$-stable family $\mathcal{A} \times \mathcal{B}$. By Theorem 5.5, $\mathcal{A} \otimes \mathcal{B} \subset \mathcal{D}_j$ for every $j \in \mathbb{N}$. Since $E_j \uparrow X \times Y$, Beppo Levi's theorem 9.6 proves (13.4) along with the measurability of the functions $\mathbf{1}_E(\bullet, y)$, $\mathbf{1}_E(x, \bullet)$, $\int \mathbf{1}_E(\bullet, y)\, \nu(dy)$ and $\int \mathbf{1}_E(x, \bullet)\, \mu(dx)$ since $\mathcal{M}$ is stable under pointwise limits.

Replacing in the above calculations E_j by $X \times Y$ finally proves that

$$E \mapsto \rho(E) := \iint \mathbf{1}_E(x, y) \, \mu(dx) \, \nu(dy)$$

is indeed a measure on $(X \times Y, \mathcal{A} \otimes \mathcal{B})$ with $\rho(A \times B) = \mu(A)\nu(B)$. ∎

13.6 Definition Let $(X, \mathcal{A}, \mu)$ and $(Y, \mathcal{B}, \nu)$ be σ-finite measure spaces. The unique measure ρ constructed in Theorem 13.5 is called the *product* of the measures μ and ν, denoted by $\mu \times \nu$. $(X \times Y, \mathcal{A} \otimes \mathcal{B}, \mu \times \nu)$ is called the *product measure space*.

Returning to the example considered at the beginning we find

13.7 Corollary *If $n > d \geqslant 1$,*

$$\left(\mathbb{R}^n, \mathcal{B}(\mathbb{R}^n), \lambda^n\right) = \left(\mathbb{R}^d \times \mathbb{R}^{n-d}, \mathcal{B}(\mathbb{R}^d) \otimes \mathcal{B}(\mathbb{R}^{n-d}), \lambda^d \times \lambda^{n-d}\right).$$

The next step is to see how we can integrate w.r.t. $\mu \times \nu$. The following two results are often stated together as the Fubini or Fubini–Tonelli theorem. We prefer to distinguish between them since the first result, Theorem 13.8, says that we can *always swap iterated integrals of positive functions* (even if we get $+\infty$), whereas 13.9 applies to more general functions but requires the (iterated) integrals to be finite.

13.8 Theorem (Tonelli) *Let $(X, \mathcal{A}, \mu)$ and $(Y, \mathcal{B}, \nu)$ be σ-finite measure spaces and let $u : X \times Y \to [0, \infty]$ be $\mathcal{A} \otimes \mathcal{B}$-measurable. Then*

(i) $x \mapsto u(x, y)$, $y \mapsto u(x, y)$ *are $\mathcal{A}$, resp. $\mathcal{B}$-measurable for all $y \in Y$, resp. $x \in X$;*

(ii) $x \mapsto \displaystyle\int_Y u(x, y) \, \nu(dy)$, $y \mapsto \displaystyle\int_X u(x, y) \, \mu(dx)$ *are $\mathcal{A}$, resp. $\mathcal{B}$-measurable;*

(iii) $\displaystyle\int_{X \times Y} u \, d(\mu \times \nu) = \iint_{YX} u(x, y) \, \mu(dx) \, \nu(dy) = \iint_{XY} u(x, y) \, \nu(dy) \, \mu(dx)$
with values in $[0, \infty]$.

Proof Since u is positive and $\mathcal{A} \otimes \mathcal{B}$-measurable, we find an increasing sequence of simple functions $f_j \in \mathcal{E}^+(\mathcal{A} \otimes \mathcal{B})$ with $\sup_{j \in \mathbb{N}} f_j = u$. Each f_j is of the form $f_j(x, y) = \sum_{k=0}^{N(j)} \alpha_k \mathbf{1}_{E_k}(x, y)$, where $\alpha_k \geqslant 0$ and the $E_k \in \mathcal{A} \otimes \mathcal{B}$, $0 \leqslant k \leqslant N(j)$, are disjoint. By Theorem 13.5, the fact that $\mathcal{M}(\mathcal{A} \otimes \mathcal{B})$ is a vector space and the linearity of the integral we conclude that

$$x \mapsto f_j(x, y), \quad y \mapsto f_j(x, y), \quad x \mapsto \int_Y f_j(x, y)\,\nu(dy), \quad y \mapsto \int_X f_j(x, y)\,\mu(dx)$$

are measurable functions and (i), (ii) follow from the usual Beppo-Levi argument since $\mathcal{M}(\mathcal{A})$ and $\mathcal{M}(\mathcal{B})$ are stable under increasing limits, cf. C8.9. Linearity of the integral and Theorem 13.5 also show

$$\int_{X \times Y} f_j\,d(\mu \times \nu) = \iint_{YX} f_j\,d\mu\,d\nu = \iint_{XY} f_j\,d\nu\,d\mu \qquad \forall j \in \mathbb{N},$$

and (iii) follows from several applications of Beppo Levi's theorem 9.6.   ∎

13.9 Corollary (Fubini's theorem) *Let $(X, \mathcal{A}, \mu)$ and $(Y, \mathcal{B}, \nu)$ be σ-finite measure spaces and let $u : X \times Y \to \bar{\mathbb{R}}$ be $\mathcal{A} \otimes \mathcal{B}$-measurable. If at least one of the following three integrals is finite*

$$\int_{X \times Y} |u|\,d(\mu \times \nu), \quad \iint_{YX} |u(x, y)|\,\mu(dx)\,\nu(dy), \quad \iint_{XY} |u(x, y)|\,\nu(dy)\,\mu(dx),$$

then all three integrals are finite, $u \in \mathcal{L}^1(\mu \times \nu)$, and

(i) *$x \mapsto u(x, y)$ is in $\mathcal{L}^1(\mu)$ for ν-a.e. $y \in Y$;*
(ii) *$y \mapsto u(x, y)$ is in $\mathcal{L}^1(\nu)$ for μ-a.e. $x \in X$;*
(iii) *$y \mapsto \displaystyle\int_X u(x, y)\,\mu(dx)$ is in $\mathcal{L}^1(\nu)$;*
(iv) *$x \mapsto \displaystyle\int_Y u(x, y)\,\nu(dy)$ is in $\mathcal{L}^1(\mu)$;*
(v) *$\displaystyle\int_{X \times Y} u\,d(\mu \times \nu) = \iint_{YX} u(x, y)\,\mu(dx)\,\nu(dy) = \iint_{XY} u(x, y)\,\nu(dy)\,\mu(dx)$.*

Proof Tonelli's theorem 13.8 shows that in $[0, \infty]$

$$\int_{X \times Y} |u|\,d(\mu \times \nu) = \iint_{YX} |u|\,d\mu\,d\nu = \iint_{XY} |u|\,d\nu\,d\mu. \qquad (13.5)$$

If one of the integrals is finite, all of them are finite and $u \in \mathcal{L}^1(\mu \times \nu)$ follows. Again by Tonelli's theorem, $x \mapsto u^{\pm}(x, y)$ is $\mathcal{A}$-measurable and $y \mapsto \int u^{\pm}(x, y)\,\mu(dx)$ is $\mathcal{B}$-measurable. Since $u^{\pm} \leqslant |u|$, (13.5) and C10.13 show that

$$\int_X u^{\pm}(x, y)\,\mu(dx) \leqslant \int_X |u(x, y)|\,\mu(dx) < \infty \qquad \text{for } \nu\text{-a.e. } y \in Y$$

and

$$\iint\limits_{Y\,X} u^{\pm}(x,y)\,\mu(dx)\,\nu(dy) \leqslant \iint\limits_{Y\,X} |u(x,y)|\,\mu(dx)\,\nu(dy) < \infty.$$

This proves (i) and (iii); (ii) and (iv) are shown in a similar way. Finally, (v) follows for u^+ and u^- from Theorem 13.8 and for $u = u^+ - u^-$ by linearity, since (i)–(iv) exclude the possibility of '$\infty - \infty$'. ∎

More on measurable functions

There is an alternative way to introduce the product σ-algebra $\mathcal{A} \otimes \mathcal{B}$. Recall that the coordinate projections

$$\pi_j : X_1 \times X_2 \to X_j, \quad (x_1, x_2) \mapsto x_j, \quad j = 1, 2,$$

induce the σ-algebra $\sigma(\pi_1, \pi_2)$ on $X_1 \times X_2$ which is by Definition 7.5 the smallest σ-algebra such that both π_1 and π_2 are measurable maps.

13.10 Theorem *Let $(X_j, \mathcal{A}_j)$, $j = 1, 2$, and $(Z, \mathcal{C})$ be measurable spaces. Then*

(i) $\mathcal{A}_1 \otimes \mathcal{A}_2 = \sigma(\pi_1, \pi_2)$;

(ii) *$T : Z \to X_1 \times X_2$ is $\mathcal{C}/\mathcal{A}_1 \otimes \mathcal{A}_2$-measurable if, and only if, $\pi_j \circ T$ is $\mathcal{C}/\mathcal{A}_j$-measurable $(j = 1, 2)$;*

(iii) *if $S : X_1 \times X_2 \to Z$ is measurable, then $S(x_1, \cdot)$ and $S(\cdot, x_2)$ are $\mathcal{A}_2/\mathcal{C}$- resp. $\mathcal{A}_1/\mathcal{C}$-measurable for every $x_1 \in X_1$, resp. $x_2 \in X_2$.*

Proof **(i)** Since $\pi_1^{-1}(x) = \{x\} \times X_2$, $\pi_2^{-1}(y) = X_1 \times \{y\}$ and $A_1 \times A_2 = (A_1 \times Y) \cap (X \times A_2)$, we have

$$\sigma(\pi_1, \pi_2) \overset{7.5}{=} \sigma(\pi_1^{-1}(\mathcal{A}_1), \pi_2^{-1}(\mathcal{A}_2)) = \sigma\left(\{A_1 \times X_2, X_1 \times A_2 : A_j \in \mathcal{A}_j\}\right),$$

which shows that $\mathcal{A}_1 \times \mathcal{A}_2 \subset \sigma(\pi_1, \pi_2) \subset \mathcal{A}_1 \otimes \mathcal{A}_2$, hence $\sigma(\pi_1, \pi_2) = \mathcal{A}_1 \otimes \mathcal{A}_2$.

(ii) If $T : Z \to X_1 \times X_2$ is measurable, then so is $\pi_j \circ T$ by part (i) and T7.4. Conversely, if $\pi_j \circ T$, $j = 1, 2$, are measurable we find

$$\begin{aligned} T^{-1}(A_1 \times A_2) &= T^{-1}\left(\pi_1^{-1}(A_1) \cap \pi_2^{-1}(A_2)\right) \\ &= T^{-1}\left(\pi_1^{-1}(A_1)\right) \cap T^{-1}\left(\pi_2^{-1}(A_2)\right) \\ &= (\pi_1 \circ T)^{-1}(A_1) \cap (\pi_2 \circ T)^{-1}(A_2) \in \mathcal{C}. \end{aligned}$$

Since $\mathcal{A}_1 \times \mathcal{A}_2$ generates $\mathcal{A}_1 \otimes \mathcal{A}_2$, T is measurable by L7.2.

(iii) Fix $x_1 \in X_1$ and consider $y \mapsto S(x_1, y)$. Then $S(x_1, \cdot) = S \circ i_{x_1}(\cdot)$, where $i_{x_1} : X_2 \to X_1 \times X_2$, $y \mapsto (x_1, y)$. By part (ii), i_{x_1} is $\mathcal{A}_2/\mathcal{A}_1 \otimes \mathcal{A}_2$-measurable since the maps $\pi_j \circ i_{x_1}(x_2) = x_j$ are $\mathcal{A}_j/\mathcal{A}_j$-measurable ($j = 1, 2$). The claim follows now from T7.4. ∎

Distribution functions

Let $(X, \mathcal{A}, \mu)$ be a σ-finite measure space. For $u \in \mathcal{M}^+(\mathcal{A})$ the decreasing, left-continuous[✓] numerical function

$$\mathbb{R} \ni t \mapsto \mu(\{u \geqslant t\})$$

is called the *distribution function* of u (under μ).

The next theorem shows that Lebesgue integrals still represent the area between the graph of a function and the abscissa.

13.11 Theorem *Let $(X, \mathcal{A}, \mu)$ be a σ-finite measure space and $u : X \to [0, \infty)$ be $\mathcal{A}$-measurable. Then*

$$\int u \, d\mu = \int_{(0,\infty)} \mu(\{u \geqslant t\}) \, \lambda^1(dt) \in [0, \infty]. \tag{13.6}$$

Proof Consider the function $U(x, t) := (u(x), t)$ on $X \times [0, \infty)$. By Theorem 13.10(ii), U is $\mathcal{A} \otimes \mathcal{B}[0, \infty)$-measurable, thus

$$E := \{(x, t) : u(x) \geqslant t\} \in \mathcal{A} \otimes \mathcal{B}[0, \infty).$$

An application of Tonelli's theorem 13.8 shows

$$\int u(x) \, \mu(dx) = \iint \mathbf{1}_{(0, u(x)]}(t) \, \lambda^1(dt) \, \mu(dx)$$

$$= \iint_{X \times (0,\infty)} \mathbf{1}_E(x, t) \, \lambda^1(dt) \, \mu(dx)$$

$$= \iint_{(0,\infty) \times X} \mathbf{1}_E(x, t) \, \mu(dx) \, \lambda^1(dt)$$

$$= \int_{(0,\infty)} \mu(\{u \geqslant t\}) \, \lambda^1(dt). \qquad ∎$$

If $\phi : [0, \infty) \to [0, \infty)$ is continuously differentiable, increasing and $\phi(0) = 0$, we even have in the setting of Theorem 13.11

$$\int \phi \circ u \, d\mu = \int_{(0,\infty)} \mu(\{\phi(u) \geqslant t\}) \, \lambda^1(dt)$$

$$\overset{(*)}{=} \int_0^\infty \mu(\{\phi(u) \geqslant t\}) \, dt$$

$$\overset{t=\phi(s)}{=} \int_0^\infty \phi'(s) \, \mu(\{\phi(u) \geqslant \phi(s)\}) \, ds$$

$$= \int_0^\infty \phi'(s) \, \mu(\{u \geqslant s\}) \, ds.$$

The problem with this calculation is the step marked $(*)$ where we equate a Lebesgue integral with a Riemann integral. By Theorem 11.8(ii) we can do this if $t \mapsto \mu(\{\phi(u) \geqslant t\})$ is Lebesgue a.e. continuous and bounded. Boundedness is not a problem since we may consider $\mu(\{\phi(u) \geqslant t\}) \wedge N$, $N \in \mathbb{N}$, and let $N \to \infty$ using T9.6. For the a.e. continuity we need

13.12 Lemma *Every monotone function* $\phi : \mathbb{R} \to \mathbb{R}$ *has at most countably many discontinuities and is, in particular, Lebesgue a.e. continuous.*

Proof Without loss of generality we may assume that ϕ increases. Therefore, the one-sided limits $\lim_{s \uparrow t} \phi(s) =: \phi(t-) \leqslant \phi(t+) := \lim_{s \downarrow t} \phi(s)$ exist in $\mathbb{R}$, so that ϕ can only have jump discontinuities where $\phi(t-) < \phi(t+)$. Define for all $\epsilon > 0$

$$J^\epsilon := \{t \in \mathbb{R} : \Delta\phi(t) := \phi(t+) - \phi(t-) \geqslant \epsilon\}.$$

Since on every compact interval $[a, b]$ and for every $\epsilon > 0$

$$0 \leqslant \phi(b) - \phi(a) = \frac{\phi(b) - \phi(a)}{\epsilon} \epsilon < \infty,$$

we can have at most $\left[\frac{\phi(b)-\phi(a)}{\epsilon}\right]$ jumps of size ϵ or larger in the interval $[a, b]$, that is $\#([a, b] \cap J^\epsilon) < \infty$. Therefore, the set of all discontinuities of ϕ

$$J := \{t \in \mathbb{R} : \Delta\phi(t) > 0\} = \bigcup_{j,k \in \mathbb{N}} [-j, j] \cap J^{1/k}$$

is a countable set, hence a Lebesgue null set. $\blacksquare$

Since $t \mapsto \mu(\{\phi(u) \geqslant t\})$ is decreasing, we finally have

13.13 Corollary *Let* $(X, \mathcal{A}, \mu)$ *be* σ-*finite and let* $\phi : [0, \infty) \to [0, \infty)$ *with* $\phi(0) = 0$ *be increasing and continuously differentiable. Then*

$$\int \phi \circ u \, d\mu = \int_0^\infty \phi'(s) \mu(\{u \geqslant s\}) \, ds \tag{13.7}$$

holds for all $u \in \mathcal{M}^+(\mathcal{A})$; *the right-hand side is an improper Riemann integral. Moreover,* $\phi \circ u \in \mathcal{L}^1(\mu)$ *if, and only if, this Riemann integral is finite.*

In the important special case where $\phi(t) = t^p$, $p \geqslant 1$, (13.7) reads

$$\|u\|_p^p = \int |u|^p \, d\mu = \int_0^\infty p s^{p-1} \mu(|u| \geqslant s) \, ds. \tag{13.8}$$

Minkowski's inequality for integrals

The following inequality is a generalization of Minkowski's inequality C12.4 to double integrals. In some sense it is also a theorem on the change of the order of iterated integrals, but equality is only obtained if $p = 1$.

13.14 Theorem (Minkowski's inequality for integrals) *Let* $(X, \mathcal{A}, \mu)$ *and* $(Y, \mathcal{B}, \mu)$ *be* σ*-finite measure spaces and* $u : X \times Y \to \bar{\mathbb{R}}$ *be* $\mathcal{A} \otimes \mathcal{B}$*-measurable. Then*

$$\left(\int_X \left(\int_Y |u(x, y)| \, \nu(dy) \right)^p \mu(dx) \right)^{1/p} \leqslant \int_Y \left(\int_X |u(x, y)|^p \, \mu(dx) \right)^{1/p} \nu(dy)$$

holds for all $p \in [1, \infty)$*, with equality for* $p = 1$*.*

Proof If $p = 1$, the assertion follows directly from Tonelli's theorem 13.8. If $p > 1$ we set

$$U_k(x) := \left(\int_Y |u(x, y)| \, \nu(dy) \wedge k \right) \mathbf{1}_{A_k}(x)$$

where $A_k \in \mathcal{A}$ is a sequence with $A_k \uparrow X$ and $\mu(A_k) < \infty$. Without loss of generality we may assume that $U_k(x) > 0$ on a set of positive μ-measure, otherwise the left-hand side of the above inequality would be 0 (using Beppo Levi's theorem 9.6) and there would be nothing to prove. By Tonelli's theorem and Hölder's inequality T12.2 with $\frac{1}{p} + \frac{1}{q} = 1$ or $q = \frac{p}{p-1}$, we find

$$\int_X U_k^p(x) \, \mu(dx) \leqslant \int_X U_k^{p-1}(x) \left(\int_Y |u(x, y)| \, \nu(dy) \right) \mu(dx)$$

$$= \int_Y \int_X U_k^{p-1}(x) \, |u(x, y)| \, \mu(dx) \, \nu(dy)$$

$$\leqslant \int_Y \left(\int_X U_k^p(x) \, \mu(dx) \right)^{1-1/p} \left(\int_X |u(x, y)|^p \, \mu(dx) \right)^{1/p} \nu(dy).$$

The claim follows upon dividing both sides by $\left(\int_X U_k^p(x) \, \mu(dx) \right)^{1-1/p}$ and letting $k \to \infty$ with Beppo Levi's theorem 9.6. ∎

Problems

13.1. Prove the rules (13.2) for Cartesian products.

13.2. Let $(X, \mathcal{A}, \mu)$ and $(Y, \mathcal{B}, \nu)$ be two σ-finite measure spaces. Show that $A \times N$, where $A \in \mathcal{A}$ and $N \in \mathcal{B}$, $\nu(N) = 0$, is a $\mu \times \nu$-null set.

13.3. Denote by λ Lebesgue measure on $(0, \infty)$. Prove that the following iterated integrals exist and that

$$\int_{(0,\infty)} \int_{(0,\infty)} e^{-xy} \sin x \sin y \, \lambda(dx)\lambda(dy) = \int_{(0,\infty)} \int_{(0,\infty)} e^{-xy} \sin x \sin y \, \lambda(dy)\lambda(dx).$$

Does this imply that the double integral exists?

13.4. Denote by λ Lebesgue measure on $(0, 1)$. Show that the following iterated integrals exist, but yield different values:

$$\int_{(0,1)} \int_{(0,1)} \frac{x^2 - y^2}{(x^2 + y^2)^2} \, \lambda(dx)\lambda(dy) \neq \int_{(0,1)} \int_{(0,1)} \frac{x^2 - y^2}{(x^2 + y^2)^2} \, \lambda(dy)\lambda(dx).$$

What does this tell about the double integral?

13.5. Denote by λ Lebesgue measure on $(-1, 1)$. Show that the iterated integrals exist, coincide,

$$\int_{(-1,1)} \int_{(-1,1)} \frac{xy}{(x^2 + y^2)^2} \, \lambda(dx)\lambda(dy) = \int_{(-1,1)} \int_{(-1,1)} \frac{xy}{(x^2 + y^2)^2} \, \lambda(dy)\lambda(dx)$$

but that the double integral does not exist.

13.6. (i) Prove that $\int_{(0,\infty)} e^{-tx} \lambda(dt) = \frac{1}{x}$ for all $x > 0$.
(ii) Use (i) and Fubini's theorem to show that the *sine integral*

$$\lim_{n \to \infty} \int_{(0,n)} \frac{\sin x}{x} \lambda(dx) = \frac{\pi}{2}.$$

13.7. Let $\mu(A) := \#A$ be the counting measure and λ be Lebesgue measure on the measurable space $([0, 1], \mathcal{B}[0, 1])$. Denote by $\Delta := \{(x, y) \in [0, 1]^2 : x = y\}$ the diagonal in $[0, 1]^2$. Check that

$$\int_{[0,1]} \int_{[0,1]} \mathbf{1}_\Delta(x, y) \, \lambda(dx)\mu(dy) \neq \int_{[0,1]} \int_{[0,1]} \mathbf{1}_\Delta(x, y) \, \mu(dy)\lambda(dx).$$

Does this contradict Tonelli's theorem?

13.8. (i) State Tonelli's and Fubini's theorems for spaces of sequences, i.e. for the measure space $(\mathbb{N}, \mathcal{P}(\mathbb{N}), \mu)$ where $\mu := \sum_{j \in \mathbb{N}} \delta_j$, and obtain criteria when one can interchange two infinite summations.
(ii) Using similar considerations as in part (i) deduce the following.
Lemma *Let $(A_j)_j$ be countably many (i.e. a finite or countably infinite number of) mutually disjoint sets whose union is $\mathbb{N}$, and let $(x_k)_{k \in \mathbb{N}} \subset \mathbb{R}$ be a sequence. Then*

$$\sum_{k \in \mathbb{N}} x_k = \sum_j \sum_{k \in A_j} x_k$$

in the sense that if either side converges absolutely, so does the other, in which case both sides are equal.

13.9. Let $u : \mathbb{R}^2 \to [0, \infty]$ be a Borel measurable function. Denote by $S[u] := \{(x, y) : 0 \leq y \leq u(x)\}$ the set above the abscissa and below the graph $\Gamma[u] := \{(x, u(x)) : x \in \mathbb{R}\}$ of u.

(i) Show that $S[u] \in \mathcal{B}(\mathbb{R}^2)$.

(ii) Is it true that $\lambda^2(S[u]) = \int u \, d\lambda^1$?

(iii) Show that $\Gamma[u] \in \mathcal{B}(\mathbb{R}^2)$ and that $\lambda^2(\Gamma[u]) = 0$.

[Hint: (i) – use T8.8 to approximate u by simple functions $f_j \uparrow u$. Thus $S[u] = \bigcup_j S[f_j]$ and $S[f_j] \in \mathcal{B}(\mathbb{R}^2)$ is easy to see; alternatively, use T13.10, set $U(x, y) := (u(x), y)$ and observe that $S[u] = U^{-1}(C)$ for the closed set $C := \{(x, y) : x \geqslant y\}$; (ii) – use Tonelli's theorem; (iii) – use $\Gamma[u] \subset S[u] \setminus S[(u - \epsilon)^+]$ or $\Gamma[u] = U^{-1}(\{(x, y) : x = y\})$; show first that $\lambda^2(\Gamma[u] \cap [-n, n]^2) = 0$ for every $n \in \mathbb{N}$ and observe that $\Gamma[u] \cap [-n, n]^2 = \Gamma[(u \mathbf{1}_{[-n,n]}) \wedge n]$.]

13.10. Let $(X, \mathcal{A}, \mu)$ be a σ-finite measure space and let $u \in \mathcal{M}^+(\mathcal{A})$ be a $[0, \infty]$-valued measurable function. Show that the set

$$Y := \{y \in \mathbb{R} : \mu(\{x : u(x) = y\}) \neq 0\} \subset \mathbb{R}$$

is countable.

[Hint: assume that $u \in \mathcal{L}^1_+(\mu)$. Set $Y_{\epsilon, \eta} := \{y > \eta : \mu(\{u = y\}) > \epsilon\}$ and observe that for $t_1, \ldots, t_N \in Y_{\epsilon, \eta}$ we have $N \epsilon \eta \leqslant \sum_{j=1}^N t_j \mu(\{u = t_j\}) \leqslant \int u \, d\mu$. Thus $Y_{\epsilon, \eta}$ is a finite set, and $Y = \bigcup_{k,n \in \mathbb{N}} Y_{\frac{1}{n}, \frac{1}{k}}$ is countable. If u is not integrable, consider $(u \wedge m) \mathbf{1}_{A_m}$, $m \in \mathbb{N}$, where $A_m \uparrow X$ is an exhaustion.]

13.11. **Completion (5).** Let $(X, \mathcal{A}, \mu)$ and $(Y, \mathcal{B}, \nu)$ be any two measure spaces such that $\mathcal{A} \neq \mathcal{P}(X)$ and such that $\mathcal{B}$ contains non-empty null sets.

(i) Show that $\mu \times \nu$ on $(X \times Y, \mathcal{A} \otimes \mathcal{B})$ is not complete, even if both μ and ν were complete.

(ii) Conclude from (i) that neither $(\mathbb{R}^2, \mathcal{B}(\mathbb{R}) \otimes \mathcal{B}(\mathbb{R}), \lambda \times \lambda)$ nor the product of the completed spaces $(\mathbb{R}^2, \mathcal{B}^*(\mathbb{R}) \otimes \mathcal{B}^*(\mathbb{R}), \bar{\lambda} \times \bar{\lambda})$ are complete.

[Hint: you may assume in (ii) that $\mathcal{B}(\mathbb{R}), \mathcal{B}^*(\mathbb{R}) \neq \mathcal{P}(\mathbb{R})$.]

13.12. Let μ be a bounded measure on the measure space $([0, \infty), \mathcal{B}[0, \infty))$.

(i) Show that $A \in \mathcal{B}[0, \infty) \otimes \mathcal{P}(\mathbb{N})$ if, and only if, $A = \bigcup_{j \in \mathbb{N}} B_j \times \{j\}$, where $(B_j)_{j \in \mathbb{N}} \subset \mathcal{B}[0, \infty)$.

(ii) Show that there exists a unique measure π on $\mathcal{B}[0, \infty) \otimes \mathcal{P}(\mathbb{N})$ satisfying

$$\pi(B \times \{n\}) = \int_B e^{-t} \frac{t^n}{n!} \mu(dt).$$

13.13. **Stieltjes measure (2). Stieltjes integrals.** This continues Problem 7.9. Let μ and ν be two measures on $(\mathbb{R}, \mathcal{B}(\mathbb{R}))$ such that $\mu((-n, n]), \nu((-n, n]) < \infty$ for all $n \in \mathbb{N}$, and denote by

$$F(x) := \begin{cases} \mu((0, x]), & \text{if } x > 0 \\ 0, & \text{if } x = 0 \\ -\mu((x, 0]), & \text{if } x < 0 \end{cases} \quad \text{and} \quad G(x) := \begin{cases} \nu((0, x]), & \text{if } x > 0 \\ 0, & \text{if } x = 0 \\ -\nu((x, 0]), & \text{if } x < 0 \end{cases}$$

the associated *right-continuous* distribution functions (in Problem 7.9 we considered *left*-continuous distribution functions). Moreover, set $\Delta F(x) = F(x) - F(x-)$ and $\Delta G(x) = G(x) - G(x-)$.

(i) Show that F, G are increasing, right-continuous and that $\Delta F(x) = 0$ if, and only if, $\mu(\{x\}) = 0$. Moreover, F and μ are in one-to-one correspondence.

(ii) Since measures and distribution functions are in one-to-one correspondence, it is customary to write $\int u \, d\mu = \int u \, dF$, etc.

If $a < b$ we set $B := \{(x, y) : a < x \leqslant b, \ x \leqslant y \leqslant b\}$. Show that B is measurable and that

$$\mu \times \nu(B) = \int_{(a,b]} F(s) \, dG(s) - F(a)(G(b) - G(a)).$$

(iii) **Integration by parts.** Show that

$$F(b)G(b) - F(a)G(a)$$

$$= \int_{(a,b]} F(s) \, dG(s) + \int_{(a,b]} G(s-) \, dF(s)$$

$$= \int_{(a,b]} F(s-) \, dG(s) + \int_{(a,b]} G(s-) \, dF(s) + \sum_{a<s\leqslant b} \Delta F(s)\Delta G(s).$$

[Hint: expand $\mu \times \nu((a, b]^2)$ in two different ways, using (ii). Note that the sum in the second part of the formula is at most countable because of L13.12.]

(iv) **Change of variable formula.** Let ϕ be a C^1-function. Then

$$\phi(F(b)) - \phi(F(a))$$

$$= \int_{(a,b]} \phi'(F(s-)) \, dF(s) + \sum_{a<s\leqslant b} \Big[\phi(F(s)) - \phi(F(s-)) - \phi'(F(s))\Delta F(s)\Big].$$

[Hint: use (iii) to show the change of variable formula for polynomials and then use the fact that continuous functions can be uniformly approximated by a sequence of polynomials – cf. Weierstraß' approximation theorem 24.6.]

13.14. **Rearrangements.** Let $(X, \mathcal{A}, \mu)$ be a σ-finite measure space and let $f \in \mathcal{L}^p(\mu)$ for some $p \in [1, \infty)$. The distribution function of f is given by $\mu_f(\{f \geqslant t\})$ and the *decreasing rearrangement* of f is the generalized inverse of μ_f,

$$f^*(\xi) := \inf\{t : \mu_f(t) \leqslant \xi\}, \qquad \xi \geqslant 0, \qquad (\inf \emptyset = +\infty).$$

(i) Let $f = 2\,\mathbf{1}_{[1,3]} + 4\,\mathbf{1}_{[4,5]} + 3\,\mathbf{1}_{[6,9]}$. Make a sketch of the graphs of $f(x)$, $\mu_f(t)$ and $f^*(\xi)$.

(ii) Show that for $f \in \mathcal{L}^p(\mu)$

$$\int_{\mathbb{R}} |f|^p \, d\mu = p \int_0^\infty t^{p-1} \mu_f(t) \, dt = \int_{(0,\infty)} (f^*)^p \, d\lambda.$$

In other words: $\|f\|_p = \|f^*\|_p$. Because of this the space $\mathcal{L}^p$ is said to be *rearrangement invariant*.

14

Integrals with respect to image measures

Let $(X, \mathcal{A}, \mu)$ be a measure space and $(X', \mathcal{A}')$ be a measurable space. As we have seen in T7.6 and D7.7, any $\mathcal{A}/\mathcal{A}'$-measurable map $T: X \to X'$ can be used to transport the measure μ, defined on $(X, \mathcal{A})$, to a measure on $(X', \mathcal{A}')$:

$$T(\mu)(A') := \mu(T^{-1}(A')) \qquad \forall\, A' \in \mathcal{A}'. \tag{14.1}$$

Let us see how (14.1) extends to integrals.

To make sense of the integral $\int \ldots dT(\mu)$ w.r.t. the image measure $T(\mu)$, we use again the recipe from Chapters 9, 10 when we introduced the integral. First, we calculate the image integrals for indicator functions and, by linearity, for (positive) simple functions. By monotone convergence T9.6 we extend the resulting formula to all positive measurable functions and, finally, considering positive and negative parts, to the whole class $\mathcal{L}^1(T(\mu))$. This is the blueprint for the proof of

14.1 Theorem *Let $T: X \to X'$ be a measurable map between the measure space $(X, \mathcal{A}, \mu)$ and the measurable space $(X', \mathcal{A}')$. For every $\mathcal{A}'/\bar{\mathcal{B}}$-measurable and $T(\mu)$-integrable function $u: X' \to \bar{\mathbb{R}}$ we find that $u \circ T: X \to \bar{\mathbb{R}}$ is $\mathcal{A}/\bar{\mathcal{B}}$-measurable, μ-integrable and satisfies*

$$\int_{X'} u\, dT(\mu) = \int_X u \circ T\, d\mu. \tag{14.2}$$

If $u \geqslant 0$ is positive, (14.2) remains valid without assuming $T(\mu)$-integrability.

Proof Since u and T are measurable, so is $u \circ T$, see T7.4, and the integrals in (14.2) are well-defined.

134

Let us assume that $u \geqslant 0$ but not necessarily integrable, i.e. $\int u\, dT(\mu) = +\infty$ is allowed. For a simple function $f \in \mathcal{E}^+(\mathcal{A}')$,

$$f = \sum_{j=0}^{M} y_j \mathbf{1}_{A'_j}, \qquad A'_j \in \mathcal{A}',\ y_j \geqslant 0,$$

the identity (14.2) follows from (14.1) by linearity:

$$
\begin{aligned}
\int f\, dT(\mu) &= \sum_{j=0}^{M} y_j \int_{X'} \mathbf{1}_{A'_j}\, dT(\mu) \\
&= \sum_{j=0}^{M} y_j\, T(\mu)(A'_j) \\
&\overset{(14.1)}{=} \sum_{j=0}^{M} y_j \mu(T^{-1}(A'_j)) \\
&= \sum_{j=0}^{M} y_j \int_X \mathbf{1}_{T^{-1}(A'_j)}(x)\, \mu(dx) \\
&= \sum_{j=0}^{M} y_j \int_X \mathbf{1}_{A'_j}(T(x))\, \mu(dx) \ = \ \int_X f(T(x))\, \mu(dx)
\end{aligned}
\tag{14.3}
$$

where we used that $\mathbf{1}_{T^{-1}(A')}(x) = \mathbf{1}_{A'}(T(x))$ for all $A' \in \mathcal{A}'$. Since every $u \in \mathcal{M}^+(\mathcal{A}')$ is the limit of an increasing sequence of positive simple functions $f_j \in \mathcal{E}^+(\mathcal{A}')$, see T8.8, we can use Beppo Levi's theorem 9.6 and (14.3) to get

$$
\int_{X'} u\, dT(\mu) \overset{9.6}{=} \sup_{j \in \mathbb{N}} \int_{X'} f_j\, dT(\mu) \overset{(14.3)}{=} \sup_{j \in \mathbb{N}} \int_X f_j \circ T\, d\mu
$$

$$
\overset{9.6}{=} \int_X u \circ T\, d\mu.
$$

If u is $T(\mu)$-integrable, we write $u = u^+ - u^-$ and apply (14.2) to $u^\pm$ separately. All we have to do is to observe that $u^\pm \circ T = (u \circ T)^\pm$ and that, due to the integrability assumption, $\int (u \circ T)^\pm\, d\mu \overset{(14.2)}{=} \int u^\pm\, dT(\mu) < \infty$. ∎

Often we are in the situation where $T : X \to X'$ is invertible with an $\mathcal{A}'/\mathcal{A}$-measurable inverse map $T^{-1} : X' \to X$. In this case we can strengthen Theorem 14.1.

14.2 Corollary *If in the situation of Theorem 14.1 the measurable map $T : X \to X'$ has a measurable inverse $T^{-1} : X' \to X$, then $u : X' \to \bar{\mathbb{R}}$ is $T(\mu)$ integrable (and,* *a fortiori, $\mathcal{A}'/\bar{\mathcal{B}}$-measurable) if, and only if, $u \circ T$ is μ integrable (and, a fortiori, $\mathcal{A}/\bar{\mathcal{B}}$-measurable). In this case (14.2) holds.*

Proof Apply Theorem 14.1 to u and $u \circ T$ using the measurable transformations T and T^{-1} respectively. ∎

14.3 Examples We will frequently encounter the following particular situation of Corollary 14.2: let $(X, \mathcal{A}, \mu)$ be $(\mathbb{R}^n, \mathcal{B}(\mathbb{R}^n), \lambda^n)$ where λ^n is n-dimensional Lebesgue measure and let $(X', \mathcal{A}') = (\mathbb{R}^n, \mathcal{B}(\mathbb{R}^n))$. The maps

$$\sigma : \mathbb{R}^n \to \mathbb{R}^n \qquad \text{and} \qquad \tau_x : \mathbb{R}^n \to \mathbb{R}^n$$
$$y \mapsto -y \qquad\qquad\qquad y \mapsto y - x$$

are continuous, hence measurable (Example 7.3) and so are their inverses $\sigma^{-1} = \sigma$ and $\tau_x^{-1} = \tau_{-x}$.

(i) By Corollary 7.11, Lebesgue measure λ^n is invariant under reflections and translations, so that $\lambda^n = \sigma(\lambda^n)$ and $\lambda^n = \tau_x(\lambda^n)$ for all $x \in \mathbb{R}^n$. But then (14.2) becomes

$$\int u(-y)\,\lambda^n(dy) = \int u(\sigma(y))\,\lambda^n(dy) = \int u(y)\,\sigma(\lambda^n)(dy)$$
$$= \int u(y)\,\lambda^n(dy), \tag{14.4}$$

and, for all $x \in \mathbb{R}^n$,

$$\int u(y \mp x)\,\lambda^n(dy) = \int u(\tau_{\pm x}y)\,\lambda^n(dy) = \int u(y)\,\tau_{\pm x}(\lambda^n)(dy)$$
$$= \int u(y)\,\lambda^n(dy). \tag{14.5}$$

(ii) If we consider Lebesgue measure with a density $f \geqslant 0$, $f\,\lambda^n$, cf. L10.8, we find

$$\int u(y)\,\tau_x(f\,\lambda^n)(dy) = \int u(\tau_x y)f(y)\,\lambda^n(dy)$$
$$= \int u(\tau_x y)f(\tau_x \tau_{-x}y)\,\lambda^n(dy) \tag{14.6}$$
$$= \int u(y)f(y + x)\,\lambda^n(dy),$$

which also proves that $\tau_x(f\,\lambda^n) = (f \circ \tau_{-x})\,\lambda^n$.

Convolutions

The *convolution* or *Faltung* of functions and measures on $(\mathbb{R}^n, \mathcal{B}(\mathbb{R}^n))$ appears naturally in functional analysis, Fourier analysis, probability theory and other branches of mathematics. One can understand it as an averaging process that respects translations and results in a gain of smoothness.

14.4 Definition Let μ and ν be measures on $(\mathbb{R}^n, \mathcal{B}(\mathbb{R}^n))$ and $u, v : \mathbb{R}^n \to \bar{\mathbb{R}}$ be measurable numerical functions. The *convolution* of ...

- ...two functions u and v is the *function*

$$u \star v(x) := \int_{\mathbb{R}^n} u(x - y)v(y)\,\lambda^n(dy) \tag{14.7}$$

  provided $u(x - \bullet)v$ is positive or contained in $\mathcal{L}^1(\lambda^n)$;
- ...of the function u and the measure μ is the *function*

$$u \star \mu(x) := \int_{\mathbb{R}^n} u(x - y)\,\mu(dy) \tag{14.8}$$

  provided $u(x - \bullet)$ is positive or contained in $\mathcal{L}^1(\mu)$;
- ...of two measures μ and ν is the *measure*

$$\mu \star \nu(B) := \iint \mathbf{1}_B(x + y)\,\mu(dx)\,\nu(dy), \qquad B \in \mathcal{B}(\mathbb{R}^n). \tag{14.9}$$

14.5 Remarks (i) The convolution of two functions (or of a function with a measure or of two measures) is linear in each of its arguments, e.g.

$$(\alpha u + \beta v) \star w = \alpha(u \star w) + \beta(v \star w), \qquad \alpha, \beta \in \mathbb{R}.$$

Similar formulae hold in the second argument and for the other cases.

  (ii) The function $\alpha : \mathbb{R}^{2n} \to \mathbb{R}^n$, $\alpha(x, y) := x + y$, is Borel measurable and

$$\mu \star \nu(B) = \iint \mathbf{1}_B(x + y)\,\mu(dx)\,\nu(dy)$$

$$= \iint \mathbf{1}_{\alpha^{-1}(B)}(x, y)\,d(\mu \times \nu)(x, y) = \alpha(\mu \times \nu).$$

If μ and ν have densities $u, v \geqslant 0$ w.r.t. Lebesgue measure, that is $\mu = u\,\lambda^n$ and $\nu = v\,\lambda^n$, we find for all $B \in \mathcal{B}(\mathbb{R}^n)$ that

$$(u\,\lambda^n) \star (v\,\lambda^n)(B) = \iint \mathbf{1}_B(x + y)u(x)v(y)\,\lambda^n(dx)\,\lambda^n(dy)$$

$$= \int_B \left(\int u(x - y)v(y)\,\lambda^n(dy) \right) \lambda^n(dx)$$

where we used Tonelli's theorem 13.8 and (14.6). Thus $(u\,\lambda^n)\star(v\,\lambda^n)=(u\star v)\,\lambda^n$; in a similar way one shows that $(u\,\lambda^n)\star\mu=(u\star\mu)\,\lambda^n$.

Interpreting (14.9) as $\int \mathbf{1}_B\,d(\mu\star v)=\iint \mathbf{1}_B(x+y)\,\mu(dx)\,v(dy)$, we easily see that

$$\int \Phi\,d(\mu\star v)=\iint \Phi(x+y)\,\mu(dx)\,v(dy),$$

first for simple functions, then by T9.6 for positive measurable functions, and finally by linearity for general $\Phi\in\mathcal{L}^1(\mu\times v)$.

Note that the definition of $u\star v$ is not really straightforward since we require $u(x-\bullet)v$ to be positive or integrable. Here is a much handier criterion due to W.H. Young.

14.6 Theorem (Young's inequality) *Let $u\in\mathcal{L}^1(\lambda^n)$ and $v\in\mathcal{L}^p(\lambda^n)$, $p\in[1,\infty)$. Then the convolution $u\star v$ defines a function in $\mathcal{L}^p(\lambda^n)$, satisfies $u\star v=v\star u$ and*

$$\|u\star v\|_p\leqslant\|u\|_1\cdot\|v\|_p. \tag{14.10}$$

Proof Assume first that $u,v\geqslant 0$. Let $\beta(x,y):=x-y$. Then β is Borel measurable and so are $(x,y)\mapsto u(x-y)v(y)$ and $(x,y)\mapsto u(y)v(x-y)$. Since λ^n is invariant under translations, we see using (14.4)–(14.6)

$$u\star v(x)=\int u(x-y)v(y)\,\lambda^n(dy)=\int u(y)v(x-y)\,\lambda^n(dy)=v\star u(x).$$

Moreover,

$$
\begin{aligned}
\|u\star v\|_p^p &= \int\left(\int u(y)v(x-y)\,\lambda^n(dy)\right)^p\lambda^n(dx)\\[4pt]
&= \|u\|_1^p\int\left(\int v(x-y)\frac{u(y)}{\|u\|_1}\,\lambda^n(dy)\right)^p\lambda^n(dx)\\[4pt]
&\overset{12.14}{\leqslant} \|u\|_1^p\iint v(x-y)^p\frac{u(y)}{\|u\|_1}\,\lambda^n(dy)\,\lambda^n(dx)\\[4pt]
&\overset{13.8}{=} \|u\|_1^p\int\underbrace{\left(\int v(x-y)^p\,\lambda^n(dx)\right)}_{=\|v\|_p^p\ \text{by (14.5)}}\frac{u(y)}{\|u\|_1}\,\lambda^n(dy)\\[4pt]
&= \|u\|_1^p\|v\|_p^p,
\end{aligned}
$$

which implies that $u\star v\in\mathcal{L}^p(\lambda^n)$.

The general case follows now from considering $u=u^+-u^-$ and $v=v^+-v^-$ and the fact that

$$(u^+-u^-)\star v^\pm=u^+\star v^\pm-u^-\star v^\pm,$$

where the difference is a.e. defined since $u^\pm\star v^\pm\in\mathcal{L}^p(\lambda^n)$ is a.e. finite. ∎

The convolution $u \star v$ is a hybrid of u and v which inherits those properties which are preserved under translations and averages, cf. Problems 14.6–14.8. In general, $u \star v$ is smoother than u and v. To see this, we need the following result which, although similar to Corollary 12.11, is much deeper and uses the topological structure of $(\mathbb{R}^n, \mathcal{B}(\mathbb{R}^n), \lambda^n)$.

14.7 Lemma *The continuous functions with compact support $C_c(\mathbb{R}^n)$ are a dense subset of $\mathcal{L}^p(\lambda^n)$, $p \in [1, \infty)$.*

Proof Postponed to Chapter 15, Theorem 15.17. ∎

14.8 Theorem *Let $u \in \mathcal{L}^p(\lambda^n)$, $p \in [1, \infty)$.*

(i) *The map $x \mapsto \int |u(x+y) - u(y)|^p \lambda^n(dy)$ is uniformly continuous.*

(ii) *If $u \in \mathcal{L}^1(\lambda^n)$, $v \in \mathcal{L}^\infty(\lambda^n)$, then $u \star v$ is bounded and continuous.*

Proof (i) Because of Lemma 14.7 we find for every $\epsilon > 0$ some $\phi_\epsilon \in C_c(\mathbb{R}^n)$ such that $\|u - \phi_\epsilon\| \leqslant \epsilon$.

By the lower triangle inequality for $\|\cdot\|_p$ and the translation invariance of λ^n we find for any two $x, x' \in \mathbb{R}^n$

$$\left| \|u(x+\bullet) - u\|_p - \|u(x'+\bullet) - u\|_p \right| \leqslant \|u(x+\bullet) - u(x'+\bullet)\|_p$$

$$\overset{(14.5)}{=} \|u(x-x'+\bullet) - u\|_p.$$

Using again the triangle inequality and translation invariance we get for every $R > 0$ and all x, x' with $|x - x'| < R/2$

$$\|u(x-x'+\bullet) - u\|_p$$

$$\leqslant \left\| (u(x-x'+\bullet) - u) \mathbf{1}_{B_R(0)} \right\|_p + \left\| (u(x-x'+\bullet) - u) \mathbf{1}_{B_R^c(0)} \right\|_p$$

$$\overset{[\checkmark]}{\leqslant} \left\| (u(x-x'+\bullet) - u) \mathbf{1}_{B_R(0)} \right\|_p + 2 \left(\int_{B_{R/2}^c(0)} |u|^p \, d\lambda^n \right)^{1/p}.$$

Since $u \in \mathcal{L}^p(\lambda^n)$, it follows from the monotone convergence theorem 11.1 that $\lim_{R \to \infty} \int_{B_R^c(0)} |u|^p \, d\lambda^n = 0$, so that we can achieve

$$\left(\int_{B_{R/2}^c(0)} |u|^p \, d\lambda^n \right)^{1/p} \leqslant \epsilon \qquad \forall R > R_\epsilon.$$

Since ϕ_ϵ is continuous with compact support, it is uniformly continuous, which means that there is a $\delta = \delta_{\epsilon, R} > 0$ such that for all $y \in \mathbb{R}^n$, $|x| < \delta$, and any fixed $R > R_\epsilon$ we have $|\phi_\epsilon(x+y) - \phi_\epsilon(y)| \leqslant \epsilon / \lambda^n(B_R(0))^{1/p}$.

Another application of the triangle inequality for $\|\cdot\|_p$ and translation invariance yields

$$\left\| \left(u(x - x' + \cdot) - u \right) \mathbf{1}_{B_R(0)} \right\|_p$$

$$\leqslant \left\| \left(u(x - x' + \cdot) - \phi_\epsilon(x - x' + \cdot) \right) \mathbf{1}_{B_R(0)} \right\|_p + \left\| (u - \phi_\epsilon) \mathbf{1}_{B_R(0)} \right\|_p$$

$$+ \left\| \left(\phi_\epsilon(x - x' + \cdot) - \phi_\epsilon \right) \mathbf{1}_{B_R(0)} \right\|_p$$

$$\leqslant 2\|u - \phi_\epsilon\|_p + \left(\int_{B_R(0)} \underbrace{\left| \phi_\epsilon(x - x' + y) - \phi_\epsilon(y) \right|^p}_{\leqslant\, \epsilon^p / \lambda^n(B_R(0)) \ \text{if} \ |x - x'| < \delta} \lambda^n(dy) \right)^{1/p}$$

$$\leqslant 3\epsilon.$$

Combining the above estimates, we get

$$\|u(x + \cdot) - u(x' + \cdot)\|_p \leqslant 5\epsilon \qquad \forall \, x, x' : |x - x'| < \delta \quad (< R/2),$$

which is but uniform continuity.

(ii) We have for any $x, x' \in \mathbb{R}^n$

$$\left| u \star v(x) - u \star v(x') \right| \leqslant \int \left| v(y)u(x - y) - v(y)u(x' - y) \right| \lambda^n(dy)$$

$$\leqslant \|v\|_\infty \|u(x - \cdot) - u(x' - \cdot)\|_1$$

$$\overset{(14.4)}{=} \|v\|_\infty \|u(x + \cdot) - u(x' + \cdot)\|_1,$$

and continuity follows from part (i). The boundedness of $u \star v$ is proved with a similar calculation. ∎

Problems

14.1. Let $(X, \mathcal{A}, \mu)$ be a measure space and $T : X \to X$ be a bijective measurable map whose inverse $T^{-1} : X \to X$ is again measurable. Show that for every $f \in \mathcal{M}^+(\mathcal{A})$ one has

$$\int u \, d(T(f\mu)) = \int u \circ T \, f \, d\mu = \int u \, f \circ T^{-1} \, dT(\mu) = \int u \, d(f \circ T^{-1} \, T(\mu)).$$

14.2. Let $(X, \mathcal{A}, \mu)$ be a measure space and $(Y, \mathcal{B})$ be a measurable space. Assume that $T : A \to B$, $A \in \mathcal{A}$, $B \in \mathcal{B}$, is an invertible measurable map. Show that

$$T(\mu)|_B = T(\mu|_A)$$

with the restrictions $\mu|_A(\cdot) := \mu(A \cap \cdot)$ and $T(\mu)|_B := T(\mu)(B \cap \cdot)$.

14.3. Let μ be a measure on $(\mathbb{R}^n, \mathcal{B}(\mathbb{R}^n))$ and $x, y, z \in \mathbb{R}^n$. Find $\delta_x \star \delta_y$ and $\delta_z \star \mu$.

14.4. Let μ, ν be two σ-finite measures on $(\mathbb{R}^n, \mathcal{B}(\mathbb{R}^n))$. Show that $\mu \star \nu$ has no atoms (cf. Problem 6.5) if μ has no atoms.

14.5. Let P be a probability measure on $(\mathbb{R}^n, \mathcal{B}(\mathbb{R}^n))$ and denote by $P^{\star n}$ the n-fold convolution product $P \star P \star \cdots \star P$. Show that

$$\int |\omega| \, P^{\star n}(d\omega) \leqslant n \int |\omega| \, P(d\omega);$$

if $\int |\omega| \, P(d\omega) < \infty$, then

$$\int \omega \, P^{\star n}(d\omega) = n \int \omega \, P(d\omega);$$

14.6. Let $p : \mathbb{R} \to \mathbb{R}$ be a polynomial and $u \in C_c(\mathbb{R})$. Show that $u \star p$ exists and is again a polynomial.

14.7. Let $w : \mathbb{R} \to \mathbb{R}$ be an increasing (hence, measurable, by Problem 8.18) and bounded function. Show that for every $u \in \mathcal{L}^1(\lambda^1)$ the convolution $u \star w$ is again increasing, bounded and continuous.

14.8. Assume that $u \in C_c(\mathbb{R})$ and $w \in C^\infty(\mathbb{R})$. Show that $u \star w$ exists, is of class C^∞ and satisfies $\partial_j(u \star w) = u \star (\partial_j w)$.

14.9. **Young's inequality.** Adapt the proof of Theorem 14.6 and show that

$$\|u \star w\|_r \leqslant \|u\|_p \cdot \|w\|_q$$

for all $p, q, r \in [1, \infty)$, $u \in \mathcal{L}^p$, $w \in \mathcal{L}^q$ and $r^{-1} + 1 = p^{-1} + q^{-1}$.

14.10. **Friedrichs mollifiers.** Let $\phi : \mathbb{R}^n \to \mathbb{R}^+$ be a C^∞-function such that $\int \phi \, d\lambda^n = 1$ and $\operatorname{supp} \phi = \overline{B_1(0)}$. For $\epsilon > 0$ define the function $\phi_\epsilon(x) := \epsilon^{-n} \phi(x/\epsilon)$. The function $\phi_\epsilon \star u$ is called the *Friedrichs mollifier* of $u \in \mathcal{L}^p$, $1 \leqslant p < \infty$.

   (i) Show that $\phi(x) := \kappa \exp(1/(|x|^2 - 1)) \, \mathbf{1}_{B_1(0)}(x)$ has, for a suitable $\kappa > 0$, the properties mentioned above. Determine κ.

   (ii) Show that $\phi_\epsilon \in C_c^\infty(\mathbb{R}^n)$, $\operatorname{supp} \phi_\epsilon = \overline{B_\epsilon(0)}$, and $\|\phi_\epsilon\|_1 = 1$.

   (iii) Show that $\operatorname{supp} u \star \phi_\epsilon \subset \operatorname{supp} u + \operatorname{supp} \phi_\epsilon = \{y : \forall x \in \operatorname{supp} u : |x - y| \leqslant \epsilon\}$.

   (iv) Show that $\phi_\epsilon \star u$ is in $C^\infty \cap \mathcal{L}^p$ and

$$\|\phi_\epsilon \star u\|_p \leqslant \|u\|_p \qquad \forall \epsilon > 0.$$

   (v) Show that $L^p\text{-}\lim_{\epsilon \to 0} \phi_\epsilon \star u = u$.
   [Hint: split the region of integration as in the proof Theorem 14.8 and use the uniform boundedness shown in (iv).]

14.11. Define $\phi : \mathbb{R} \to \mathbb{R}$ by $\phi(x) := (1 - \cos x) \, \mathbf{1}_{[0, 2\pi)}(x)$ and let $u(x) := 1$, $v(x) := \phi'(x)$, and $w(x) := \int_{(-\infty, x)} \phi(t) \, dt$. Then

   (i) $u \star v(x) = 0$ for all $x \in \mathbb{R}$;

   (ii) $u \star w(x) = \phi \star \phi(x) > 0$ for all $x \in (0, 4\pi)$;

   (iii) $(u \star v) \star w \equiv 0 \neq u \star (v \star w)$.

Does this contradict the commutativity of the convolution which was used in Theorem 14.6?

15

Integrals of images and Jacobi's transformation rule[1]

The previous chapter dealt with image measures and, by their definition, with measures of pre-images of sets. Sometimes one needs to know the measure of the direct image of a set under $T: X \to Y$. If T^{-1} exists and is measurable, we can apply the results of Chapter 14 to $S := T^{-1}$ and we are done. If, however, T^{-1} is not measurable, the direct image $T(A)$ of a set $A \in \mathcal{A}$ need not be $\mathcal{A}'$-measurable; in particular, an expression of the type $\mu'(T(A))$ – here μ' is any measure on $(X', \mathcal{A}')$ – may not be well-defined, let alone a measure. Let us consider this problem in a very particular setting, where

$$X \subset \mathbb{R}^n, \ X' \subset \mathbb{R}^d \quad \text{and} \quad \mu, \mu' \text{ are Lebesgue measures } \lambda^n, \text{ resp. } \lambda^d.$$

We need some *notation*: if $\Phi: X \to \mathbb{R}^d$, we write $\Phi = (\Phi_1, \Phi_2, \ldots, \Phi_d)$ for its components and we set for vectors $x = (x_1, \ldots, x_n) \in \mathbb{R}^n$ and matrices $A = (a_{jk})_{\substack{j=1,\ldots,n \\ k=1,\ldots,d}}$

$$|x|_\infty := \max_{1 \leqslant j \leqslant n} |x_j|, \qquad |A|_\infty := \max_{\substack{1 \leqslant j \leqslant n \\ 1 \leqslant k \leqslant d}} |a_{jk}|. \tag{15.1}$$

A set $F \subset \mathbb{R}^n$ $[G \subset \mathbb{R}^n]$ is called an F_σ-set $[G_\delta$-set$]$ if it is the *countable union of closed sets* [*countable intersection of open sets*], i.e. if

$$F = \bigcup_{\ell \in \mathbb{N}} C_\ell \qquad \left[G = \bigcap_{\ell \in \mathbb{N}} U_\ell \right] \tag{15.2}$$

for closed sets C_ℓ [open sets U_ℓ]. Obviously, both F_σ- and G_δ-sets are Borel sets; but, in general, neither are F_σ-sets closed nor are G_δ-sets open.

---

[1] The proofs in this chapter can be left out at first reading.

15.1 Theorem *Let* $F \subset \mathbb{R}^n$ *be an* F_σ*-set and* $\Phi : F \to \mathbb{R}^d$ *be an* α*-Hölder continuous map, that is*

$$|\Phi(x) - \Phi(y)|_\infty \leqslant L |x - y|_\infty^\alpha \qquad \forall x, y \in F \qquad (15.3)$$

with constant L *and exponent* $\alpha \in (0, 1]$. *For every* F_σ*-set* $E \subset \mathbb{R}^n$, $\Phi(F \cap E)$ *is an* F_σ*-set in* $\mathbb{R}^d$, *hence Borel measurable. If* $d \geqslant \alpha n$, *we have*

$$\lambda^d(\Phi(F \cap E)) \leqslant L^d \, \lambda^n(E). \qquad (15.4)$$

Proof Since E, F are F_σ-sets, they have representations $E = \bigcup_{j \in \mathbb{N}} \Gamma_j$ and $F = \bigcup_{j \in \mathbb{N}} C_j$ with closed sets $\Gamma_j, C_j \subset \mathbb{R}^n$. Moreover,

$$E = \bigcup_{k \in \mathbb{N}} E \cap \overline{B_k(0)} = \bigcup_{k,j \in \mathbb{N}} \Gamma_j \cap \overline{B_k(0)} =: \bigcup_{\ell \in \mathbb{N}} K_\ell,$$

where $(K_\ell)_{\ell \in \mathbb{N}}$ is an enumeration of the family $(\Gamma_j \cap \overline{B_k(0)})_{j,k \in \mathbb{N}}$ of closed and bounded, hence compact, sets. Thus $C_j \cap K_\ell$ is a compact set, and since images of compact sets under continuous maps are compact, we see that $\Phi(C_j \cap K_\ell)$ is compact and, in particular, closed. So,

$$\Phi(F \cap E) \overset{(2.4)}{=} \bigcup_{j,\ell \in \mathbb{N}} \Phi(C_j \cap K_\ell)$$

is an F_σ-set.

Assume now that $d \geqslant \alpha n$. If $\lambda^n(E) = \infty$, (15.4) is trivial and we will consider only the case $\lambda^n(E) < \infty$. The proof of Carathéodory's extension theorem 6.1 – in particular (6.1) – for $\mu = \lambda^n$ and the semi-ring $\mathcal{S} = \mathcal{J}^n$ of n-dimensional half-open rectangles (cf. P6.4) shows that we can find for every $\epsilon > 0$ a sequence $(J_j^\epsilon)_{j \in \mathbb{N}} \subset \mathcal{J}^n$ with

$$E \subset \bigcup_{j \in \mathbb{N}} J_j^\epsilon \qquad \text{and} \qquad \sum_{j \in \mathbb{N}} \lambda^n(J_j^\epsilon) \leqslant \lambda^n(E) + \epsilon. \qquad (15.5)$$

Without loss of generality we can assume that all J_j^ϵ are *squares*, i.e. have sides of equal length $s_j^\epsilon < s < 1$, otherwise we could subdivide each J_j^ϵ into finitely many non-overlapping squares of this type.[✓] So,

$$\lambda^d(\Phi(F \cap E)) \leqslant \lambda^d \Big(\bigcup_{j \in \mathbb{N}} \Phi(F \cap J_j^\epsilon) \Big) \overset{4.6}{\leqslant} \sum_{j \in \mathbb{N}} \lambda^d\big(\Phi(F \cap J_j^\epsilon)\big), \qquad (15.6)$$

which means that it is enough to check (15.4) for a square $E = J$ of side-length $s < 1$ and centre $c \in \mathbb{R}^n$. Because of (15.3),

$$\Phi(J) = \Phi\Big(\underset{k=1}{\overset{n}{\bigtimes}} \big[c_k - \tfrac{1}{2}s, c_k + \tfrac{1}{2}s\big) \Big) \subset \underset{k=1}{\overset{d}{\bigtimes}} \big[\Phi_k(c) - \tfrac{L}{2}s^{1/\alpha}, \Phi_k(c) + \tfrac{L}{2}s^{1/\alpha}\big)$$

and (notice that $\lambda^n(J) \leqslant 1$ and $d/\alpha n \geqslant 1$)

$$\lambda^d(\Phi(F \cap J)) \leqslant (Ls^{1/\alpha})^d = L^d \left(\lambda^n(J)\right)^{d/(\alpha n)} \leqslant L^d \lambda^n(J).$$

From (15.5), (15.6) we conclude

$$\lambda^d(\Phi(F \cap E)) \leqslant \sum_{j \in \mathbb{N}} L^d \lambda^n(J_j^\epsilon) \leqslant L^d \left(\lambda^n(E) + \epsilon\right)$$

and the claim follows upon letting $\epsilon \to 0$.   ∎

Theorem 15.1 can be improved if we use the completed Borel-σ-algebra $\mathcal{B}^*(\mathbb{R}^n)$, cf. Problems 4.13, 6.2, 10.11 and 10.12. Recall that

$$B^* \in \mathcal{B}^*(\mathbb{R}^n) \iff \begin{array}{l} B^* = B \cup N \quad \text{for some } B \in \mathcal{B}(\mathbb{R}^n) \\ \text{and a subset } N \text{ of a Borel null set.} \end{array}$$

The advantage of $\mathcal{B}^*(\mathbb{R}^n)$ over $\mathcal{B}(\mathbb{R}^n)$ is that α-Hölder continuous maps $\Phi : \mathbb{R}^n \to \mathbb{R}^d$ map $\mathcal{B}^*(\mathbb{R}^n)$-measurable sets into $\mathcal{B}^*(\mathbb{R}^d)$-sets if $d \geqslant \alpha n$; this is not true for $\mathcal{B}(\mathbb{R}^n)$. To see this we need a few preparations.

15.2 Lemma *Let $B \in \mathcal{B}(\mathbb{R}^n)$ be a Borel set. Then there exists an F_σ-set F and a G_δ-set G such that*

$$F \subset B \subset G \quad and \quad \lambda^n(F) = \lambda^n(B) = \lambda^n(G).$$

Proof The proof consists of three stages:

Step 1: Construction of the set G. If $\lambda^n(B) = \infty$, we take $G = \mathbb{R}^n$. If $\lambda^n(B) < \infty$, we find as in the proof of Theorem 15.1 (or as in Carathéodory's extension theorem 6.1, (6.1), with $\mu = \lambda^n$ and $\mathcal{S} = \mathcal{J}^n$) for every $k \in \mathbb{N}$ a sequence of half-open squares $(J_j^k)_{j \in \mathbb{N}} \subset \mathcal{J}^n$ of side-length s_j such that

$$B \subset \bigcup_{j \in \mathbb{N}} J_j^k \quad and \quad \sum_{j \in \mathbb{N}} \lambda^n(J_j^k) \leqslant \lambda^n(B) + \frac{1}{k}.$$

We can now enlarge J_j^k by moving the lower left corner by $\epsilon_j := (s_j^n + 2^{-j}/k)^{1/n} - s_j$ units 'to the left' in each coordinate direction. The new *open* square $\tilde{J}_j^k$ has volume

$$\lambda^n(\tilde{J}_j^k) = \lambda^n(J_j^k) + \frac{1}{k} 2^{-j},$$

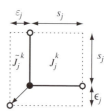

and we see that the open sets $G^k := \bigcup_{j\in\mathbb{N}} \tilde{J}_j^k \supset B$ satisfy

$$\lambda^n(G^k) \overset{4.6}{\leqslant} \sum_{j\in\mathbb{N}} \lambda^n(\tilde{J}_j^k) = \sum_{j\in\mathbb{N}} \lambda^n(J_j^k) + \frac{1}{k} \leqslant \left(\lambda^n(B) + \frac{1}{k}\right) + \frac{1}{k}.$$

Thus $G := \bigcap_{k\in\mathbb{N}} G^k$ is a G_δ-set with $G \supset B$, and

$$\lambda^n(B) \leqslant \lambda^n(G) \overset{4.4}{=} \lim_{k\to\infty} \lambda^n(G^k) \leqslant \lim_{k\to\infty} \left(\lambda^n(B) + \frac{2}{k}\right) = \lambda^n(B).$$

Step 2: Construction of the set F if $\lambda^n(B) < \infty$. Denote by $\bar{B}$ the closure[2] of B. Since $\bar{B} \setminus B$ is a Borel set, we find as in step 1 open sets U^k with

$$\bar{B} \setminus B \subset U^k \qquad \text{and} \qquad \lambda^n(U^k) \leqslant \lambda^n(\bar{B} \setminus B) + \frac{1}{k}. \qquad (15.7)$$

Observe that

$$B \subset \left(B \setminus U^k\right) \cup \left(U^k \cap B\right) \subset \left(\bar{B} \setminus U^k\right) \cup \left(U^k \setminus (\bar{B} \setminus B)\right),$$

so that by the subadditivity of measures

$$\begin{aligned}
\lambda^n(B) \ &\leqslant\ \lambda^n\left(\bar{B} \setminus U^k\right) + \lambda^n\left(U^k \setminus (\bar{B} \setminus B)\right) \\
&=\ \lambda^n(\bar{B} \setminus U^k) + \lambda^n(U^k) - \lambda^n(\bar{B} \setminus B) \\
&\overset{(15.7)}{\leqslant}\ \lambda^n(\bar{B} \setminus U^k) + \frac{1}{k}.
\end{aligned}$$

By construction, $C_k := \bar{B} \setminus U^k \subset \bar{B} \setminus (\bar{B} \setminus B) = B$ is a closed set and $F := \bigcup_{k\in\mathbb{N}} C_k \subset B$ is an F_σ-set satisfying

$$\lambda^n(B) - \frac{1}{k} \leqslant \lambda^n(C_k) \leqslant \lambda^n\left(\bigcup_{j\in\mathbb{N}} C_j\right) = \lambda^n(F) \leqslant \lambda^n(B).$$

The claim follows as $k \to \infty$.

Step 3: Construction of the set F if $\lambda^n(B) = \infty$. Setting

$$B_j := B \cap \left(B_j(0) \setminus B_{j-1}(0)\right), \qquad j \in \mathbb{N},$$

we get a disjoint partitioning of $B = \biguplus_{j\in\mathbb{N}} B_j$ where each set B_j is a Borel set with finite volume. Applying step 2 to each B_j, we find F_σ-sets $F_j \subset B_j$ with

---

[2] i.e. the smallest closed set containing B, cf. Appendix B, Definition B.3(iii).

$\lambda^n(F_j) = \lambda^n(B_j)$, $j \in \mathbb{N}$. Since the B_j are mutually disjoint, so are the F_j, and since $F := \bigcup_{j \in \mathbb{N}} F_j$ is again an F_σ-set (cf. Problem 15.1) we end up with $F \subset B$ and

$$\lambda^n(F) = \sum_{j \in \mathbb{N}} \lambda^n(F_j) = \sum_{j \in \mathbb{N}} \lambda^n(B_j) = \lambda^n(B).$$

The proof of the lemma is now complete.   ∎

15.3 Lemma *Let* $\Phi : \mathbb{R}^n \to \mathbb{R}^d$ *be an α-Hölder continuous map with $\alpha \in (0, 1]$ and $d \geqslant \alpha n$. If N^* is a subset of a Borel null set $N \in \mathcal{B}(\mathbb{R}^n)$, then $\Phi(N^*)$ is a subset of a Borel null set $M \in \mathcal{B}(\mathbb{R}^d)$.*

Proof Since $N^* \subset N \in \mathcal{B}(\mathbb{R}^n)$ where $\lambda^n(N) = 0$, we can repeat the argument of the proof of Theorem 15.1 to find for $k \in \mathbb{N}$ a covering of N by half-open squares $J_j^k \in \mathcal{J}^n$ such that

$$N \subset \bigcup_{j \in \mathbb{N}} J_j^k \qquad \text{and} \qquad \lambda^n\Big(\bigcup_{j \in \mathbb{N}} J_j^k\Big) \leqslant \sum_{j \in \mathbb{N}} \lambda^n(J_j^k) \leqslant \frac{1}{k}.$$

Since $\lambda^n(J_j^k) = \lambda^n(\bar{J}_j^k)$, $\bar{J}_j^k$ is the closed square, we have also

$$\lambda^n\Big(\bigcup_{j \in \mathbb{N}} \bar{J}_j^k\Big) \leqslant \sum_{j \in \mathbb{N}} \lambda^n(\bar{J}_j^k) \leqslant \frac{1}{k}.$$

Applying T15.1 to the F_σ-set $F^k := \bigcup_{j \in \mathbb{N}} \bar{J}_j^k$ shows that $\bigcap_{k \in \mathbb{N}} \Phi(F^k) \in \mathcal{B}(\mathbb{R}^n)$ as well as

$$\lambda^d(\Phi(F^k)) \leqslant L^d \lambda^n(F^k) \leqslant \frac{L^d}{k}.$$

Since $\bigcap_{k \in \mathbb{N}} \Phi(F^k) \supset \Phi(N) \supset \Phi(N^*)$, we conclude

$$\lambda^d\Big(\bigcap_{\ell \in \mathbb{N}} \Phi(F^\ell)\Big) \leqslant \lambda^d(\Phi(F^k)) \leqslant \frac{L^d}{k} \xrightarrow{k \to \infty} 0. \qquad ∎$$

Lemma 15.3 is just a special case of the following theorem which has already been announced above.

15.4 Theorem *Let* $F \subset \mathbb{R}^n$ *be an F_σ-set, $\Phi : F \to \mathbb{R}^d$ be an α-Hölder continuous map with exponent $\alpha \in (0, 1]$. If $d \geqslant \alpha n$, then Φ maps the completed Borel σ-algebra $F \cap \mathcal{B}^*(\mathbb{R}^n)$ into $\mathcal{B}^*(\mathbb{R}^d)$, and the inequality (15.4) holds for all $B \in \mathcal{B}^*(\mathbb{R}^n)$ with the completed Lebesgue measures*[3] *$\bar{\lambda}^n$ and $\bar{\lambda}^d$.*

---

[3] See Problems 4.13, 6.2, 10.11, 10.12, 13.3 for the completion of measures and their properties.

Proof Pick $B^* \in \mathcal{B}^*(\mathbb{R}^n)$ and write $B^* = B \cup N^*$ where $B \in \mathcal{B}(\mathbb{R}^n)$ and N^* is a subset of a Borel null set $N \in \mathcal{B}(\mathbb{R}^n)$. According to L15.2 we have $B^* = E \cup M^* \cup N^* =: E \cup N^{**}$ where E is an F_σ-set, $\lambda^n(E) = \lambda^n(B)$, and $M^*, N^{**} := N^* \cup M^*$ are subsets of Borel null sets. Thus

$$\Phi(B^*) = \Phi(E \cup N^{**}) = \Phi(E) \cup \Phi(N^{**})$$

and $\Phi(E)$ is an F_σ-set, see T15.1, and $\Phi(N^{**})$ is contained in a Borel null set $\subset \mathbb{R}^d$, see L15.3, hence $\Phi(B^*) \in \mathcal{B}^*(\mathbb{R}^d)$. Finally, by T15.1,

$$
\begin{aligned}
\bar{\lambda}^d\big(\Phi(F \cap B^*)\big) &= \bar{\lambda}^d\big(\Phi(F \cap (E \cup N^{**}))\big) \\[2mm]
&\leqslant \bar{\lambda}^d\big(\Phi(F \cap E) \cup \Phi(N^{**})\big) \\[2mm]
&= \lambda^d\big(\Phi(F \cap E)\big) \\[2mm]
&\overset{15.1}{\leqslant} L^d \lambda^n(E) = L^d \lambda^n(B) = L^d \bar{\lambda}^n(B^*).
\end{aligned}
$$
∎

Let us stress that both Hölder continuity of Φ and the condition $d \geqslant \alpha n$ are crucial for Theorem 15.4; one can find counterexamples if we have only $\Phi \in C(F, \mathbb{R}^d)$ or $d < \alpha n$.

Jacobi's transformation formula

One of the most interesting situations arises if $\Phi = (\Phi_1, \dots, \Phi_n) : \mathbb{R}^n_x \to \mathbb{R}^n_y$ (we write $\mathbb{R}^n_x$ if we want to indicate the generic variable in order to distinguish between the domain and range of Φ) is a C^1-map with everywhere defined inverse $\Phi^{-1} : \mathbb{R}^n_y \to \mathbb{R}^n_x$ which is again a C^1-map. Such maps are called $C^1(\mathbb{R}^n, \mathbb{R}^n)$-*diffeomorphisms*. As usual, we write $D\Phi(x) := \left(\frac{\partial}{\partial x_j} \Phi_k(x)\right)_{j,k=1,\dots,n}$ for the *Jacobian* at the point $x \in \mathbb{R}^n_x$. By Taylor's theorem we find for all $x, x' \in K$ from a compact set $K \subset \mathbb{R}^n_x$

$$
\begin{aligned}
|\Phi_k(x) - \Phi_k(x')| &\leqslant \sum_{j=1}^{n} \left|\frac{\partial}{\partial x_j} \Phi_k(\xi)\right| \cdot |x_j - x'_j| \\[2mm]
&\leqslant n \sup_{\xi \in K} |D\Phi(\xi)|_\infty \cdot |x - x'|_\infty,
\end{aligned}
\tag{15.8}
$$

i.e. Φ is *locally Lipschitz (1-Hölder) continuous* with Lipschitz constant $L = L_K = n \sup_{\xi \in K} |D\Phi(\xi)|_\infty$.

15.5 Theorem (Jacobi's transformation theorem) *Let* $\Phi : \mathbb{R}_x^n \to \mathbb{R}_y^n$ *be a* C^1-*diffeomorphism. Then*

$$\lambda^n(\Phi(B)) = \int_B |\det D\Phi(x)| \, \lambda^n(dx) \tag{15.9}$$

holds for all Borel sets $B \in \mathcal{B}(\mathbb{R}_x^n)$.

The proof of Theorem 15.5 is based on two auxiliary results.

15.6 Lemma *Let* μ *and* ν *be two measures on the measurable space* $(X, \mathcal{A})$ *and let* $\mathcal{S}$ *be a semi-ring such that* $\sigma(\mathcal{S}) = \mathcal{A}$. *If* $\mu|_{\mathcal{S}} \leqslant \nu|_{\mathcal{S}}$ [4] *and if there is a sequence* $(S_j)_{j \in \mathbb{N}} \subset \mathcal{S}$ *with* $S_j \uparrow X$, *then* $\mu \leqslant \nu$.

Proof It is clear from the properties of μ and ν that $\rho := \nu - \mu : \mathcal{S} \to [0, \infty)$ is a pre-measure. By T6.1, ρ has a unique extension to a measure $\tilde{\rho}$ on $\mathcal{A}$ and

$$\nu(S) = (\rho + \mu)(S) = \widetilde{\rho + \mu}(S) \qquad \forall S \in \mathcal{S}$$

where $\widetilde{\rho + \mu}$ is the unique extension of the pre-measure $(\rho + \mu)|_{\mathcal{S}}$ to a measure on $\mathcal{A}$. But the measures $\tilde{\rho} + \mu$ and ν satisfy

$$\nu(S) = \tilde{\rho}(S) + \mu(S) = \rho(S) + \mu(S) = \widetilde{\rho + \mu}(S) \qquad \forall S \in \mathcal{S}$$

and we conclude from the uniqueness of the extensions that $\nu = \tilde{\rho} + \mu$ on $\mathcal{A}$, i.e. $\nu(A) - \mu(A) = \tilde{\rho}(A) \geqslant 0$ for all $A \in \mathcal{A}$. ∎

Caution: Lemma 15.6 fails if $\mathcal{S}$ is not a semi-ring; see Problem 15.4.

15.7 Lemma *For every* C^1-*diffeomorphism* $\Phi : \mathbb{R}_x^n \to \mathbb{R}_y^n$ *we have*

$$\lambda^n(\Phi(J)) \leqslant \int_J |\det D\Phi(x)| \, \lambda^n(dx) \qquad \forall J \in \mathcal{J}(\mathbb{R}_x^n).$$

Proof Let $J = [\![a, b)\!)$, $a, b \in \mathbb{R}_x^n$, and note that $\bar{J} = [\![a, b]\!]$ is a compact set. Since $D(\Phi^{-1})$ is continuous, we find on the compact set $\Phi(\bar{J})$

$$L := \sup_{x \in J} |[D\Phi(x)]^{-1}|_\infty \leqslant \sup_{y \in \Phi(\bar{J})} |D(\Phi^{-1})(y)|_\infty, \tag{15.10}$$

where we used the inverse function theorem.[✓] Since $D\Phi$ is uniformly continuous on $\bar{J}$, we find for a given $\epsilon > 0$ some $\delta > 0$ such that

$$\sup_{\substack{x,x' \in [\![a,b]\!] \\ |x-x'|_\infty \leqslant \delta}} |D\Phi(x) - D\Phi(x')|_\infty \leqslant \frac{\epsilon}{L}. \tag{15.11}$$

---

[4] This is short for $\mu(S) \leqslant \nu(S)$ for all $S \in \mathcal{S}$.

Partition J into N disjoint half-open squares $J_1, \ldots, J_N \in \mathcal{J}(\mathbb{R}^n_x)$ of the same side-length $< \delta$. Since $D\Phi$ and $\det D\Phi$ are continuous functions[✓], we can find for each $\ell = 1, 2, \ldots, N$ a point

$$x_\ell \in \bar{J}_\ell \quad \text{such that} \quad |\det D\Phi(x_\ell)| = \inf_{x \in \bar{J}_\ell} |\det D\Phi(x)|.$$

Set $T_\ell := (D\Phi)(x_\ell) \in \mathbb{R}^{n \times n}$ and observe that

$$D(T_\ell^{-1} \circ \Phi)(x) = T_\ell^{-1} \circ (D\Phi)(x) = \mathrm{id}_n + T_\ell^{-1} \circ (D\Phi(x) - D\Phi(x_\ell))$$

(id_n is the identity matrix in $\mathbb{R}^{n \times n}$). The estimates (15.10), (15.11) show that

$$\sup_{x \in \bar{J}_\ell} |D(T_\ell^{-1} \circ \Phi)(x)|_\infty \leqslant 1 + L \frac{\epsilon}{L} = 1 + \epsilon \qquad \forall 1 \leqslant \ell \leqslant N,$$

i.e. $T_\ell^{-1} \circ \Phi$ is Lipschitz (1-Hölder) continuous with constant $1 + \epsilon$, see (15.8). Therefore, the special transformation rule T6.10 for Lebesgue measure and T15.1 show

$$\lambda^n\big(\Phi(J_\ell)\big) = \lambda^n\big(T_\ell \circ T_\ell^{-1} \circ \Phi(J_\ell)\big)$$

$$= |\det T_\ell| \cdot \lambda^n\big(T_\ell^{-1} \circ \Phi(J_\ell)\big)$$

$$\leqslant |\det T_\ell| (1+\epsilon)^n \lambda^n(J_\ell).$$

Since $J = \biguplus_{\ell=1}^N J_\ell$ and $|\det T_\ell| \leqslant |\det D\Phi(x)|$ for all $x \in J_\ell$, we get

$$\lambda^n(\Phi(J)) \leqslant \sum_{\ell=1}^N \lambda^n(\Phi(J_\ell)) \leqslant (1+\epsilon)^n \sum_{\ell=1}^N |\det T_\ell| \lambda^n(J_\ell)$$

$$\leqslant (1+\epsilon)^n \sum_{\ell=1}^N \int_{J_\ell} |\det D\Phi(x)| \lambda^n(dx)$$

$$= (1+\epsilon)^n \int_J |\det D\Phi(x)| \lambda^n(dx),$$

and the proof is finished by letting $\epsilon \to 0$. ∎

We can finally proceed to the proof of Theorem 15.5.

Proof (of Theorem 15.5) Set $\Psi := \Phi^{-1}$. Since Ψ is continuous, $\mu := \lambda^d \circ \Phi = \Psi(\lambda^d)$ is a measure on $\mathcal{B}(\mathbb{R}^n_x)$, compare T7.6 and D7.7. The determinant $\det D\Phi$ is also continuous, thus $\nu(A) := \int_A |\det D\Phi(x)| \lambda^n(dx)$ defines a measure on $\mathcal{B}(\mathbb{R}^n_x)$, see L10.8. From Lemma 15.7 we know that $\mu(J) \leqslant \nu(J) < \infty$ for all rectangles $J \in \mathcal{J}(\mathbb{R}^n_x)$, and Lemma 15.6 shows that $\mu \leqslant \nu$ holds on the whole of $\mathcal{B}(\mathbb{R}^n_x)$, i.e.

$$\lambda^n(\Phi(X)) \leqslant \int_X |\det D\Phi(x)| \, \lambda^n(dx) \qquad \forall X \in \mathcal{B}(\mathbb{R}^n_x). \tag{15.12}$$

This proves '$\leqslant$' of (15.9). For the other direction our strategy is to apply Lemma 15.7 to the inverse function $\Psi = \Phi^{-1}$. If $X = \Phi^{-1}(Y)$, $Y \in \mathcal{B}(\mathbb{R}^n_y)$, (15.12) becomes

$$\int \mathbf{1}_Y(y) \, \lambda^n(dy) = \lambda^n(Y) \leqslant \int \mathbf{1}_{\Phi^{-1}(Y)}(x) \, |\det D\Phi(x)| \, \lambda^n(dx)$$

$$= \int \mathbf{1}_Y(\Phi(x)) \, |\det D\Phi(x)| \, \lambda^n(dx),$$

and with exactly the same arguments which we used to prove Theorem 14.1, this inequality is easily extended from indicator functions to all $u \in \mathcal{M}^+(\mathcal{B}(\mathbb{R}^n_y))$:

$$\int_{\mathbb{R}^n_y} u(y) \, \lambda^n(dy) \leqslant \int_{\mathbb{R}^n_x} u(\Phi(x)) \, |\det D\Phi(x)| \, \lambda^n(dx). \tag{15.13}$$

Switching in (15.13) the rôles of $\mathbb{R}^n_x \leftrightarrow \mathbb{R}^n_y$, $x \leftrightarrow y$ and considering the C^1-diffeomorphism $\Psi : \mathbb{R}^n_y \to \mathbb{R}^n_x$ (instead of Φ) and the measurable[✓] function $u(x) := \mathbf{1}_{\Phi(A)} \circ \Phi(x) \, |\det D\Phi(x)|$ for some $A \in \mathcal{B}(\mathbb{R}^n_x)$ yields

$$\int_{\mathbb{R}^n_x} \mathbf{1}_{\Phi(A)} \circ \Phi(x) \, |\det D\Phi(x)| \, \lambda^n(dx)$$

$$\leqslant \int_{\mathbb{R}^n_y} \left(\mathbf{1}_{\Phi(A)} \circ \Phi \, |\det D\Phi| \right) \circ \Phi^{-1}(y) \, |\det D(\Phi^{-1})(y)| \, \lambda^n(dy)$$

$$= \int_{\mathbb{R}^n_y} \mathbf{1}_{\Phi(A)}(y) \cdot |\det(D\Phi) \circ \Phi^{-1}(y)| \cdot |\det D(\Phi^{-1})(y)| \, \lambda^n(dy)$$

$$= \int_{\mathbb{R}^n_y} \mathbf{1}_{\Phi(A)}(y) \cdot \left| \det \left[\underbrace{(D\Phi) \circ \Phi^{-1}(y) \cdot D(\Phi^{-1})(y)}_{\mathrm{id}_n = D(\mathrm{id}_n) = D(\Phi \circ \Phi^{-1}) = (D\Phi) \circ \Phi^{-1} \cdot D(\Phi^{-1})} \right] \right| \, \lambda^n(dy)$$

$$= \int_{\mathbb{R}^n_y} \mathbf{1}_{\Phi(A)}(y) \, \lambda^n(dy) \; = \; \lambda^n(\Phi(A)).$$

This proves that for all $A \in \mathcal{B}(\mathbb{R}^n_x)$

$$\int_{\mathbb{R}^n_x} \mathbf{1}_A(x) \, |\det D\Phi(x)| \, \lambda^n(dx) = \int_{\mathbb{R}^n_x} \mathbf{1}_{\Phi(A)} \circ \Phi(x) \, |\det D\Phi(x)| \, \lambda^n(dx)$$

$$\leqslant \lambda^n(\Phi(A)),$$

and, together with the converse inequality (15.12), the theorem follows. ∎

If $X \subset \mathbb{R}_x^n$, $Y \subset \mathbb{R}_y^n$ are open sets and $\Phi : X \to Y$ is a C^1-diffeomorphism, we still can apply Theorem 15.5 to $A = \Phi^{-1}(B)$, $A \in X \cap \mathcal{B}(\mathbb{R}_x^n)$, $B \in Y \cap \mathcal{B}(\mathbb{R}_y^n)$ to get

$$\lambda^n\big|_Y = \Phi(\mu)\big|_Y = \Phi(\mu\big|_X), \qquad \mu(\cdot) := \int_{\cdot \cap X} |\det D\Phi(x)| \, \lambda^n(dx), \qquad (15.14)$$

i.e. Theorem 14.1 yields the following important result.

15.8 Corollary (General transformation theorem) *Let $X, Y \subset \mathbb{R}^n$ be open sets and $\Phi : X \to Y$ be a C^1-diffeomorphism. A function $u : Y \to \bar{\mathbb{R}}$ is integrable w.r.t. λ^n if, and only if, the function $u \circ \Phi \cdot |\det D\Phi| : X \to \bar{\mathbb{R}}$ is integrable w.r.t. λ^n. In this case*

$$\int_Y u(y) \, \lambda^n(dy) = \int_X u(\Phi(x)) \, |\det D\Phi(x)| \, \lambda^n(dx). \qquad (15.15)$$

For many applications we need a somewhat reinforced version of C15.8 since Φ is often only *almost everywhere* a diffeomorphism. The following simple generalization takes care of that. Recall that $\bar{\lambda}^n$ is the completed Lebesgue measure, cf. Problems 4.13, 6.2, 10.11, 10.12, 13.11.

15.9 Corollary *Let $\Phi : X \to \mathbb{R}_y^n$ be a C^1-map on a measurable set $X \in \mathcal{B}^*(\mathbb{R}_x^n)$ whose open interior is denoted by X°. If $X \setminus X^\circ$ is a $\bar{\lambda}^n$-null set[5] and $\Phi\big|_{X^\circ}$ is a C^1-diffeomorphism onto $\Phi(X^\circ)$, then*

$$\int_{\Phi(X)} u(y) \, \bar{\lambda}^n(dy) = \int_X u \circ \Phi(x) \, |\det D\Phi(x)| \, \bar{\lambda}^n(dx) \qquad (15.16)$$

holds for all $\mathcal{B}^*$-measurable positive functions $u : \Phi(X) \to [0, \infty]$. Moreover, $u : \Phi(X) \to \mathbb{R}$ is $\bar{\lambda}^n$ integrable if, and only if, $u \circ \Phi \cdot |\det D\Phi| : X \to \bar{\mathbb{R}}$ is $\bar{\lambda}^n$ integrable; in this case (15.16) remains valid.

Proof The argument proving C15.8 remains literally valid for $\bar{\lambda}^n$, i.e. the difficulty of C15.9 is not the completion of the measure but the fact that Φ is only almost everywhere a diffeomorphism.

Since $\bar{\lambda}^n(X \setminus X^\circ) = 0$, we get $\Phi(X) \setminus \Phi(X^\circ) \subset \Phi(X \setminus X^\circ)$, cf. Chapter 2, which is again a $\bar{\lambda}^n$-null set by Lemma 15.3. In view of C10.10 we can alter $\mathcal{L}^1$-functions on null sets, which means that the equality

$$\int_{\Phi(X^\circ)} u \, d\bar{\lambda}^n = \int \left(\mathbf{1}_{\Phi(X^\circ)} \cdot u\right) \circ \Phi \cdot |\det D\Phi| \, d\bar{\lambda}^n$$

from C15.8 immediately implies (15.16). ∎

---

[5] i.e. a subset of a Borel null set.

15.10 Remark Formulae (15.9) and (15.15) have the following interesting interpretation in connection with the Radon–Nikodým theorem 19.2 and Lebesgue's
differentiation theorem for measures T19.20, in particular C19.21:

$$\frac{d\lambda^n \circ \Phi}{d\lambda^n}(x) = |\det D\Phi(x)| = \lim_{r \to 0} \frac{\lambda^n(\Phi(B_r(x)))}{\lambda^n(B_r(x))}.$$

Spherical coordinates and the volume of the unit ball

Some of the most interesting applications of Corollaries 15.8 and 15.9 are coordinate changes.

15.11 Example (Planar polar coordinates) Consider the map

$$P : (0, \infty) \times (0, 2\pi) \to \mathbb{R}^2 \setminus [0, \infty) \times \{0\}, \qquad P(r, \theta) := (r \cos \theta, r \sin \theta)$$

which introduces *polar coordinates* (r, θ) in $\mathbb{R}^2$. It is not hard to see that P
is bijective and even a C^1-diffeomorphism. The determinant of the Jacobian is
given by

$$\det\left(\frac{\partial P(r, \theta)}{\partial(r, \theta)}\right) = \left|\begin{pmatrix} \cos\theta & -r\sin\theta \\ \sin\theta & r\cos\theta \end{pmatrix}\right| = r\cos^2\theta + r\sin^2\theta = r.$$

Since $[0, \infty) \times \{0\}$ is a λ^2-null set, we can apply Corollary 15.8 (or 15.9) and find
for every $u : \mathbb{R}^2 \to \mathbb{R}$, $u \in \mathcal{L}^1(\mathbb{R}^2, \lambda^2)$

$$\int_{\mathbb{R}^2} u(x, y)\, d\lambda^2(x, y) = \int_{(0,\infty)\times(0,2\pi)} r\, u(r\cos\theta, r\sin\theta)\, d\lambda^2(r, \theta)$$

$$= \int_{(0,\infty)} \int_{(0,2\pi)} r\, u(r\cos\theta, r\sin\theta)\, d\lambda^1(\theta)\, d\lambda^1(r),$$

where we used Fubini's theorem 13.9 for the last equality. This shows, in
particular, that

$$u \in \mathcal{L}^1(\mathbb{R}^2) \iff (r, \theta) \mapsto r\, u(r\cos\theta, r\sin\theta) \in \mathcal{L}^1\big((0, \infty) \times (0, 2\pi)\big).$$

A simple but quite interesting application of planar polar coordinates is the
following formula which plays a central rôle in probability theory: this is where
the norming factor $\frac{1}{\sqrt{2\pi}}$ for the Gaussian distribution comes from.

15.12 Example We have

$$\int_{\mathbb{R}} e^{-x^2} \, d\lambda^1(x) = \sqrt{\pi}. \tag{15.17}$$

Proof: We use the following trick: by Tonelli's theorem 13.8

$$\left(\int_{\mathbb{R}} e^{-x^2} \, d\lambda^1(x) \right)^2 = \int_{\mathbb{R}} \int_{\mathbb{R}} e^{-x^2} e^{-y^2} \, d\lambda^1(x) \, d\lambda^1(y)$$

$$= \int_{\mathbb{R}^2} e^{-(x^2+y^2)} \, d\lambda^2(x, y)$$

$$= \int_{(0,\infty)} \int_{(0,2\pi)} r \, e^{-r^2} \, d\lambda^1(r) \, d\lambda^1(\theta).$$

Since re^{-r^2} is positive and improperly Riemann integrable[✓], we know that Lebesgue and Riemann integrals coincide (cf. 11.8, 11.18), and therefore

$$\left(\int_{\mathbb{R}} e^{-x^2} \, d\lambda^1(x) \right)^2 = \lambda^1(0, 2\pi) \int_0^\infty r \, e^{-r^2} \, dr = 2\pi \left[-\tfrac{1}{2} e^{-r^2} \right]_0^\infty = \pi.$$

Polar coordinates also exist in higher dimensions but, unfortunately, the formulae become quite messy. The idea here is that we parametrize $\mathbb{R}^n$ by the radius $r \in [0, \infty)$, and $n-1$ angles $\theta \in [0, 2\pi)$ and $\omega \in [-\pi/2, \pi/2)^{n-2}$, so that $x = P(r, \theta, \omega)$. The Jacobian is now of the form $r^{n-1} J(\theta, \omega)$ and, if we denote by $v = u \circ P$ the function u expressed in polar coordinates, the transformation formula gives

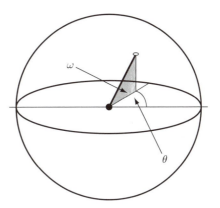

$$\int_{\mathbb{R}^n} u \, d\lambda^n = \iiint_{\substack{(0,\infty)\times(0,2\pi)\times \\ \times(-\pi/2,\pi/2)^{n-2}}} r^{n-1} v(r, \theta, \omega) \, |\det J(\theta, \omega)| \, d\lambda^1(r) \, d\lambda^1(\theta) \, d\lambda^{n-2}(\omega).$$

We will not give further details but settle for the slightly simpler case of *spherical coordinates* which will lead to a similar formula. Let $S^{n-1} := \{x \in \mathbb{R}^n : |x|^2 = 1\}$ be the unit sphere of $\mathbb{R}^n$ ($|x|^2 = x_1^2 + \cdots + x_n^2$ is the Euclidean norm) and set

$$\Phi : \mathbb{R}^n \setminus \{0\} \to (0, \infty) \times S^{n-1}, \quad x \mapsto (|x|, \phi(x))$$

where $\phi(x) := x/|x| \in S^{n-1}$ is the directional unit vector for x. Obviously, Φ is bijective, differentiable and has a differentiable inverse $\Phi^{-1}(r, s) = r \cdot s$.

15.13 Theorem *On $\mathcal{B}(S^{n-1}) = S^{n-1} \cap \mathcal{B}(\mathbb{R}^n)$ there exists a measure σ^{n-1} which is invariant under rotations and satisfies*

$$\int_{\mathbb{R}^n} u\, d\lambda^n = \iint_{(0,\infty) \times S^{n-1}} r^{n-1}\, u(rs)\, \lambda^1(dr)\, \sigma^{n-1}(ds) \qquad (15.18)$$

for all $u \in \mathcal{L}^1(\lambda^n)$. In other words, $\Phi(\lambda^n) = \mu \times \sigma^{n-1}$ where $\mu(dr) = r^{n-1}\, \mathbf{1}_{(0,\infty)}(r)\, \lambda^1(dr)$; in particular

$$u \in \mathcal{L}^1(\mathbb{R}^n, \lambda^n) \iff r^{n-1}\, u(rs) \in \mathcal{L}^1\big((0,\infty) \times S^{n-1}, \lambda^1 \times \sigma^{n-1}\big).$$

Proof We define σ^{n-1} by

$$\sigma^{n-1}(A) := n\, \lambda^n\big(\phi^{-1}(A) \cap B_1(0)\big) \quad \forall A \in \mathcal{B}(S^{n-1})$$

which is an image measure, hence a measure, cf. T7.6. Since ϕ^{-1} and λ^n are invariant w.r.t. rotations around the origin, see T7.9, it is obvious that σ^{n-1} inherits this property, too.

Both Φ and Φ^{-1} are continuous, hence measurable. Therefore,

$$\Phi^{-1}\big(\mathcal{B}(\mathbb{R}) \otimes \mathcal{B}(S^{n-1})\big) \subset \mathcal{B}(\mathbb{R}^n) \quad \text{and} \quad \Phi\big(\mathcal{B}(\mathbb{R}^n)\big) \subset \mathcal{B}(\mathbb{R}) \otimes \mathcal{B}(S^{n-1}),$$

which shows that $\mathcal{B}(\mathbb{R}^n) = \Phi^{-1}(\mathcal{B}(\mathbb{R}) \otimes \mathcal{B}(S^{n-1}))$. To see (15.18), fix $A \in \mathcal{B}(S^{n-1})$ and consider first the set $B := \{x \in \mathbb{R}^n : |x| \in [a,b),\ \phi(x) \in A\} = \phi^{-1}(A) \cap \{x : a \leqslant |x| < b\}$, which is clearly a Borel set of $\mathbb{R}^n$. Thus

$$\begin{aligned}
\lambda^n(B) &= \lambda^n\big(\phi^{-1}(A) \cap \{x : a \leqslant |x| < b\}\big) \\
&= \lambda^n\big(\phi^{-1}(A) \cap B_b(0)\big) - \lambda^n\big(\phi^{-1}(A) \cap B_a(0)\big) \\
&= b^n\, \lambda^n\big(\phi^{-1}(A) \cap B_1(0)\big) - a^n\, \lambda^n\big(\phi^{-1}(A) \cap B_1(0)\big) \\
&= (b^n - a^n)\, \lambda^n\big(\phi^{-1}(A) \cap B_1(0)\big),
\end{aligned}$$

where we used that $\lambda^n(a \cdot B) = a^n\, \lambda^n(B)$, cf. T7.10 or Problems 5.8, 7.7, and that ϕ^{-1} is invariant under dilations. This shows

$$\begin{aligned}
\lambda^n(B) = \tfrac{1}{n}\, (b^n - a^n)\, \sigma^{n-1}(A) &= \int_{[a,b)} r^{n-1}\, \sigma^{n-1}(A)\, \lambda^1(dr) \\
&= \mu \times \sigma^{n-1}\big([a,b) \times A\big).
\end{aligned}$$

Since the family $\{[a, b) \times A : a < b,\ A \in \mathcal{B}(S^{n-1})\}$ generates $\mathcal{B}(\mathbb{R}) \otimes \mathcal{B}(S^{n-1})$, see Lemma 13.3, and satisfies the conditions of the uniqueness theorem 5.7, the above relation extends to all sets $B' \in \mathcal{B}(\mathbb{R}) \otimes \mathcal{B}(S^{n-1})$. Since $\mathcal{B}(\mathbb{R}^n) = \Phi^{-1}(\mathcal{B}(\mathbb{R}) \otimes \mathcal{B}(S^{n-1}))$, we have $B' = \Phi(B)$ for some $B \in \mathcal{B}(\mathbb{R}^n)$, so that

$$\lambda^n(B) = \lambda^n(\Phi^{-1} \circ \Phi(B)) = \lambda^n(\Phi^{-1}(B')) = \mu \times \sigma^{n-1}(B').$$

All other assertions follow now from Theorem 14.1 on image integrals and Fubini's theorem 13.9. ∎

Let us note the particularly interesting case where $u(x) = f(|x|)$ is rotationally invariant.

15.14 Corollary *If $u(x) = f(|x|)$ is a rotationally invariant function, then $u \in \mathcal{L}^1(\mathbb{R}^n, \lambda^n)$ if, and only if, $r \mapsto r^{n-1} f(r) \in \mathcal{L}^1((0, \infty), \lambda^1)$. In this case*

$$\int_{\mathbb{R}^n} f(|x|)\, \lambda^n(dx) = n\,\omega_n \int_{(0,\infty)} r^{n-1}\, f(r)\, \lambda^1(dr),$$

where $\omega_n = \lambda^n(B_1(0))$ denotes the volume of the unit ball in $\mathbb{R}^n$. In particular, we get for the functions $f_\alpha(x) := |x|^\alpha$, $\alpha \in \mathbb{R}$,

$$f_\alpha \in \mathcal{L}^1(B_1(0) \setminus \{0\}) \iff \alpha > -n,$$
$$f_\alpha \in \mathcal{L}^1(\mathbb{R}^n \setminus B_1(0)) \iff \alpha < -n.$$

Proof The integral formula follows from (15.18) where the constant $n\,\omega_n := \sigma^{n-1}(S^{n-1})$.[6] That ω_n must be the volume of $B_1(0)$ is immediately clear if we choose $u(x) = \mathbf{1}_{B_1(0)}(x)$. The integrability of f_α follows now from Example 11.12. ∎

Let us finally determine ω_n, the volume of the unit ball in $\mathbb{R}^n$. For this we use the same method which we employed in Example 15.12:

$$(\sqrt{\pi})^n \overset{(15.17)}{=} \left(\int e^{-t^2} \lambda^1(dt) \right)^n = \int \cdots \int e^{-(x_1^2 + \cdots + x_n^2)}\, \lambda^1(dx_1) \ldots \lambda^2(dx_n)$$

$$\overset{15.14}{=} n\,\omega_n \int_{(0,\infty)} r^{n-1}\, e^{-r^2}\, \lambda^1(dr).$$

---

[6] This is, actually, the surface area of the unit ball $B_1(0)$ in $\mathbb{R}^n$.

Since $r^{n-1}e^{-r^2}$ is positive and improperly Riemann integrable[✓], Riemann and Lebesgue integrals coincide (use 11.8, 11.18), and we find after a change of variables according to $s = r^2$

$$(\sqrt{\pi})^n = n\,\omega_n \int_0^\infty r^{n-1}\,e^{-r^2}\,dr = \omega_n\,\frac{n}{2}\int_0^\infty s^{n/2-1}\,e^{-s}\,ds = \omega_n\,\frac{n}{2}\,\Gamma(\tfrac{n}{2}),$$

see Example 11.14. Since $\frac{n}{2}\,\Gamma(\frac{n}{2}) = \Gamma(\frac{n}{2}+1)$, we have finally established

15.15 Corollary $\omega_n = \lambda^n(B_1(0)) = \dfrac{\pi^{n/2}}{\Gamma(\frac{n}{2}+1)}.$

Continuous functions are dense in $\mathcal{L}^p(\lambda^n)$

We will now establish a result that is closely related to Lemma 15.2: we show that the *continuous functions with compact support* $C_c(\mathbb{R}^n)$ are dense in the space of Lebesgue p-integrable functions $\mathcal{L}^p(\lambda^n)$, $1 \leqslant p < \infty$, that is, if $u \in \mathcal{L}^p(\lambda^n)$, then

$$\forall\epsilon > 0 \quad \exists\phi = \phi_{\epsilon,u} \in C_c(\mathbb{R}^n) : \|u - \phi\|_p \leqslant \epsilon.$$

Since every compact set $K \subset \mathbb{R}^n$ is bounded, we find for some sufficiently large $R > 0$ that $K \subset [-R, R]^n$, hence $\lambda^n(K) \leqslant (2R)^n$. Thus for $\phi \in C_c(\mathbb{R}^n)$ with support $\operatorname{supp}\phi := \overline{\{\phi \neq 0\}} \subset K$,

$$\|\phi\|_p^p = \int |\phi|^p\,d\lambda^n = \int_K |\phi|^p\,d\lambda^n \leqslant \sup_{x\in K}|\phi(x)|^p\,(2R)^n < \infty,$$

so that $C_c(\mathbb{R}^n) \subset \mathcal{L}^p(\lambda^n)$ (measurability is clear because of continuity).

Our strategy will be to approximate first indicator functions of Borel sets and simple functions. For this we need the following

15.16 Lemma (Urysohn) *Let $K \subset \mathbb{R}^n$ be a compact set and $U \supset K$ be an open set. Then there exists a continuous function $\phi = \phi_{K,U} \in C(\mathbb{R}^n)$ such that $\mathbf{1}_K \leqslant \phi \leqslant \mathbf{1}_U$.*

Proof Let $d(x, A) := \inf_{y\in A}|x - y|$ be the distance of the point $x \in \mathbb{R}^n$ from the set $A \subset \mathbb{R}^n$. For $x, x' \in \mathbb{R}^n$ we have

$$d(x, A) = \inf_{y\in A}|x - y| \leqslant \inf_{y\in A}\big(|x - x'| + |x' - y|\big) = |x - x'| + d(x', A),$$

which shows, due to the symmetry in x and x', that $|d(x, A) - d(x', A)| \leqslant |x - x'|$, or, in other words, that $x \mapsto d(x, A)$ is continuous. It is now easy to see that the function

$$\phi(x) := \frac{d(x, U^c)}{d(x, K) + d(x, U^c)}$$

is continuous and satisfies $\mathbf{1}_K \leqslant \phi \leqslant \mathbf{1}_U$. ∎

15.17 Theorem $C_c(\mathbb{R}^n)$ *is a dense subset of* $\mathcal{L}^p(\lambda^n)$, $1 \leqslant p < \infty$.

Proof We have already verified that $C_c(\mathbb{R}^n) \subset \mathcal{L}^p(\lambda^n)$.

Step 1: $C(\mathbb{R}^n) \cap \mathcal{L}^p(\lambda^n)$ *is dense in* $\mathcal{E}(\mathcal{B}(\mathbb{R}^n)) \cap \mathcal{L}^p(\lambda^n)$. Let $B \in \mathcal{B}(\mathbb{R}^n)$ such that $\mathbf{1}_B \in \mathcal{L}^p(\lambda^n)$ (i.e. $\lambda^n(B) < \infty$). In steps 1,2 of the proof of Lemma 15.2 we constructed for such sets open sets U_ϵ and closed sets C_ϵ such that

$$C_\epsilon \subset B \subset U_\epsilon \quad \text{and} \quad (\lambda^n(U_\epsilon) - \lambda^n(B)) + (\lambda^n(B) - \lambda^n(C_\epsilon)) \leqslant \epsilon^p.$$

By the continuity of measures T4.4(iii) we find for the closed and bounded, hence compact, sets $\overline{B_j(0)} \cap C_\epsilon \uparrow C_\epsilon$ that $\lim_{j \to \infty} \lambda^n(\overline{B_j(0)} \cap C_\epsilon) = \lambda^n(C_\epsilon)$. This means that we can replace C_ϵ by a compact set $K_\epsilon \subset C_\epsilon$ and still have

$$K_\epsilon \subset B \subset U_\epsilon \quad \text{and} \quad (\lambda^n(U_\epsilon) - \lambda^n(K_\epsilon)) \leqslant (2\epsilon)^p.$$

Using Lemma 15.16 we find a continuous function $\phi_\epsilon := \phi_{U_\epsilon, K_\epsilon} \in C(\mathbb{R}^n)$ with $\mathbf{1}_{K_\epsilon} \leqslant \phi_\epsilon \leqslant \mathbf{1}_{U_\epsilon}$. As $\mathbf{1}_{K_\epsilon} \leqslant \mathbf{1}_B \leqslant \mathbf{1}_{U_\epsilon}$ we have, in particular,

$$\|\mathbf{1}_B - \phi_\epsilon\|_p \leqslant \|\mathbf{1}_B - \mathbf{1}_{K_\epsilon}\|_p + \|\mathbf{1}_{K_\epsilon} - \phi_\epsilon\|_p \leqslant 2\|\mathbf{1}_{U_\epsilon} - \mathbf{1}_{K_\epsilon}\|_p \leqslant 4\epsilon,$$

which also shows that $\phi_\epsilon \in \mathcal{L}^p(\lambda^n)$. Since any $f \in \mathcal{E}(\mathcal{B}(\mathbb{R}^n)) \cap \mathcal{L}^p(\lambda^n)$ has a standard representation of the form $f = \sum_{j=0}^M y_j \mathbf{1}_{B_j}$ where $y_0 = 0$ and $B_1, \ldots, B_M$ are Borel sets of finite volume, it is clear that $C(\mathbb{R}^n) \cap \mathcal{L}^p(\lambda^n)$ is dense in the set of all pth power integrable simple functions.

Step 2 : $C(\mathbb{R}^n) \cap \mathcal{L}^p(\lambda^n)$ *is dense in* $\mathcal{L}^p(\lambda^n)$. Fix $\epsilon > 0$. Since $\mathcal{E}(\mathcal{B}(\mathbb{R}^n)) \cap \mathcal{L}^p(\lambda^n)$ is dense in $\mathcal{L}^p(\lambda^n)$, cf. C12.11, there exists some $f_\epsilon \in \mathcal{E}(\mathcal{B}(\mathbb{R}^n)) \cap \mathcal{L}^p(\lambda^n)$ such that

$$\|f_\epsilon - u\|_p \leqslant \epsilon.$$

Using step 1 we find some $\phi_\epsilon \in C(\mathbb{R}^n) \cap \mathcal{L}^p(\lambda^n)$ with

$$\|\phi_\epsilon - f_\epsilon\|_p \leqslant \epsilon,$$

and the claim follows from Minkowski's inequality for $\|\cdot\|_p$

$$\|\phi_\epsilon - u\|_p \leqslant \|\phi_\epsilon - f_\epsilon\|_p + \|f_\epsilon - u\|_p \leqslant 2\epsilon.$$

Step 3 : $C_c(\mathbb{R}^n)$ *is dense in* $\mathcal{L}^p(\lambda^n)$. Let $\phi_\epsilon \in C(\mathbb{R}^n)$ be the function constructed in step 2. Using Lemma 15.16 we obtain a sequence of functions χ_j such that $\mathbf{1}_{\overline{B_j(0)}} \leqslant \chi_j \leqslant \mathbf{1}_{B_{j+1}(0)}$. Obviously, $\chi_j \phi_\epsilon \xrightarrow{j \to \infty} \phi_\epsilon$, $|\chi_j \phi_\epsilon| \leqslant |\phi_\epsilon|$ and $\chi_j \phi_\epsilon \in C_c(\mathbb{R}^n)$. Lebesgue's dominated convergence theorem 11.2 (or 12.9) therefore shows that

$$\lim_{j \to \infty} \|u - \chi_j \phi_\epsilon\|_p = \|u - \phi_\epsilon\|_p \leqslant 2\epsilon,$$

and the theorem is proved.   ∎

Regular measures

The seemingly innocuous question whether the continuous functions are a dense subset of $\mathcal{L}^p$ is – even for Lebesgue measure in $\mathbb{R}^n$ – quite hard to answer, as we have seen in Theorem 15.17. In general measure spaces, such results require a connection between *measure* and *topology* that reaches further than just considering the Borel (= topological) σ-algebra on a topological space $(X, \mathcal{O})$. This connection is made in the following

15.18 Definition Let $(X, \mathcal{O})$ be a topological space, denote by $\mathcal{K}$ the compact subsets of X and let μ be a measure on $(X, \mathcal{B})$, $\mathcal{B} = \sigma(\mathcal{O})$. The measure μ is called *outer regular* if

$$\mu(B) = \inf\{\mu(U) : U \in \mathcal{O}, \ U \supset B\} \qquad \forall B \in \mathcal{B},$$

and *(compact) inner regular* if

$$\mu(B) = \sup\{\mu(K) : K \in \mathcal{K}, \ K \subset B\} \qquad \forall B \in \mathcal{B}.$$

For Lebesgue measure λ^n on $(\mathbb{R}^n, \mathcal{B}(\mathbb{R}^n))$ we have proved outer and inner regularity in Lemma 15.2, see also step 1 in the proof of Theorem 15.17 and Problem 15.2. Let us note, without proof, the following characterization of outer regular measures.

15.19 Theorem *Let* (X, ρ) *be a complete separable metric space*[7] *and denote the open sets by* $\mathcal{O}$ *and the compact sets by* $\mathcal{K}$. *Every measure* μ *on* $(X, \mathcal{B}(X))$ *which is locally finite, i.e. every* $x \in X$ *has an open neighbourhood* $U = U(x)$ *of finite measure* $\mu(U) < \infty$, *is both outer regular and inner regular, i.e.*

$$\mu(B) = \inf\{\mu(U) : U \in \mathcal{O}, \ U \supset B\} = \sup\{\mu(K) : K \in \mathcal{K}, \ K \subset B\}.$$

---

[7] cf. Appendix B.

A proof can be found in Bauer [6, §26]. Note the analogy to Lemma 15.2 and the proof of Theorem 15.17 where we (essentially) verified Theorem 15.19 for Lebesgue measure. Also note that the measure μ in Theorem 15.19 is σ-finite: since X is separable, there is a countable dense subset $D \subset X$, and the collection

$$\Phi = \left\{ B_r(d) : r \in \mathbb{Q}^+, \ d \in D, \ B_r(d) \subset U(d), \quad U(d) \text{ as in T15.19} \right\}$$

is a countable family of open balls with finite μ-measure. Moreover, since every $U \in \mathcal{O}$ can be written in the form[8]

$$U = \bigcup_{\Phi \ni B_r(d) \subset U} B_r(d)$$

we find that $X = \bigcup_{N=1}^{\infty} \bigcup_{j=1}^{N} B_{r_j}(d_j)$ with $\mu(B_{r_j}(d_j)) < \infty$.

Almost the same argument that was used in the proof of Theorem 15.17 is valid in the abstract setting.

15.20 Theorem *Let $(X, \mathcal{O})$ be a topological space and μ be an outer regular measure on $(X, \mathcal{B}(X))$. Then the set*

$$C_{fin}(X) := \{u : X \to \mathbb{R} : u \text{ is continuous}, \quad \mu(\{u \neq 0\}) < \infty\}$$

is dense in $L^p(X, \mathcal{B}, \mu)$, $1 \leqslant p < \infty$.

Proof Let $A \in \mathcal{B}$ be a set with $\mu(A) < \infty$. Since μ is outer regular, we find for every $\epsilon > 0$ some $U \in \mathcal{O}$ such that

$$A \subset U \quad \text{and} \quad \mu(U) - \epsilon^p \leqslant \mu(A) \leqslant \mu(U).$$

Literally as in step 2 of the proof of Lemma 15.2 we can find some closed set F with

$$F \subset A \quad \text{and} \quad \mu(F) \leqslant \mu(A) \leqslant \mu(F) + \epsilon^p,$$

and, consequently, $\mu(U) - \mu(F) \leqslant 2\epsilon^p$. The rest of the proof is now as in T15.17. ∎

Problems

15.1. Let $F, F_1, F_2, F_3, \ldots$ be F_σ-sets in $\mathbb{R}^n$. Show that

(i) $F_1 \cap F_2 \cap \ldots \cap F_N$ is for every $N \in \mathbb{N}$ an F_σ-set;

(ii) $\bigcup_{j \in \mathbb{N}} F_j$ is an F_σ-set;

---

[8] This is similar to (3.2) in the proof of T3.8: the inclusion '$\subset$' is obvious, for '$\supset$' fix $x \in U$. Then there exists some $r \in \mathbb{Q}^+$ with $B_r(x) \subset U$. Since D is dense, $x \in B_{r/2}(d)$ for some $d \in D$ with $\rho(d, x) < r/4$, so that $x \in B_{r/2}(d) \subset U$.

(iii) F^c and $\bigcap_{j \in \mathbb{N}} F_j^c$ are G_δ-sets;

(iv) all closed sets are F_σ-sets.

15.2. Prove the following corollary to Lemma 15.2: *Lebesgue measure λ^n on $\mathbb{R}^n$ is outer regular, i.e.*

$$\lambda^n(B) = \inf \left\{ \lambda^n(U) : U \supset B, \ U \ \ open \right\} \qquad \forall B \in \mathcal{B}(\mathbb{R}^n),$$

and inner regular, i.e.

$$\lambda^n(B) = \sup \left\{ \lambda^n(F) : F \subset B, \ F \ \ closed \right\} \qquad \forall B \in \mathcal{B}(\mathbb{R}^n)$$
$$= \sup \left\{ \lambda^n(K) : K \subset B, \ K \ \ compact \right\} \qquad \forall B \in \mathcal{B}(\mathbb{R}^n).$$

15.3. **Completion (6).** Combine Problems 15.2 and 10.12 to show that the completion $\bar{\lambda}^n$ of n-dimensional Lebesgue measure is again inner and outer regular.

15.4. Consider the Borel σ-algebra $\mathcal{B}[0, \infty)$ and write $\lambda = \lambda^1|_{[0,\infty)}$ for Lebesgue measure on the half-line $[0, \infty)$.

(i) Show that $\mathcal{G} := \{[a, \infty) : a \geqslant 0\}$ generates $\mathcal{B}[0, \infty)$.

(ii) Show that $\mu(B) := \int_B \mathbf{1}_{[2,4]} \lambda(dx)$ and $\rho(B) := \mu(5 \cdot B)$, $B \in \mathcal{B}[0, \infty)$ are measures on $\mathcal{B}[0, \infty)$ such that $\rho|_{\mathcal{G}} \leqslant \mu|_{\mathcal{G}}$ but not $\rho \leqslant \mu$ in general.

Why does this not contradict Lemma 15.6?

15.5. Use Jacobi's transformation formula to recover Theorem 5.8(i), Problem 5.8 and Theorem 7.10. Show, in particular, that for all integrable functions $u : \mathbb{R}^n \to [0, \infty)$

$$\int u(x+y) \, \lambda^n(dx) = \int u(x) \, \lambda^n(dx) \qquad \forall y \in \mathbb{R}^n,$$

$$\int u(t\,x) \, \lambda^n(dx) = \frac{1}{t^n} \int u(x) \, \lambda^n(dx) \qquad \forall t > 0,$$

$$\int u(Ax) \, \lambda^n(dx) = \frac{1}{|\det A|} \int u(x) \, \lambda^n(dx) \qquad \forall A \in GL(n, \mathbb{R}).$$

In particular, the l.h.s. of the above equalities exists and is finite if, and only if, the r.h.s. exists and is finite.

Why can't we use 15.5 and 15.8 to *prove* these formulae?

15.6. **Arc-length.** Let $f : \mathbb{R} \to \mathbb{R}$ be a twice continuously differentiable function and denote by $\Gamma_f := \{(t, f(t)) : t \in \mathbb{R}\}$ its graph. Define a function $\Phi : \mathbb{R} \to \mathbb{R}^2$ by $\Phi(x) := (x, f(x))$. Then

(i) $\Phi : \mathbb{R} \to \Gamma_f$ is a C^1-diffeomorphism and $\det D\Phi(x) = 1 + (f'(x))^2$.

(ii) $\sigma := \Phi(|\det D\Phi| \, \lambda^1)$ is a measure on Γ_f.

(iii) $\int_{\Gamma_f} u(x, y) \, d\sigma(x, y) = \int_{\mathbb{R}} u(t, f(t)) \sqrt{1 + (f'(t))^2} \, d\lambda^1(t)$ with the understanding that whenever one side of the equality makes sense (measurability!) and is finite, so does the other.

The measure σ is called *canonical surface measure* on Γ_f. This name is justified by the following compatibility property w.r.t. λ^2: Let $n(x)$ be a unit normal vector to Γ_f at the point $(x, f(x))$ and define a map $\tilde{\Phi} : \mathbb{R} \times \mathbb{R} \to \mathbb{R}^2$ by $\tilde{\Phi}(x, r) := \Phi(x) + r\, n(x)$. Then

(iv) $n(x) = (-f'(x), 1)/\sqrt{1 + (f'(x))^2}$ and $\det D\tilde{\Phi}(x, r) = 1 + (f'(x))^2 - r f''(x)$.
Conclude that for every compact interval $[c, d]$ there exists some $\epsilon > 0$ such that $\tilde{\Phi}|_{(c,d) \times (-\epsilon, \epsilon)}$ is a C^1-diffeomorphism.

(v) Let $C \subset \Gamma_{f|(c,d)}$ and $r < \epsilon$ with ϵ as in (iv). Make a sketch of the set $C(r) := \tilde{\Phi}(\Phi^{-1}(C) \times (-r, r))$ and show that it is Borel measurable.

(vi) Use dominated convergence to show that for every $x \in (c, d)$

$$\lim_{r \downarrow 0} \frac{1}{2r} \int_{(-r,r)} \left| \det D\tilde{\Phi}(x, r) \right| \lambda^1(dr) = \left| \det D\tilde{\Phi}(x, 0) \right|.$$

(vii) Use the general transformation theorem 15.8, Tonelli's theorem 13.8, (vi) and dominated convergence to show that

$$\lim_{r \downarrow 0} \lambda^2(C(r)) = \int_{\Phi^{-1}(C)} \left| \det D\Phi(x) \right| \lambda^1(dx).$$

(viii) Conclude that $\int \sqrt{1 + (f'(t))^2}\, dt$ is the arc-length of the graph of Γ_f.

15.7. Let $\Phi : \mathbb{R}^d \to M \subset \mathbb{R}^n$, $d \leqslant n$, be a C^1-diffeomorphism.

(i) Show that $\lambda_M := \Phi(|\det D\Phi| \lambda^d)$ is a measure on M. Find a formula for $\int_M u\, d\lambda_M$.

(ii) Show that for a dilation $\theta_r : \mathbb{R}^n \to \mathbb{R}^n$, $x \mapsto rx$, $r > 0$, we have

$$\int_M u(r\xi)\, r^n\, d\lambda_M(\xi) = \int_{\theta_r(M)} u(\xi)\, d\lambda_M(\xi).$$

(iii) Let $M = \{\|x\| = 1\} = S^{n-1}$ be the unit sphere in $\mathbb{R}^n$, so that $d = n - 1$. Show that for every integrable $u \in \mathcal{L}^1(\mathbb{R}^n)$ and $\sigma := \lambda_M$

$$\int u(x) \lambda^n(dx) = \int_{(0,\infty)} \int_{\{\|x\| = r\}} u(x)\, \sigma(dx)\, \lambda^1(dr)$$

$$= \int_{(0,\infty)} \int_{\{\|x\| = 1\}} u(rx)\, \sigma(dx)\, \lambda^1(dr).$$

Remark. With somewhat more effort it is possible to show the analogue of the approximation formula in Problem 15.6(vii) for λ_M; all that changes are technical details, the idea of the proof is the same, cf. Stroock [50, pp. 94–101] for a nice presentation.

15.8. In Example 11.14 we introduced Euler's Gamma function:

$$\Gamma(t) = \int_{(0,\infty)} x^{t-1} e^{-x} \lambda^1(dx).$$

Show that $\Gamma(\frac{1}{2}) = \sqrt{\pi}$.

15.9.   **3-d polar coordinates.** Define $\Phi : [0, \infty) \times [0, 2\pi) \times [-\pi/2, \pi/2) \to \mathbb{R}^3$ by

$$\Phi(r, \theta, \omega) := \left(r \cos\theta \cos\omega, \, r \sin\theta \cos\omega, \, r \sin\omega \right).$$

Show that $|\det D\Phi(r, \theta, \omega)| = r^2 \cos\omega$ and find the integral formula for the coordinate change from Cartesian to polar coordinates $(x, y, z) \rightsquigarrow (r, \theta, \omega)$.

15.10.   Compute for $m, n \in \mathbb{N}$ the integral $\displaystyle\iint\limits_{B_1(0)} x^m y^n \, d\lambda^2(x, y)$.

16

Uniform integrability and Vitali's convergence theorem

Lebesgue's dominated convergence theorem 11.2 gives sufficient conditions which allow us to interchange limits and integrals. A crucial ingredient is the assumption that $|u_j| \leqslant w$ a.e. for all $j \in \mathbb{N}$ and some $w \in \mathcal{L}^1_+(\mu)$. This condition is not necessary, but a slightly weaker one is indeed necessary and sufficient in order to swap limits and integrals. The key idea is to control the size of the sets where the u_j exceed a given reference function. This is the rationale behind the next definition.

16.1 Definition Let $(X, \mathcal{A}, \mu)$ be a measure space and $\mathcal{F} \subset \mathcal{M}(\mathcal{A})$ be a family of measurable functions. We call $\mathcal{F}$ *uniformly integrable* (also: *equi-integrable*) if

$$\forall \epsilon > 0 \quad \exists w_\epsilon \in \mathcal{L}^1_+(\mu) : \sup_{u \in \mathcal{F}} \int_{\{|u| > w_\epsilon\}} |u| \, d\mu < \epsilon. \tag{16.1}$$

Note that there are other (but for $\mu(X) < \infty$ usually equivalent) definitions of uniform integrability, see Theorem 16.8 below for a discussion. We follow the universal formulation due to G. A. Hunt [21, p. 33].

The other key assumption in Theorem 11.2 was that $u_j(x) \xrightarrow{j \to \infty} u(x)$ for (almost) all $x \in X$; we can weaken this assumption, too.

16.2 Definition Let $(X, \mathcal{A}, \mu)$ be a measure space. A sequence of $\mathcal{A}$-measurable numerical functions $u_j : X \to \bar{\mathbb{R}}$ *converges in measure*[1] if

$$\forall \epsilon > 0, \ \forall A \in \mathcal{A}, \ \mu(A) < \infty : \lim_{j \to \infty} \mu\left(\{|u_j - u| > \epsilon\} \cap A\right) = 0 \tag{16.2}$$

holds for some $u \in \mathcal{M}(\mathcal{A})$. We write $\mu\text{-}\lim_{j \to \infty} u_j = u$ or $u_j \xrightarrow{\mu} u$.

---

[1] If μ is a probability measure one usually speaks of *convergence in probability*.

163

16.3 Example Convergence in measure is strictly weaker than pointwise convergence. To see this, take $(X, \mathcal{A}, \mu) = ([0, 1], \mathcal{B}[0, 1], \lambda^1|_{[0,1]})$ and set

$$u_n(x) := \mathbf{1}_{[j2^{-k}, (j+1)2^{-k}]}(x), \quad n = j + 2^k, \ 0 \leqslant j < 2^k.$$

This is a sequence of rectangular functions of width 2^{-k} moving in 2^k steps through $[0, 1]$, jump back to $x = 0$, halve their width and start moving again. Obviously,

$$\lambda^1(\{|u_n| > \epsilon\}) = 2^{-k} \xrightarrow{n = n(k) \to \infty} 0 \qquad \forall \epsilon \in (0, 1),$$

so that $u_n \xrightarrow{\lambda^1} 0$ in measure, but the pointwise limit $\lim_{n \to \infty} u_n(x)$ does not exist anywhere.[✓]

16.4 Lemma Let $(u_j)_{j \in \mathbb{N}} \subset \mathcal{L}^p(\mu)$, $p \in [1, \infty)$, and $(w_k)_{k \in \mathbb{N}} \subset \mathcal{M}(\mathcal{A})$. Then

(i) $\lim_{j \to \infty} \|u_j - u\|_p = 0$ implies $u_j \xrightarrow{\mu} u$;

(ii) $\lim_{k \to \infty} w_k(x) = w(x)$ a.e. implies $w_k \xrightarrow{\mu} w$.

Proof (i) follows immediately from the Markov inequality P10.12,

$$\mu\left(\{|u_j - u| > \epsilon\} \cap A\right) \leqslant \mu\left(\{|u_j - u| > \epsilon\}\right) = \mu\left(\{|u_j - u|^p > \epsilon^p\}\right)$$

$$\leqslant \frac{1}{\epsilon^p} \|u_j - u\|_p^p.$$

(ii) Observe that for all $\epsilon > 0$

$$\{|w_k - w| > \epsilon\} \subset \{\epsilon \wedge |w_k - w| \geqslant \epsilon\}.$$

An application of the Markov inequality P10.12 yields

$$\mu(\{|w_k - w| > \epsilon\} \cap A) \leqslant \mu(\{\epsilon \wedge |w_k - w| \geqslant \epsilon\} \cap A)$$

$$\leqslant \frac{1}{\epsilon} \int_A \epsilon \wedge |w_k - w| \, d\mu = \frac{1}{\epsilon} \int (\epsilon \wedge |w_k - w|) \mathbf{1}_A \, d\mu.$$

If $\mu(A) < \infty$, the function $\epsilon \mathbf{1}_A \in \mathcal{L}^1_+(\mu)$ is integrable, dominates the integrand $(\epsilon \wedge |w_k - w|) \mathbf{1}_A$, and Lebesgue's dominated convergence theorem 11.2 implies that $\lim_{k \to \infty} \int_A (\epsilon \wedge |w_k - w|) \, d\mu = 0$. ∎

16.5 Lemma Assume that $(X, \mathcal{A}, \mu)$ is σ-finite and that $(u_j)_{j \in \mathbb{N}} \subset \mathcal{M}(\mathcal{A})$ converges in measure to u. Then u is a.e. unique.

Proof Let $(A_k)_{k\in\mathbb{N}} \subset \mathcal{A}$ be a sequence with $A_k \uparrow X$ and $\mu(A_k) < \infty$. Suppose that u and w are two measurable functions such that $u_j \xrightarrow{\mu} u$ and $u_j \xrightarrow{\mu} w$. Because of $|u - w| \leqslant |u - u_j| + |u_j - w|$ we find for all $j, n \in \mathbb{N}$ that

$$\left\{ |u - w| > \tfrac{2}{n} \right\} \subset \left\{ |u - u_j| > \tfrac{1}{n} \right\} \cup \left\{ |u_j - w| > \tfrac{1}{n} \right\}.$$

Therefore,

$$\mu\left(A_k \cap \left\{ |u - w| > \tfrac{2}{n} \right\}\right)$$

$$\leqslant \mu\left(A_k \cap \left\{ |u - u_j| > \tfrac{1}{n} \right\}\right) + \mu\left(A_k \cap \left\{ |u_j - w| > \tfrac{1}{n} \right\}\right) \xrightarrow{j\to\infty} 0$$

holds for all $k, n \in \mathbb{N}$, i.e. $A_k \cap \left\{ |u - w| > \tfrac{2}{n} \right\}$ is a null set for all $k, n \in \mathbb{N}$; but then $\{u \neq w\} \subset \bigcup_{n\in\mathbb{N}} \left\{ |u - w| > \tfrac{2}{n} \right\} = \bigcup_{k,n\in\mathbb{N}} \left(A_k \cap \left\{ |u - w| > \tfrac{2}{n} \right\}\right)$ is also a null set, and we are done. ∎

Caution: Limits in measure on a non-σ-finite measure space $(X, \mathcal{A}, \mu)$ need not be unique, see Problem 16.6.

We are now ready for the main result of this chapter, which generalizes Lebesgue's dominated convergence theorem 11.2.

16.6 Theorem (Vitali) *Let $(X, \mathcal{A}, \mu)$ be σ-finite and let $(u_j)_{j\in\mathbb{N}} \subset \mathcal{L}^p(\mu)$, $p \in [1, \infty)$, be a sequence which converges in measure to some measurable function $u \in \mathcal{M}(\mathcal{A})$. Then the following assertions are equivalent:*

(i) $\lim\limits_{j\to\infty} \|u_j - u\|_p = 0$;

(ii) $\left(|u_j|^p\right)_{j\in\mathbb{N}}$ *is a uniformly integrable family;*

(iii) $\lim\limits_{j\to\infty} \int |u_j|^p \, d\mu = \int |u|^p \, d\mu.$

Proof (iii)⇒(ii): Since $\lim_{j\to\infty} \int |u_j|^p \, d\mu = \int |u|^p \, d\mu$, there exists some constant $C < \infty$ such that $\sup_{j\in\mathbb{N}} \int |u_j|^p \, d\mu \leqslant C$, and for every $\epsilon > 0$ there is some $N_\epsilon \in \mathbb{N}$ such that

$$\left| \int |u_j|^p \, d\mu - \int |u|^p \, d\mu \right| \leqslant \epsilon^p \qquad \forall j \geqslant N_\epsilon.$$

Setting $w_\epsilon := \max\{|u_1|, |u_2|, \ldots, |u_{N_\epsilon}|, |u|\}$, we have $w_\epsilon \in \mathcal{L}^p_+(\mu)^{[\checkmark]}$ and we see for every $\epsilon \in (0, 1)$ that

$$\left\{ |u_j| > \tfrac{1}{\epsilon} w_\epsilon \right\} = \emptyset \quad \forall j \leqslant N_\epsilon, \qquad \left\{ |u_j| > \tfrac{1}{\epsilon} w_\epsilon \right\} \subset \left\{ \epsilon |u_j| > |u| \right\} \quad \forall j \in \mathbb{N}.$$

This implies for all $j \in \mathbb{N}$ that

$$\int_{\{|u_j| > \frac{1}{\epsilon} w_\epsilon\}} |u_j|^p \, d\mu \leqslant \left| \int_{\{|u_j| > \frac{1}{\epsilon} w_\epsilon\}} \left(|u_j|^p - |u|^p \right) d\mu \right| + \int_{\{|u_j| > \frac{1}{\epsilon} w_\epsilon\}} |u|^p \, d\mu$$

$$\leqslant \epsilon^p + \int_{\{\epsilon |u_j| > |u|\}} |u|^p \, d\mu$$

$$\leqslant \epsilon^p + \epsilon^p \sup_{j \in \mathbb{N}} \int |u_j|^p \, d\mu \leqslant (1 + C) \, \epsilon^p.$$

Since

$$w_\epsilon \in \mathcal{L}_+^p(\mu) \Leftrightarrow w_\epsilon^p \in \mathcal{L}_+^1(\mu) \quad \text{and} \quad \{|u_j| > \tfrac{1}{\epsilon} w_\epsilon\} = \{|u_j|^p > \tfrac{1}{\epsilon^p} w_\epsilon^p\}, \quad (16.3)$$

we have established the uniform integrability of $\left(|u_j|^p \right)_{j \in \mathbb{N}}$.

(ii)⇒(i): Let us first check that the double sequence $\left(|u_j - u_k|^p \right)_{j,k \in \mathbb{N}}$ is again uniformly integrable. In view of (16.3), our assumption reads

$$\int_{\{|u_j| > w\}} |u_j|^p \, d\mu < \epsilon \qquad \forall \, j \in \mathbb{N} \qquad\qquad (16.4)$$

for some suitable $w = w_\epsilon \in \mathcal{L}_+^p(\mu)$. From $|a - b| \leqslant |a| + |b| \leqslant 2 \max\{|a|, |b|\}$ we deduce

$$\int_{\{|u_j - u_k| > 2w\}} |u_j - u_k|^p \, d\mu \leqslant 2^p \int_{\{|u_j - u_k| > 2w\}} \left(|u_j| \vee |u_k| \right)^p d\mu,$$

and since $|u_j - u_k| \leqslant |u_j| + |u_k|$ we get

$$\{|u_j - u_k| > 2w\} \subset \{|u_j| > w\} \cup \{|u_k| > w\}.$$

Consequently,

$$\int_{\{|u_j - u_k| > 2w\}} |u_j - u_k|^p \, d\mu$$

$$\leqslant 2^p \left\{ \int_{\{|u_j| > w\} \cap \{|u_k| > w\}} + \int_{\{|u_j| > w \geqslant |u_k|\}} + \int_{\{|u_k| > w \geqslant |u_j|\}} \right\} |u_j|^p \vee |u_k|^p \, d\mu$$

$$\leqslant 2^p \left\{ \int_{\{|u_j| > w\} \cap \{|u_k| > w\}} |u_j|^p \, d\mu + \int_{\{|u_j| > w\} \cap \{|u_k| > w\}} |u_k|^p \, d\mu \right\}$$

$$+ 2^p \int_{\{|u_j| > w\}} |u_j|^p \, d\mu + 2^p \int_{\{|u_k| > w\}} |u_k|^p \, d\mu$$

$$\overset{(16.4)}{\leqslant} 4 \cdot 2^p \, \epsilon = 2^{p+2} \, \epsilon.$$

From this we conclude that for $W = 2w \in \mathcal{L}^p_+(\mu)$ and large $R > 0$

$$\int |u_j - u_k|^p \, d\mu$$

$$= \int_{\{|u_j-u_k|>W\}} |u_j - u_k|^p \, d\mu + \int_{\{|u_j-u_k|\leqslant W\}} |u_j - u_k|^p \, d\mu$$

$$\leqslant 2^{p+2}\epsilon + \int_{\{|u_j-u_k|\leqslant W \wedge \epsilon\}} |u_j - u_k|^p \, d\mu + \int_{\{\epsilon < |u_j-u_k|\leqslant W\}} |u_j - u_k|^p \, d\mu$$

$$\leqslant 2^{p+2}\epsilon + \int \epsilon^p \wedge W^p \, d\mu + \left\{ \int_{\substack{\{|u_j-u_k|>\epsilon\} \\ \cap\{W>R\}}} W^p \, d\mu + \int_{\substack{\{|u_j-u_k|>\epsilon\} \\ \cap\{\epsilon<W\leqslant R\}}} W^p \, d\mu \right\}$$

$$\leqslant 2^{p+2}\epsilon + \int \epsilon^p \wedge W^p \, d\mu + \int_{\{W>R\}} W^p \, d\mu$$

$$+ R^p \mu\big(\{|u_j - u_k| > \epsilon\} \cap \{\epsilon < W \leqslant R\}\big).$$

Letting first $j, k \to \infty$ we find because of $u_j \xrightarrow{\mu} u$ that[✓]

$$\limsup_{j,k\to\infty} \int |u_j - u_k|^p \, d\mu \leqslant 2^{p+2}\epsilon + \int \epsilon^p \wedge W^p \, d\mu + \int_{\{W>R\}} W^p \, d\mu.$$

The last two terms vanish as $\epsilon \to 0$ and $R \to \infty$ by the dominated convergence theorem 12.9, so that $\lim_{j,k\to\infty} \int |u_j - u_k|^p \, d\mu = 0$. Since $\mathcal{L}^p(\mu)$ is complete (cf. T12.7), $(u_j)_{j\in\mathbb{N}}$ converges in $\mathcal{L}^p(\mu)$ to a limit $\tilde{u} \in \mathcal{L}^p(\mu)$.

Due to Lemma 16.4, $\mathcal{L}^p$-convergence also implies $u_j \xrightarrow{\mu} \tilde{u}$ and, by Lemma 16.5, we have $u = \tilde{u}$ a.e., hence $\mathcal{L}^p\text{-}\lim_{j\to\infty} u_j = u$.

(i)$\Rightarrow$(iii): is a consequence of the lower triangle inequality for the $\mathcal{L}^p$-norm, cf. the first part of the proof of Theorem 12.10 ∎

16.7 Remark Vitali's theorem 16.6 still holds for measure spaces $(X, \mathcal{A}, \mu)$ which are not σ-finite. In this case, however, we can no longer identify the $\mathcal{L}^p$-limit and the theorem reads: If $u_j \xrightarrow{\mu} u$, then the following are equivalent:

(i) $(u_j)_{j\in\mathbb{N}}$ converges in $\mathcal{L}^p$;
(ii) $(u_j)_{j\in\mathbb{N}}$ is uniformly integrable;
(iii) $(\|u_j\|_p)_{j\in\mathbb{N}}$ converges in $\mathbb{R}$.

The reason for this is evident from the proof of T16.6: the last few lines of the step (ii)$\Rightarrow$(i) require σ-finiteness of $(X, \mathcal{A}, \mu)$.

Different forms of uniform integrability[2]

In view of Vitali's convergence theorem 16.6 one is led to suspect that uniform integrability is essentially a sufficient (and also necessary, if $(X, \mathcal{A}, \mu)$ is σ-finite) condition for weak sequential relative compactness in $\mathcal{L}^1(\mu)$, i.e.

$$(16.1) \implies \begin{cases} \text{every } (u_j)_{j \in \mathbb{N}} \subset \mathcal{F} \text{ has a subsequence } (u_{j(k)})_{k \in \mathbb{N}} \text{ such} \\ \text{that } \lim_{k \to \infty} \int u_{j(k)} \cdot \phi \, d\mu \text{ exists for all } \phi \in \mathcal{L}^\infty(\mu). \end{cases}$$

(see Dunford and Schwartz [15, pp. 289–90, 386–7]). In $\mathcal{L}^p(\mu)$, $1 < p < \infty$, uniform boundedness of $\mathcal{F} \subset \mathcal{L}^p(\mu)$ is enough for this:

$$\sup_{u \in \mathcal{F}} \|u\|_p < \infty \iff \begin{cases} \text{every } (u_j)_{j \in \mathbb{N}} \subset \mathcal{F} \text{ has a subsequence} \\ (u_{j(k)})_{k \in \mathbb{N}} \text{ such that } \lim_{k \to \infty} \int u_{j(k)} \cdot \phi \, d\mu \\ \text{exists for all } \phi \in \mathcal{L}^q(\mu), \ \frac{1}{p} + \frac{1}{q} = 1. \end{cases}$$

This is a consequence of the reflexivity of the spaces $\mathcal{L}^p(\mu)$, $p > 1$.

Let us give various equivalent conditions for uniform integrability.

16.8 Theorem *Let $(X, \mathcal{A}, \mu)$ be some measure space and $\mathcal{F} \subset \mathcal{L}^1(\mathcal{A})$. Then the following statements* **(i)–(iv)** *are equivalent:*

(i) $\mathcal{F}$ *is uniformly integrable, i.e. (16.1) holds;*

(ii) a) $\sup\limits_{u \in \mathcal{F}} \int |u| \, d\mu < \infty$;

     b) $\forall \epsilon > 0 \ \ \exists w_\epsilon \in \mathcal{L}^1_+(\mathcal{A}), \ \delta > 0 \ \ \forall B \in \mathcal{A} : \int_B w_\epsilon \, d\mu < \delta$

               $\implies \sup\limits_{u \in \mathcal{F}} \int_B |u| \, d\mu < \epsilon$;

(iii) a) $\sup\limits_{u \in \mathcal{F}} \int |u| \, d\mu < \infty$;

     b) $\forall \epsilon > 0 \ \ \exists K_\epsilon \in \mathcal{A}, \ \mu(K_\epsilon) < \infty : \sup\limits_{u \in \mathcal{F}} \int_{K_\epsilon^c} |u| \, d\mu < \epsilon$;

     c) $\forall \epsilon > 0 \ \ \exists \delta > 0 \ \ \forall B \in \mathcal{A} : \mu(B) < \delta \implies \sup\limits_{u \in \mathcal{F}} \int_B |u| \, d\mu < \epsilon$;

(iv) a) $\forall \epsilon > 0 \ \ \exists K_\epsilon \in \mathcal{A}, \ \mu(K_\epsilon) < \infty : \sup\limits_{u \in \mathcal{F}} \int_{K_\epsilon^c} |u| \, d\mu < \epsilon$;

     b) $\lim\limits_{R \to \infty} \sup\limits_{u \in \mathcal{F}} \int_{\{|u| > R\}} |u| \, d\mu = 0$.

If $(X, \mathcal{A}, \mu)$ is a **σ-finite** *measure space,* **(i)–(iv)** *are also equivalent to*

---

[2] This section can be left out at first reading.

(v) a) $\displaystyle\sup_{u\in\mathcal{F}} \int |u|\,d\mu < \infty;$

    b) $\displaystyle\lim_{j\to\infty} \sup_{u\in\mathcal{F}} \int_{A_j} u\,d\mu = 0$ *for every decreasing sequence* $(A_j)_{j\in\mathbb{N}} \subset \mathcal{A}$, $A_j \downarrow$

    $\emptyset$. *[Note:* $\mu(A_j) < \infty$ *is* **not** *assumed.]*

If $(X, \mathcal{A}, \mu)$ *is a* **finite** *measure space,* **(i)–(v)** *are also equivalent to*

(vi) $\displaystyle\lim_{R\to\infty} \sup_{u\in\mathcal{F}} \int_{\{|u|>R\}} |u|\,d\mu = 0;$

(vii) $\displaystyle\sup_{u\in\mathcal{F}} \int \Phi(|u|)\,d\mu < \infty$ *for some increasing, convex function* $\Phi : [0, \infty)$

    $\to [0, \infty)$ *such that* $\displaystyle\lim_{t\to\infty} \frac{\Phi(t)}{t} = \infty.$

16.9 Remark Almost any combination of the above criteria appears in the literature as uniform integrability or under different names. Here is a short list:

(ii-a) – uniform boundedness

(iii-b) – tightness

(iii-c) – uniform absolute continuity

(v-b) – uniform σ-additivity

(vii) – de la Vallée Poussin's condition

(iii) – Dieudonné's condition (weak seq. relative compactness)

(v) – Dunford–Pettis condition (weak seq. relative compactness)

Proof (of Theorem 16.8) First we show **(iv)**$\Rightarrow$**(iii)**$\Rightarrow$**(ii)**$\Rightarrow$**(i)**$\Rightarrow$**(iv)** for general measure spaces, then **(ii)**$\Rightarrow$**(v)**$\Rightarrow$**(i)** for σ-finite measure spaces and, finally, for finite measure spaces **(iv)**$\Rightarrow$**(vi)**$\Rightarrow$**(vii)**$\Rightarrow$**(i)**.

    (iv)$\Rightarrow$**(iii):** Condition **(iii-b)** is clear. Given $\epsilon > 0$ we can pick $K = K_{\epsilon/2} \in \mathcal{A}$ and $R = R_{\epsilon/2} > 0$ such that

$$\int |u|\,d\mu = \int_{K\cap\{|u|>R\}} |u|\,d\mu + \int_{K\cap\{|u|\leqslant R\}} |u|\,d\mu + \int_{K^c} |u|\,d\mu$$

$$\leqslant \frac{\epsilon}{2} + R\mu(K) + \frac{\epsilon}{2} < \infty,$$

uniformly for all $u \in \mathcal{F}$. Setting $\delta := \frac{\epsilon}{2R}$ we see for every $B \in \mathcal{A}$ with $\mu(B) < \delta$ that

$$\int_B |u|\,d\mu = \int_{B\cap\{|u|>R\}} |u|\,d\mu + \int_{B\cap\{|u|\leqslant R\}} |u|\,d\mu$$

$$\leqslant \int_{\{|u|>R\}} |u|\,d\mu + R\mu(B) \leqslant \frac{\epsilon}{2} + R\delta = \epsilon,$$

and **(iii)** follows.

(iii)⇒(ii): Condition **(ii-a)** is clear. Given $\epsilon > 0$ we pick $K = K_\epsilon \in \mathcal{A}$ with $\mu(K) < \infty$ and $\delta = \delta_\epsilon > 0$ and set $w_\epsilon := \mathbf{1}_{K_\epsilon}$. If $B \in \mathcal{A}$ is such that $\mu(B \cap K_\epsilon) = \int_B w_\epsilon \, d\mu < \delta$, we get from **(iii-c)** and **(iii-b)** that

$$\int_B |u| \, d\mu = \int_{B \cap K_\epsilon} |u| \, d\mu + \int_{B \cap K_\epsilon^c} |u| \, d\mu \leqslant \epsilon + \epsilon,$$

uniformly for all $u \in \mathcal{F}$ which is just **(ii-b)**.

(ii)⇒(i): Take $w = w_\epsilon$ and $\delta = \delta_\epsilon > 0$ as in **(ii)**. If $R > \frac{1}{\delta} \sup_{u \in \mathcal{F}} \int |u| \, d\mu$ we see

$$\int |u| \, d\mu \geqslant \int_{\{|u| > Rw\}} |u| \, d\mu \geqslant R \int_{\{|u| > Rw\}} w \, d\mu,$$

and so

$$\int_{\{|u| > Rw\}} w \, d\mu \leqslant \frac{1}{R} \sup_{u \in \mathcal{F}} \int |u| \, d\mu \leqslant \delta.$$

From **(ii-b)** we infer that $\sup_{u \in \mathcal{F}} \int_{\{|u| > Rw\}} |u| \, d\mu \leqslant \epsilon$.

(i)⇒(iv): Let $w = w_\epsilon$ be as in **(i)** resp. (16.1). Since $\{|u| \leqslant w\} \cap \{|u| \geqslant R\} \subset \{w \geqslant R\}$, we have

$$\int_{\{|u| > R\}} |u| \, d\mu = \int_{\{|u| > w\} \cap \{|u| > R\}} |u| \, d\mu + \int_{\{|u| \leqslant w\} \cap \{|u| > R\}} |u| \, d\mu$$

$$\leqslant \int_{\{|u| > w\}} |u| \, d\mu + \int_{\{|w| > R\}} w \, d\mu \qquad (16.5)$$

$$\leqslant \epsilon + \int w \mathbf{1}_{\{|w| > R\}} \, d\mu.$$

From the dominated convergence theorem 11.2 we see that the right-hand side tends (uniformly for all $u \in \mathcal{F}$) to ϵ as $R \to \infty$ and **(iv-b)** follows. To see **(iv-a)** we choose $r = r_\epsilon > 0$ so small that $\int_{\{w \leqslant r\}} w \, d\mu \leqslant \int w \wedge r \, d\mu \leqslant \epsilon$; this is possible since by Lebesgue's convergence theorem 11.2 $\lim_{r \to 0} \int |w| \wedge r \, d\mu = 0$. By the Markov inequality P10.12 we see $\mu(\{w > r\}) \leqslant \frac{1}{r} \int w \, d\mu < \infty$, and we get for $K := \{w > r\}$

$$\sup_{u \in \mathcal{F}} \int_{K^c} |u| \, d\mu = \sup_{u \in \mathcal{F}} \left(\int_{\{w \leqslant r\} \cap \{|u| > w\}} |u| \, d\mu + \int_{\{w \leqslant r\} \cap \{|u| \leqslant w\}} |u| \, d\mu \right)$$

$$\overset{(16.1)}{\leqslant} \epsilon + \sup_{u \in \mathcal{F}} \int_{\{w \leqslant r\} \cap \{|u| \leqslant w\}} |u| \, d\mu$$

$$\leqslant \epsilon + \int_{\{w \leqslant r\}} w \, d\mu$$

$$\leqslant 2\epsilon.$$

This proves **(iv)**.

Assume for the rest of the proof that μ is σ-finite

(ii)$\Rightarrow$(v): **(v-a)** is clear. If $A_j \downarrow \emptyset$ we see from the monotone convergence theorem 11.1 that $\lim_{j\to\infty} \int_{A_j} w\, d\mu = 0$, so that for we have by **(ii-b)** $\sup_{u\in\mathscr{F}} \int_{A_j} u\, d\mu \leqslant \sup_{u\in\mathscr{F}} \int_{A_j} |u|\, d\mu < \epsilon$ for sufficiently large $j \in \mathbb{N}$.

(v)$\Rightarrow$(i): Note that for the positive, resp. negative parts $u^\pm$ of u

$$\int_{A_j} u^\pm \, d\mu = \int_{A_j\cap\{\pm u\geqslant 0\}} (\pm u)\, d\mu \qquad \text{and} \qquad A_j\cap\{\pm u\geqslant 0\}\downarrow\emptyset,$$

which implies that we may replace u in **(v-b)** by $|u|$. Since μ is σ-finite, we can find an exhausting sequence $E_k \in \mathcal{A}$, $E_k \uparrow X$, $\mu(E_k) < \infty$. The function

$$w := \sum_{k\in\mathbb{N}} \frac{2^{-k}}{1+\mu(E_k)} \mathbf{1}_{E_k}$$

is clearly positive and $\in \mathcal{L}^1_+(\mu)$. Assume **(i)** false; in particular,

$$\exists\, \eta > 0 \quad \forall j \in \mathbb{N} : \sup_{u\in\mathscr{F}} \int_{\{|u|>jw\}} |u|\, d\mu > \eta.$$

But $A_j := \{|u| > jw\} \downarrow \emptyset$ and **(v)** (with the above discussed modification) will then lead to a contradiction.

Assume for the rest of the proof that μ is finite

(iv)$\Rightarrow$(vi): is trivial.

(vi)$\Rightarrow$(vii): For $u \in \mathscr{F}$ we set $\alpha_n := \alpha_n(u) := \mu(\{|u| > n\})$ and define

$$\Phi(t) := \int_{[0,t)} \phi(s)\, \lambda(ds), \qquad \phi(s) := \sum_{n=1}^{\infty} \gamma_n \mathbf{1}_{[n,n+1)}(s).$$

We will now determine the numbers $\gamma_1, \gamma_2, \gamma_3, \dots$. Clearly,

$$\Phi(t) = \sum_{n=1}^{\infty} \gamma_n \int_{[0,t)} \mathbf{1}_{[n,n+1)}(s)\, \lambda(ds) = \sum_{n=1}^{\infty} \gamma_n [(t-n)^+ \wedge 1]$$

and

$$\int \Phi(|u|)\, d\mu = \sum_{n=1}^{\infty} \gamma_n \int \left[(|u|-n)^+ \wedge 1 \right] d\mu \leqslant \sum_{n=1}^{\infty} \gamma_n \, \mu(|u| > n). \tag{16.6}$$

If we can construct $(\gamma_n)_{n\in\mathbb{N}}$ such that it increases to ∞ and (16.6) is finite (uniformly for all $u \in \mathcal{F}$), then we are done: $\phi(s)$ will increase to ∞, $\Phi(t)$ will be convex[3] and satisfy

$$\frac{\Phi(t)}{t} = \frac{1}{t}\int_{[0,t)}\phi(s)\,\lambda(ds) \geqslant \frac{1}{t}\int_{[t/2,t)}\phi(s)\,\lambda(ds) \geqslant \frac{1}{2}\phi\left(\frac{t}{2}\right) \uparrow \infty.$$

By assumption we can find an increasing sequence $(r_j)_{j\in\mathbb{N}} \subset \mathbb{R}$ such that $\lim_{j\to\infty} r_j = \infty$ and $\int_{\{|u|>r_j\}}|u|\,d\mu \leqslant 2^{-j}$. Thus

$$\sum_{k=r_j}^{\infty}\mu(\{|u|>k\}) = \sum_{k=r_j}^{\infty}\sum_{\ell=k}^{\infty}\mu(\{\ell < |u| \leqslant \ell+1\})$$

$$= \sum_{\ell=r_j}^{\infty}\sum_{k=r_j}^{\ell}\mu(\{\ell < |u| \leqslant \ell+1\})$$

$$\leqslant \sum_{\ell=r_j}^{\infty}\ell\,\mu(\{\ell < |u| \leqslant \ell+1\}).$$

Now sum the above inequality over $j = 1, 2, 3, \ldots$ to get

$$\sum_{j=1}^{\infty}\sum_{k=r_j}^{\infty}\mu(\{|u|>k\}) \leqslant \sum_{j=1}^{\infty}\sum_{\ell=r_j}^{\infty}\ell\,\mu(\{\ell < |u| \leqslant \ell+1\})$$

$$\leqslant \sum_{j=1}^{\infty}\sum_{\ell=r_j}^{\infty}\int_{\{\ell<|u|\leqslant\ell+1\}}|u|\,d\mu$$

$$= \sum_{j=1}^{\infty}\underbrace{\int_{\{|u|>r_j\}}|u|\,d\mu}_{\leqslant 2^{-j}\text{ by assumption}} \leqslant 1,$$

---

[3] Usually one argues that $\Phi'' \geqslant 0$ a.e., but for this we need to know that the monotone function $\phi = \Phi'$ is almost everywhere differentiable – and this requires Lebesgue's differentiation theorem 19.20. Here is an alternative elementary argument: it is not hard to see that $\Phi : (a,b) \to \mathbb{R}$ is convex if, and only if, $\frac{\Phi(y)-\Phi(x)}{y-x} \leqslant \frac{\Phi(z)-\Phi(x)}{z-x}$ holds for all $a < x < y < z < b$, use e.g. the technique of the proof of Lemma 12.13. Since $\Phi(x) = \int_0^x \phi(s)\,ds$ (by L13.12 and T11.8), this is the same as

$$\frac{1}{y-x}\int_x^y \phi(s)\,ds \leqslant \frac{1}{z-x}\int_x^z \phi(s)\,ds \iff \frac{1}{y-x}\int_x^y \phi(s)\,ds \leqslant \frac{1}{z-y}\int_y^z \phi(s)\,ds$$

$$\iff \int_0^1 \phi(s(y-x)+x)\,ds \leqslant \int_0^1 \phi(s(z-y)+y)\,ds.$$

The latter inequality follows from the fact that ϕ is increasing and $s(y-x)+x \in [x,y]$ while $s(z-y)+y \in [y,z]$ for $0 \leqslant s \leqslant 1$.

and interchange the order of summation in the first double sum on the left:

$$\sum_{j=1}^{\infty} \sum_{k=r_j}^{\infty} \mu(\{|u| > k\}) = \sum_{k=1}^{\infty} \underbrace{\left(\sum_{j=1}^{\infty} \mathbf{1}_{[1,k]}(r_j) \right) \mu(\{|u| > k\})}_{=: \ \gamma_k} \leqslant 1.$$

This finishes the construction of the sequence $(\gamma_k)_{k\in\mathbb{N}}$.

(vii)⇒(i): Since $\mu(X) < \infty$, constants are integrable and we may take $w_\epsilon(x) := r_\epsilon$ for all $x \in X$. Fix $\epsilon > 0$ and choose r_ϵ so big that $t^{-1} \Phi(t) > 1/\epsilon$ for all $t > r_\epsilon$. Then

$$\int_{\{|u|>r_\epsilon\}} |u| \, d\mu \leqslant \int_{\{|u|>r_\epsilon\}} \epsilon \, \Phi(|u|) \, d\mu \leqslant \epsilon \int \Phi(|u|) \, d\mu,$$

and **(i)** follows. ∎

Problems

16.1. Let $(X, \mathcal{A}, \nu)$ be a finite measure space and $(u_j)_{j\in\mathbb{N}} \subset \mathcal{M}(\mathcal{A})$. Prove that

$$\lim_{k\to\infty} \mu\left(\left\{ \sup_{j\geqslant k} |u_j| > \epsilon \right\} \right) = 0 \quad \forall \epsilon > 0 \quad \Longrightarrow \quad f_j \xrightarrow{j\to\infty} 0 \text{ a.e.}$$

[Hint: $|u_j| \to 0$ a.e. if, and only if, $\mu\left(\bigcup_{j\geqslant k} \{|u_j| > \epsilon\} \right)$ is small for all $\epsilon > 0$ and big $k \geqslant k_\epsilon$.]

16.2. Show that for a sequence $(u_j)_{j\in\mathbb{N}}$ of measurable functions on a finite measure space

$$\lim_{k\to\infty} \mu\left(\left\{ \sup_{j\geqslant k} |u_j| > \epsilon \right\} \right) = \mu\left(\limsup_{j\to\infty} \{|u_j| > \epsilon\} \right) \quad \forall \epsilon > 0,$$

and combine this with Problem 16.1 to give a new criterion for a.e. convergence.

16.3. Let $(X, \mathcal{A}, \mu)$ be a measure space and $(u_j)_{j\in\mathbb{N}} \subset \mathcal{M}(\mathcal{A})$. Show that $u_j \xrightarrow{j\to\infty} u$ in measure if, and only if, $u_j - u_k \xrightarrow{j,k\to\infty} 0$ in measure.

16.4. Consider one-dimensional Lebesgue measure λ on $([0,1], \mathcal{B}[0,1])$. Compare the convergence behaviour (a.e., $\mathcal{L}^p$, in measure) of the following sequences:

(i) $f_{n,j} := n \mathbf{1}_{[(j-1)/n, j/n]}$, $n \in \mathbb{N}$, $1 \leqslant j \leqslant n$ run through in lexicographical order;
(ii) $g_n := n \mathbf{1}_{(0,1/n)}$, $n \in \mathbb{N}$;
(iii) $h_n := a_n(1 - nx)^+$, $n \in \mathbb{N}$, $x \in [0, 1]$ and a sequence $(a_n)_{n\in\mathbb{N}} \subset \mathbb{R}^+$.

16.5. Let $(u_j)_{j\in\mathbb{N}}$, $(w_j)_{j\in\mathbb{N}}$ be two sequences of measurable functions on $(X, \mathcal{A}, \mu)$. Suppose that $u_j \xrightarrow{\mu} u$ and $w_j \xrightarrow{\mu} w$. Show that $au_j + bw_j$, $a, b \in \mathbb{R}$, $\max\{u_j, w_j\}$, $\min\{u_j, w_j\}$ and $|u_j|$ converge in measure and find their limits.

16.6. Let $(X, \mathcal{A}, \mu)$ be a measure space which is not σ-finite. Construct an example of a sequence $(u_j)_{j\in\mathbb{N}} \subset \mathcal{M}(\mathcal{A})$ which converges in measure but whose limit is not unique. Can this happen in a σ-finite measure space?

[Hint: let $X_{\sigma f} := \bigcup \{F : \mu(F) < \infty\}$ be the σ-finite part of X. Show that $X \setminus X_{\sigma f} \neq \emptyset$, that every measurable $E \subset X \setminus X_{\sigma f}$ satisfies $\mu(E) = \infty$ and that we can change every limit of $(u_j)_{j \in \mathbb{N}}$ outside $X_{\sigma f}$.]

16.7. (i) Prove, without using Vitali's convergence theorem, the following

Theorem (Bounded convergence). *Let $(X, \mathcal{A}, \mu)$ be a measure space, $A \in \mathcal{A}$ be a set with $\mu(A) < \infty$ and $(u_j)_{j \in \mathbb{N}}$ be a sequence of measurable functions. Suppose that all u_j vanish on A^c, that $|u_j| \leqslant C$ for all $j \in \mathbb{N}$ and some constant $C > 0$ and that $u_j \xrightarrow{\mu} u$. Then L^1-$\lim_j u_j = u$.*

(ii) Use one-dimensional Lebesgue measure and the sequence $u_j = \mathbb{1}_{[j,j+1)}$ to show that the assumption $\mu(A) < \infty$ is really needed in (i).

(iii) As L^1-limit the function u is unique but, as we have seen in Problem 16.6, this is not the case for limits in measure. Why does the uniqueness of the limit in (i) not contradict Problem 16.6?

16.8. Let $(\Omega, \mathcal{A}, P)$ be a probability space. Define for two random variables X, Y

$$\rho_\mu(X, Y) := \inf \left\{ \epsilon > 0 : P(\{|X - Y| \geqslant \epsilon\}) \leqslant \epsilon \right\}.$$

(i) ρ_μ is a pseudo-metric on the space of random variables $\mathcal{M}(\mathcal{A})$, i.e. ρ_μ satisfies properties (d_2), (d_3) of a metric, cf. Appendix B, Definition B.15.

(ii) A sequence $(X_j)_{j \in \mathbb{N}} \subset \mathcal{M}(\mathcal{A})$ converges in probability to a random variable X if, and only if, $\rho_\mu(X_j, X) \xrightarrow{j \to \infty} 0$.

(iii) ρ_μ is a complete pseudo-metric on $\mathcal{M}(\mathcal{A})$, i.e. every ρ_μ-Cauchy sequence converges in probability to some limit in $\mathcal{M}(\mathcal{A})$.

(iv) Show that

$$g_\mu(X, Y) := \int \frac{|X - Y|}{1 + |X - Y|}\, dP \quad \text{and} \quad \delta_\mu(X, Y) := \int [|X - Y| \wedge 1]\, dP$$

are pseudo-metrics on $\mathcal{M}(\mathcal{A})$ which have the same Cauchy sequences as ρ_μ.

16.9. Let $(X, \mathcal{A}, \mu)$ be a σ-finite measure space. Suppose that $(A_j)_{j \in \mathbb{N}} \subset \mathcal{A}$ satisfies $\mu(A_j) \xrightarrow{j \to \infty} 0$. Show that

$$\lim_{j \to \infty} \int_{A_j} u\, d\mu = 0 \qquad \forall u \in \mathcal{L}^1(\mu).$$

[Hint: use Vitali's convergence theorem 16.6.]

16.10. Let $(X, \mathcal{A}, \mu)$ be a measure space and $(u_n)_{n \in \mathbb{N}} \subset \mathcal{M}(\mathcal{A})$.

(i) Let $(x_n)_{n \in \mathbb{N}} \subset \mathbb{R}$. Show that $x_n \xrightarrow{n \to \infty} 0$ if, and only if, every subsequence $(x_{n_k})_{k \in \mathbb{N}}$ satisfies $x_{n_k} \xrightarrow{k \to \infty} 0$.

(ii) Show that $u_n \xrightarrow{\mu} u$ if, and only if, every subsequence $(u_{n_k})_{k \in \mathbb{N}}$ has a sub-subsequence $(\tilde{u}_{n_k})_{k \in \mathbb{N}}$ which converges a.e. to u on every set $A \in \mathcal{A}$ of finite μ-measure.

[Hint: use L16.4 for necessity. For sufficiency show that $\tilde{u}_{n_k} \to u$ in measure, hence the sequence of reals $\mu(A \cap \{|u_{n_k} - u| > \epsilon\})$ has a subsequence converging to 0; use (i) to conclude that $\mu(A \cap \{|u_n - u| > \epsilon\}) \to 0$.]

(iii) Use part (ii) to show that $u_n \xrightarrow{\mu} u$ entails that $\Phi \circ u_n \xrightarrow{\mu} \Phi \circ u$ for every continuous function $\Phi : \mathbb{R} \to \mathbb{R}$.

16.11. Let $\mathcal{F}$ and $\mathcal{G}$ be two families of uniformly integrable functions on an arbitrary measure space $(X, \mathcal{A}, \mu)$. Show that

(i) every finite collection of functions $\{f_1, \ldots, f_n\} \subset \mathcal{L}^1(\mu)$ is uniformly integrable.

(ii) $\mathcal{F} \cup \{f_1, \ldots, f_n\}$, $f_1, \ldots, f_n \in \mathcal{L}^1(\mu)$, is uniformly integrable.

(iii) $\mathcal{F} + \mathcal{G} := \{f + g : f \in \mathcal{F}, g \in \mathcal{G}\}$ is uniformly integrable.

(iv) c.h.$(\mathcal{F}) := \{tf + (1-t)\phi : f, \phi \in \mathcal{F}, 0 \leqslant t \leqslant 1\}$ ('c.h.' stands for convex hull) is uniformly integrable.

(v) the closure of c.h.$(\mathcal{F})$ in the space $\mathcal{L}^1$ is uniformly integrable.

16.12. Assume that $(u_j)_{j \in \mathbb{N}}$ is uniformly integrable. Show that

$$\lim_{k \to \infty} \frac{1}{k} \int \sup_{j \leqslant k} u_j \, d\mu = 0.$$

16.13. Let $(\Omega, \mathcal{A}, P)$ be a probability space. Adapt the proof of Theorem 16.8 to show that a sequence $(u_j)_{j \in \mathbb{N}} \subset \mathcal{L}^1(\mu)$ is uniformly integrable if it is bounded in some space $\mathcal{L}^p(P)$ with $p > 1$, i.e. if $\sup_{j \in \mathbb{N}} \|u_j\|_p < \infty$.

Use Vitali's convergence theorem 16.6 to construct an example illustrating that $\mathcal{L}^1$-boundedness of $(u_j)_{j \in \mathbb{N}}$ does not guarantee uniform integrability.

16.14. Let $(X, \mathcal{A}, \mu)$ be a finite measure space and $\mathcal{F} \subset \mathcal{L}^1(\mu)$ be a family of integrable functions. Show that $\mathcal{F}$ is uniformly integrable if, and only if, $\sum_{j=1}^{\infty} j \mu(\{j < |f| \leqslant j+1\})$ converges uniformly for all $f \in \mathcal{F}$.

[Hint: compare (vi)$\Rightarrow$(vii) of the proof of Theorem 16.8.]

17

Martingales

Martingales are a key tool of modern probability theory, in particular, when it comes to a.e. convergence assertions and related limit theorems. The origins of martingale techniques can be traced back to analysis papers by Kac, Marcinkiewicz, Paley, Steinhaus, Wiener and Zygmund from the early 1930s on independent (or orthogonal) functions and the convergence of certain series of functions, see e.g. the paper by Marcinkiewicz and Zygmund [28] which contains many references. The theory of martingales as we know it now goes back to Doob and most of the material of this and the following chapter can be found in his seminal monograph [13] from 1953.

We want to understand martingales as an analysis tool which will be useful for the study of L^p- and almost everywhere convergence and, in particular, for the further development of measure and integration theory. Our presentation differs somewhat from the standard way to introduce martingales – conditional expectations will be defined later in Chapter 22 – but the results and their proofs are pretty much the usual ones. The only difference is that we develop the theory for σ-finite measure spaces rather than just for probability spaces. Those readers who are familiar with martingales and the language of conditional expectations we ask for patience until Chapter 23, in particular Theorem 23.9, when we catch up with these notions.

Throughout this chapter $(X, \mathcal{A}, \mu)$ is a measure space which admits a *filtration*, i.e. an increasing sequence

$$\mathcal{A}_0 \subset \mathcal{A}_1 \subset \ldots \subset \mathcal{A}_j \subset \ldots \subset \mathcal{A}$$

of sub-σ-algebras of $\mathcal{A}$. If $(X, \mathcal{A}_0, \mu)$ is σ-finite[1] we call $(X, \mathcal{A}, \mathcal{A}_j, \mu)$ a *σ-finite filtered measure space*. This will always be the case from now on. Finally,

---

[1] i.e. $(A_j)_{j \in \mathbb{N}} \subset \mathcal{A}_0$ with $A_j \uparrow X$ and $\mu(A_j) < \infty$.

we write $\mathcal{A}_\infty := \sigma(\mathcal{A}_j, j = 0, 1, 2, \ldots)$ for the smallest σ-algebra generated by all $\mathcal{A}_j$.

17.1 Definition Let $(X, \mathcal{A}, \mathcal{A}_j, \mu)$ be a σ-finite filtered measure space. A sequence of $\mathcal{A}$-measurable functions $(u_j)_{j\in\mathbb{N}}$ is called a *martingale* (w.r.t. the filtration $(\mathcal{A}_j)_{j\in\mathbb{N}}$), if $u_j \in \mathcal{L}^1(\mathcal{A}_j)$ for each $j \in \mathbb{N}$ and if

$$\int_A u_{j+1} \, d\mu = \int_A u_j \, d\mu \qquad \forall A \in \mathcal{A}_j. \tag{17.1}$$

We say that $(u_j)_{j\in\mathbb{N}}$ is a *submartingale* (w.r.t. $(\mathcal{A}_j)_{j\in\mathbb{N}}$) if $u_j \in \mathcal{L}^1(\mathcal{A}_j)$ and

$$\int_A u_{j+1} \, d\mu \geqslant \int_A u_j \, d\mu \qquad \forall A \in \mathcal{A}_j, \tag{17.2}$$

and a *supermartingale* (w.r.t. $(\mathcal{A}_j)_{j\in\mathbb{N}}$) if $u_j \in \mathcal{L}^1(\mathcal{A}_j)$ and

$$\int_A u_{j+1} \, d\mu \leqslant \int_A u_j \, d\mu \qquad \forall A \in \mathcal{A}_j. \tag{17.3}$$

If we want to emphasize the underlying filtration, we write $(u_j, \mathcal{A}_j)_{j\in\mathbb{N}}$.

17.2 Remark **(i)** It is enough to assume instead of (17.1) that $\int_G u_{j+1} \, d\mu = \int_G u_j \, d\mu$ for all $G \in \mathcal{G}_j$ where $\mathcal{G}_j$ is a generator of $\mathcal{A}_j$ containing an exhausting sequence $(G_k)_{k\in\mathbb{N}} \subset \mathcal{G}_j$ with $G_k \uparrow X$. This follows from the fact that

$$\int_A u_{j+1} \, d\mu = \int_A u_j \, d\mu \quad \Longleftrightarrow \quad \underbrace{\int_A (u_{j+1}^+ + u_j^-) \, d\mu}_{=:\,\nu(A)} = \underbrace{\int_A (u_{j+1}^- + u_j^+) \, d\mu}_{=:\,\rho(A)}$$

where ν, ρ are finite measures on $\mathcal{A}_j$ and from the uniqueness theorem 5.7: $\nu|_{\mathcal{G}_j} = \rho|_{\mathcal{G}_j}$ implies – under our assumptions on $\mathcal{G}_j$ – that $\nu = \rho$ on $\mathcal{A}_j$.

(For sub- or supermartingales we need, in addition, that $\mathcal{G}_j$ is a semi-ring, cf. Lemma 15.6.)

(ii) Set $\mathcal{S}_j := \{A \in \mathcal{A}_j : \mu(A) < \infty\}$. It is not hard to see that $\mathcal{S}_j$ is a semi-ring and that, because of σ-finiteness, $\sigma(\mathcal{S}_j) = \mathcal{A}_j$. Therefore (ii) means that it is enough to assume (17.1)–(17.3) for all sets in $\mathcal{S}_j$, i.e. for all sets with *finite* μ-measure.

(iii) Condition (17.2) in Definition 17.1 is equivalent to

$$\int \phi u_{j+1} \, d\mu \geqslant \int \phi u_j \, d\mu \qquad \forall \phi \in \mathcal{L}_+^\infty(\mathcal{A}_j). \tag{17.2'}$$

Indeed: Since $\phi := \mathbf{1}_A \in \mathcal{L}_+^\infty(\mathcal{A}_j)$ for all $A \in \mathcal{A}_j$, (17.2$'$) implies (17.2). Conversely, if $\phi \in \mathcal{E}^+(\mathcal{A}_j)$ is a simple function, (17.2$'$) follows from (17.2) by linearity. For general $\phi \in \mathcal{L}_+^\infty(\mathcal{A}_j)$, we find by T8.8 a sequence of $\mathcal{A}_j$-measurable simple functions ϕ_k such that $\phi_k \leqslant \phi$ and $\phi_k \uparrow \phi$. Since $\phi u_j, \phi u_{j+1} \in \mathcal{L}^1(\mathcal{A})$, we can use Lebesgue's dominated convergence theorem 11.2 and get

$$\int \phi u_{j+1} \, d\mu = \lim_{k \to \infty} \int \phi_k u_{j+1} \, d\mu \overset{(17.2')}{\geqslant} \lim_{k \to \infty} \int \phi_k u_j \, d\mu = \int \phi u_j \, d\mu.$$

Similar statements hold for martingales (17.1) and supermartingales (17.3).

(iv) With some obvious (notational) changes in Definiton 17.1 we can also consider other index sets such as $\mathbb{N}_0$, $\mathbb{Z}$ or $-\mathbb{N}$.

17.3 Examples Let $(X, \mathcal{A}, \mathcal{A}_j, \mu)$ be a σ-finite filtered measure space.

(i) $(u_j)_{j \in \mathbb{N}}$ is a martingale if, and only if, it is both a sub- and a supermartingale.

(ii) $(u_j)_{j \in \mathbb{N}}$ is a supermartingale if, and only if, $(-u_j)_{j \in \mathbb{N}}$ is a submartingale.

(iii) Let $(u_j)_{j \in \mathbb{N}}$ and $(w_j)_{j \in \mathbb{N}}$ be [sub-]martingales and let α, β be [positive] real numbers. Then $(\alpha u_j + \beta w_j)_{j \in \mathbb{N}}$ is a [sub-]martingale.

(iv) Let $(u_j)_{j \in \mathbb{N}}$ be a submartingale. Then $(u_j^+)_{j \in \mathbb{N}}$ is a submartingale.

Indeed: Take $A \in \mathcal{A}_j$ and observe that $\{u_j \geqslant 0\} \in \mathcal{A}_j$. Then

$$\int_A u_{j+1}^+ \, d\mu \geqslant \int_{A \cap \{u_j \geqslant 0\}} u_{j+1}^+ \, d\mu \geqslant \int_{A \cap \{u_j \geqslant 0\}} u_{j+1} \, d\mu$$

$$\overset{(17.2)}{\geqslant} \int_{A \cap \{u_j \geqslant 0\}} u_j \, d\mu = \int_A u_j^+ \, d\mu.$$

(v) Let $(u_j)_{j \in \mathbb{N}}$ be a martingale. Then $(|u_j|)_{j \in \mathbb{N}}$ is a submartingale. This follows from $|u_j| = 2u_j^+ - u_j$, (iii) and (iv).

(vi) Let $(u_j)_{j \in \mathbb{N}}$ be a martingale. If $u_j \in \mathcal{L}^p(\mathcal{A}_j)$ for some $p \in [1, \infty)$, then $(|u_j|^p)_{j \in \mathbb{N}}$ a submartingale.

Indeed: Note that $|y|^p - |x|^p = \int_{|x|}^{|y|} p \, t^{p-1} \, dt \geqslant p|x|^{p-1}(|y| - |x|)$ for all $x, y \in \mathbb{R}$ where we set, as usual, $\int_{|x|}^{|y|} = -\int_{|y|}^{|x|}$ if $|x| > |y|$. If we take $y = u_{j+1}$ and $x = u_j$ and integrate over $A \in \mathcal{A}_j$, we find by dominated convergence T11.2

$$\int_A \left(|u_{j+1}|^p - |u_j|^p \right) d\mu \geqslant p \int \left(\mathbf{1}_A |u_j|^{p-1} \right) \left(|u_{j+1}| - |u_j| \right) d\mu$$

$$= \lim_{N \to \infty} p \int \underbrace{\left[\left(\mathbf{1}_A |u_j|^{p-1} \right) \wedge N \right]}_{\in \mathcal{L}_+^\infty(\mathcal{A}_j)} \left(|u_{j+1}| - |u_j| \right) d\mu$$

$$\overset{(17.2'),(v)}{\geqslant} \quad 0,$$

since $(|u_j|)_{j \in \mathbb{N}}$ is, by (v), a submartingale.

(vii) Let $u_j \in \mathcal{L}^1(\mathcal{A}_j)$, $j \in \mathbb{N}$, and $u_1 \leqslant u_2 \leqslant u_3 \leqslant \dots$. Then $(u_j)_{j \in \mathbb{N}}$ is a submartingale.

(viii) Let $(X, \mathcal{A}, \mu) = \big([0, 1), \mathcal{B}[0, 1), \lambda := \lambda^1|_{[0,1)}\big)$ and consider the finite (σ-) algebras generated by all *dyadic intervals* of $[0, 1)$ of length 2^{-j}, $j \in \mathbb{N}_0$:

$$\mathcal{A}_j^\Delta := \sigma\big([0, 2^{-j}), \dots, [k2^{-j}, (k+1)2^{-j}), \dots, [(2^j-1)2^{-j}, 1)\big).$$

Obviously, $\mathcal{A}_0^\Delta \subset \mathcal{A}_1^\Delta \subset \dots \subset \mathcal{B}[0, 1)$ and $\big([0, 1), \mathcal{B}[0, 1), \mathcal{A}_j^\Delta, \lambda\big)$ is a (σ-) finite filtered measure space. *Then* $(u_j)_{j \in \mathbb{N}_0}$, $u_j := 2^j \mathbf{1}_{[0, 2^{-j})}$, *is a martingale.*

Indeed: Since the sets $[k2^{-j}, (k+1)2^{-j})$, $k = 0, 1, \dots, 2^j - 1$ are a disjoint partition of $[0, 1)$, every $A \in \mathcal{A}$ consists of a (finite) disjoint union of such sets. If $[0, 2^{-j}) \subset A$, we have

$$\int_A u_{j+1} \, d\lambda = \int 2^{j+1} \mathbf{1}_{A \cap [0, 2^{-(j+1)})} \, d\lambda = 2^{j+1} 2^{-(j+1)}$$

$$= 2^j 2^{-j} = \int 2^j \mathbf{1}_{A \cap [0, 2^{-j})} \, d\lambda = \int_A u_j \, d\lambda$$

and, otherwise,

$$\int_A u_{j+1} \, d\lambda = \int_A 2^{j+1} \mathbf{1}_{[0, 2^{-(j+1)})} \, d\lambda = 0 = \int_A 2^j \mathbf{1}_{[0, 2^{-j})} \, d\lambda = \int_A u_j \, d\lambda.$$

(ix) Let $(X, \mathcal{A}, \mu) = \big([0, \infty)^n, \mathcal{B}([0, \infty)^n), \lambda = \lambda^n|_{[0,\infty)^n}\big)$ and consider the σ-algebras $\mathcal{A}_j$ generated by the lattice of half-open *dyadic squares* of side-length 2^{-j}, $j \in \mathbb{N}_0$,

$$\mathcal{A}_j^\Delta := \sigma\big(z + [0, 2^{-j})^n : z \in 2^{-j}\mathbb{N}_0^n\big), \quad j \in \mathbb{N}_0.$$

Then $\mathcal{A}_0^\Delta \subset \mathcal{A}_1^\Delta \subset \dots \subset \mathcal{B}([0, \infty)^n)$, and $\big([0, \infty)^n, \mathcal{B}([0, \infty)^n), \mathcal{A}_j^\Delta, \lambda\big)$ is a σ-finite filtered measure space.

For every real-valued function $u \in \mathcal{L}^1([0, \infty)^n, \lambda)$ we can define an $\mathcal{A}_j^\Delta$-measurable step function u_j on the dyadic squares in $\mathcal{A}_j^\Delta$ by

$$u_j(x) := \sum_{z \in 2^{-j}\mathbb{N}_0^n} \frac{\int_{z+[0,2^{-j})^n} u \, d\lambda}{\lambda\big(z + [0, 2^{-j})^n\big)} \mathbf{1}_{z+[0,2^{-j})^n}(x)$$

$$(17.4)$$

$$= \sum_{z \in 2^{-j}\mathbb{N}_0^n} \left\{ \int u \frac{\mathbf{1}_{z+[0,2^{-j})^n}}{\lambda\big(z + [0, 2^{-j})^n\big)} \, d\lambda \right\} \mathbf{1}_{z+[0,2^{-j})^n}(x).$$

Then $(u_j, \mathcal{A}_j^\Delta)_{j \in \mathbb{N}}$ *is a martingale.*

Indeed: Since the sets $z+[0,2^{-j})^n$ are disjoint for different $z \in 2^{-j}\mathbb{N}_0^n$, the sums in (17.4) are actually finite sums.

That $u_j \in \mathcal{L}^1(\mathcal{A}_j^{\Delta})$ is clear from the construction. To see (17.1), fix $z' \in 2^{-j}\mathbb{N}_0^n$ and $j \in \mathbb{N}_0$ and observe that for all $k = j, j+1, j+2, \dots$

$$\int_{z'+[0,2^{-j})^n} u_k(x)\,\lambda(dx)$$

$$= \sum_{z \in 2^{-k}\mathbb{N}_0^n} \left\{ \int u \frac{\mathbf{1}_{z+[0,2^{-k})^n}}{\lambda(z+[0,2^{-k})^n)}\,d\lambda \right\} \cdot \int \mathbf{1}_{z+[0,2^{-k})^n}\mathbf{1}_{z'+[0,2^{-j})^n}\,d\lambda$$

$$= \sum_{\substack{z \in 2^{-k}\mathbb{N}_0^n \\ z+[0,2^{-k})^n \subset z'+[0,2^{-j})^n}} \int u \frac{\mathbf{1}_{z+[0,2^{-k})^n}}{\lambda(z+[0,2^{-k})^n)}\,d\lambda \cdot \lambda(z+[0,2^{-k})^n)$$

$$= \sum_{\substack{z \in 2^{-k}\mathbb{N}_0^n \\ z+[0,2^{-k})^n \subset z'+[0,2^{-j})^n}} \int_{z+[0,2^{-k})^n} u(x)\,\lambda(dx)$$

$$= \int_{z'+[0,2^{-j})^n} u(x)\,\lambda(dx).$$

The r.h.s. is independent of k and, therefore, we get

$$\int_{z'+[0,2^{-j})^n} u_j\,d\lambda = \int_{z'+[0,2^{-j})^n} u\,d\lambda = \int_{z'+[0,2^{-j})^n} u_{j+1}\,d\lambda.$$

Since $\mathcal{A}_j^{\Delta}$ is generated by (disjoint unions of) squares of the form $z' + [0,2^{-j})^n$, $z' \in 2^{-j}\mathbb{N}_0^n$, the claim follows from Remark 17.2(i).

(x) Assume that $(X, \mathcal{A}, \mu)$ is a probability space, i.e. a measure space where $\mu(X) = 1$. A family of real functions $(u_j)_{j \in \mathbb{N}} \subset \mathcal{L}^1(\mathcal{A})$ is called *independent*, if

$$\mu\left(\bigcap_{j=1}^M u_j^{-1}(B_j)\right) = \prod_{j=1}^M \mu(u_j^{-1}(B_j)) \tag{17.5}$$

holds for all $M \in \mathbb{N}$ and any choice of $B_1, B_2, \dots, B_M \in \mathcal{B}(\mathbb{R})$. If $\mathcal{A}_k := \sigma(u_1, u_2, \dots, u_k)$ is the σ-algebra generated by $u_1, u_2, \dots, u_k$, then the sequence of partial sums

$$s_k := u_1 + u_2 + \cdots + u_k, \qquad k \in \mathbb{N},$$

is an $(\mathcal{A}_k)_{k \in \mathbb{N}}$-submartingale if, and only if, $\int u_j\,d\mu \geqslant 0$ for all j.

To see this we need an *auxiliary result* which is of some interest on its own: *If $u_1, u_2, \ldots, u_{k+1}$ are independent integrable functions, then*

$$\int_A u_{k+1} \, d\mu = \mu(A) \int u_{k+1} \, d\mu \qquad \forall A \in \sigma(u_1, u_2, \ldots, u_k) \qquad (17.6)$$

and

$$\int \phi \, u_{k+1} \, d\mu = \int \phi \, d\mu \cdot \int u_{k+1} \, d\mu \quad \forall \phi \in \mathcal{L}^1(\sigma(u_1, \ldots, u_k)). \qquad (17.7)$$

In particular, integrable independent functions satisfy

$$\int \prod_{j=1}^k u_j \, d\mu = \prod_{j=1}^k \int u_j \, d\mu.$$

The proof of (17.6) and (17.7) will be given in Scholium 17.4 below. Returning to the original problem, we find for all $A \in \mathcal{A}_k$ that

$$\int_A s_{k+1} \, d\mu = \int_A (s_k + u_{k+1}) \, d\mu \;\; = \;\; \int_A s_k \, d\mu + \int_A u_{k+1} \, d\mu$$

$$\overset{(17.6)}{=} \int_A s_k \, d\mu + \mu(A) \int u_{k+1} \, d\mu.$$

Thus $\int u_{k+1} \, d\mu \geqslant 0$ is necessary and sufficient for $(s_k)_{k \in \mathbb{N}}$ to be a submartingale.

(xi) Let $(u_j)_{j \in \mathbb{N}} \subset \mathcal{L}^1_+(\mathcal{A}) \cap \mathcal{L}^\infty_+(\mathcal{A})$ be independent functions (in the sense of (x)). Then $p_k := u_0 \cdot u_1 \cdot \ldots \cdot u_k$, $k \in \mathbb{N}$, is a submartingale w.r.t. the filtration $\mathcal{A}_k := \sigma(u_0, u_1, \ldots, u_k)$ if, and only if, $\int u_j \, d\mu \geqslant 1$ for all j. This follows directly from

$$\int_A p_{k+1} \, d\mu = \int \mathbf{1}_A \, p_k \, u_{k+1} \, d\mu \overset{(17.7)}{=} \int \mathbf{1}_A \, p_k \, d\mu \cdot \int u_{k+1} \, d\mu$$

$$= \int_A p_k \, d\mu \cdot \int u_{k+1} \, d\mu \quad \forall A \in \mathcal{A}_k.$$

17.4 Scholium (on independent functions) (i) Let $u_1, u_2, \ldots, u_{k+1}$ be independent integrable functions on the probability space $(X, \mathcal{A}, \mu)$. Then

$$\int_A u_{k+1} \, d\mu = \mu(A) \int u_{k+1} \, d\mu \qquad \forall A \in \sigma(u_1, u_2, \ldots, u_k) \qquad (17.6)$$

and

$$\int \phi\, u_{k+1}\, d\mu = \int \phi\, d\mu \cdot \int u_{k+1}\, d\mu \qquad \forall \phi \in \mathcal{L}^1(\sigma(u_1, \ldots, u_k)). \qquad (17.7)$$

Proof. We begin with (17.6). Pick a set $A_M := \bigcap_{j=1}^{M} u_j^{-1}(B_j)$, $B_1, \ldots, B_M \in \mathcal{B}(\mathbb{R})$, $M \leqslant k$, from the generator of $\mathcal{A}_k = \sigma(u_1, u_2, \ldots, u_k)$. Because of Theorem 8.8 (and Problem 8.10) we find a sequence of simple functions $(f_\ell)_{\ell \in \mathbb{N}} \subset \mathcal{E}(\sigma(u_{k+1}))$ such that $|f_\ell| \leqslant |u_{k+1}|$ and $\lim_{\ell \to \infty} f_\ell = u_{k+1}$. For the standard representations $f_\ell = \sum_{j=0}^{N(\ell)} y_j^\ell \mathbf{1}_{H_j^\ell}$, $H_j^\ell \in \sigma(u_{k+1})$, we get using dominated convergence T11.2

$$\int_{A_M} u_{k+1}\, d\mu \overset{11.2}{=} \lim_{\ell \to \infty} \int_{A_M} \sum_{j=0}^{N(\ell)} y_j^\ell \mathbf{1}_{H_j^\ell}\, d\mu$$

$$= \lim_{\ell \to \infty} \sum_{j=0}^{N(\ell)} y_j^\ell\, \mu(A_M \cap H_j^\ell)$$

$$\overset{(17.5)}{=} \lim_{\ell \to \infty} \sum_{j=0}^{N(\ell)} y_j^\ell\, \mu(A_M) \mu(H_j^\ell)$$

$$\overset{11.2}{=} \mu(A_M) \int u_{k+1}\, d\mu,$$

where we applied (17.5) for $H_j^\ell \in \sigma(u_{k+1})$ ($\Longleftrightarrow H_j^\ell = u_{k+1}^{-1}(C_j^\ell)$ with some suitable $C_j^\ell \in \mathcal{B}(\mathbb{R})$) and A_M. This proves (17.6) for a generator of $\mathcal{A}_k$ which satisfies the conditions stated in Remark 17.2(i); a similar argument as the one in this remark now proves that (17.6) holds for all $A \in \mathcal{A}_k$.

For (17.7) let us first assume that ϕ is bounded. Set $\mathcal{A}_k := \sigma(u_1, \ldots, u_k)$. By Theorem 8.8 (and Problem 8.10) we find a sequence of simple functions $(f_\ell)_{\ell \in \mathbb{N}} \subset \mathcal{E}(\mathcal{A}_k)$ such that $|f_\ell| \leqslant |\phi|$ and $\lim_{\ell \to \infty} f_\ell = \phi$. For the standard representations $f_\ell = \sum_{j=0}^{N(\ell)} y_j^\ell \mathbf{1}_{A_j^\ell}$, $A_j^\ell \in \mathcal{A}_k$, we get using dominated convergence T11.2 and (17.6)

$$\int \phi\, u_{k+1}\, d\mu \overset{11.2}{=} \lim_{\ell \to \infty} \int \sum_{j=0}^{N(\ell)} y_j^\ell \mathbf{1}_{A_j^\ell}\, u_{k+1}\, d\mu$$

$$= \lim_{\ell \to \infty} \sum_{j=0}^{N(\ell)} y_j^\ell\, \mu(A_j^\ell) \int u_{k+1}\, d\mu$$

$$= \lim_{\ell \to \infty} \int \sum_{j=0}^{N(\ell)} y_j^\ell \mathbf{1}_{A_j^\ell} \, d\mu \cdot \int u_{k+1} \, d\mu$$

$$\overset{11.2}{=} \int \phi \, d\mu \cdot \int u_{k+1} \, d\mu.$$

If ϕ is integrable but not bounded, we apply the previous calculation to the bounded functions $|\phi_\ell| := |\phi| \wedge \ell$ and use dominated convergence on the right and monotone convergence on the left to get

$$\int |\phi| \cdot |u_{k+1}| \, d\mu \overset{9.6}{=} \lim_{\ell \to \infty} \int |\phi_\ell| \cdot |u_{k+1}| \, d\mu = \lim_{\ell \to \infty} \int |\phi_\ell| \, d\mu \cdot \int |u_{k+1}| \, d\mu$$

$$\overset{11.2}{=} \int |\phi| \, d\mu \cdot \int |u_{k+1}| \, d\mu.$$

This shows, in particular, that $\phi \, u_{k+1} \in \mathcal{L}^1(\mu)$. We can therefore apply dominated convergence to $\psi_\ell := (-\ell) \vee \phi \wedge \ell$ to derive

$$\int \phi \, u_{k+1} \, d\mu = \lim_{\ell \to \infty} \int \psi_\ell \, u_{k+1} \, d\mu = \lim_{\ell \to \infty} \int \psi_\ell \, d\mu \cdot \int u_{k+1} \, d\mu$$

$$= \int \phi \, d\mu \cdot \int u_{k+1} \, d\mu.$$

(ii) In Example 17.3(x) we assumed the existence of infinitely many independent functions. As a matter of fact, this is a not completely trivial matter. If we want to construct *finitely many* independent functions $u_1, u_2, \ldots, u_n$, we can proceed as follows. Replace the probability space $(X, \mathcal{A}, \mu)$ by the n-fold product measure space $\left(X^n, \mathcal{A}^{\otimes n}, \mu^{\times n}\right)$ (which is again a probability space[✓]) and define $\tilde{u}_j(x_1, \ldots, x_n) := u_j(x_j)$ for $j = 1, 2, \ldots, n$. Since each of the new functions $\tilde{u}_j$ depends only on the variable x_j, their independence follows from a simple Fubini-type argument. A similar argument can be applied to countably many functions – provided we know how to construct infinite-dimensional products.

We will not follow this route but construct instead countably many independent functions $(X_j)_{j \in \mathbb{N}}$ on the probability space $([0, 1], \mathcal{B}[0, 1], \lambda := \lambda^1|_{[0,1]})$ which are *identically distributed*, i.e. the image measures satisfy $X_1(\lambda) = X_j(\lambda)$ for all $j \in \mathbb{N}$ with a *Bernoulli distribution* $X_1(\lambda) = p \, \delta_1 + (1 - p) \, \delta_0$, $p \in (0, 1)$.

Consider the interval map $\beta_p : [0, 1) \to [0, 1)$

$$\beta_p(x) := \frac{x}{p} \mathbf{1}_{[0, p)}(x) + \frac{x - p}{1 - p} \mathbf{1}_{[p, 1)}(x),$$

and its iterates $\beta_p^n := \underbrace{\beta_p \circ \cdots \circ \beta_p}_{n \text{ times}}$, see the pictures for the graphs of β_p and β_p^2.

Define

$$X_n(x) := \mathbf{1}_{[0,p)}(\beta_p^{n-1}(x)), \quad n \in \mathbb{N}.$$

In the first step the interval $[0, 1)$ is split according to $p : (1-p)$ into two intervals $[0, p)$ and $[p, 1)$ and X_1 is 1 on the left segment and 0 on the right. The subsequent iterations split each of the intervals of the previous step – say, step $n-1$ – into two new sub-intervals according to the ratio $p : (1-p)$, and we define X_n to be 1 on each new left sub-interval and 0 otherwise, see the picture for $n = 1, 2$. Thus $\lambda(\{X_n = 1\}) = p$ and $\lambda(\{X_n = 0\}) = 1 - p$, which means that the X_n are identically Bernoulli distributed.

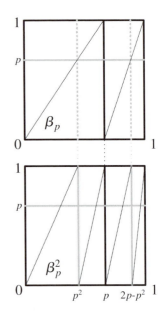

To see independence, fix $\epsilon_j \in \{0, 1\}$, and observe that $\{X_1 = \epsilon_1\} \cap \{X_2 = \epsilon_2\} \cap \ldots \cap \{X_{n-1} = \epsilon_{n-1}\}$ exactly determines the segment before the nth split. Since each split preserves the proportion between p and $1 - p$, we find

$$\lambda(\{X_1 = \epsilon_1\} \cap \ldots \cap \{X_{n-1} = \epsilon_{n-1}\} \cap \{X_n = 1\})$$
$$= \lambda(\{X_1 = \epsilon_1\} \cap \ldots \cap \{X_{n-1} = \epsilon_{n-1}\}) \cdot p,$$

so that

$$\lambda(\{X_1 = \epsilon_1\} \cap \ldots \cap \{X_{n-1} = \epsilon_{n-1}\} \cap \{X_n = \epsilon_n\})$$
$$= p^{\epsilon_1 + \cdots + \epsilon_n}(1-p)^{n - \epsilon_1 - \cdots - \epsilon_n} = \prod_{j=1}^{n} \lambda(\{X_j = \epsilon_j\}).$$

This shows that the X_j are all independent.

For later reference purposes let us derive some formulae for the arithmetic means $\frac{1}{n} S_n := \frac{1}{n}(X_1 + X_2 + \cdots + X_n)$. The *mean value* is

$$\frac{1}{n} \int S_n \, d\lambda = \frac{1}{n} \int (X_1 + \cdots + X_n) \, d\lambda = \int X_1 \, d\lambda = 1 \cdot p + 0 \cdot (1 - p) = p,$$

while the *variance* is given by

$$\int \left[\tfrac{1}{n}(S_n - np)\right]^2 d\lambda = \frac{1}{n^2} \int \left(\sum_{j=1}^{n}(X_j - p)\right)^2 d\lambda$$

$$= \frac{1}{n^2} \sum_{j,k=1}^{n} \int (X_j - p)(X_k - p)\, d\lambda$$

$$= \frac{1}{n^2} \sum_{j=1}^{n} \int (X_j - p)^2\, d\lambda \qquad \text{(independence)}$$

$$= \frac{1}{n} \int (X_1 - p)^2\, d\lambda \qquad \text{(identical distr.)}$$

$$= \frac{1}{n}\big((1-p)^2 p + p^2(1-p)\big)$$

$$= \frac{1}{n}\, p(1-p). \qquad\qquad \blacksquare$$

In the next chapter we study the convergence behaviour of a martingale $(u_j)_{j\in\mathbb{N}}$; therefore, it is natural to ask questions of the type *from which index j onwards does $u_j(x)$ exceed a certain threshold*, etc. This means that we must be able to admit indices τ which may depend on the argument x of $u_j(x)$: $u_{\tau(x)}(x)$. The problem is measurability.

17.5 Definition Let $(X, \mathcal{A}, \mathcal{A}_j, \mu)$ be a σ-finite filtered measure space. A *stopping time* is a map $\tau : X \to \mathbb{N} \cup \{\infty\}$ which satisfies $\{\tau \leqslant j\} \in \mathcal{A}_j$ for all $j \in \mathbb{N}$. The associated σ-algebra is given by

$$\mathcal{A}_\tau := \big\{A \in \mathcal{A} : A \cap \{\tau \leqslant j\} \in \mathcal{A}_j \ \forall j \in \mathbb{N}\big\}.$$

As usual, we write $u_\tau(x)$ instead of the more correct $u_{\tau(x)}(x)$.

17.6 Lemma *Let σ, τ be stopping times on a σ-finite filtered measure space $(X, \mathcal{A}, \mathcal{A}_j, \mu)$.*

(i) $\sigma \wedge \tau$, $\sigma \vee \tau$, $\sigma + k$, $k \in \mathbb{N}_0$ *are stopping times.*

(ii) $\{\sigma < \tau\} \in \mathcal{A}_\sigma \cap \mathcal{A}_\tau$ *and* $\mathcal{A}_\sigma \subset \mathcal{A}_\tau$ *if* $\sigma \leqslant \tau$.

(iii) *If u_j is a sequence of real functions such that $u_j \in \mathcal{M}(\mathcal{A}_j)$, then u_σ is $\mathcal{A}_\sigma/\mathcal{B}(\mathbb{R})$-measurable.*

Proof (i) follows immediately from the identities

$$\{\sigma \wedge \tau \leqslant j\} = \{\sigma \leqslant j\} \cup \{\tau \leqslant j\} \in \mathcal{A}_j,$$

$$\{\sigma \vee \tau \leqslant j\} = \{\sigma \leqslant j\} \cap \{\tau \leqslant j\} \in \mathcal{A}_j,$$

$$\{\sigma + k \leqslant j\} = \{\sigma \leqslant j - k\} \in \mathcal{A}_{(j-k)\vee 0} \subset \mathcal{A}_j.$$

(ii) Since for all $j \in \mathbb{N}$

$$\{\sigma < \tau\} \cap \{\sigma \leqslant j\} = \bigcup_{k=1}^{j} \{\sigma = k\} \cap \{k < \tau\}$$

$$= \bigcup_{k=1}^{j} \underbrace{\{\sigma \leqslant k\}}_{\in \mathcal{A}_k} \cap \underbrace{\{\sigma \leqslant k-1\}^c}_{\in \mathcal{A}_k} \cap \underbrace{\{\tau \leqslant k\}^c}_{\in \mathcal{A}_k} \in \mathcal{A}_j,$$

we find that $\{\sigma < \tau\} \in \mathcal{A}_\sigma$, while a similar calculation for $\{\sigma < \tau\} \cap \{\tau \leqslant j\}$ yields $\{\sigma < \tau\} \in \mathcal{A}_\tau$.

If $\sigma \leqslant \tau$ we find for $A \in \mathcal{A}_\sigma$

$$A \cap \{\tau \leqslant j\} = A \cap \underbrace{\{\sigma \leqslant \tau\}}_{=\Omega} \cap \{\tau \leqslant j\} = A \cap \underbrace{\{\sigma \leqslant j\}}_{\in \mathcal{A}_j} \cap \underbrace{\{\tau \leqslant j\}}_{\in \mathcal{A}_j} \in \mathcal{A}_j,$$

i.e. $A \in \mathcal{A}_\tau$, hence $\mathcal{A}_\sigma \subset \mathcal{A}_\tau$.

(iii) We have for all $B \in \mathcal{B}(\mathbb{R})$ and $j \in \mathbb{N} \cup \{\infty\}$

$$\{u_\sigma \in B\} \cap \{\sigma \leqslant j\} = \bigcup_{k=1}^{j} \{u_k \in B\} \cap \{\sigma = k\}$$

$$= \bigcup_{k=1}^{j} \underbrace{\{u_k \in B\}}_{\in \mathcal{A}_k} \cap \underbrace{\{\sigma \leqslant k\}}_{\in \mathcal{A}_k} \cap \underbrace{\{\sigma \leqslant k-1\}^c}_{\in \mathcal{A}_k} \in \mathcal{A}_j. \qquad \blacksquare$$

The next result is a very useful characterization of (sub-)martingales.

17.7 Theorem *Let $(X, \mathcal{A}, \mathcal{A}_j, \mu)$ be a σ-finite filtered measure space. For a sequence $(u_j)_{j\in\mathbb{N}}$, $u_j \in \mathcal{L}^1(\mathcal{A}_j)$, the following assertions are equivalent:*

(i) $(u_j)_{j\in\mathbb{N}}$ *is a submartingale;*
(ii) $\int u_\sigma \, d\mu \leqslant \int u_\tau \, d\mu$ *for all bounded stopping times $\sigma \leqslant \tau$;*
(iii) $\int_A u_\sigma \, d\mu \leqslant \int_A u_\tau \, d\mu$ *for all bounded stopping times $\sigma \leqslant \tau$ and $A \in \mathcal{A}_\sigma$.*

Proof (i)⇒(ii): Let $\sigma \leqslant \tau \leqslant N$ be two stopping times. By Lemma 17.6 u_τ is measurable, and since

$$\int |u_\tau| \, d\mu = \sum_{j=1}^{N} \int_{\{\tau=j\}} |u_j| \, d\mu \leqslant \sum_{j=1}^{N} \int |u_j| \, d\mu < \infty,$$

we find that $u_\tau, u_\sigma \in \mathcal{L}^1(X, \mathcal{A}, \mu)$.

Step 1: $\tau - \sigma \leqslant 1$. In this case

$$\{\sigma < \tau\} \cap \{\sigma = j\} = \{\tau > j\} \cap \{\sigma = j\} = \{\tau \leqslant j\}^c \cap \{\sigma = j\} \in \mathcal{A}_j,$$

and we see

$$\int u_\sigma \, d\mu = \int_{\{\tau = \sigma\}} u_\sigma \, d\mu + \sum_{j=1}^{N-1} \int_{\{\sigma < \tau\} \cap \{\sigma = j\}} u_j \, d\mu$$

$$\overset{(17.2)}{\leqslant} \int_{\{\tau = \sigma\}} u_\tau \, d\mu + \sum_{j=1}^{N-1} \int_{\{\sigma < \tau\} \cap \{\sigma = j\}} u_{j+1} \, d\mu$$

$$= \int_{\{\tau = \sigma\}} u_\tau \, d\mu + \int_{\{\sigma < \tau\}} u_\tau \, d\mu \qquad \text{(use } \tau - \sigma \leqslant 1\text{)}$$

$$= \int u_\tau \, d\mu.$$

Step 2: if $\sigma \leqslant \tau \leqslant N$ we introduce (at most N) intermediate stopping times $\rho_j := (\sigma + j) \wedge \tau$, $j = 0, 1, 2, \ldots, k \leqslant N$. For some $k \leqslant N$ we get $\sigma = \rho_0 \leqslant \rho_1 \leqslant \ldots \leqslant \rho_k = \tau$ while $\rho_{j+1} - \rho_j \leqslant 1$. Repeating step 1 from above k times yields

$$\int u_\sigma \, d\mu = \int u_{\rho_0} \, d\mu \leqslant \int u_{\rho_1} \, d\mu \leqslant \ldots \leqslant \int u_{\rho_k} \, d\mu = \int u_\tau \, d\mu.$$

(ii)$\Rightarrow$(iii): Note that for any $A \in \mathcal{A}_\sigma$ the function $\rho := \rho_A := \sigma \mathbf{1}_A + \tau \mathbf{1}_{A^c}$ is again a bounded stopping time. This follows from

$$\{\rho \leqslant j\} = (\{\sigma \leqslant j\} \cap A) \cup (\{\tau \leqslant j\} \cap A^c) \in \mathcal{A}_j, \qquad j \in \mathbb{N},$$

where we used that $A \in \mathcal{A}_\sigma \subset \mathcal{A}_\tau$, cf. Lemma 17.6. Since $\rho \leqslant \tau$, (ii) shows

$$\int (u_\sigma \mathbf{1}_A + u_\tau \mathbf{1}_{A^c}) \, d\mu = \int u_\rho \, d\mu \leqslant \int u_\tau \, d\mu,$$

which is but $\int_A u_\sigma \, d\mu \leqslant \int_A u_\tau \, d\mu$.

(iii)$\Rightarrow$(i): Take $\sigma = j$ and $\tau = j + 1$. ∎

17.8 Remark One should read Theorem 17.7(iii) in the following way: *Let $\tau_1 \leqslant \tau_2 \leqslant \ldots \leqslant \tau_k \leqslant N$ be bounded stopping times. Then*

$$\begin{array}{ccc} (u_j, \mathcal{A}_j)_{j \in \mathbb{N}} \text{ is a} & \Longrightarrow & (u_{\tau_j}, \mathcal{A}_{\tau_j})_{j=1,\ldots,k} \text{ is a} \\ \text{submartingale} & & \text{submartingale.} \end{array}$$

This statement is often called the *optional sampling theorem*.

Problems

Unless otherwise stated $(X, \mathcal{A}, \mathcal{A}_j, \mu)$ will be a σ-finite filtered measure space.

17.1. Let $(X, \mathcal{A}, \mu)$ be a finite measure space and let $(u_j, \mathcal{A}_j)_{j\in\mathbb{N}}$ be a martingale. Set $\mathcal{A}_0 := \{\emptyset, X\}$. Show that $(u_j, \mathcal{A}_j)_{j\in\mathbb{N}_0}$ is a martingale if, and only if, $u_0 = \int u_1\, d\mu$.

17.2. Let $(u_j, \mathcal{A}_j)_{j\in\mathbb{N}}$ be a (sub-, super-)martingale and let $(\mathcal{B}_j)_{j\in\mathbb{N}}$ and $(\mathcal{C}_j)_{j\in\mathbb{N}}$ be filtrations in $\mathcal{A}$ which are smaller resp. larger than $(\mathcal{A}_j)_{j\in\mathbb{N}}$, i.e. such that $\mathcal{B}_j \subset \mathcal{A}_j \subset \mathcal{C}_j$.

   (a) Show that $(u_j, \mathcal{B}_j)_{j\in\mathbb{N}}$ is again a (sub-, super-)martingale.

   (b) Show that $(u_j, \mathcal{C}_j)_{j\in\mathbb{N}}$ is, in general, no longer a (sub-, super-)martingale.

17.3. **Completion (7).** Let $(u_j, \mathcal{A}_j)_{j\in\mathbb{N}}$ be a submartingale and denote by $\mathcal{A}_j^*$ the completion of $\mathcal{A}_j$. Then $(u_j, \mathcal{A}_j^*)_{j\in\mathbb{N}}$ is still a submartingale.

17.4. Show that $(u_j)_{j\in\mathbb{N}}$ is a submartingale if, and only if, $u_j \in \mathcal{L}^1(\mathcal{A}_j)$ for all $j \in \mathbb{N}$ and

$$\int_A u_j\, d\mu \leqslant \int_A u_k\, d\mu \qquad \forall j < k, \ \forall A \in \mathcal{A}_j.$$

   Find similar statements for martingales and supermartingales.

17.5. Prove the assertion made in Remark 17.2(ii).

17.6. Let $(u_j, \mathcal{A}_j)_{j\in\mathbb{N}}$ be a martingale with $u_j \in \mathcal{L}^2(\mathcal{A}_j)$. Show that

$$\int u_j u_k\, d\mu = \int u_{j\wedge k}^2\, d\mu.$$

   [Hint: assume that $j < k$. Approximate u_j by simple functions from $\mathcal{E}(\mathcal{A}_j)$, use dominated convergence and (17.1).]

17.7. **Martingale transform** Let $(u_j, \mathcal{A}_j)_{j\in\mathbb{N}}$ be a martingale and let $(f_j)_{j\in\mathbb{N}}$ be a sequence of bounded functions such that $f_j \in \mathcal{M}(\mathcal{A}_j)$ for every $j \in \mathbb{N}$. Set $f_0 := 0$ and $u_0 := \int u_1\, d\mu$. Then the so-called *martingale transform*

$$(f \bullet u)_k := \sum_{j=1}^{k} f_{j-1} \cdot (u_j - u_{j-1}), \qquad k \in \mathbb{N},$$

   is again a martingale w.r.t. $(\mathcal{A}_j)_{j\in\mathbb{N}}$.

17.8. Let $(\Omega, \mathcal{A}, P)$ be a probability space and let $(X_j)_{j\in\mathbb{N}}$ be a sequence of independent identically distributed random variables with $X_j \in \mathcal{L}^2(\mathcal{A})$ and $\int X_j\, dP = 0$. Set $\mathcal{A}_j := \sigma(X_1, X_2, \ldots, X_j)$.

   (i) Show, without using Example 17.3(vi), that $S_n^2 := (X_1 + X_2 + \cdots + X_n)^2$ is a submartingale w.r.t. $(\mathcal{A}_n)_{n\in\mathbb{N}}$.

   (ii) Show that there exists a constant κ such that $S_n^2 - \kappa n$ is a martingale w.r.t. $(\mathcal{A}_n)_{n\in\mathbb{N}}$.

17.9. Let $(\Omega, \mathcal{A}, P)$ be a probability space and let $(X_j)_{j\in\mathbb{N}}$ be a sequence of independent random variables with $X_j \in \mathcal{L}^2(\mathcal{A})$, $\int X_j\, dP = 0$ and $\int X_j^2\, dP = \sigma_j^2$. Set $\mathcal{A}_j := \sigma(X_1, X_2, \ldots, X_j)$ and $A_j := \sigma_1^2 + \cdots + \sigma_j^2$. Show that

$$M_n := S_n^2 - A_n = \left(\sum_{j=1}^{n} X_j\right)^2 - \sum_{j=1}^{n} \sigma_j^2$$

   is a martingale.

   [Hint: use formulae (17.6), (17.7) and Remark 17.2(ii).]

17.10. **Martingale difference sequence** Let $(d_j)_{j \in \mathbb{N}}$ be a sequence in $\mathcal{L}^2(\mathcal{A}) \cap \mathcal{L}^1(\mathcal{A})$ and define $\mathcal{A}_0 := \{\varnothing, X\}$ and $\mathcal{A}_j := \sigma(d_1, d_2, \dots, d_j)$. Suppose that for each $j \in \mathbb{N}$

$$\int_A d_j \, d\mu = 0 \qquad \forall A \in \mathcal{A}_{j-1}.$$

Show that $(u_n^2)_{n \in \mathbb{N}}$ where $u_n := d_1 + \dots + d_n$ is a submartingale which satisfies

$$\int u_n^2 \, d\mu = \sum_{j=1}^n \int d_j^2 \, d\mu.$$

Show that on $(\mathbb{R}, \mathcal{B}(\mathbb{R}), \lambda^1)$ the sequence $d_j(x) := \operatorname{sgn} \sin(2^j \pi x)$, $x \in \mathbb{R}$, $j \in \mathbb{N}$, is a martingale difference sequence. (See Chapter 24, in particular pp. 299 and 302 for more details.)

17.11. Let $(\Omega, \mathcal{A}, P)$ be a probability space and let $(X_j)_{j \in \mathbb{N}}$ be a sequence of independent identically Bernoulli $(p, 1-p)$-distributed random variables with values ± 1, i.e. such that $P(X_j = 1) = p$ and $P(X_j = -1) = 1 - p$ – this can be constructed as in Scholium 17.4. Set $S_n := X_1 + \dots + X_n$. Then $(\frac{1-p}{p})^{S_n}$ is a martingale w.r.t. the filtration given by $\mathcal{A}_n := \sigma(X_1, \dots, X_n)$.

17.12. Let $(X, \mathcal{A}, \mu)$ be a σ-finite measure space, let ν be a further measure on $\mathcal{A}$ and let $(A_{n,j})_{j \in \mathbb{N}} \subset \mathcal{A}$ be for each $n \in \mathbb{N}$ a sequence of mutually disjoint sets such that $X = \bigcup_{j \in \mathbb{N}} A_{n,j}$. Assume, moreover, that each set $A_{n,j}$ is the union of finitely many sets from the sequence $(A_{n+1,k})_{k \in \mathbb{N}}$. Show that

(i) the σ-algebras $\mathcal{A}_n := \sigma(A_{n,j} : j \in \mathbb{N})$ form a filtration;
(ii) if $\mu(A_{n,j}) > 0$ for all $n, j \in \mathbb{N}$, then

$$u_n := \sum_{j=1}^\infty \frac{\nu(A_{n,j})}{\mu(A_{n,j})} \mathbf{1}_{A_{n,j}}$$

is a martingale w.r.t. $(\mathcal{A}_n)_{n \in \mathbb{N}}$.

17.13. Let $(u_j, \mathcal{A}_j)_{j \in \mathbb{N}}$ be a supermartingale and $u_j \geqslant 0$ a.e. Prove that $u_k = 0$ a.e. implies that $u_{k+j} = 0$ a.e. for all $j \in \mathbb{N}$.

17.14. Verify that the family $\mathcal{A}_\tau$ defined in Definition 17.5 is indeed a σ-algebra.

17.15. Show that τ is a stopping time if, and only if, $\{\tau = j\} \in \mathcal{A}_j$ for all $j \in \mathbb{N}$.

17.16. Show that, in the notation of Lemma 17.6, $\mathcal{A}_{\sigma \wedge \tau} = \mathcal{A}_\sigma \cap \mathcal{A}_\tau$ for any two stopping times σ, τ.

18

Martingale convergence theorems

Throughout this chapter $(X, \mathcal{A}, \mathcal{A}_j, \mu)$ is a σ-finite filtered measure space.

One of the foremost applications of martingales is to convergence theorems. Let us begin with the following simple observation for a sequence $(u_j)_{j \in \mathbb{N}}$ of real numbers. If $(u_j)_{j \in \mathbb{N}}$ has a limit $\ell = \lim_{j \to \infty} u_j$ and if we know that $\ell \in (a, b)$, only *finitely many* of the u_j can be outside of (a, b). In particular, if infinitely many u_j are bigger than b and infinitely many smaller than a, then the sequence has no limit at all. We call any occurrence of

$$u_j \leqslant a \qquad \text{and} \qquad u_{j+k} \geqslant b \quad \text{(for some } k \in \mathbb{N})$$

an *upcrossing of* $[a, b]$ – the picture below shows three such upcrossings if $j = 0, 1, \ldots, N$ – and we have just observed that, if for *some* pair $a, b \in \mathbb{R}$, $a < b$,

$$\#\{\text{upcrossings of } [a, b]\} = \infty \quad \Longrightarrow \quad (u_j)_{j \in \mathbb{N}} \text{ has no limit.} \tag{18.1}$$

For a submartingale we can estimate the average number of upcrossings over any interval:

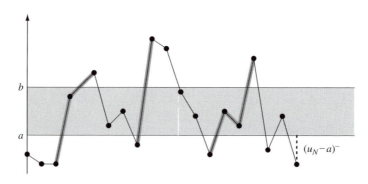

18.1 Lemma (Doob's upcrossing estimate) *Let $(u_j)_{j \in \mathbb{N}}$ be a submartingale and denote by $U([a, b]; N; x)$ the number of upcrossings of $(u_j(x))_{j \in \mathbb{N}}$ across $[a, b]$ which occur for $1 \leqslant j \leqslant N$. Then*

$$\int_A U([a, b]; N)\, d\mu \leqslant \frac{1}{b - a} \int_A (u_N - a)^+\, d\mu \qquad \forall A \in \mathcal{A}_0.$$

Proof In order to keep track of the upcrossings we introduce the following stopping times[✓], cf. Problem 18.1: $\tau_0 := 0$ and

$$\sigma_k := \inf\{j > \tau_{k-1} : u_j \leqslant a\} \wedge N, \qquad \tau_k := \inf\{j > \sigma_k : u_j \geqslant b\} \wedge N,$$

(as usual we set $\inf \emptyset = +\infty$). Then

$$\tau_0 = 0 < \sigma_1 \leqslant \tau_1 \leqslant \sigma_2 \leqslant \ldots \leqslant \sigma_N = \tau_N = N.$$

By the very definition of an upcrossing we find

$$(b - a)\, U([a, b]; N) \leqslant \underbrace{(u_{\tau_1} - a)}_{\geqslant b - a} + \underbrace{(u_{\tau_2} - u_{\sigma_2})}_{\geqslant b - a} + \cdots + (u_{\tau_N} - u_{\sigma_N}),$$

and integrating both sides of this inequality over $A \in \mathcal{A}_0$ yields, after some simple rearrangements,

$$(b - a) \int_A U([a, b]; N)\, d\mu$$

$$\leqslant -\int_A a\, d\mu + \overbrace{\int_A (u_{\tau_1} - u_{\sigma_2})\, d\mu}^{\leqslant 0} + \cdots + \overbrace{\int_A (u_{\tau_{N-1}} - u_{\sigma_N})\, d\mu}^{\leqslant 0} + \int_A u_{\tau_N}\, d\mu$$

$$\overset{17.7}{\leqslant} \int_A (u_{\tau_N} - a)\, d\mu \leqslant \int_A (u_{\tau_N} - a)^+\, d\mu. \qquad \blacksquare$$

The upcrossing lemma is the basis for all martingale convergence theorems.

18.2 Theorem (Submartingale convergence) *Let $(u_j, \mathcal{A}_j)_{j \in \mathbb{N}}$ be a submartingale on the σ-finite filtered measure space $(X, \mathcal{A}, \mathcal{A}_j, \mu)$.*

If $\sup_{j \in \mathbb{N}} \int u_j^+\, d\mu < \infty$, then $u_\infty(x) := \lim_{j \to \infty} u_j(x)$ exists for almost all $x \in \mathbb{R}$ and defines an $\mathcal{A}_\infty$-measurable function.

Before we give the details of the proof, let us note some immediate consequences.

18.3 Corollary *Under any of the following conditions the pointwise limit* $\lim_{j \to \infty} u_j$ *exists a.e. in* $\mathbb{R}$:

(i) $(u_j)_{j \in \mathbb{N}}$ *is a supermartingale and* $\sup_{j \in \mathbb{N}} \int u_j^- \, d\mu < \infty$.

(ii) $(u_j)_{j \in \mathbb{N}}$ *is a positive supermartingale.*

(iii) $(u_j)_{j \in \mathbb{N}}$ *is a martingale and* $\sup_{j \in \mathbb{N}} \int |u_j| \, d\mu < \infty$.

Proof (of Theorem 18.2) In view of (18.1) we have

$$\left\{ x : \lim_{j \to \infty} u_j(x) \text{ does not exist} \right\} = \left\{ x : \limsup_{j \to \infty} u_j(x) > \liminf_{j \to \infty} u_j(x) \right\}$$

$$= \bigcup_{\substack{a < b \\ a, b \in \mathbb{Q}}} \left\{ x : \sup_{N \in \mathbb{N}} U([a, b]; N; x) = \infty \right\}.$$

Since $\mu|_{\mathcal{A}_0}$ is σ-finite, there exists an exhausting sequence $(A_k)_{k \in \mathbb{N}} \subset \mathcal{A}_0$ such that $A_k \uparrow X$ and $\mu(A_k) < \infty$. From the inequality $(\beta - \alpha)^+ \leqslant \beta^+ + |\alpha|$ we find

$$\int_{A_k} \sup_{N \in \mathbb{N}} U([a, b]; N) \, d\mu \overset{9.6}{=} \sup_{N \in \mathbb{N}} \int_{A_k} U([a, b]; N) \, d\mu$$

$$\overset{18.1}{\leqslant} \frac{1}{b - a} \sup_{N \in \mathbb{N}} \int_{A_k} (u_N - a)^+ \, d\mu$$

$$\leqslant \frac{1}{b - a} \left(\sup_{N \in \mathbb{N}} \int_{A_k} u_N^+ \, d\mu + |a| \, \mu(A_k) \right) < \infty,$$

and a routine application of Markov's inequality P10.12 yields

$$\mu\left(\left\{ \sup_{N \in \mathbb{N}} U([a, b]; N) = \infty \right\} \cap A_k \right) = 0.$$

Since $\bigcup_{k \in \mathbb{N}} \bigcup_{\substack{a < b \\ a, b \in \mathbb{Q}}} \left(\{ x : \sup_{N \in \mathbb{N}} U([a, b]; N; x) = \infty \} \cap A_k \right)$ is a countable union of null sets, it is itself a null set; thus the limit $u_\infty(x) = \lim_{j \to \infty} u_j(x)$ exists for almost all $x \in \mathbb{R}$ and is $\mathcal{A}_\infty$-measurable. An application of Fatou's lemma T9.11 shows

$$\int |u_\infty| \, d\mu = \int \liminf_{j \to \infty} |u_j| \, d\mu \leqslant \liminf_{j \to \infty} \int |u_j| \, d\mu \leqslant \sup_{j \in \mathbb{N}} \int |u_j| \, d\mu,$$

while the submartingale property gives

$$\sup_{j \in \mathbb{N}} \int |u_j| \, d\mu = \sup_{j \in \mathbb{N}} \left(2 \int u_j^+ \, d\mu - \int u_j \, d\mu \right) \leqslant 2 \sup_{j \in \mathbb{N}} \int u_j^+ \, d\mu - \int u_1 \, d\mu.$$

The last expression is, by assumption, finite and we conclude that $u_\infty \in \mathcal{L}^1(\mathcal{A}_\infty)$ and $|u_\infty| < \infty$ a.e. ∎

We have seen in Example 17.3(v) that for a martingale $(u_j)_{j \in \mathbb{N}}$ the sequence $(|u_j|)_{j \in \mathbb{N}}$ is a submartingale. Therefore,

$$\int |u_1| \, d\mu \leqslant \int |u_2| \, d\mu \leqslant \int |u_3| \, d\mu \leqslant \dots,$$

which means that, if we had a martingale with index set *running to the left*, say, $(w_\ell)_{\ell \in -\mathbb{N}}$, condition (iii) of C18.3 would be automatically fulfilled: $\int |w_{-j}| \, d\mu \leqslant \int |w_{-1}| \, d\mu < \infty$, and the limit $\lim_{j \to +\infty} w_{-j}$ would always exist.

18.4 Definition Let $(X, \mathcal{A}, \mu)$ be a measure space and $\mathcal{A}_{-1} \supset \mathcal{A}_{-2} \supset \mathcal{A}_{-3} \supset \dots$ be a decreasing filtration of sub-σ-algebras of $\mathcal{A}$ such that $\mu|_{\mathcal{A}_{-j}}$ is σ-finite for every $j \in \mathbb{N}$. A (sub-, super-)martingale $(w_\ell, \mathcal{A}_\ell)_{\ell \in -\mathbb{N}}$ is called *reversed* or *backwards (sub-, super-)martingale*.

18.5 Corollary (Backwards convergence theorem) *Let $(w_\ell, \mathcal{A}_\ell)_{\ell \in -\mathbb{N}}$ be a backwards submartingale. If on $\mathcal{A}_{-\infty} = \bigcap_{\ell \in -\mathbb{N}} \mathcal{A}_\ell$ the measure $\mu|_{\mathcal{A}_{-\infty}}$ is σ-finite, then $\lim_{j \to +\infty} w_{-j}(x) \in [-\infty, +\infty)$ exists for almost all x and defines an $\mathcal{A}_{-\infty}/\mathcal{B}[-\infty, \infty)$-measurable function.*

Proof The proof of Doob's upcrossing lemma 18.1 obviously applies to the (finite) sequence $w_{-N}, w_{-N+1}, \dots, w_{-1}$ and yields the following variant of the upcrossing inequality:

$$(b - a) \int_A U([a, b]; -N) \, d\mu \leqslant \int_A (w_{-1} - a)^+ \, d\mu \qquad \forall \, A \in \mathcal{A}_{-N}.$$

This means that the arguments of the proof of Theorem 18.2 remain valid without further conditions on $(w_\ell)_{\ell \in -\mathbb{N}}$. But $(w_\ell^+)_{\ell \in -\mathbb{N}}$ is again a submartingale, see Example 17.3(iv), thus

$$\int w_{-\infty}^+ \, d\mu = \int \liminf_{j \to +\infty} w_{-j}^+ \, d\mu \overset{9.11}{\leqslant} \liminf_{j \to +\infty} \int w_{-j}^+ \, d\mu \leqslant \int w_{-1}^+ \, d\mu < \infty.$$

By Corollary 10.13, $\mu(\{w_{-\infty} = +\infty\}) = \mu(\{w_{-\infty}^+ = +\infty\}) = 0.$ ∎

Theorem 18.2 does not guarantee $\mathcal{L}^1$-convergence of a martingale. An example of a martingale which satisfies all conditions of T18.2 but fails to have a limit in $\mathcal{L}^1$, is given in 17.3(viii): here $u_j \overset{j \to \infty}{\longrightarrow} 0$ a.e. while $\int_{[0,1)} u_j \, d\lambda = 1 \nrightarrow 0$. Such phenomena can be avoided if we assume that the submartingale is *uniformly integrable (UI)*.

Recall from Definition 16.1 that $(u_j)_{j\in\mathbb{N}}$ is uniformly integrable if

$$\forall \epsilon > 0 \quad \exists w_\epsilon \in \mathcal{L}^1_+(\mathcal{A}, \mu) : \sup_{j\in\mathbb{N}} \int_{\{|u_j|>w_\epsilon\}} |u_j|\, d\mu < \epsilon.$$

18.6 Theorem (Convergence of UI submartingales) *Let $(u_j)_{j\in\mathbb{N}}$ be a submartingale on the σ-finite filtered measure space $(X, \mathcal{A}, \mathcal{A}_j, \mu)$. Then the following assertions are equivalent:*

(i) $u_\infty(x) = \lim\limits_{j\to\infty} u_j(x)$ *exists a.e.,* $u_\infty \in \mathcal{L}^1(\mathcal{A}_\infty, \mu)$,

$$\lim_{j\to\infty} \int u_j\, d\mu = \int u_\infty\, d\mu, \text{ and } (u_j)_{j\in\mathbb{N}\cup\{\infty\}} \text{ is a submartingale.}$$

(ii) $(u_j)_{j\in\mathbb{N}}$ *is uniformly integrable.*

(iii) $(u_j)_{j\in\mathbb{N}}$ *converges in* $\mathcal{L}^1(\mathcal{A}_\infty)$.

Proof (i)$\Rightarrow$(ii): Since $\mu|_{\mathcal{A}_0}$ is σ-finite, we can fix an exhausting sequence $(A_k)_{k\in\mathbb{N}} \subset \mathcal{A}_0$ with $A_k \uparrow X$ and $\mu(A_k) < \infty$. It is not hard to see that the function $w := \sum_{k=1}^\infty 2^{-k}(1+\mu(A_k))^{-1} \mathbf{1}_{A_k}$ is strictly positive $w > 0$ and integrable $w \in \mathcal{L}^1(\mathcal{A}_0, \mu)$. Because of $u_\infty \in \mathcal{L}^1(\mathcal{A}_\infty, \mu)$, we find for every $\epsilon > 0$ some $\kappa > 0$ and some $N \in \mathbb{N}$ such that $\int_{\{u_\infty^+ > \kappa\}} u_\infty^+\, d\mu + \int_{A_j^c} u_\infty^+\, d\mu < \epsilon$ for all $j \geqslant N$. Example 17.3(iv) shows that $(u_j^+)_{j\in\mathbb{N}\cup\{\infty\}}$ is still a submartingale, so that for every $L > 0$

$$\int_{\{u_j^+ > Lw\}} u_j^+\, d\mu \leqslant \int_{\{u_j^+ > Lw\}} u_\infty^+\, d\mu$$

$$\leqslant \int_{\{u_j^+ > Lw\}\cap\{u_\infty^+ \leqslant \kappa\}\cap A_N} u_\infty^+\, d\mu + \int_{\{u_\infty^+ > \kappa\}\cup A_N^c} u_\infty^+\, d\mu$$

$$\leqslant \kappa\mu\left(\{u_j^+ > Lw\}\cap A_N\right) + \epsilon$$

$$\leqslant \kappa\mu\left(\{u_j^+ > L2^{-N}(1+\mu(A_N))^{-1}\}\right) + \epsilon,$$

where we used that $w \geqslant 2^{-N}(1+\mu(A_N))^{-1}$ on A_N. The Markov inequality P10.12 and the submartingale property imply

$$\sup_{j\in\mathbb{N}} \int_{\{u_j^+ > Lw\}} u_j^+\, d\mu \leqslant \kappa \frac{2^N(1+\mu(A_N))}{L} \sup_{j\in\mathbb{N}} \int u_j^+\, d\mu + \epsilon$$

$$\leqslant \frac{\kappa 2^N(1+\mu(A_N))}{L} \int u_\infty^+\, d\mu + \epsilon.$$

Since we may choose $L > 0$ arbitrarily large, we have found that $(u_j^+)_{j\in\mathbb{N}}$ is uniformly integrable. From $\lim_{j\to\infty} u_j = u_\infty$ a.e., we conclude $\lim_{j\to\infty} u_j^+ = u_\infty^+$, and Vitali's convergence theorem 16.6 shows that $\lim_{j\to\infty} \int u_j^+ \, d\mu = \int u_\infty^+ \, d\mu$. Thus

$$\int |u_j| \, d\mu = \int (2u_j^+ - u_j) \, d\mu \xrightarrow{j\to\infty} \int (2u_\infty^+ - u_\infty) \, d\mu = \int |u_\infty| \, d\mu,$$

and another application of Vitali's theorem proves that $(u_j)_{j\in\mathbb{N}}$ is uniformly integrable.

(ii)$\Rightarrow$(iii): Because of uniform integrability we have for some $\epsilon > 0$ and a suitable $w_\epsilon \in \mathcal{L}^1(\mathcal{A})$

$$\int |u_j| \, d\mu = \int_{\{|u_j| > w_\epsilon\}} |u_j| \, d\mu + \int_{\{|u_j| \leqslant w_\epsilon\}} |u_j| \, d\mu$$

$$\leqslant \epsilon + \int w_\epsilon \, d\mu \; < \; \infty,$$

and the martingale convergence theorem 18.2 guarantees that the pointwise limit $u_\infty = \lim_{j\to\infty} u_j$ exists a.e.; $\mathcal{L}^1$-convergence follows from Vitali's convergence theorem 16.6.

(iii)$\Rightarrow$(i): Since $\mathcal{L}^1\text{-}\lim_{j\to\infty} u_j = u$ exists we find (e.g. as in Theorem 12.10) that $\sup_{j\in\mathbb{N}} \int |u_j| \, d\mu < \infty$. By the martingale convergence theorem 18.2, the pointwise limit $u_\infty = \lim_{j\to\infty} u_j$ exists a.e. On the other hand, by Corollary 12.8, $u = \lim_{k\to\infty} u_{j(k)}$ a.e. for some subsequence. This implies that $u = u_\infty$ a.e. and, in particular, that $u_\infty = \mathcal{L}^1\text{-}\lim_{j\to\infty} u_j$; this entails $\lim_{j\to\infty} \int_A u_j \, d\mu = \int_A u_\infty \, d\mu$ for all $A \in \mathcal{A}$.[✓] Since $(u_j)_{j\in\mathbb{N}}$ is a submartingale, we find for all $k > j$ and $A \in \mathcal{A}_j$

$$\int_A u_j \, d\mu \leqslant \int_A u_k \, d\mu \xrightarrow{k\to\infty} \int_A u_\infty \, d\mu,$$

so that $(u_j)_{j\in\mathbb{N}\cup\{\infty\}}$ is also a submartingale. ∎

Again, $\mathcal{L}^1$-convergence of backwards (sub-)martingales holds under much weaker assumptions.

18.7 Theorem *Let $(w_\ell, \mathcal{A}_\ell)_{\ell\in-\mathbb{N}}$ be a backwards submartingale and assume that $\mu|_{\mathcal{A}_{-\infty}}$ is σ-finite. Then*

(i) $\lim\limits_{j\to+\infty} w_{-j} = w_{-\infty} \in [-\infty, \infty)$ *exists a.e.*

(ii) $\mathcal{L}^1\text{-}\lim\limits_{j\to+\infty} w_{-j} = w_{-\infty}$ *if, and only if,* $\inf_{j\in\mathbb{N}} \int w_{-j} \, d\mu > -\infty$. *In this case, $(w_\ell, \mathcal{A}_\ell)_{\ell\in-\mathbb{N}\cup\{-\infty\}}$ is a submartingale and $w_{-\infty}$ is a.e. real-valued.*

For a backwards martingale, the condition in (ii) *is automatically satisfied.*

Proof Part **(i)** has already been proved in Corollary 18.5. For **(ii)** we start with the observation that for a backwards submartingale

$$\sup_{j \in \mathbb{N}} \int |w_{-j}| \, d\mu < \infty \iff \inf_{j \in \mathbb{N}} \int w_{-j} \, d\mu > -\infty \iff \lim_{j \to +\infty} \int w_{-j} \, d\mu \in \mathbb{R}.$$

Indeed: the second equivalence follows from the submartingale property,

$$\int w_{-j-1} \, d\mu \leqslant \int w_{-j} \, d\mu \leqslant \int w_{-1} \, d\mu,$$

while '$\Leftarrow$' of the first equivalence derives from the fact that $(w_\ell^+)_{\ell \in -\mathbb{N}}$ is again a submartingale, cf. Example 17.3(iv), and

$$\int |w_{-j}| \, d\mu = \int (2w_{-j}^+ - w_{-j}) \, d\mu \leqslant 2 \int w_{-1}^+ \, d\mu - \int w_{-j} \, d\mu;$$

the other direction '$\Rightarrow$' is obvious. With exactly the same reasoning which was used in the proof of T18.6, (i)$\Rightarrow$(ii), we can now show that $(w_\ell^+)_{\ell \in -\mathbb{N}}$ and $(w_\ell)_{\ell \in -\mathbb{N}}$ are uniformly integrable (of course, the function w_ϵ used as a bound for uniform integrability is now $\mathcal{A}_{-\infty}$-measurable). The submartingale property of $(w_\ell)_{\ell \in -\mathbb{N} \cup \{-\infty\}}$ follows literally with the same arguments as the corresponding assertion in (iii)$\Rightarrow$(i) of T18.6. ∎

We close this chapter with a simple but far-reaching application of the (backwards) martingale convergence theorem.

18.8 Example (Kolmogorov's strong law of large numbers) For every sequence $(X_j)_{j \in \mathbb{N}}$ of identically distributed independent random variables on the probability space $(\Omega, \mathcal{A}, P)$ – that is, all $X_j : \Omega \to \mathbb{R}$ are measurable, independent functions (in the sense of Example 17.3(x) and Scholium 17.4) such that $X_j(P) = X_1(P)$ for all $j \in \mathbb{N}$ – the *strong law of large numbers* holds, i.e. the limit

$$\lim_{n \to \infty} \frac{1}{n} (X_1(\omega) + \cdots + X_n(\omega)) \quad \text{exists and is finite for a.e. } \omega \in \Omega,$$

if, and only if, the X_j are integrable. If this is the case, the above limit is given by $\int X_1 \, dP$.

Sufficiency: Suppose the X_j are integrable. Then $Y_j := X_j - \int X_j \, dP$ are again independent identically distributed random variables with zero mean: $\int Y_j \, dP = 0$. Set

$$S_n := Y_1 + Y_2 + \cdots + Y_n \quad \text{and} \quad \mathcal{A}_{-n} := \sigma(S_n, S_{n+1}, S_{n+2}, \ldots),$$

and $\left(\frac{1}{n} S_n, \mathcal{A}_{-n}\right)_{n \in \mathbb{N}}$ is a backwards martingale. In fact, any function of $(Y_1, Y_2, \ldots, Y_n, S_n)$ is independent of $(Y_{n+1}, Y_{n+2}, \ldots)$, and (17.6) yields for every set of the form $A = \bigcap_{j=1}^{N}\{Y_{n+j} \in B_j\} \cap \{S_n \in B_0\}$, $B_0, \dots, B_N \in \mathcal{B}(\mathbb{R})$, $N \in \mathbb{N}$, and all $k = 1, 2, \dots, n$

$$\int_A Y_k \, dP = \int_{\bigcap_{j=1}^{N}\{Y_{n+j} \in B_j\}} \mathbb{1}_{\{S_n \in B_0\}} Y_k \, dP$$

$$= \int_{\{S_n \in B_0\}} Y_k \, dP \cdot P\left(\bigcap_{j=1}^{N}\{Y_{n+j} \in B_j\}\right) \qquad \text{(by (17.6))}$$

$$= \int_{\{S_n \in B_0\}} Y_1 \, dP \cdot P\left(\bigcap_{j=1}^{N}\{Y_{n+j} \in B_j\}\right),$$

noting that the Y_k are identically distributed. Summing over $k = 1, \dots, n$ gives

$$\int_A S_n \, dP = n \int_{\{S_n \in B_0\}} Y_1 \, dP \cdot P\left(\bigcap_{j=1}^{N}\{Y_{n+j} \in B_j\}\right) = n \int_A Y_1 \, dP.$$

This means that $\int_A Y_1 \, dP = \int_A \frac{1}{n} S_n \, dP$ for all $n \in \mathbb{N}$ and all sets A from a generator of $\mathcal{A}_{-n}$ which clearly satisfies the conditions of Remark 17.2(i), proving that $\left(\frac{1}{n} S_n, \mathcal{A}_{-n}\right)_{n \in \mathbb{N}}$ is a backwards martingale. Theorem 18.7 now guarantees that

$$L := \lim_{n \to \infty} \frac{S_n}{n} = \lim_{n \to \infty} \frac{S_{n^2}}{n^2} \quad \text{exists a.e. and in } \mathcal{L}^1.$$

It remains to show that $L = 0$ a.e. Note that $\lim_{n \to \infty} S_n/n^2 = 0$ a.e.; since $e^{-|x|} \leqslant 1$ and since constants are integrable, the dominated convergence theorem 11.2 and independence (17.7) show

$$\int \left[e^{-|L|}\right]^2 dP = \int \lim_{n \to \infty} \left[\exp\left(-\left|\tfrac{S_n}{n}\right|\right) \exp\left(-\left|\tfrac{S_{n^2} - S_n}{n^2}\right|\right)\right] dP$$

$$= \lim_{n \to \infty} \int \exp\left(-\left|\tfrac{S_n}{n}\right|\right) \exp\left(-\left|\tfrac{S_{n^2} - S_n}{n^2}\right|\right) dP$$

$$= \lim_{n \to \infty} \left(\int \exp\left(-\left|\tfrac{S_n}{n}\right|\right) dP \int \exp\left(-\left|\tfrac{S_{n^2} - S_n}{n^2}\right|\right) dP\right)$$

$$= \left(\int e^{-|L|} \, dP\right)^2.$$

Thus

$$\int \left(e^{-|L|} - \int e^{-|L|} \, dP\right)^2 dP = \int \left[e^{-|L|}\right]^2 dP - \left(\int e^{-|L|} \, dP\right)^2 = 0,$$

and we conclude with Theorem 10.9(i) that $e^{-|L|} = \int e^{-|L|}\,dP$ a.e.; as a consequence, L is almost everywhere constant. Using $L = L^1\text{-}\lim_{n\to\infty} S_n/n$, we get

$$L = \int L\,dP = \lim_{n\to\infty} \underbrace{\int \frac{S_n}{n}\,dP}_{=0} = 0 \quad \text{a.e.}$$

Necessity: Suppose the a.e. limit $L = \lim_{n\to\infty} \frac{1}{n}(X_1(\omega) + \cdots + X_n(\omega))$ exists and is finite. If all X_j were positive, we could argue as follows: the truncated random variables $X_j^c := X_j \wedge c$ are still independent and identically distributed. Since they are also integrable, the sufficiency direction of Kolmogorov's law shows that for all $c > 0$

$$\int X_1^c\,dP = \lim_{n\to\infty} \frac{X_1^c + \cdots + X_n^c}{n} \leqslant \lim_{n\to\infty} \frac{X_1 + \cdots + X_n}{n} = L.$$

Letting $c \to \infty$, Beppo Levi's theorem 9.6 proves $\int X_1\,dP < \infty$.

Such a simple argument is not available in the general case. For this we need the converse or 'difficult' half of the Borel – Cantelli lemma (cf. Problem 6.9).

18.9 Theorem (Borel–Cantelli) *Let $(\Omega, \mathcal{A}, P)$ be a probability space and $(A_j)_{j\in\mathbb{N}} \subset \mathcal{A}$. Then*

$$\sum_{j=1}^{\infty} P(A_j) < \infty \quad \Longrightarrow \quad P(\limsup_{j\to\infty} A_j) = 0;$$

if the sets A_j are pairwise independent,[1] then

$$\sum_{j=1}^{\infty} P(A_j) = \infty \quad \Longrightarrow \quad P(\limsup_{j\to\infty} A_j) = 1.$$

Proof Recall that $\limsup_j A_j = \bigcap_k \bigcup_{j\geqslant k} A_j$. Thus $\omega \in \limsup_j A_j$ if, and only if, ω appears in infinitely many of the A_j. This shows that $\limsup_j A_j = \{\sum_{j=1}^{\infty} \mathbf{1}_{A_j} = \infty\}$.

The first of the two implications follows thus: by the Beppo Levi theorem for series C9.9, we see

$$\int \sum_{j=1}^{\infty} \mathbf{1}_{A_j}\,dP = \sum_{j=1}^{\infty} \int \mathbf{1}_{A_j}\,dP = \sum_{j=1}^{\infty} P(A_j) < \infty.$$

Corollary 10.13 then shows $\sum_{j=1}^{\infty} \mathbf{1}_{A_j} < \infty$ a.e., and $P(\limsup_{j\to\infty} A_j) = 0$ follows.

---

[1] i.e. $P(A_j \cap A_k) = P(A_j)P(A_k)$ for all $j \neq k$.

For the second implication we set $S_n := \sum_{j=1}^n \mathbf{1}_{A_j}$ and $S := \sum_{j=1}^\infty \mathbf{1}_{A_j}$. Then $m_n := \int S_n \, dP = \sum_{j=1}^n P(A_n)$ and, by pairwise independence,

$$\int (S_n - m_n)^2 \, dP = \sum_{j,k=1}^n \int (\mathbf{1}_{A_j} - P(A_j))(\mathbf{1}_{A_k} - P(A_k)) \, dP$$

$$= \sum_{j=1}^n \int (\mathbf{1}_{A_j} - P(A_j))^2 \, dP$$

$$= \sum_{j=1}^n P(A_j)(1 - P(A_j)) \leqslant m_n.$$

Since $S_n \leqslant S$, we can use Markov's inequality P10.12 to get

$$P\left(S \leqslant \tfrac{1}{2} m_n\right) \leqslant P\left(S_n \leqslant \tfrac{1}{2} m_n\right) = P\left(S_n - m_n \leqslant -\tfrac{1}{2} m_n\right)$$

$$\leqslant P\left(|S_n - m_n| \geqslant \tfrac{1}{2} m_n\right) = P\left([S_n - m_n]^2 \geqslant \tfrac{1}{4} m_n^2\right)$$

$$\leqslant \frac{4}{m_n^2} \int [S_n - m_n]^2 \, dP \leqslant \frac{4}{m_n}.$$

By assumption $m_n \xrightarrow{n \to \infty} \infty$, hence $P(S < \infty) = \lim_{n \to \infty} P(S \leqslant \tfrac{1}{2} m_n) = 0$. ∎

18.8 Example (continued) We can now continue with the proof of the *necessity* part of Kolmogorov's strong law of large numbers. Since the a.e. limit exists, we get

$$\frac{X_n}{n} = \frac{S_n}{n} - \frac{n-1}{n} \frac{S_{n-1}}{n-1} \xrightarrow{n \to \infty} 0,$$

which shows that $\omega \in A_n := \{|X_n| > n\}$ happens only for finitely many n. In other words, $P(\sum_{j=1}^\infty \mathbf{1}_{A_j} = \infty) = 0$; since the A_n are all independent, the Borel–Cantelli lemma T18.9 shows that $\sum_{j=1}^\infty P(A_j) < \infty$. Thus

$$\int |X_1| \, dP = \sum_{j=1}^\infty \int_{\{j-1 \leqslant |X_1| < j\}} |X_1| \, dP \leqslant \sum_{j=1}^\infty j \, P(j-1 \leqslant |X_1| < j)$$

$$= \sum_{j=1}^\infty \sum_{n=1}^j P(j-1 \leqslant |X_1| < j) = \sum_{n=1}^\infty \sum_{j=n}^\infty P(j-1 \leqslant |X_1| < j)$$

$$= 1 + \sum_{n=1}^\infty P(|X_1| \geqslant n) = 1 + \sum_{n=1}^\infty P(|X_n| \geqslant n) < \infty,$$

since X_1 and X_n have the same distribution. ∎

We will see more applications of the martingale convergence theorems in the following chapters.

Problems

Unless otherwise stated $(X, \mathcal{A}, \mathcal{A}_j, \mu)$ will be a σ-finite filtered measure space.

18.1. Verify that the random times σ_k and τ_k defined in the proof of Lemma 18.1 are stopping times.

18.2. Let $(\mathcal{A}_{-j})_{j \in \mathbb{N}}$ be a decreasing filtration such that $\mu|_{\mathcal{A}_{-\infty}}$ is σ-finite. Assume that $(u_{-j}, \mathcal{A}_{-j})_{j \in \mathbb{N}}$ is a backwards supermartingale which converges a.e. to a real-valued function $u_{-\infty} \in \mathcal{L}^1(\mu)$ which closes the supermartingale to the left, i.e. such that $(u_{-j}, \mathcal{A}_{-j})_{j \in \mathbb{N} \cup \{\infty\}}$ is still a supermartingale. Then

$$\lim_{j \to \infty} \int u_{-j} \, d\mu = \int u_{-\infty} \, d\mu.$$

18.3. Let $(u_j, \mathcal{A}_j)_{j \in \mathbb{N}}$ be a supermartingale such that $u_j \geq 0$ and $\lim_{j \to \infty} \int u_j \, d\mu = 0$. Then $u_j \xrightarrow{j \to \infty} 0$ pointwise a.e. and in $\mathcal{L}^1$.

   Remark: Positive supermartingales with $\lim_{j \to \infty} \int u_j \, d\mu = 0$ are called *potentials*.

18.4. Let $(u_j, \mathcal{A}_j)_{j \in \mathbb{N}}$ be a martingale. If $\mathcal{L}^1$-$\lim_{j \to \infty} u_j$ exists, then the pointwise limit $\lim_{j \to \infty} u_j(x)$ exists for almost every x.

18.5. Let $(\Omega, \mathcal{A}, P)$ be a probability space. Find a martingale $(u_j)_{j \in \mathbb{N}}$ for which $0 < P(u_j \text{ converges}) < 1$.
   [Hint: take a sequence $(X_k)_{k \in \mathbb{N}_0}$ of independent Bernoulli $(\frac{1}{2}, \frac{1}{2})$-distributed random variables with values ± 1; try $u_j := \frac{1}{2}(X_0 + 1)(X_1 + X_2 + \cdots + X_j)$.]

18.6. The following exercise furnishes an example of a martingale $(M_j)_{j \in \mathbb{N}}$ on the probability space $\big([0, 1], \mathcal{B}[0, 1], \lambda = \lambda^1|_{[0,1]}\big)$ such that λ-$\lim_{j \to \infty} M_j$ exists but the pointwise limit $\lim_{j \to \infty} M_j(x)$ doesn't. Compare this with Problem 18.4.

   (i) Construct a sequence $(X_j)_{j \in \mathbb{N}}$ of independent, identically Bernoulli distributed random variables with $\lambda(\{X_1 = 1\}) = \lambda(\{X_1 = -1\}) = \frac{1}{2}$.

   (ii) Let $\mathcal{A}_n = \sigma(X_1, \ldots, X_{2^n})$. Show that $A_n := \{X_{2^{n-1}+1} + \cdots + X_{2^n} = 0\}$ is for each $n \in \mathbb{N}$ contained in $\mathcal{A}_n$ and

$$\lim_{n \to \infty} \lambda(A_n) = 0 \qquad \text{and} \qquad \lambda(\limsup_{n \to \infty} A_n) = 1.$$

   Conclude that the set of all x for which $\lim_n \mathbf{1}_{A_n}(x)$ exists is a null set.
   [Hint: use the 'difficult' direction of the Borel–Cantelli lemma T18.9. Moreover, Stirling's formula $n! \sim \sqrt{2\pi n}\,(n/e)^n$ might come in handy.]

   (iii) The sequence $M_0 := 0$ and $M_{n+1} := M_n(1 + X_{2^n+1}) + \mathbf{1}_{A_n} X_{2^n+1}$, $n \in \mathbb{N}_0$, defines a martingale $(M_n, \mathcal{A}_n)_{n \geq 1}$.

   (iv) Show that $\lambda(M_{n+1} \neq 0) \leq \frac{1}{2}\lambda(M_n \neq 0) + \lambda(A_n)$.

   (v) Show that for every $x \in \{\lim_n M_n \text{ exists}\}$ the limit $\lim_n \mathbf{1}_{A_n}(x)$ exists, too. Conclude that $\lim_n \lambda(M_n = 0) = 1$ and that $\lambda(\{\lim_n M_n \text{ exists}\}) = 0$.

18.7. Consider the probability space $(\mathbb{N}, \mathcal{P}(\mathbb{N}), P)$ with $P(\{j\}) := \frac{1}{j} - \frac{1}{j+1}$. Set

$$\mathcal{A}_n := \sigma\big(\{1\}, \{2\}, \ldots, \{n\}, [n+1, \infty) \cap \mathbb{N}\big)$$

and show that $X_n := (n+1)\mathbf{1}_{[n+1,\infty) \cap \mathbb{N}}$, $n \in \mathbb{N}$, is a positive martingale such that $\int X_n \, dP = 1$, $\lim_{n \to \infty} X_n = 0$ but $\sup_{n \in \mathbb{N}} X_n = \infty$.

18.8. $\mathcal{L}^2$-**bounded martingales** A martingale $(u_j, \mathcal{A}_j)_{j \in \mathbb{N}}$ is called $\mathcal{L}^2$-*bounded*, if $\sup_{j \in \mathbb{N}} \int u_j^2 \, d\mu < \infty$. For ease of notation set $u_0 := 0$.

(i) Show that $(u_j)_{j \in \mathbb{N}}$ is $\mathcal{L}^2$-bounded if, and only if, $\sum_{j=1}^{\infty} \int (u_j - u_{j-1})^2 \, d\mu < \infty$.

[Hint: use Problem 17.6.]

Assume from now on that $(u_j)_{j \in \mathbb{N}}$ is $\mathcal{L}^2$-bounded.

(ii) Show that $\lim_{j \to \infty} u_j = u$ exists a.e.

[Hint: argue that $(u_j^2)_{j \in \mathbb{N}}$ is a submartingale.]

(iii) Show that $\lim_{j \to \infty} \int (u - u_j)^2 \, d\mu = 0$.

[Hint: check that $\int (u_{j+k} - u_j)^2 \, d\mu = \sum_{\ell=j+1}^{j+k} \int (u_\ell - u_{\ell-1})^2 \, d\mu$ and apply Fatou's lemma T9.11.]

(iv) Assume now that $\mu(X) < \infty$. Show that $(u_j)_{j \in \mathbb{N}}$ is uniformly integrable, that $u_j \xrightarrow{j \to \infty} u$ in $\mathcal{L}^1$ and that $u_\infty := u$ closes the martingale to the right, i.e. that $(u_j)_{j \in \mathbb{N} \cup \{\infty\}}$ is again a martingale.

18.9. Let $(\Omega, \mathcal{A}, P)$ be a probability space.

(i) Let $(\varepsilon_j)_{j \in \mathbb{N}}$ be a sequence of independent identically Bernoulli $(\frac{1}{2}, \frac{1}{2})$-distributed random variables with values ± 1. Show that for any sequence $(y_j)_{j \in \mathbb{N}}$

$$\sum_{j=1}^{\infty} y_j^2 < \infty \quad \Longleftrightarrow \quad \sum_{j=1}^{\infty} \varepsilon_j y_j \quad \text{converges a.e.}$$

(ii) Generalize (i) to a sequence of independent random variables $(X_j)_{j \in \mathbb{N}}$ with zero mean $\int X_j \, dP = 0$ and finite variances $\int X_j^2 \, dP = \sigma_j^2 < \infty$ and prove

$$\sum_{j=1}^{\infty} \sigma_j^2 < \infty. \quad \Longrightarrow \quad \sum_{j=1}^{\infty} X_j \quad \text{converges a.e.}$$

[Hint: consider the martingale $S_n = X_1 + \cdots + X_n$ and use Problem 18.8.]

(iii) If $|X_j| \leqslant C$ for all $j \in \mathbb{N}$, the converse of (ii) is also true, i.e.

$$\sum_{j=1}^{\infty} \sigma_j^2 < \infty. \quad \Longleftrightarrow \quad \sum_{j=1}^{\infty} X_j \quad \text{converges a.e.}$$

[Hint: show that $M_n := (X_1 + \cdots + X_n)^2 - (\sigma_1^2 + \cdots + \sigma_n^2) =: S_n^2 - A_n$ is a martingale, use optional sampling 17.8 for M_n with $\tau_\kappa := \inf\{j : |M_j| > \kappa\}$, observe that $|M_{n \wedge \tau_\kappa}| \leqslant C + \kappa$ and that $\int A_{n \wedge \tau_\kappa} \, dP \leqslant (K + c)^2$.]

19

The Radon–Nikodým theorem and other applications of martingales

After our excursion into the theory of martingales we want to apply martingales to continue the development of measure and integration theory. The central topics of this chapter are

- the Radon–Nikodým theorem 19.2 and Lebesgue's decomposition theorem 19.9;
- the Hardy–Littlewood maximal theorem 19.17;
- Lebesgue's differentiation theorem 19.20.

For the last two we need (maximal) inequalities for martingales. These will be treated in a short interlude which is also of independent interest.

The Radon–Nikodým theorem

Let $(X, \mathcal{A}, \mu)$ be a measure space. We have seen in Lemma 10.8 that for any $f \in \mathcal{L}^1_+(\mathcal{A})$ – or indeed for $f \in \mathcal{M}^+(\mathcal{A})$ – the set-function $\nu := f\mu$ given by $\nu(A) := \int_A f(x)\,\mu(dx)$ is again a measure. From Theorem 10.9(ii) we know that

$$N \in \mathcal{A}, \quad \mu(N) = 0 \quad \Longrightarrow \quad \nu(N) = 0. \tag{19.1}$$

This observation motivates the following

19.1 Definition Let μ, ν be two measures on the measurable space $(X, \mathcal{A})$. If (19.1) holds, we call ν *absolutely continuous* w.r.t. μ and write $\nu \ll \mu$.

Measures with densities are always absolutely continuous w.r.t. their base measure: $f\mu \ll \mu$. Remarkably, the converse is also true.

19.2 Theorem (Radon–Nikodým). *Let μ, ν be two measures on the measurable space $(X, \mathcal{A})$. If μ is σ-finite, then the following assertions are equivalent*

(i) $\nu(A) = \int_A f(x)\,\mu(dx)$ *for some a.e. unique* $f \in \mathcal{M}^+(\mathcal{A})$;

(ii) $\nu \ll \mu$.

The unique function f is called the Radon–Nikodým derivative and (traditionally) denoted by $f = d\nu/d\mu$.

Above we have just verified that **(i)**$\Rightarrow$**(ii)**. The converse direction is less obvious and we want to use a martingale argument for its proof. For this we need a few more preparations which extend the notion of martingale to directed index sets.

Let $(I, \leqslant)$ be any partially ordered index set. We call I *upwards filtering* or *upwards directed* if

$$\alpha, \beta \in I \quad \Longrightarrow \quad \exists \gamma \in I : \alpha \leqslant \gamma, \ \beta \leqslant \gamma. \tag{19.2}$$

A family $(\mathcal{A}_\alpha)_{\alpha \in I}$ of sub-σ-algebras of $\mathcal{A}$ is called a *filtration* if

$$\alpha, \beta \in I, \ \alpha \leqslant \beta \quad \Longrightarrow \quad \mathcal{A}_\alpha \subset \mathcal{A}_\beta;$$

as before, we set $\mathcal{A}_\infty := \sigma\left(\bigcup_{\alpha \in I} \mathcal{A}_\alpha\right)$, and we treat ∞ as the biggest element of $I \cup \{\infty\}$, i.e. $\alpha < \infty$ for all $\alpha \in I$. If a σ-algebra $\mathcal{A}_0 \subset \mathcal{A}_\alpha$ for all $\alpha \in I$ and if $\mu|_{\mathcal{A}_0}$ is σ-finite, we call $(X, \mathcal{A}, \mathcal{A}_\alpha, \mu)$ a *σ-finite filtered measure space*.

19.3 Definition Let $(X, \mathcal{A}, \mathcal{A}_\alpha, \mu)$ be a σ-finite filtered measure space. A family of measurable functions $(u_\alpha)_{\alpha \in I}$ is called a *martingale* (w.r.t. the filtration $(\mathcal{A}_\alpha)_{\alpha \in I}$), if $u_\alpha \in \mathcal{L}^1(\mathcal{A}_\alpha)$ for each $\alpha \in I$ and if

$$\int_A u_\beta \, d\mu = \int_A u_\alpha \, d\mu \qquad \forall \alpha \leqslant \beta, \ \forall A \in \mathcal{A}_\alpha. \tag{19.3}$$

The notion of convergence *along an upwards filtering set* is slightly more complicated than for the index set $\mathbb{N}$. We say

$$u = \mathcal{L}^1\text{-}\lim_{\alpha \in I} u_\alpha \iff \forall \epsilon > 0 \ \exists \gamma_\epsilon \in I \ \forall \alpha \geqslant \gamma_\epsilon : \ \|u - u_\alpha\|_1 < \epsilon.$$

We can now extend Theorem 18.6.

19.4 Theorem *Let I be an upwards filtering index set, $(X, \mathcal{A}, \mathcal{A}_\alpha, \mu)$ be a σ-finite measure space and $(u_\alpha, \mathcal{A}_\alpha)_{\alpha \in I}$ be a martingale. Then the following assertions are equivalent.*

(i) *There exists a unique $u_\infty \in \mathcal{L}^1(\mathcal{A}_\infty)$ such that $(u_\alpha, \mathcal{A}_\alpha)_{\alpha \in I \cup \{\infty\}}$ is a martingale. In this case $u_\infty = \mathcal{L}^1\text{-}\lim_{\alpha \in I} u_\alpha$.*

(ii) *$(u_\alpha, \mathcal{A}_\alpha)_{\alpha \in I}$ is uniformly integrable.*

Proof (i)⇒(ii): (compare with T18.6) Denote by $(A_j)_{j\in\mathbb{N}}$ an exhausting sequence in $\mathcal{A}_0$. Since $u_\infty \in \mathcal{L}^1(\mathcal{A}_\infty)$, we find for every $\epsilon > 0$ some $\kappa > 0$ and $N \in \mathbb{N}$ such that

$$\int_{\{|u_\infty|>\kappa\}} |u_\infty|\, d\mu + \int_{A_j^c} |u_\infty|\, d\mu \leqslant \epsilon \qquad \forall j \geqslant N.$$

Clearly, the function $w(x) := \sum_{j\in\mathbb{N}} 2^{-j}\,(1+\mu(A_j))^{-1}\,\mathbf{1}_{A_j}(x)$ is in $\mathcal{L}^1_+(\mathcal{A}_0)$, $w > 0$ and, as $(|u_\alpha|)_{\alpha\in I\cup\{\infty\}}$ is a submartingale (cf. Example 17.3(v)), we find for every $L > 0$

$$\sup_{\alpha\in I} \int_{\{|u_\alpha|>Lw\}} |u_\alpha|\, d\mu \leqslant \sup_{\alpha\in I} \int_{\{|u_\alpha|>Lw\}} |u_\infty|\, d\mu$$

$$\leqslant \sup_{\alpha\in I} \int_{\{|u_\alpha|>Lw\}\cap A_N\cap\{|u_\infty|\leqslant\kappa\}} |u_\infty|\, d\mu + \int_{A_N^c} |u_\infty|\, d\mu + \int_{\{|u_\infty|>\kappa\}} |u_\infty|\, d\mu$$

$$\leqslant \kappa \sup_{\alpha\in I} \mu\big(\{|u_\alpha| > L\,2^{-N}\,(1+\mu(A_N))^{-1}\}\big) + \epsilon$$

(use for the last step that $w(x) \geqslant 2^{-N}\,(1+\mu(A_N))^{-1}$ for $x \in A_N$). By Markov's inequality P10.12 and the submartingale property we get

$$\sup_{\alpha\in I} \int_{\{|u_\alpha|>Lw\}} |u_\alpha|\, d\mu \leqslant \kappa\, \frac{2^{-N}\,(1+\mu(A_N))}{L} \sup_{\alpha\in I} \int |u_\alpha|\, d\mu + \epsilon$$

$$\leqslant \frac{\kappa\, 2^{-N}\,(1+\mu(A_N))}{L} \int |u_\infty|\, d\mu + \epsilon,$$

and (ii) follows since we can choose $L > 0$ as large as we want.

(ii)⇒(i): *Step 1: uniqueness.* Assume that $u, w \in \mathcal{L}^1(\mathcal{A}_\infty)$ are two functions which close the martingale $(u_\alpha)_{\alpha\in I}$, i.e. functions satisfying

$$\int_A u\, d\mu = \int_A w\, d\mu = \int_A u_\alpha\, d\mu \qquad \forall A \in \mathcal{A}_\alpha,\ \alpha \in I.$$

Since u and w are integrable functions, the family

$$\Sigma := \left\{A \in \mathcal{A}_\infty : \int_A u\, d\mu = \int_A w\, d\mu\right\}$$

is a σ-algebra which satisfies $\bigcup_{\alpha\in I} \mathcal{A}_\alpha \subset \Sigma \subset \mathcal{A}_\infty$. Since $\mathcal{A}_\infty$ is generated by the $\mathcal{A}_\alpha$, we get $\Sigma = \mathcal{A}_\infty$, which means that $\int_A u\, d\mu = \int_A w\, d\mu$ holds for all $A \in \mathcal{A}_\infty$. Now Corollary 10.14 applies and we get $u = w$ almost everywhere.

Step 2: existence of the limit. We claim that

$$\forall \epsilon > 0\ \exists \gamma_\epsilon \in I\ \forall \alpha, \beta \geqslant \gamma_\epsilon : \int |u_\alpha - u_\beta|\, d\mu < \epsilon. \tag{19.4}$$

Otherwise we could find a sequence $(\alpha_j)_{j\in\mathbb{N}} \subset I$ such that $\int |u_{\alpha_{j+1}} - u_{\alpha_j}|\, d\mu > \epsilon$ for all $j \in \mathbb{N}$. Since I is upwards filtering, we can assume that $(\alpha_j)_{j\in\mathbb{N}}$ is an increasing sequence.[✓] Because of (ii), $(u_{\alpha_j}, \mathcal{A}_{\alpha_j})_{j\in\mathbb{N}}$ is a uniformly integrable martingale with index set $\mathbb{N}$ which is, by construction, not an $\mathcal{L}^1$-Cauchy sequence. This contradicts Theorem 18.6.

We will now prove the existence of the $\mathcal{L}^1$-limit. Pick in (19.4) $\epsilon = \frac{1}{n}$ and choose $\gamma_{1/n}$. Since I is upwards directed, we can assume that $\gamma_{1/n}$ increases as $n \to \infty$;[✓] thus $(u_{\gamma_{1/n}})_{n\in\mathbb{N}} \subset \mathcal{L}^1(\mathcal{A}_\infty)$ is an $\mathcal{L}^1$-Cauchy sequence. By Theorem 18.6 it converges in $\mathcal{L}^1(\mathcal{A}_\infty)$ and a.e. to some $u_\infty := \lim_{n\to\infty} u_{\gamma_{1/n}} \in \mathcal{L}^1(\mathcal{A}_\infty)$. Moreover, for all $A \in \mathcal{A}_\infty$ and $\alpha > \gamma_{1/n}$ we have

$$\int_A |u_\alpha - u_\infty|\, d\mu \leqslant \underbrace{\int_A |u_\alpha - u_{\gamma_{1/n}}|\, d\mu}_{\leqslant 1/n \text{ by } (19.4)} + \int_A |u_{\gamma_{1/n}} - u_\infty|\, d\mu \leqslant \frac{2}{n}.$$

This shows, in particular, that $\mathbf{1}_A u_\alpha \xrightarrow{\mathcal{L}^1} \mathbf{1}_A u_\infty$ for all $A \in \mathcal{A}_\infty$, and in view of step 1, u_∞ is the only possible limit. The same argument that we used in (iii)⇒(i) of T18.6 now yields that $(u_\alpha)_{\alpha\in I\cup\{\infty\}}$ is still a martingale. ∎

Theorem 19.4 does not claim that $u_\alpha \xrightarrow{\text{a.e. along } I} u_\infty$. This is, in general, *false* for non-linearly ordered index sets I, see e.g. Dieudonné [12].

That uncountable, partially ordered index sets are not at all artificial is shown by the following example which will be essential for the proof of Theorem 19.2.

19.5 Example Let $(X, \mathcal{A}, \mu)$ be a finite measure space and assume that ν is a measure such that $\nu \ll \mu$. Set

$$I := \left\{ \alpha = \{A_1, A_2, \ldots, A_n\} : n \in \mathbb{N},\ A_j \in \mathcal{A}\ \text{and}\ \overset{n}{\underset{j=1}{\biguplus}} A_j = X \right\}$$

and define an order relation '$\leqslant$' on I through

$$\alpha \leqslant \alpha' \iff \forall A \in \alpha : A = A_1' \uplus \ldots \uplus A_\ell'\ \text{where}\ A_k' \in \alpha',\ \ell \in \mathbb{N}.$$

Since the common refinement β of any two elements $\alpha, \alpha' \in I$,

$$\beta := \{A \cap A' : A \in \alpha,\ A' \in \alpha'\},$$

is again in I and satisfies $\alpha \leqslant \beta$ and $\alpha' \leqslant \beta$, it is clear that $(I, \leqslant)$ is upwards filtering. In particular,

$$(\mathcal{A}_\alpha)_{\alpha\in I} \quad \text{where} \quad \mathcal{A}_\alpha := \sigma(A : A \in \alpha)$$

is a filtration as $\mathcal{A}_\alpha \subset \mathcal{A}_{\alpha'}$ whenever $\alpha \leqslant \alpha'$. Moreover, $(f_\alpha, \mathcal{A}_\alpha)_{\alpha \in I}$ defined by

$$f_\alpha := \sum_{A \in \alpha} \frac{\nu(A)}{\mu(A)} \mathbf{1}_A, \qquad \left(\frac{\nu(A)}{\mu(A)} := 0 \quad \text{if} \quad \mu(A) = 0 \right)$$

is a martingale. *Indeed*, if $\alpha \leqslant \beta$, $\alpha, \beta \in I$, then

$$\int_A f_\alpha \, d\mu = \frac{\nu(A)}{\mu(A)} \mu(A) = \left\{ \begin{array}{ll} \nu(A) & \text{if } \mu(A) > 0 \\ 0 & \text{if } \mu(A) = 0 \end{array} \right\} = \nu(A)$$

as $\nu \ll \mu$. Similarly, for $A \in \alpha$ with $A = B_1 \uplus \ldots \uplus B_\ell$ and $B_1, \ldots, B_\ell \in \beta$

$$\int_A f_\beta \, d\mu = \sum_{k=1}^{\ell} \int_{B_k} f_\beta \, d\mu = \sum_{k=1}^{\ell} \frac{\nu(B_k)}{\mu(B_k)} \mu(B_k)$$

$$= \sum_{k : \mu(B_k) > 0} \nu(B_k)$$

$$\stackrel{(*)}{=} \sum_{k=1}^{\ell} \nu(B_k) = \nu(A),$$

where we used in $(*)$ that $\nu \ll \mu$, i.e. $\nu(B_k) = 0$ if $\mu(B_k) = 0$. Thus $\int_A f_\alpha \, d\mu = \int_A f_\beta \, d\mu$ for all $A \in \alpha$, hence on $\mathcal{A}_\alpha$ since all $A \in \alpha$ are disjoint and generate $\mathcal{A}_\alpha$ ([✓], cf. also Remark 17.2(i)).

What Example 19.5 really says is that

$$\nu(A) = \int_A f_\alpha \, d\mu \quad \forall A \in \mathcal{A}_\alpha, \tag{19.5}$$

or $\nu|_{\mathcal{A}_\alpha} \ll \mu|_{\mathcal{A}_\alpha}$ and $d(\nu|_{\mathcal{A}_\alpha})/d(\mu|_{\mathcal{A}_\alpha}) = f_\alpha$. Heuristically we should expect that, if $f_\alpha \xrightarrow{\alpha \to \infty} f_\infty$ exists, f_∞ is the Radon–Nikodým derivative $d\nu/d\mu = f_\infty$. This idea can be made rigorous and is the basis for the

Proof (of Theorem 19.2 (ii)⇒(i)) Let us first

assume that μ and ν are finite measures

Denote by $(f_\alpha, \mathcal{A}_\alpha)_{\alpha \in I}$ the martingale of Example 19.5. It is enough to show that

$$f_\infty = \mathcal{L}^1\text{-}\lim_{\alpha \in I} f_\alpha \quad \text{exists and that} \quad \mathcal{A} = \mathcal{A}_\infty. \tag{19.6}$$

Indeed, (19.6) combined with (19.5) implies

$$\nu(A) = \int_A f_\infty \, d\mu \qquad \forall A \in \bigcup_{\alpha \in I} \mathcal{A}_\alpha,$$

and the uniqueness theorem 5.7 for measures extends this equality to $\mathcal{A}_\infty = \sigma\left(\bigcup_{\alpha \in I} \mathcal{A}_\alpha\right)$. Since $A \in \mathcal{A}$ is trivially contained in $\mathcal{A}_\alpha$ where $\alpha := \{A, A^c\}$ – at this point we use the finiteness of the measure μ – we see

$$\mathcal{A} \supset \mathcal{A}_\infty = \sigma\left(\bigcup_{\alpha \in I} \mathcal{A}_\alpha\right) \supset \bigcup_{\alpha \in I} \mathcal{A}_\alpha \supset \mathcal{A},$$

and all that remains is to prove the existence of the limit in (19.6). In view of Theorem 19.4 we have to show that $(f_\alpha, \mathcal{A}_\alpha)_{\alpha \in I}$ is uniformly integrable.

We claim that $\sup_{\alpha \in I} \nu(\{f_\alpha > R\}) \leqslant \epsilon$ for all large enough $R = R_\epsilon > 0$. Otherwise we could find some $\epsilon_0 > 0$ with $\nu(\{f_\alpha > n\}) > \epsilon_0$ for all $n \in \mathbb{N}$, so that $\nu\left(\bigcap_{n \in \mathbb{N}} \{f_\alpha > n\}\right) > 0$ by the continuity of measures, T4.4. Since ν is a finite measure,

$$\mu\left(\bigcap_{n \in \mathbb{N}} \{f_\alpha > n\}\right) \overset{4.4}{=} \inf_{n \in \mathbb{N}} \mu(\{f_\alpha > n\}) \overset{10.12}{\leqslant} \lim_{n \to \infty} \frac{1}{n} \int f_\alpha \, d\mu = \lim_{n \to \infty} \frac{\nu(X)}{n} = 0,$$

which contradicts the fact that $\nu \ll \mu$. Finally,

$$\int_{\{|f_\alpha| > R\}} |f_\alpha| \, d\mu = \int_{\{f_\alpha > R\}} f_\alpha \, d\mu = \nu(\{f_\alpha > R\}) \leqslant \epsilon$$

if $R = R_\epsilon > 0$ is sufficiently large, and uniform integrability follows since the constant function $R \in \mathcal{L}^1(\mu)$. The uniqueness of f_∞ follows also from Theorem 19.4.

> **Assume that μ is finite and $\nu(X) = \infty$**

Denote by $\mathcal{F} := \{F \in \mathcal{A} : \nu(F) < \infty\}$ the sets with finite ν-measure. Obviously, $\mathcal{F}$ is $\cup$-stable, and the constant

$$c := \sup_{F \in \mathcal{F}} \mu(F) \leqslant \mu(X) < \infty$$

can be approximated by an increasing sequence $(F_j)_{j \in \mathbb{N}} \subset \mathcal{F}$ such that $c = \mu\left(\bigcup_{j \in \mathbb{N}} F_j\right) = \sup_{j \in \mathbb{N}} \mu(F_j)$.[✓] When restricted to the set $F_\infty := \bigcup_{j \in \mathbb{N}} F_j$, ν is by definition σ-finite, while for $A \subset F_\infty^c$, $A \in \mathcal{A}$, we have

$$\text{either} \quad \mu(A) = \nu(A) = 0 \quad \text{or} \quad 0 < \mu(A) < \nu(A) = \infty. \tag{19.7}$$

In fact, if $\nu(A) < \infty$, then $F_j \cup A \in \mathcal{F}$ for all $j \in \mathbb{N}$, which implies that

$$c \geqslant \mu\left(\bigcup_{j \in \mathbb{N}} F_j \cup A\right) = \mu(F_\infty \cup A) = \mu(F_\infty) + \mu(A) = c + \mu(A),$$

that is $\mu(A) = 0$, hence $\nu(A) = 0$ by absolute continuity; if, however, $\nu(A) = \infty$ we have again by absolute continuity that $\mu(A) > 0$. Define now

$$\nu_j := \nu(\cdot \cap (F_j \setminus F_{j-1})), \quad \mu_j := \mu(\cdot \cap (F_j \setminus F_{j-1})), \tag{$F_0 := \emptyset$}$$

and it is clear that $\nu_j \ll \mu_j$ for every $j \in \mathbb{N}$. Since μ_j, ν_j are finite measures, the first part of this proof shows that $\nu_j = f_j \mu_j$. Obviously, the function

$$f(x) := \begin{cases} f_j(x) & \text{if } x \in F_j \setminus F_{j-1}, \\ \infty & \text{if } x \in F_\infty^c, \end{cases} \tag{19.8}$$

fulfils $\nu = f \mu$. By construction, f is unique on the set F_∞. But since *every* density $\tilde{f}$ of ν with respect to μ satisfies

$$\nu\big(\{\tilde{f} \leqslant n\} \cap F_\infty^c\big) = \int_{\{\tilde{f} \leqslant n\} \cap F_\infty^c} \tilde{f}\, d\mu \leqslant n\, \mu\big(\{\tilde{f} \leqslant n\} \cap F_\infty^c\big) < \infty,$$

the alternative (19.7) reveals that $\nu\big(\{\tilde{f} \leqslant n\} \cap F_\infty^c\big) = \mu\big(\{\tilde{f} \leqslant n\} \cap F_\infty^c\big) = 0$ for all $n \in \mathbb{N}$, i.e. that $\tilde{f}\big|_{F_\infty^c} = \infty$. In other words: f, as defined in (19.8), is also unique on F_∞^c.

$$\boxed{\textbf{Assume that } \mu \textbf{ is } \sigma\textbf{-finite and } \nu(X) \leqslant \infty}$$

Let $(A_j)_{j\in\mathbb{N}} \subset \mathcal{A}$ be an exhausting sequence with $A_j \uparrow X$ and $\mu(A_j) < \infty$. Then the measures

$$h\mu \text{ and } \mu \qquad \text{where} \qquad h(x) := \sum_{j=1}^{\infty} \frac{2^{-j}}{1+\mu(A_j)}\, \mathbf{1}_{A_j}(x)$$

have the same null sets.[✓] Therefore $\nu \ll \mu$ if, and only if, $\nu \ll h\mu$. Since $h\mu$ is a finite measure[✓], the first two parts of the proof show that $\nu = f \cdot (h\mu) = (fh)\mu$ for a suitable density $f \in \mathcal{M}^+(\mathcal{A})$. The last equality needs proof: if $f = \sum_{j=0}^{M} y_j \mathbf{1}_{A_j}$ is a positive simple function,

$$\nu(A) = \int_A \sum_{j=0}^{M} y_j \mathbf{1}_{A_j}\, d(h\mu) = \sum_{j=0}^{M} y_j \int \mathbf{1}_{A_j \cap A}\, h\, d\mu = \int_A (fh)\, d\mu,$$

and the general case follows from Beppo Levi's theorem 9.6. Uniqueness is clear as f is $(h\mu)$-a.e. unique, which implies that fh is μ-a.e. unique since $h > 0$. ■

19.6 Corollary *Let $(X, \mathcal{A}, \mu)$ be a σ-finite measure space and $\nu = f\mu$. Then*

(i) $\nu(X) < \infty \quad \Longleftrightarrow \quad f \in \mathcal{L}^1(\mu)$;

(ii) ν *is σ-finite* $\quad \Longleftrightarrow \quad \mu(\{f = \infty\}) = 0$.

Proof The first assertion (i) is obvious. For (ii) assume first that $\mu(\{f = \infty\}) = 0$. Since μ is σ-finite, we find an exhausting sequence $(A_j)_{j\in\mathbb{N}} \subset \mathcal{A}$ with $A_j \uparrow X$ and $\mu(A_j) < \infty$. The sets

$$B_k := \{0 \leqslant f \leqslant k\}, \qquad B_\infty := \{f = \infty\}$$

obviously satisfy $\bigcup_{k\in\mathbb{N}}(B_k\cup B_\infty)=X$ as well as $\nu(B_\infty)=0$ and

$$\nu(B_k\cap A_j)=\int_{B_k\cap A_j}f\,d\mu\leqslant k\int_{A_j}d\mu=k\,\mu(A_j)<\infty.$$

This shows that $\left(A_j\cap(B_k\cup B_\infty)\right)_{j,k\in\mathbb{N}}$ is an exhausting sequence for ν which means that ν is σ-finite.

Conversely, let ν be σ-finite and assume that $\mu(\{f=\infty\})>0$. As we can find *one* exhausting sequence $(C_k)_{k\in\mathbb{N}}\subset\mathcal{A}$ for both μ and $\nu^{[\vee]}$, we see that

$$\{f=\infty\}=\bigcup_{k\in\mathbb{N}}\left(\{f=\infty\}\cap C_k\right)\supset\{f=\infty\}\cap C_{k_0}$$

for some fixed $k_0\in\mathbb{N}$ with $\mu(C_{k_0})>0$. But then

$$\nu(C_{k_0})\geqslant\int_{\{f=\infty\}\cap C_{k_0}}f\,d\mu=\infty,$$

which is impossible. ∎

It is clear that not all measures are absolutely continuous with respect to each other. In some sense, the next notion is the opposite of absolute continuity.

19.7 Definition Two measures μ,ν on a measurable space $(X,\mathcal{A})$ are called (mutually) *singular* if there is a set $N\in\mathcal{A}$ such that $\nu(N)=0=\mu(N^c)$. We write in this case $\mu\perp\nu$ (or $\nu\perp\mu$ as '$\perp$' is symmetric).

19.8 Examples Let $(X,\mathcal{A})=(\mathbb{R}^n,\mathcal{B}(\mathbb{R}^n))$. Then

(i) $\delta_x\perp\lambda^n$ for all $x\in\mathbb{R}^n$;
(ii) $f\mu\perp g\mu$ if $\operatorname{supp}f\cap\operatorname{supp}g=\emptyset$.[1]

The measures μ and ν are singular, if they have disjoint 'supports', that is, if μ lives in a region of X which is not charged by ν and *vice versa*. In this sense, Example 19.8(ii) is the model case for singular measures. In general, however, two measures are neither purely absolutely continuous nor purely singular, but are a mixture of both.

19.9 Theorem (Lebesgue decomposition) *Let* μ,ν *be two* σ-*finite measures on a measurable space* $(X,\mathcal{A})$. *Then there exists a (up to null sets) unique decomposition* $\nu=\nu^\circ+\nu^\perp$ *where* $\nu^\circ\ll\mu$ *and* $\nu^\perp\perp\mu$.

---

[1] $\operatorname{supp}f:=\overline{\{f\neq 0\}}$.

Proof Obviously $\mu + \nu$ is still a σ-finite measure[✓], and $\nu \ll (\nu + \mu)$. In this situation Theorem 19.2 applies and shows that

$$\nu = f(\mu + \nu) = f\mu + f\nu. \tag{19.9}$$

For any $\epsilon > 0$ we conclude, in particular, that

$$\nu(\{f \geq 1 + \epsilon\}) = \int_{\{f \geq 1+\epsilon\}} f \, d(\mu + \nu)$$

$$\geq (1 + \epsilon)\mu(\{f \geq 1 + \epsilon\}) + (1 + \epsilon)\nu(\{f \geq 1 + \epsilon\}),$$

i.e. $\mu(\{f \geq 1 + \epsilon\}) = \nu(\{f \geq 1 + \epsilon\}) = 0$ for all ϵ, hence $\mu(\{f > 1\}) = \nu(\{f > 1\}) = 0$. Without loss of generality we may therefore assume that $0 \leq f \leq 1$. In this case (19.9) can be rewritten as

$$(1 - f)\nu = f\mu, \tag{19.10}$$

and on the set $N := \{f = 1\}$ we have

$$\mu(N) = \int_{\{f=1\}} d\mu = \int_{\{f=1\}} f \, d\mu \overset{(19.10)}{=} \int_{\{f=1\}} (1 - f) \, d\nu = 0.$$

Therefore, $\mu \perp \nu^{\perp}$ where $\nu^{\perp} := \nu(\bullet \cap \{f = 1\})$, and for $\nu^{\circ} := \nu(\bullet \cap \{f < 1\})$ we get from (19.10)

$$\nu^{\circ}(A) = \nu(A \cap \{f < 1\}) = \int_{A \cap \{f<1\}} d\nu = \int_{A \cap \{f<1\}} \frac{f}{1 - f} \, d\mu \quad \forall A \in \mathcal{A},$$

showing that $\nu^{\circ} \ll \mu$.

The uniqueness (up to null sets) of this decomposition follows directly from the uniqueness of the Radon–Nikodým derivative $f/(1 - f)\mathbf{1}_{\{f<1\}}$. ∎

19.10 Remark We have used the martingale convergence theorem to prove the Radon–Nikodým theorem. But the connection between these two theorems is much deeper. For measures with values in a Banach space ('*vector measures*') the Radon–Nikodým theorem holds if, and only if, the pointwise martingale convergence theorem is valid. One should add that the Radon–Nikodým theorem for Banach spaces is intimately connected with the geometry of Banach spaces. Note, however, that the techniques required in the theory of vector measures are distinctly different from those in the real case. For more on this see Diestel-Uhl [11, Chapter V.2], Benyamini-Lindenstrauss [7, Chapter 5.2] or Métivier [29, § 11].

Martingale inequalities

Martingales will allow us to prove maximal inequalities which are useful and important both in analysis and probability theory. In order to ease the exposition we introduce the following (quite common) shorthand notation:

$$u_N^*(x) := \max_{1 \leqslant j \leqslant N} |u_j(x)| \qquad \text{and} \qquad u_\infty^*(x) := \lim_{N \to \infty} u_N^*(x) = \sup_{j \in \mathbb{N}} |u_j(x)|.$$

The following simple lemma is the key to all maximal inequalities.

19.11 Lemma *Let $(X, \mathcal{A}, \mathcal{A}_j, \mu)$ be a σ-finite filtered measure space and let $(u_j)_{j \in \mathbb{N}}$ be a submartingale. Then we have for all $s > 0$*

$$\mu\left(\left\{ \max_{1 \leqslant j \leqslant N} u_j \geqslant s \right\}\right) \leqslant \frac{1}{s} \int_{\left\{ \max_{1 \leqslant j \leqslant N} u_j \geqslant s \right\}} u_N \, d\mu \leqslant \frac{1}{s} \int u_N^+ \, d\mu. \qquad (19.11)$$

If $u_j \in \mathcal{L}_+^p(\mu)$ or if $(u_j)_{j \in \mathbb{N}} \subset \mathcal{L}^p(\mu)$, $p \in [1, \infty)$, is a martingale, then

$$\mu(\{u_N^* \geqslant s\}) \leqslant \frac{1}{s^p} \int_{\{u_N^* \geqslant s\}} |u_N|^p \, d\mu \leqslant \frac{1}{s^p} \int |u_N|^p \, d\mu. \qquad (19.12)$$

Proof Consider the stopping time when u_j exceeds the level s for the first time:

$$\sigma := \inf\{j \leqslant N : u_j \geqslant s\} \wedge (N+1), \qquad (\inf \emptyset = +\infty)$$

and set $A := \left\{ \max_{1 \leqslant j \leqslant N} u_j \geqslant s \right\} = \bigcup_{j=1}^N \{u_j \geqslant s\} = \{\sigma \leqslant N\} \in \mathcal{A}_\sigma$, where we used Lemma 17.6. From Theorem 17.7(iii) and the fact that $u_\sigma \geqslant s$ on A, we conclude

$$\mu\bigg(\underbrace{\bigcup_{j=1}^N \{u_j \geqslant s\}}_{=A}\bigg) \leqslant \int_A \frac{u_\sigma}{s} \, d\mu = \frac{1}{s} \int_A u_\sigma \, d\mu \leqslant \frac{1}{s} \int_A u_N \, d\mu \leqslant \frac{1}{s} \int u_N^+ \, d\mu.$$

The second inequality (19.12) follows along the same lines since, under our assumptions, $(|u_j|^p)_{j \in \mathbb{N}}$ is a submartingale, cf. Example 17.3(vi). ∎

The next theorem is commonly referred to as *Doob's maximal inequality*.

19.12 Theorem (Doob's maximal L^p-inequality) *Let $(X, \mathcal{A}, \mathcal{A}_j, \mu)$ be a σ-finite filtered measure space, $1 < p < \infty$ and let $(u_j)_{j \in \mathbb{N}}$ be a martingale or $(|u_j|^p)_{j \in \mathbb{N}}$ be a submartingale. Then we have*

$$\|u_N^*\|_p \leqslant \frac{p}{p-1} \|u_N\|_p \leqslant \frac{p}{p-1} \max_{1 \leqslant j \leqslant N} \|u_j\|_p.$$

Proof It is enough to consider the case where $(u_j)_{j\in\mathbb{N}}$ is a martingale; the situation where $(|u_j|^p)_{j\in\mathbb{N}}$ is a submartingale is similar and simpler.

If $\|u_N\|_p = \infty$, the inequality is trivial; if $u_N \in \mathcal{L}^p(\mu)$, then $u_1, \ldots, u_{N-1} \in \mathcal{L}^p(\mu)$ since $(|u_j|^p)_{j\in\mathbb{N}}$ is a submartingale by 17.3(vi). Thus

$$u_N^* \leqslant |u_1| + |u_2| + \cdots + |u_N| \quad\Longrightarrow\quad u_N^* \in \mathcal{L}^p(\mu)$$

and using (13.8) of Corollary 13.13 and Tonelli's theorem 13.8 we find

$$\int (u_N^*)^p \, d\mu \stackrel{(13.8)}{=} p \int_0^\infty s^{p-1} \mu\left(\{u_N^* \geqslant s\}\right) ds$$

$$\stackrel{(13.12)}{\leqslant} p \int_0^\infty s^{p-2}\left(\int |u_N| \, \mathbf{1}_{\{u_N^* \geqslant s\}} \, d\mu\right) ds$$

$$\stackrel{13.8}{=} p \int |u_N|\left(\int_0^{u_N^*} s^{p-2} \, ds\right) d\mu$$

$$\stackrel{(13.8)}{=} \frac{p}{p-1} \int |u_N| \, (u_N^*)^{p-1} \, d\mu.$$

Hölder's inequality T12.2 with $\frac{1}{p} + \frac{1}{q} = 1$, i.e. $q = \frac{p}{p-1}$, yields

$$\int (u_N^*)^p \, d\mu \leqslant \frac{p}{p-1}\left(\int |u_N|^p \, d\mu\right)^{1/p}\left(\int (u_N^*)^p \, d\mu\right)^{1-1/p},$$

and the claim follows. ∎

Using the continuity of measures T4.4, resp. Beppo Levi's Theorem 9.6 we derive from (19.11), resp. Theorem 19.12 the following result.

19.13 Corollary *Let $(u_j)_{j\in\mathbb{N}}$ be a martingale on the σ-finite filtered measure space $(X, \mathcal{A}, \mathcal{A}_j, \mu)$. Then*

$$\mu(\{u_\infty^* \geqslant s\}) \leqslant \frac{1}{s} \sup_{j\in\mathbb{N}} \|u_j\|_1; \tag{19.13}$$

$$\|u_\infty^*\|_p \leqslant \frac{p}{p-1} \sup_{j\in\mathbb{N}} \|u_j\|_p, \quad p \in (1, \infty). \tag{19.14}$$

If $(u_j)_{j\in\mathbb{N}\cup\{\infty\}}$ is a martingale, we may replace $\sup_{j\in\mathbb{N}} \|u_j\|_p$, $p \in [1, \infty)$, in (19.13) and (19.14) by $\|u_\infty\|_p$.

An inequality of the form (19.13) is a so-called *weak-type maximal* inequality opposed to the strong-type (p, p) inequalities of the form (19.14).

If $p = 1$ and $(u_j)_{j \in \mathbb{N} \cup \{\infty\}}$ is a martingale, we cannot expect a $(1, 1)$ strong-type inequality like (19.14) and we have to settle for the weak-type maximal inequality (19.13) instead. Otherwise, the best we can hope for is

$$\|u_\infty^*\|_1 \leqslant \frac{e}{e-1}\left(\mu(X) + \int u_\infty (\log u_\infty)^+ \, d\mu\right) \qquad \text{if } \mu(X) < \infty$$

or $\displaystyle\int_{\{u_\infty^* \geqslant \alpha\}} u_\infty^* \, d\mu \leqslant \frac{e\alpha}{e\alpha - 1}\left(\|u_\infty\|_1 + \int u_\infty (\log u_\infty)^+ \, d\mu\right) \qquad \text{else.}$

Details can be found in Doob [13, pp. 313–4] apart from some obvious modifications if $\mu(X) = \infty$.

The Hardy–Littlewood maximal theorem

Doob's martingale inequalities T19.12 and C19.13 can be seen as abstract versions of the classical Hardy–Littlewood estimates for maximal functions in $\mathbb{R}^n$. To prepare the ground we begin with a dyadic example.

19.14 Example Consider in $\mathbb{R}^n$ the half-open squares

$$Q_k(z) := z + [0, 2^{-k})^n, \quad k \in \mathbb{Z}, \ z \in 2^{-k}\mathbb{Z}^n,$$

with lower left corner z and side-length 2^{-k}. Then

$$\mathcal{A}_k^{[0]} := \sigma\big(Q_k(z) : z \in 2^{-k}\mathbb{Z}^n\big), \quad k \in \mathbb{Z},$$

defines a (two-sided infinite) filtration

$$\ldots \subset \mathcal{A}_{-2}^{[0]} \subset \mathcal{A}_{-1}^{[0]} \subset \mathcal{A}_0^{[0]} \subset \mathcal{A}_1^{[0]} \subset \mathcal{A}_2^{[0]} \subset \ldots$$

of sub-σ-algebras of $\mathcal{B}(\mathbb{R}^n)$. The superscript '[0]' indicates that the square lattice in each $\mathcal{A}_k^{[0]}$ contains some square with the origin $0 \in \mathbb{R}^n$ as lower left corner. Just as in Example 17.3(ix) one sees that for a function $f \in \mathcal{L}^1(\lambda^n)$

$$f_k(x) := \sum_{z \in 2^{-k}\mathbb{Z}^n} \frac{1}{\lambda^n(Q_k(z))} \int_{Q_k(z)} f \, d\lambda^n \, \mathbf{1}_{Q_k(z)}(x), \qquad k \in \mathbb{Z}, \qquad (19.15)$$

is a martingale – if you are unhappy about the two-sided infinite index set, then think of $\big(f_k, \mathcal{A}_k^{[0]}\big)_{k \in \mathbb{N}_0}$ as a martingale and of $\big(f_{-k}, \mathcal{A}_{-k}^{[0]}\big)_{k \in \mathbb{N}_0}$ as backwards martingale.

For the *square maximal function*

$$f_{[0]}^*(x) := \sup_{k \in \mathbb{Z}} |f_k(x)| = \sup \left\{ \frac{1}{\lambda^n(Q)} \int_Q |f| \, d\lambda^n \, : \, Q \in \bigcup_{k \in \mathbb{Z}} \mathcal{A}_k^{[0]}, \, x \in Q \right\}$$

and the submartingale $(|f_k|)_{k \in \mathbb{Z}}$, cf. Example 17.3(v), Doob's inequalities become

$$\lambda \left(\left\{ f_{[0]}^* \geq s \right\} \right) \leq \frac{1}{s} \sup_{k \in \mathbb{Z}} \int |f_k| \, d\lambda^n \leq \frac{1}{s} \int |f| \, d\lambda^n.$$

The classical Hardy–Littlewood maximal function is defined similar to the square maximal function from Example 19.14, the only difference being that one uses balls rather than squares.

19.15 Definition The *Hardy–Littlewood maximal function* of the function $u \in \mathcal{L}^p(\lambda^n)$, $1 \leq p < \infty$ is defined by

$$u^*(x) := \sup_{B : B \ni x} \frac{1}{\lambda^n(B)} \int_B |u| \, d\lambda^n,$$

where $B \subset \mathbb{R}^n$ stands for a generic (open or closed) ball of any radius.

From the Hölder inequality we see that for all sets with finite Lebesgue measure

$$\int_A |u| \, d\lambda^n \leq (\lambda^n(A))^{1-1/p} \left(\int_A |u|^p \, d\lambda^n \right)^{1/p}, \quad 1 \leq p < \infty,$$

so that u^* is well-defined. However, since u^* is given by a (possibly uncountable) supremum, it is not obvious whether u^* is Borel measurable.

19.16 Lemma *Let* $u \in \mathcal{L}^p(\lambda^n)$, $1 \leq p < \infty$. *The Hardy–Littlewood maximal function satisfies*

$$u^*(x) = \sup \left\{ \frac{1}{\lambda^n(B_r(c))} \int_{B_r(c)} |u| \, d\lambda^n \, : \, r \in \mathbb{Q}_+, \, c \in \mathbb{Q}^n, \, x \in B_r(c) \right\}.$$

In particular, u^* *is Borel measurable.*

Proof Since $\mathbb{Q}_+ \times \mathbb{Q}^n$ is countable, the formula shows that u^* arises from a countable supremum of Borel measurable functions and is, by Corollary 8.9, again Borel measurable.

The inequality '$\geq$' is clear since every ball with rational centre and radius is admissible in the definition of the maximal function u^*. To see '$\leq$', we fix $x \in \mathbb{R}^n$

and pick some generic (open or closed) ball B with $x \in B$. Given some $\epsilon > 0$ we can find $r \in \mathbb{Q}_+$ and $c \in \mathbb{Q}^n$ such that $B' := B_r(c) \subset B$, $\frac{1}{2}\lambda^n(B) \leqslant \lambda^n(B')$ and

$$\lambda^n(B \setminus B')^{1-1/p} \leqslant \frac{\lambda^n(B)}{\|u\|_p} \frac{\epsilon}{2} \leqslant \frac{\lambda^n(B')}{\|u\|_p} \epsilon.$$

Then

$$\frac{1}{\lambda^n(B)} \int_B |u| \, d\lambda^n \leqslant \frac{1}{\lambda^n(B')} \int_B |u| \, d\lambda^n$$

$$= \frac{1}{\lambda^n(B')} \int_{B \setminus B'} |u| \, d\lambda^n + \frac{1}{\lambda^n(B')} \int_{B'} |u| \, d\lambda^n$$

$$\leqslant \frac{1}{\lambda^n(B')} \lambda^n(B \setminus B')^{1-1/p} \|u\|_p + \frac{1}{\lambda^n(B')} \int_{B'} |u| \, d\lambda^n$$

$$\leqslant \epsilon + \sup_{B' : x \in B'} \frac{1}{\lambda^n(B')} \int_{B'} |u| \, d\lambda^n$$

(the supremum ranges over all balls B' with rational radius and centre s.t. $x \in B'$), where we again used Hölder's inequality in the penultimate line. Since ϵ and B were arbitrary, the inequality '$\leqslant$' follows by considering the supremum over all balls with $x \in B$ and then letting $\epsilon \to 0$. ∎

We will see now that u^* is in $\mathcal{L}^p$ if $1 < p < \infty$.

19.17 Theorem (Hardy, Littlewood) *Let* $u \in \mathcal{L}^p(\lambda^n)$, $1 \leqslant p < \infty$, *and write* u^* *for the maximal function. Then*

$$\lambda^n(\{u^* \geqslant s\}) \leqslant \frac{c_n}{s} \|u\|_1, \qquad\qquad s > 0, \ p = 1, \qquad (19.16)$$

$$\|u^*\|_p \leqslant \frac{p \, c_n}{p-1} \|u\|_p, \qquad\qquad 1 < p < \infty, \qquad (19.17)$$

with the universal constant $c_n = \left(\frac{16}{\sqrt{\pi}}\right)^n \Gamma(\frac{n}{2}+1)$.

Proof If we could show that the square maximal function $u^*_{[0]}$ satisfies $u^*_{[0]} \geqslant u^*$, then (19.16), (19.17) would immediately follow from Doob's inequalities C19.13, compare Example 19.14. The problem, however, is that a ball B_r of radius $r \in \left[\frac{1}{4} 2^{-k-1}, \frac{1}{4} 2^{-k}\right)$, $k \in \mathbb{Z}$, need not entirely fall into any single square of our lattice $\mathcal{A}^{[0]}_k$:

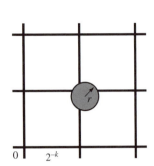
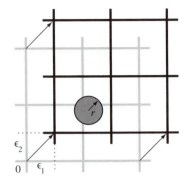

But if we move our lattice by $2 \cdot \frac{1}{4} 2^{-k} = \frac{1}{2} 2^{-k}$ in certain (combinations of) coordinate directions, we can 'catch' B_r inside a single cube Q' of the shifted lattice.[✓] More precisely, if

$$\mathbf{e} := (\epsilon_1, \ldots, \epsilon_n), \qquad \epsilon_j \in \left\{0, \tfrac{1}{2} 2^{-k}\right\},$$

then

$$\mathcal{A}_k^{[\mathbf{e}]} := \sigma\big(\mathbf{e} + Q_k(z) : z \in 2^{-k}\mathbb{Z}^n\big), \ k \in \mathbb{Z}, \quad \big(u_k^{[\mathbf{e}]}\big)_{k\in\mathbb{Z}}, \quad u_{[\mathbf{e}]}^*,$$

are 2^n filtrations with corresponding martingales and square maximal functions. As in Example 19.14 we find that

$$\lambda^n\big(\{u_{[\mathbf{e}]}^* \geqslant s\}\big) \leqslant \frac{1}{s}\|u\|_1, \qquad s > 0. \tag{19.18}$$

Combining Corollary 15.15 with the translation invariance and scaling behaviour of Lebesgue measure we see that the volume of a ball B_r of radius $\frac{1}{4} 2^{-k-1} \leqslant r < \frac{1}{4} 2^{-k}$ and arbitrary centre is

$$\lambda^n(B_r) = r^n \, \lambda^n(B_1) \overset{15.15}{=} \frac{\pi^{n/2}\, r^n}{\Gamma(\frac{n}{2}+1)} \geqslant \frac{\pi^{n/2} \left(\frac{1}{4}2^{-k-1}\right)^n}{\Gamma(\frac{n}{2}+1)},$$

hence we get from $x \in B_r \subset Q'$ and $\lambda^n(Q') = (2^{-k})^n$ that

$$\frac{1}{\lambda^n(B_r)} \int_{B_r} |u| \, d\lambda^n \leqslant \frac{\lambda^n(Q')}{\lambda^n(B_r)} \frac{1}{\lambda^n(Q')} \int_{Q'} |u| \, d\lambda^n$$

$$\leqslant \frac{(2^{-k})^n \, \Gamma\left(\frac{n}{2}+1\right)}{\pi^{n/2}\left(\frac{1}{8}2^{-k}\right)^n} \frac{1}{\lambda^n(Q')} \int_{Q'} |u| \, d\lambda^n$$

$$\leqslant \underbrace{\left(\frac{8}{\sqrt{\pi}}\right)^n \Gamma\left(\tfrac{n}{2}+1\right)}_{=:\,\gamma_n} \max_{\mathbf{e}} u_{[\mathbf{e}]}^*(x).$$

This shows that $u^* \leqslant \gamma_n \max_{\mathbf{e}} u^*_{[\mathbf{e}]}$ and

$$\lambda^n(\{u^* \geqslant s\}) \leqslant \lambda^n\left(\left\{\max_{\mathbf{e}} u^*_{[\mathbf{e}]} \geqslant \frac{s}{\gamma_n}\right\}\right)$$

$$\leqslant \sum_{\mathbf{e}} \lambda^n\left(\left\{u^*_{[\mathbf{e}]} \geqslant \frac{s}{\gamma_n}\right\}\right)$$

$$\overset{(19.18)}{\leqslant} 2^n \gamma_n \frac{1}{s} \int |u| \, \lambda^n(ds).$$

A very similar argument yields $\|u^*_{[\mathbf{e}]}\|_p \leqslant \frac{p}{p-1} \|u\|_p$ for all shifts $\mathbf{e}$, and Doob's inequality (19.14) applied to each $u^*_{[\mathbf{e}]}$ finally shows

$$\|u^*\|_p \leqslant \gamma_n \left\|\max_{\mathbf{e}} u^*_{[\mathbf{e}]}\right\|_p \leqslant \gamma_n \sum_{\mathbf{e}} \|u^*_{[\mathbf{e}]}\|_p \leqslant 2^n \gamma_n \frac{p}{p-1} \|u\|_p.$$

All that remains to be done is to call $c_n := 2^n \gamma_n$.   ∎

The proof of Theorem 19.17 extends with very little effort to maximal functions of finite measures.

19.18 Definition Let μ be a locally finite[2] measure on $(\mathbb{R}^n, \mathcal{B}(\mathbb{R}^n))$. The *maximal function* is given by

$$\mu^*(x) := \sup_{B : B \ni x} \frac{\mu(B)}{\lambda^n(B)}$$

where $B \subset \mathbb{R}^n$ stands for a generic open ball of any radius.

If we replace in the proof of Theorem 19.17 the expression $\int_B |u| \, d\lambda^n$ by $\mu(B)$ and $u^*_{[\mathbf{e}]}(x)$ by

$$\mu^*_{[\mathbf{e}]}(x) := \sup\left\{\frac{\mu(Q)}{\lambda^n(Q)} : Q \in \bigcup_{k \in \mathbb{Z}} \mathcal{A}_k^{[\mathbf{e}]}, \, x \in Q\right\},$$

we arrive at the following generalization of (19.16)

19.19 Corollary *Let μ be a finite measure on $(\mathbb{R}^n, \mathcal{B}(\mathbb{R}^n))$ with total mass $\|\mu\|$ and maximal function μ^*. Then*

$$\lambda^n(\{\mu^* \geqslant s\}) \leqslant \frac{c_n}{s} \|\mu\|, \quad s > 0, \tag{19.19}$$

with the universal constant $c_n = \left(\frac{16}{\sqrt{\pi}}\right)^n \Gamma\left(\frac{n}{2} + 1\right)$.

---

[2] i.e. every point $x \in \mathbb{R}^n$ has a neighbourhood $U = U(x)$ such that $\mu(U) < \infty$. In $\mathbb{R}^n$ this is clearly equivalent to saying that $\mu(B) < \infty$ for every open ball B.

Lebesgue's differentiation theorem

Let us return once again to the Radon–Nikodým theorem 19.2. There we have seen that $\nu \ll \mu$ implies $\nu = f\,\mu$. The proof, though, shows even more, namely

$$\nu|_{A_\alpha} = f_\alpha\,\mu|_{A_\alpha} \qquad \text{and} \qquad \mathcal{L}^1\text{-}\lim_{\alpha \in I} f_\alpha = f$$

(notation as in T19.2). Let us consider a concrete measure space $(X, \mathcal{A}, \mu) = (\mathbb{R}^n, \mathcal{B}(\mathbb{R}^n), \lambda^n)$. In this case we could reduce our consideration to a *countable sequence* of σ-algebras (instead of $(\mathcal{A}_\alpha)_{\alpha \in I}$) – cf. Problem 19.1 – and use Theorem 18.6 instead of 19.4. In fact, this would even allow us to get $f(x)$ as *pointwise* limit. This is one way to prove *Lebesgue's differentiation theorem*.

19.20 Theorem (Lebesgue) *Let $u \in \mathcal{L}^1(\lambda^n)$. Then*

$$\lim_{r \to 0} \frac{1}{\lambda^n(B_r(x))} \int_{B_r(x)} |u(y) - u(x)|\,\lambda^n(dy) = 0 \qquad (19.20)$$

for (Lebesgue) almost all $x \in \mathbb{R}^n$. In particular,

$$u(x) = \lim_{r \to 0} \frac{1}{\lambda^n(B_r(x))} \int_{B_r(x)} u(y)\,\lambda^n(dy). \qquad (19.21)$$

We will not follow the route laid out above, but use instead the Hardy–Littlewood maximal theorem 19.17 to prove T19.20. The reason is mainly a didactic one since this is a beautiful example of how weak-type maximal inequalities (i.e. inequalities like (19.16) or (19.13)) can be used to get a.e. convergence. More on this theme can be found in Krantz [25, pp. 27–30] and Garsia [16, pp. 1–4]. Our proof will also show that the limits in (19.20) and (19.21) can be strengthened to $B \downarrow \{x\}$ where B is any ball containing x and, in the limit, shrinking to $\{x\}$.

Proof (of Theorem 19.20) We know from Theorem 15.17 that the continuous functions with compact support $C_c(\mathbb{R}^n)$ are dense in $\mathcal{L}^1(\lambda^n)$. Since $\phi \in C_c(\mathbb{R}^n)$ is uniformly continuous, we find for every $\epsilon > 0$ some $\delta > 0$ such that

$$|\phi(x) - \phi(y)| < \epsilon \qquad \forall\,|x - y| \leqslant r,\ r < \delta.$$

Thus

$$\lim_{r \to 0} \frac{1}{\lambda^n(B_r(x))} \int_{B_r(x)} |\phi(y) - \phi(x)|\,\lambda^n(dy) \leqslant \epsilon \qquad \forall\,\phi \in C_c(\mathbb{R}^n), \qquad (19.22)$$

and (19.20) is true for $C_c(\mathbb{R}^n)$.

For a general $u \in \mathcal{L}^1(\lambda^n)$ we pick a sequence $(\phi_j)_{j \in \mathbb{N}} \subset C_c(\mathbb{R}^n)$ with $\lim_{j \to \infty} \|u - \phi_j\|_1 = 0$. Denote by

$$w_\delta^\sharp(x) := \sup_{0 < r < \delta} \frac{1}{\lambda^n(B_r(x))} \int_{B_r(x)} |w|\,d\lambda^n$$

the restricted maximal function. Since the supremum is subadditive and since $(u - \phi_j)_\delta^\natural \leqslant (u - \phi_j)^*$, we get

$$\lambda^n\left(\left\{x : \limsup_{r \to 0} \frac{1}{\lambda^n(B_r(x))} \int_{B_r(x)} |u(y) - u(x)| \, \lambda^n(dy) > 3\epsilon\right\}\right)$$

$$= \lambda^n\left(\left\{x : \inf_{\delta > 0} (u - u(x))_\delta^\natural(x) > 3\epsilon\right\}\right)$$

$$\leqslant \lambda^n\left(\left\{x : (u - u(x))_\delta^\natural(x) > 3\epsilon\right\}\right)$$

$$= \lambda^n\left(\left\{x : \left[(u - \phi_j) + (\phi_j - \phi_j(x)) + (\phi_j(x) - u(x))\right]_\delta^\natural(x) > 3\epsilon\right\}\right)$$

$$\leqslant \lambda^n\left(\left\{(u - \phi_j)^* > \epsilon\right\}\right) + \lambda^n\left(\left\{x : (\phi_j - \phi_j(x))_\delta^\natural(x) > \epsilon\right\}\right)$$

$$+ \lambda^n\left(\left\{x : |\phi_j(x) - u(x)| > \epsilon\right\}\right)$$

$$\leqslant \frac{c_n}{\epsilon} \|u - \phi_j\|_1 + 0 + \frac{1}{\epsilon} \|\phi_j - u\|_1,$$

where we used Theorem 19.17, resp., (19.22) with $\delta \to 0$, resp., the Markov inequality 10.12 to deal with each of the above three terms respectively. The assertion now follows by letting first $j \to \infty$ and then $\epsilon \to 0$. ∎

Let us now investigate the connection between ordinary derivatives and the Radon–Nikodým derivative. For this the following auxiliary notation will be useful. If μ is a measure on $(\mathbb{R}^n, \mathcal{B}(\mathbb{R}^n))$ that assigns finite volume to any ball, we set

$$\bar{D}\mu(x) := \limsup_{r \to 0} \frac{\mu(B_r(x))}{\lambda^n(B_r(x))} = \lim_{k \to \infty} \sup_{0 < r < 1/k} \frac{\mu(B_r(x))}{\lambda^n(B_r(x))}$$

and, whenever the limit exists and is finite,

$$D\mu(x) := \lim_{r \to 0} \frac{\mu(B_r(x))}{\lambda^n(B_r(x))}.$$

Note that by Lemma 19.16, both $\bar{D}\mu$ and $D\mu$ are Borel measurable functions.

19.21 Corollary *Let μ be a locally finite[3] measure on $(\mathbb{R}^n, \mathcal{B}(\mathbb{R}^n))$ which is absolutely continuous w.r.t. Lebesgue measure: $\mu \ll \lambda^n$. Then $D\mu$ exists Lebesgue a.e. and coincides a.e. with the Radon-Nikodým derivative $d\mu/d\lambda^n$, that is, $\mu = D\mu \, \lambda^n$.*

---

[3] See the footnote on page 217. Note that μ is automatically σ-finite.[✓]

Proof Assume first that μ is a finite measure. By the Radon–Nikodým Theorem 19.2 we know that there is a unique function $f \in \mathcal{L}^1(\lambda^n)$ with

$$\frac{\mu(B_r(x))}{\lambda^n(B_r(x))} = \frac{1}{\lambda^n(B_r(x))} \int_{B_r(x)} f(y)\,\lambda^n(dy).$$

By Lebesgue's differentiation theorem the right-hand side of the above equality tends for λ^n-almost all x to $f(x)$ as $r \to 0$, so that $D\mu = f$ almost everywhere.

If μ is not finite, we choose an exhausting sequence of open balls $B_k(0)$, $k \in \mathbb{N}$, and set $\mu_k(\bullet) := \mu(B_k(0) \cap \bullet)$ and $\lambda_k^n(\bullet) := \lambda^n(B_k(0) \cap \bullet)$. Since the measures μ_k and λ_k^n are finite, the previous argument applies and shows that $D\mu_k = d\mu_k/d\lambda_k^n$ a.e. for every $k \in \mathbb{N}$. By the very definition of the Radon-Nikodým derivative, we find that $d\mu_k/d\lambda_k^n = d\mu_j/d\lambda_j^n$ on $B_j(0)$ whenever $j < k$ and the same is true for $D\mu_k$, resp. $D\mu_j$. Thus

$$D\mu(x) := D\mu_k(x), \quad x \in B_k(0) \qquad \text{and} \qquad \frac{d\mu}{d\lambda^n}(x) := \frac{d\mu_k}{d\lambda_k^n}(x), \quad x \in B_k(0)$$

are well-defined functions which satisfy $D\mu = d\mu/d\lambda^n$ λ^n-almost everywhere. ∎

19.22 Corollary *Let ν be a locally finite[4] measure on $(\mathbb{R}^n, \mathcal{B}(\mathbb{R}^n))$ which is singular w.r.t. λ^n, i.e. $\nu \perp \lambda^n$. Then $D\nu = 0$ λ^n-almost everywhere.*

Proof Assume first that ν is a finite measure. Write $\|\nu\|$ for the total mass of ν. It is enough to show that $\bar{D}\nu = 0$ a.e. Since $\nu \perp \lambda^n$, there is some λ^n-null set N with $\nu(N) = \|\nu\|$. From Theorem 15.19 we know that ν is inner regular. Thus for every $\epsilon > 0$, there is some compact set $K = K_\epsilon \subset N$ such that $\nu(K) > \|\nu\| - \epsilon$. Setting

$$\nu_1(\bullet) := \nu(K \cap \bullet) \quad \text{and} \quad \nu_2 := \nu - \nu_1$$

we obtain two measures ν_1, ν_2 with $\nu = \nu_1 + \nu_2$ and $\|\nu_2\| \leqslant \epsilon$. Since K^c is open, we conclude from the definition of the derivative that $\bar{D}\nu_1(x) = 0$ for all $x \in K^c$, so that

$$\bar{D}\nu(x) = \bar{D}\nu_1(x) + \bar{D}\nu_2(x) = \bar{D}\nu_2(x) \leqslant \nu_2^*(x) \qquad \forall x \in K^c,$$

where ν_2^* denotes the maximal function for the measure ν_2. This shows that

$$\{\bar{D}\nu > s\} \subset K \cup \{\nu_2^* > s\} \qquad \forall s > 0.$$

Using that $\lambda^n(K) \leqslant \lambda^n(N) = 0$ and the maximal inequality Corollary 19.19 we find

$$\lambda^n(\{\bar{D}\nu > s\}) \leqslant \lambda^n(\{\nu_2^* > s\}) \leqslant \frac{c_n}{s}\|\nu_2\| \leqslant \frac{c_n}{s}\epsilon.$$

---

[4] See the footnote on p. 217.

Since $\epsilon > 0$ and $s > 0$ were arbitrary, we conclude that $\bar{D}\nu = 0$ Lebesgue a.e.

If ν is not finite, we choose an exhausting sequence of open balls $B_k(0)$, $k \in \mathbb{N}$, and set $\nu_k(\,\boldsymbol{\cdot}\,) := \nu(B_k(0) \cap \boldsymbol{\cdot}\,)$. Obviously, $\bar{D}\nu = \bar{D}\nu_k$ on $B_k(0)$, and therefore the first part of the proof shows that $\bar{D}\nu(x) = 0$ for Lebesgue almost all $x \in B_k(0)$. Denoting the exceptional set by M_k we see that $\bar{D}\nu(x) = 0$ for all $x \notin M := \bigcup_{k\in\mathbb{N}} M_k$; the latter, however, is an λ^n-null set, and the theorem follows. $\blacksquare$

The Calderón–Zygmund lemma

Our last topic is the famous *Calderón–Zygmund decomposition* lemma which is the heart of many further developments in the theory of singular integral operators. We take the proof from Stein's book [47, p. 17] and rephrase it a little to bring out the martingale connection.

19.23 Lemma (Calderón–Zygmund decomposition) *Let $u \in \mathcal{L}^1_+(\lambda^n)$ and $\alpha > 0$. Then there exists a decomposition of $\mathbb{R}^n$ such that*

(i) $\mathbb{R}^n = F \cup \Omega$ *and* $F \cap \Omega = \emptyset$;

(ii) $u \leqslant \alpha$ *almost everywhere on F;*

(iii) $\Omega = \bigcup_{k\in\mathbb{N}} Q_k$ *with mutually disjoint half-open axis-parallel squares Q_k such that for each Q_k*

$$\alpha < \frac{1}{\lambda^n(Q_k)} \int_{Q_k} u \, d\lambda^n \leqslant 2^n \alpha.$$

Proof Let $\mathcal{A}_k := \mathcal{A}_k^{[0]}$, $k \in \mathbb{Z}$, be the dyadic filtration of Example 19.14 and let $(u_k)_{k\in\mathbb{Z}}$ be the corresponding martingale (19.15). Introduce a stopping time

$$\tau := \inf\{k \in \mathbb{Z} : u_k > \alpha\}, \qquad \inf \emptyset := +\infty,$$

and set $F := \{\tau = +\infty\} \cup \{\tau = -\infty\}$ and $\Omega := \{-\infty < \tau < +\infty\}$. By the very definition of the martingale $(u_k)_{k\in\mathbb{N}}$ we see

$$u_k(x) \leqslant \frac{1}{\lambda^n(Q_k)} \int u \, d\lambda^n = 2^{nk} \|u\|_1 \xrightarrow{k\to-\infty} 0,$$

so that $\lim_{k\to-\infty} u_k(x) = 0$ and $\{\tau = -\infty\} = \emptyset$. If $x \in \{\tau = +\infty\}$, we have $u_k(x) \leqslant \alpha$ and so $u(x) = \lim_{k\to\infty} u_k(x) \leqslant \alpha$ a.e., as the almost everywhere pointwise limit exists by Corollary 18.3 (note that $\int u_k \, d\lambda^n = \int u \, d\lambda^n < \infty$). This settles (i) and (ii).

Since τ is a stopping time, $\{\tau = k\} = \{\tau \leqslant k\} \setminus \{\tau \leqslant k-1\} \in \mathcal{A}_k$, hence $\{\tau = k\}$ as well as $\Omega = \bigcup_{k\in\mathbb{Z}} \{\tau = k\}$ are unions of disjoint half-open squares. The estimate in (iii) can be written as

$$\alpha < u_\tau(x) \leqslant 2^n \alpha \quad \forall x \in \Omega.$$

From its definition, $u_\tau > \alpha$ is clear. For the upper estimate we note that every square $Q_{k-1} \in \mathcal{A}_{k-1}$ contains 2^n squares $Q_k \in \mathcal{A}_k$, so that

$$\frac{u_k}{u_{k-1}} = \frac{\displaystyle\sum_{z\in 2^{-k+1}\mathbb{Z}^n}\left(\sum_{Q_k(y)\subset Q_{k-1}(z)}\frac{1}{\lambda^n(Q_k(y))}\int_{Q_k(y)} u\,d\lambda^n\, \mathbf{1}_{Q_k(y)}\right)}{\displaystyle\sum_{z\in 2^{-k+1}\mathbb{Z}^n}\frac{1}{\lambda^n(Q_{k-1}(z))}\int_{Q_{k-1}(z)} u\,d\lambda^n\, \mathbf{1}_{Q_{k-1}(z)}}$$

$$\leqslant 2^n\, \frac{\displaystyle\sum_{z\in 2^{-k+1}\mathbb{Z}^n}\left(\sum_{Q_k(y)\subset Q_{k-1}(z)}\frac{2^{-n}}{\lambda^n(Q_k(y))}\int_{Q_k(y)} u\,d\lambda^n\right)\mathbf{1}_{Q_{k-1}(z)}}{\displaystyle\sum_{z\in 2^{-k+1}\mathbb{Z}^n}\frac{1}{\lambda^n(Q_{k-1}(z))}\int_{Q_{k-1}(z)} u\,d\lambda^n\, \mathbf{1}_{Q_{k-1}(z)}}$$

$$= 2^n.$$

Finally, by the definition of τ,

$$\mathbf{1}_{\{-\infty<\tau<+\infty\}}u_\tau = \sum_{k\in\mathbb{Z}} u_k\, \mathbf{1}_{\{\tau=k\}} \leqslant \sum_{k\in\mathbb{Z}} 2^n u_{k-1}\, \mathbf{1}_{\{\tau=k\}}$$

$$\leqslant \sum_{k\in\mathbb{Z}} 2^n \alpha\, \mathbf{1}_{\{\tau=k\}}$$

$$= 2^n \alpha\, \mathbf{1}_{\{-\infty<\tau<+\infty\}}$$

and the proof is complete. ∎

Note that all results that we have proved here for λ^n and $\mathbb{R}^n$ can be extended to *spaces of homogeneous type*, i.e. metric spaces X with a measure μ that is finite and strictly positive on balls and has the following *volume doubling* property: for some positive constant $\kappa > 0$ we have

$$\mu(B_{2r}(x)) \leqslant \kappa\mu(B_r(x)) \qquad \forall x \in X,\ r > 0,$$

see Krantz [25, §6.1, pp. 235–61].

Problems

19.1. Show that Theorem 18.6 is enough to prove the Radon–Nikodým theorem 19.2 in the situation where $\mathcal{A}$ is countably generated, i.e. where $\mathcal{A} = \sigma(\{A_j\}_{j\in\mathbb{N}})$.
[Hint: set $\mathcal{A}_n := \sigma(A_1, A_2, \dots, A_n)$ and observe that the atoms of $\mathcal{A}_n$ are of the form $C_1 \cap \dots \cap C_n$ where $C_j \in \{A_j, A_j^c\}$, $1 \leqslant j \leqslant n$.]

19.2. **A theorem of Doob.** Let $(\mu_t)_{t\geqslant 0}$ and $(\nu_t)_{t\geqslant 0}$ be two families of measures on the σ-finite measure space $(X, \mathcal{A})$ such that $\nu_t \ll \mu_t$ for all $t \geqslant 0$ and $t \mapsto \mu_t(A), \nu_t(A)$

are measurable for all $A \in \mathcal{A}$. Then there exists a measurable function $(t, x) \mapsto p(t, x)$, $(t, x) \in [0, \infty) \times X$, such that $\nu_t = p(t, \cdot)\,\mu_t$ for all $t \geqslant 0$.
[Hint: set, as in the proof of Theorem 19.2, $p_\alpha(t, x) := \sum_{A \in \alpha} \frac{\nu_t(A)}{\mu_t(A)}\, \mathbf{1}_A(x)$, and check that this function is jointly measurable in t and x. Now argue as in the proof of 19.2.]

19.3. **Conditional expectations.** Let $(X, \mathcal{A}, \mu)$ be a σ-finite measure space and let $\mathcal{F} \subset \mathcal{A}$ be a sub-σ-algebra. Use the Radon–Nikodým theorem to show that for every $u \in \mathcal{L}^1(\mathcal{A})$ there exists an – up to null sets unique – $\mathcal{F}$-measurable function $u^{\mathcal{F}} \in \mathcal{L}^1(\mathcal{F})$ such that

$$\int_F u^{\mathcal{F}}\, d\mu = \int_F u\, d\mu \qquad \forall F \in \mathcal{F}. \tag{19.23}$$

Use this result to rephrase the (sub-, super-)martingale property 17.1.

Remark. Since $u^{\mathcal{F}}$ is unique (modulo null sets) one often writes $u^{\mathcal{F}} = E^{\mathcal{F}} u$ where $E^{\mathcal{F}}$ is an operator which is called *conditional expectation*. We will introduce this operator in a different way in Chapter 22 and show in Theorem 23.9 that we could have defined $E^{\mathcal{F}}$ by (19.23).

19.4. Let $(X, \mathcal{A}, \mu)$ be a σ-finite measure space and let ν be a further measure. Show that $\nu \leqslant \mu$ entails that $\nu = f\,\mu$ for some (a.e. uniquely determined) density function f such that $0 \leqslant f \leqslant 1$.

19.5. Let μ, ν be two σ-finite measures on $(X, \mathcal{A})$ which have the same null sets. Show that $\nu = f\,\mu$ and $\mu = g\,\nu$ where $0 < f < \infty$ a.e. and $g = 1/f$ a.e.

Remark. Measures having the same null sets are called *equivalent*.

19.6. Give an example of a measure μ and a density f such that $f\,\mu$ is not σ-finite.

19.7. Let μ, ν be σ-finite measures on the measurable space $(X, \mathcal{A})$. Let $(A_j)_{j \in \mathbb{N}}$ be a filtration of sub-σ-algebras of $\mathcal{A}$ such that $\mathcal{A} = \sigma\left(\bigcup_{j \in \mathbb{N}} A_j\right)$ and denote by $\mu_j := \mu|_{A_j}$ and $\nu_j := \nu|_{A_j}$. If $\mu_j \ll \nu_j$ for all $j \in \mathbb{N}$, then $\mu \ll \nu$. Find an expression for the density $d\mu/d\nu$.

19.8. Let μ be Lebesgue measure on $[0, 2]$ and ν be Lebesgue measure on $[1, 3]$. Find the Lebesgue decomposition of ν with respect to μ.

19.9. **Stieltjes measure (3).** Let $(X, \mathcal{A}, \mu)$ be a finite measure space and denote by F the left-continuous distribution function of μ as in Problem 7.9. Use Lebesgue's decomposition theorem 19.9 to show that we can decompose $F = F_1 + F_2 + F_3$ and, accordingly, $\mu = \mu_1 + \mu_2 + \mu_3$ in such a way that

(1) F_1 is discrete, i.e. μ_1 is the countable sum of weighted Dirac δ-measures.
(2) F_2 is *absolutely continuous*, i.e. for every $\epsilon > 0$ there exists a $\delta > 0$ such that $\sum_{j=1}^N |F(y_j) - F(x_j)| \leqslant \epsilon$ for all points $x_1 < y_1 < x_2 < y_2 < \ldots < x_N < y_N$ with $\sum_{j=1}^N |y_j - x_j| < \delta$.
(3) F_3 is continuous and singular, i.e. $\mu_3 \perp \lambda^1$.

[Hint: use in (19.2) the characterization of null sets of Problem 6.1.]

19.10. **The devil's staircase.** Recall the construction of Cantor's ternary set from Problem 7.10. In each step of the construction $E_k = I_k^1 \,\dot\cup\, \ldots \,\dot\cup\, I_k^{2^k}$. Denote by $J_k^1, \ldots, J_k^{2^k - 1}$

the intervals which make up $[0, 1] \setminus E_k$ arranged in increasing order of their endpoints. We construct a sequence of functions $F_k : [0, 1] \to [0, 1]$ by

$$F_k(x) := \begin{cases} 0, & \text{if } x = 0 \\ j2^{-k} & \text{if } x \in J_k^j, \ 1 \leqslant j \leqslant 2^k - 1 \\ 1, & \text{if } x = 1 \end{cases}$$

and interpolate linearly between these values to get $F_k(x)$ for all other x.

(i) Sketch the first three functions F_1, F_2, F_3.

(ii) Show that the limit $F(x) := \lim_{k \to \infty} F_k(x)$ exists.

Remark. F is usually called the *Cantor function*.

(iii) Show that F is continuous and increasing.

(iv) Show that F' exists a.e. and equals 0.

(v) Show that F is not absolutely continuous (in the sense of Problem 19.9(2)) but singular, i.e. the corresponding measure μ with distribution function F is singular w.r.t. Lebesgue measure $\lambda^1|_{[0,1]}$.

19.11. **Kolmogorov's inequality.** Let $(X_j)_{j \in \mathbb{N}}$ be a sequence of independent, identically distributed random variables on a probability space $(\Omega, \mathcal{A}, P)$. Then we have the following generalization of Chebyshev's inequality, cf. Problem 10.5(vi),

$$P\left(\max_{1 \leqslant j \leqslant n} \left| \sum_{j=1}^n (X_j - EX_j) \right| \geqslant t \right) \leqslant \frac{1}{t^2} \sum_{j=1}^n VX_j,$$

where, in probabilistic notation, $EY = \int Y \, dP$ is the expectation or mean value and $VY = \int (Y - EY)^2 \, dP$ the variance of the random variable (i.e. measurable function) $Y : \Omega \to \mathbb{R}$.

19.12. Let $u, w \geqslant 0$ be measurable functions on a σ-finite measure space $(X, \mathcal{A}, \mu)$.

(i) Show that $t \mu(\{u \geqslant t\}) \leqslant \int_{\{u \geqslant t\}} w \, d\mu$ for all $t > 0$ implies that

$$\int u^p \, d\mu \leqslant \frac{p}{p-1} \int u^{p-1} w \, d\mu \qquad \forall p > 1.$$

(ii) Assume now that $u, w \in L^p$. Conclude from (i) that $\|u\|_p \leqslant \frac{p}{p-1} \|w\|_p$ for $p > 1$.

[Hint: use the technique of the proof of Theorem 19.12; for (ii) use Hölder's inequality.]

19.13. Show the following slight improvement of Doob's maximal inequality T19.12: Let $(u_j)_{j \in \mathbb{N}}$ be a martingale or $(|u_j|^p)_{j \in \mathbb{N}}$, $1 < p < \infty$, be a submartingale on a σ-finite filtered measure space. Then

$$\max_{j \leqslant N} \|u_j\|_p \leqslant \|u_N^*\|_p \leqslant \frac{p}{p-1} \|u_N\|_p \leqslant \frac{p}{p-1} \max_{1 \leqslant j \leqslant N} \|u_j\|_p.$$

19.14. **$\mathcal{L}^p$-bounded martingales.** A martingale $(u_j, \mathcal{A}_j)_{j\in\mathbb{N}}$ is called $\mathcal{L}^p$-bounded, if $\sup_{j\in\mathbb{N}} \int |u_j|^p \, d\mu < \infty$ for some $p > 1$. Show that the sequence $(u_j)_{j\in\mathbb{N}}$ converges a.e. and in $\mathcal{L}^p$-sense to a function $u \in \mathcal{L}^p$.
   [Hint: compare with Problem 18.8]

19.15. Use Theorem 18.6 to show that the martingale of Example 19.14 is uniformly integrable.

19.16. Let $u : [a, b] \to \mathbb{R}$ be a continuous function. Show that $x \mapsto \int_{[a,x]} u(t) \, dt$ is everywhere differentiable and find its derivative. What happens if we only assume that $u \in \mathcal{L}^1(dt)$?
   [Hint: Theorem 19.20.]

19.17. Let $f : \mathbb{R} \to \mathbb{R}$ be a bounded increasing function. Show that f' exists Lebesgue almost everywhere and that $f(b) - f(a) \geqslant \int_{(a,b)} f'(x) \, dx$. When do we have equality?
   [Hint: assume first that f is left- or right-continuous. Then you can interpret f as distribution function of a Stieltjes measure μ. Use Lebesgue's decomposition theorem 19.9 to write $\mu = \mu^\circ + \mu^\perp$ and use Corollaries 19.21 and 19.22 to find f'. If f is not one-sided continuous in the first place, use Lemma 13.12 to find a version ϕ of f which is left- or right-continuous such that $\{\phi \neq f\}$ is at most countable, hence a Lebesgue null set.]

19.18. **Fubini's 'other' theorem.** *Let $(f_j)_{j\in\mathbb{N}}$ be a sequence of monotone increasing functions $f_j : [a, b] \to \mathbb{R}$. If the series $s(x) := \sum_{j=1}^\infty f_j(x)$ converges, then $s'(x)$ exists a.e. and is given by $s'(x) = \sum_{j=1}^\infty f_j'(x)$ a.e.*
   [Hint: the partial sums $s_n(x)$ and $s(x)$ are again increasing functions and, by Problem 19.17 $s'(x)$ and $s_n'(x)$ exist a.e.; the latter can be calculated through term-by-term differentiation. Since the f_j are increasing functions, the limits of the difference quotients show that $0 \leqslant s_n' \leqslant s_{n+1}' \leqslant s'$ a.e., hence $\sum_j f_j'$ converges a.e. To identify this series with s', show that $\sum_k(s(x) - s_{n_k}(x))$ converges on $[a, b]$ for some suitable subsequence. The first part of the proof applied to this series implies that $\sum_k(s'(x) - s_{n_k}'(x))$ converges, thus $s' - s_{n_k}' \to 0$.]

20

Inner product spaces

Up to now we have only considered functions with values in $\mathbb{R}$ or $\bar{\mathbb{R}}$. Often it is necessary to admit complex-valued functions, too. In what follows $\mathbb{K}$ will stand for $\mathbb{R}$ or $\mathbb{C}$.

Recall that a $\mathbb{K}$-vector space is a set V with a vector addition '$+$' : $V \times V \to V$, $(v, w) \mapsto v + w$ and a multiplication of a vector with a scalar '$\cdot$' : $\mathbb{K} \times V \to V$, $(\alpha, v) \mapsto \alpha \cdot v$ which are defined in such a way that $(V, +)$ is an Abelian group and that for all $\alpha, \beta \in \mathbb{K}$ and $v, w \in V$ the relations

$$(\alpha + \beta)v = \alpha v + \beta v \qquad\qquad \alpha(v + w) = \alpha v + \alpha w$$

$$(\alpha\beta)v = \alpha(\beta v) \qquad\qquad 1 \cdot v = v$$

hold. Typical examples of $\mathbb{R}$-vector spaces are the spaces $\mathcal{L}^p$ or L^p (see Remark 12.5) and, in particular, the sequence spaces ℓ^p from Example 12.12. For the $\mathbb{C}$-versions we first need to know how to integrate complex functions.

20.1 Scholium (integral of complex functions) It is often necessary to consider complex-valued functions $u : X \to \mathbb{C}$ on a measurable space $(X, \mathcal{A})$. Since $\mathbb{C}$ is a normed space, we have a natural topology on $\mathbb{C}$ and we may consider the Borel σ-algebra $\mathcal{B}(\mathbb{C})$ on $\mathbb{C}$. Since we can (even topologically) identify the complex plane $\mathbb{C}$ with $\mathbb{R}^2$, the Borel sets in $\mathbb{C}$ are generated by the half-open rectangles

$$[\![z, w)\!) := \{x + iy : \mathrm{Re}\, z \leqslant x < \mathrm{Re}\, w, \ \mathrm{Im}\, z \leqslant y < \mathrm{Im}\, w\}.$$

The correspondence $\mathbb{C} \leftrightarrow \mathbb{R}^2$ is accomplished by the map $\phi : \mathbb{C} \to \mathbb{R}^2$, $z \mapsto \phi(z) := (\mathrm{Re}\, z, \mathrm{Im}\, z) = \left(\frac{1}{2}(z + \bar{z}), \frac{1}{2i}(z - \bar{z})\right)$ which is, along with its inverse $\phi^{-1} : \mathbb{R}^2 \to \mathbb{C}$, $(x, y) \mapsto \phi^{-1}(x, y) = x + iy$, continuous, hence measurable.

Consequently, we have

$$\left.\begin{array}{l} f : X \to \mathbb{C} \quad \text{is} \\[2mm] \mathcal{A}/\mathcal{B}(\mathbb{C}) \text{ measurable} \end{array}\right\} \Longleftrightarrow \left\{\begin{array}{l} \operatorname{Re} f, \operatorname{Im} f : X \to \mathbb{R} \quad \text{are} \\[2mm] \mathcal{A}/\mathcal{B}(\mathbb{R}) \text{ measurable.} \end{array}\right. \tag{20.1}$$

To see '$\Rightarrow$' note that the maps $\operatorname{Re} : z \mapsto \frac{1}{2}(z + \bar{z})$ and $\operatorname{Im} : z \mapsto \frac{1}{2i}(z - \bar{z})$ are continuous, hence measurable, and so are by Theorem 7.4 the compositions $\operatorname{Re} \circ f$ and $\operatorname{Im} \circ f$.

Conversely, '$\Leftarrow$' follows – if we write $f = u + iv$ – from the formula

$$f^{-1}\big([\![z, w)\!)\big) = \underbrace{u^{-1}\big([\operatorname{Re} z, \operatorname{Re} w)\big)}_{\in \mathcal{A}} \cap \underbrace{v^{-1}\big([\operatorname{Im} z, \operatorname{Im} w)\big)}_{\in \mathcal{A}} \in \mathcal{A}$$

and the fact that the rectangles of the form $[\![z, w)\!)$ generate $\mathcal{B}(\mathbb{C})$.

This means that we can define the integral of a $\mathbb{C}$-valued measurable function by linearity

$$\int f \, d\mu := \int \operatorname{Re} f \, d\mu + i \int \operatorname{Im} f \, d\mu, \tag{20.2}$$

and we call $f : X \to \mathbb{C}$ *integrable* and write $f \in \mathcal{L}^1_{\mathbb{C}}(\mu)$ if $\operatorname{Re} f, \operatorname{Im} f : X \to \mathbb{R}$ are integrable in the usual sense. The following rules for $f \in \mathcal{L}^1_{\mathbb{C}}(\mu)$ are readily checked:

$$\operatorname{Re} \int f \, d\mu = \int \operatorname{Re} f \, d\mu, \quad \operatorname{Im} \int f \, d\mu = \int \operatorname{Im} f \, d\mu, \quad \overline{\int f \, d\mu} = \int \bar{f} \, d\mu, \tag{20.3}$$

$$f \in \mathcal{L}^1_{\mathbb{C}}(\mu) \Longleftrightarrow f \in \mathcal{M}(\mathcal{B}(\mathbb{C})) \text{ and } |f| \in \mathcal{L}^1_{\mathbb{R}}(\mu). \tag{20.4}$$

In (20.4) the direction '$\Rightarrow$' follows since $|f| = \big((\operatorname{Re} f)^2 + (\operatorname{Im} f)^2\big)^{1/2}$ is measurable and $|f| \leqslant |\operatorname{Re} f| + |\operatorname{Im} f|$, while '$\Leftarrow$' is implied by $|\operatorname{Re} f|, |\operatorname{Im} f| \leqslant |f|$.

The equivalence (20.4) can be used to show that $\mathcal{L}^1_{\mathbb{C}}(\mu)$ is a $\mathbb{C}$-vector space: for $f, g \in \mathcal{L}^1_{\mathbb{C}}(\mu)$ and $\alpha, \beta \in \mathbb{C}$ we have $\alpha f + \beta g \in \mathcal{L}^1_{\mathbb{C}}(\mu)$, in which case

$$\int (\alpha f + \beta g) \, d\mu = \alpha \int f \, d\mu + \beta \int g \, d\mu; \tag{20.5}$$

moreover, we have the following *standard estimate*:

$$\left|\int f \, d\mu\right| \leqslant \int |f| \, d\mu. \tag{20.6}$$

Only (20.6) is not entirely straightforward. Since $\int f\,d\mu \in \mathbb{C}$, we can find some $\theta \in [0, 2\pi)$ such that

$$\left| \int f\,d\mu \right| = e^{i\theta} \int f\,d\mu \quad = \quad \mathrm{Re}\left(e^{i\theta} \int f\,d\mu \right)$$

$$\overset{(20.3),(20.5)}{=} \int \mathrm{Re}\left(e^{i\theta} f \right) d\mu$$

$$\leqslant \quad \int \left| e^{i\theta} f \right| d\mu = \int |f|\,d\mu.$$

The spaces $\mathcal{L}^p_{\mathbb{C}}(\mu)$, $1 < p \leqslant \infty$, are now defined by

$$\mathcal{L}^p_{\mathbb{C}}(\mu) := \left\{ f \in \mathcal{M}(\mathcal{B}(\mathbb{C})) : |f| \in \mathcal{L}^p_{\mathbb{R}}(\mu) \right\}, \tag{20.7}$$

and it is obvious that all assertions from Chapter 12 remain valid. In particular, $L^p_{\mathbb{C}}(\mu)$ stands for the set of all equivalence classes of $\mathcal{L}^p_{\mathbb{C}}(\mu)$-functions if we identify functions which coincide outside some μ-null set. Note also that most of our results on $\mathbb{R}$-valued integrands carry over to $\mathbb{C}$-valued functions by considering real and imaginary parts separately.

As we have seen in Chapter 12, cf. Remark 12.5, the spaces $\mathcal{L}^p_{\mathbb{K}}(\mu)$, resp. $L^p_{\mathbb{K}}(\mu)$ are semi-normed, resp. normed vector spaces. The same and more is true for $\mathbb{R}^n$ and $\mathbb{C}^n$: here we can even define a product of two vectors which, however, results in a scalar. It is this notion which we want to study in greater detail.

20.2 Definition A $\mathbb{K}$-vector space V is an *inner product space* if it supports a *scalar* or *inner product*, i.e. a map $\langle \cdot, \cdot \rangle : V \times V \to \mathbb{K}$ with the following properties: for all $u, v, w \in V$ and $\alpha, \beta \in \mathbb{K}$

(definiteness) $\qquad\qquad \langle v, v \rangle > 0 \quad \Longleftrightarrow \quad v \neq 0, \qquad\qquad (SP_1)$

(skew-symmetry) $\qquad\qquad \langle v, w \rangle \;=\; \overline{\langle w, v \rangle}, \qquad\qquad (SP_2)$

$\qquad\qquad\qquad \langle \alpha u + \beta v, w \rangle \;=\; \alpha \langle u, w \rangle + \beta \langle v, w \rangle. \qquad (SP_3)$

If $\mathbb{K} = \mathbb{R}$, (SP_2) becomes *symmetry* and (SP_2), (SP_3) together show that both $v \mapsto \langle v, w \rangle$ and $w \mapsto \langle v, w \rangle$ are $\mathbb{R}$-linear; therefore we call $(v, w) \mapsto \langle v, w \rangle$ *bilinear*. If $\mathbb{K} = \mathbb{C}$, (SP_2), (SP_3) give

$$\langle u, \alpha v + \beta w \rangle \overset{(SP_2)}{=} \overline{\langle \alpha v + \beta w, u \rangle} \overset{(SP_3)}{=} \overline{\alpha \langle v, u \rangle + \beta \langle w, u \rangle}$$

$$= \bar{\alpha} \overline{\langle v, u \rangle} + \bar{\beta} \overline{\langle w, u \rangle} \overset{(SP_2)}{=} \bar{\alpha} \langle u, v \rangle + \bar{\beta} \langle u, w \rangle,$$

i.e. $w \mapsto \langle v, w \rangle$ is *skew-linear*. We call $\langle \cdot, \cdot \rangle$ in this case a *sesqui-linear form*. Since $\mathbb{K} = \mathbb{C}$ always includes $\mathbb{K} = \mathbb{R}$, we will restrict ourselves to $\mathbb{K} = \mathbb{C}$.

20.3 Lemma (Cauchy–Schwarz inequality) *Let* $(V, \langle \cdot, \cdot \rangle)$ *be an inner product space. Then*

$$|\langle v, w \rangle|^2 \leqslant \langle v, v \rangle \langle w, w \rangle \qquad \forall\, v, w \in V. \tag{20.8}$$

Equality holds if, and only if, $v = \alpha w$ *for some* $\alpha \in \mathbb{C}$.

Proof If $v = 0$ or $w = 0$, there is nothing to show. For all other $v, w \in V$ and $\alpha \in \mathbb{C}$ we have

$$0 \leqslant \langle v - \alpha w, v - \alpha w \rangle = \langle v, v \rangle - \alpha \langle w, v \rangle - \bar{\alpha} \langle v, w \rangle + \alpha \bar{\alpha} \langle w, w \rangle$$
$$= \langle v, v \rangle - 2\operatorname{Re}\left(\alpha \langle w, v \rangle\right) + |\alpha|^2 \langle w, w \rangle,$$

where we used that $z + \bar{z} = 2\operatorname{Re} z$. If we set $\alpha = \langle v, v \rangle / \langle w, v \rangle$, we get

$$0 \leqslant \langle v, v \rangle - 2\operatorname{Re}\langle v, v \rangle + \frac{\langle v, v \rangle^2 \langle w, w \rangle}{|\langle w, v \rangle|^2},$$

which implies (20.8).

Since $\langle v - \alpha w, v - \alpha w \rangle = 0$ only if $v = \alpha w$, this is necessary for equality in (20.8), too. If, indeed, $v = \alpha w$, we see

$$|\langle v, w \rangle|^2 = |\langle \alpha w, w \rangle|^2 = \alpha \bar{\alpha} \langle w, w \rangle \langle w, w \rangle = \langle \alpha w, \alpha w \rangle \langle w, w \rangle = \langle v, v \rangle \langle w, w \rangle,$$

showing that $v = \alpha w$ is also sufficient for equality in (20.8). ∎

Lemma 20.3 is an abstract version of the Cauchy–Schwarz inequality for integrals C12.3. Just as in Chapter 12 we will use it to show that in an inner product space $(V, \langle \cdot, \cdot \rangle)$

$$\|v\| := \sqrt{\langle v, v \rangle}, \qquad v \in V, \tag{20.9}$$

defines a norm, i.e. a map $\|\cdot\| : V \to [0, \infty)$ satisfying for all $v, w \in V$ and $\alpha \in \mathbb{C}$

(*definiteness*) $\qquad\qquad\qquad \|v\| > 0 \iff v \neq 0,$ $\qquad\qquad$ (N_1)

(*pos. homogeneity*) $\qquad\qquad\qquad \|\alpha v\| = |\alpha| \cdot \|v\|,$ $\qquad\qquad$ (N_2)

(*triangle inequality*) $\qquad\qquad\qquad \|v + w\| \leqslant \|v\| + \|w\|.$ $\qquad\qquad$ (N_3)

20.4 Lemma $(V, \langle \cdot, \cdot \rangle^{1/2})$ *is a normed space.*

Proof Because of (SP_1) the map $\|\cdot\| : V \to [0, \infty)$, $\|v\| := \sqrt{\langle v, v \rangle}$, is well-defined. All we have to do is to check the properties (N_1)–(N_3). Obviously $(SP_1) \Leftrightarrow (N_1)$, (N_2) follows from $(SP_2), (SP_3)$:

$$\|\alpha v\|^2 = \langle \alpha v, \alpha v \rangle = \alpha \bar{\alpha} \langle v, v \rangle = |\alpha|^2 \cdot \|v\|^2,$$

and the triangle inequality (N_3) is a consequence of the Cauchy–Schwarz inequality (20.8):

$$
\begin{aligned}
\|v + w\|^2 = \langle v + w, v + w \rangle &= \langle v, v \rangle + \langle v, w \rangle + \langle w, v \rangle + \langle w, w \rangle \\
&= \|v\|^2 + 2\,\mathrm{Re}\,\langle v, w \rangle + \|w\|^2 \\
&\leqslant \|v\|^2 + 2\,|\langle v, w \rangle| + \|w\|^2 \\
&\overset{(20.8)}{\leqslant} \|v\|^2 + 2\,\|v\| \cdot \|w\| + \|w\|^2 \\
&= (\|v\| + \|w\|)^2.
\end{aligned}
$$

∎

20.5 Examples **(i)** The typical finite-dimensional inner product spaces are

$$\mathbb{R}^n \quad (\mathbb{R}\text{-vector space}) \qquad\qquad \mathbb{C}^n \quad (\mathbb{C}\text{-vector space})$$

$$\langle x, y \rangle = \sum_{j=1}^{n} x_j y_j \qquad\qquad \langle z, w \rangle = \sum_{j=1}^{n} z_j \bar{w}_j$$

$$\|x\| = \left(\sum_{j=1}^{n} x_j^2 \right)^{1/2} \qquad\qquad \|z\| = \left(\sum_{j=1}^{n} |z_j|^2 \right)^{1/2}.$$

(ii) The typical separable[1] infinite-dimensional inner product spaces are

$$\ell^2_{\mathbb{R}}(\mathbb{N}) \quad (\mathbb{R}\text{-vector space}) \qquad\qquad \ell^2_{\mathbb{C}}(\mathbb{N}) \quad (\mathbb{C}\text{-vector space})$$

$$x = (x_j)_{j \in \mathbb{N}},\ y = (y_j)_{j \in \mathbb{N}} \qquad\qquad z = (z_j)_{j \in \mathbb{N}},\ w = (w_j)_{j \in \mathbb{N}}$$

$$\langle x, y \rangle = (x, y)_{\ell^2} = \sum_{j=1}^{\infty} x_j y_j \qquad\qquad \langle z, w \rangle = (z, w)_{\ell^2} = \sum_{j=1}^{\infty} z_j \bar{w}_j$$

$$\|x\| = \|x\|_{\ell^2} = \left(\sum_{j=1}^{\infty} x_j^2 \right)^{1/2} \qquad\qquad \|z\| = \|z\|_{\ell^2} = \left(\sum_{j=1}^{\infty} |z_j|^2 \right)^{1/2}.$$

---

[1] *Separable* means that the space contains a countable dense subset, see Definition 21.14 below.

(iii) Let $(X, \mathcal{A}, \mu)$ be a measure space. The typical general (finite and infinite-dimensional) inner product spaces are

$$L_{\mathbb{R}}^2(\mu) \quad (\mathbb{R}\text{-vector space}) \qquad\qquad L_{\mathbb{C}}^2(\mu) \quad (\mathbb{C}\text{-vector space})$$

$$\langle u, v \rangle = (u, v)_2 = \int u\, v\, d\mu \qquad\qquad \langle f, g \rangle = (f, g)_2 = \int f\, \bar{g}\, d\mu$$

$$\|u\| = \|u\|_2 = \left(\int u^2\, d\mu \right)^{1/2} \qquad\qquad \|f\| = \|f\|_2 = \left(\int |f|^2\, d\mu \right)^{1/2}.$$

Every inner product space becomes a normed space with norm given by (20.9), but not every normed space is necessarily an inner product space. In fact, $L^p(\mu)$ or $\ell^p(\mathbb{N})$ are for all $1 \leqslant p \leqslant \infty$ normed spaces, but only for $p = 2$ inner product spaces. The reason for this is that in $L^p(\mu)$, $p \neq 2$, the parallelogram law does not hold.

20.6 Lemma (Parallelogram identity) *Let $(V, \langle \cdot, \cdot \rangle)$ be an inner product space. Then*

$$\left\| \frac{v+w}{2} \right\|^2 + \left\| \frac{v-w}{2} \right\|^2 = \frac{1}{2} \left(\|v\|^2 + \|w\|^2 \right) \qquad \forall\, v, w \in V. \qquad (20.10)$$

Proof Obvious. ∎

Geometrically $v + w$ and $v - w$ are the diagonals of the parallelogram spanned by v and w. The proof of (20.10) in $\mathbb{R}^n$ would show the *cosine law* for the angle $\sphericalangle(x, y)$ between the vectors $x, y \in \mathbb{R}^n$:

$$\frac{\langle x, y \rangle}{\|x\| \cdot \|y\|} = \cos \sphericalangle(x, y). \qquad (20.11)$$

In fact, inner products induce a natural geometry on V which resembles in many aspects the Euclidean geometry on $\mathbb{R}^n$ and $\mathbb{C}^n$.

20.7 Definition Let $(V, \langle \cdot, \cdot \rangle)$ be an inner product space. We call $v, w \in V$ *orthogonal* and write $v \perp w$ if $\langle v, w \rangle = 0$.

20.8 Remark (i) If $\|\cdot\|$ derives from a scalar product, we can recover $\langle \cdot, \cdot \rangle$ from $\|\cdot\|$ with the help of the so-called *polarization identities*: if $\mathbb{K} = \mathbb{R}$,

$$\langle v, w \rangle = \tfrac{1}{4} \left(\|v+w\|^2 - \|v-w\|^2 \right) = \tfrac{1}{2} \left(\|v+w\|^2 - \|v\|^2 - \|w\|^2 \right), \qquad (20.12)$$

and if $\mathbb{K} = \mathbb{C}$,

$$\langle v, w \rangle = \tfrac{1}{4} \left(\|v+w\|^2 - \|v-w\|^2 + i\|v-iw\|^2 - i\|v+iw\|^2 \right). \qquad (20.13)$$

(ii) One can show that a norm $\|\cdot\|$ derives from a scalar product if, and only if, $\|\cdot\|$ satisfies the parallelogram identity (20.10). For a proof we refer to Yosida [55, p. 39], see also Problem 20.2.

(iii) Let $V = V_{\mathbb{R}}$ be an $\mathbb{R}$-inner product space with scalar product $\langle \cdot, \cdot \rangle_{\mathbb{R}}$. Then we can turn V into a $\mathbb{C}$-inner product space using the following *complexification procedure*:

$$V_{\mathbb{C}} := V_{\mathbb{R}} \oplus i V_{\mathbb{R}} = \{v + iw : v, w \in V_{\mathbb{R}}\},$$

with the following addition

$$(v + iw) + (v' + iw') := (v + v') + i(w + w'), \qquad v, v', w, w' \in V_{\mathbb{R}},$$

scalar multiplication

$$(\alpha + i\beta)(v + iw) := (\alpha v - \beta w) + i(\beta v + \alpha w), \qquad \alpha, \beta \in \mathbb{R}, \; v, w \in V_{\mathbb{R}},$$

inner product

$$\langle v + iw, v' + iw' \rangle_{\mathbb{C}} := \langle v, v' \rangle + i \langle w, v' \rangle - i \langle v, w' \rangle + \langle w, w' \rangle, \qquad v, v', w, w' \in V_{\mathbb{R}},$$

and norm $\| \cdot \| := \langle \cdot, \cdot \rangle_{\mathbb{C}}^{1/2}$.

Problems

20.1. Show that the examples given in 20.5 are indeed inner product spaces.

20.2. This exercise shows the following

> **Theorem (Fréchet–von Neumann–Jordan).** *An inner product $\langle \cdot, \cdot \rangle$ on the $\mathbb{R}$-vector space V derives from a norm if, and only if, the parallelogram identity (20.10) holds.*

(i) Necessity: prove Lemma 20.6.

Assume from now on that $\|\cdot\|$ is a norm satisfying (20.10) and set

$$(v, w) := \tfrac{1}{4}\left(\|v + w\|^2 - \|v - w\|^2 \right).$$

(ii) Show that (v, w) satisfies the properties (SP_1) and (SP_2) of Definition 20.2.

(iii) Prove that $(u + v, w) = (u, w) + (v, w)$.

(iv) Use (iii) to prove that $(qv, w) = q(v, w)$ for all dyadic numbers $q = j2^{-k}$, $j \in \mathbb{Z}$, $k \in \mathbb{N}_0$ and conclude that (SP_3) holds for dyadic α, β.

(v) Prove that the maps $t \mapsto \|tv + w\|$ and $t \mapsto \|tv - w\|$ ($t \in \mathbb{R}$, $v, w \in X$) are continuous and conclude that $t \mapsto (tv, w)$ is continuous. Use this and (iv) to show that (SP_3) holds for all $\alpha, \beta \in \mathbb{R}$.

20.3. (Continuation of Problem 20.2) Assume now that W is a $\mathbb{C}$-vector space with norm $\|\cdot\|$ satisfying the parallelogram identity (20.10) and let

$$(v, w)_{\mathbb{R}} := \tfrac{1}{4}\big(\|v + w\|^2 - \|v - w\|^2\big).$$

Then $(v, w)_{\mathbb{C}} := (v, w)_{\mathbb{R}} + i(v, iw)_{\mathbb{R}}$ is a complex-valued inner product.

20.4. Does the norm $\|\cdot\|_1$ on $L^1([0, 1], \mathcal{B}[0, 1], \lambda^1|_{[0,1]})$ derive from an inner product?

20.5. Let $(V, \langle \cdot, \cdot \rangle)$ be a $\mathbb{C}$-inner product space, $n \in \mathbb{N}$ and set $\theta := e^{2\pi i/n}$.

(i) Show that $\quad \dfrac{1}{n} \displaystyle\sum_{j=1}^{n} \theta^{jk} = \begin{cases} 1 & \text{if} \quad k = 0 \\ 0 & \text{if} \quad 1 \leqslant k \leqslant n - 1. \end{cases}$

(ii) Use (i) to prove for all $n \geqslant 3$ the following generalization of (20.12) and (20.13):

$$\langle v, w \rangle = \frac{1}{n} \sum_{j=1}^{n} \theta^j \, \|v + \theta^j w\|^2.$$

(iii) Prove the following continuous version of (ii)

$$\langle v, w \rangle = \frac{1}{2\pi} \int_{(-\pi, \pi]} e^{i\phi} \, \|v + e^{i\phi} w\|^2 \, d\phi.$$

20.6. Let V be an inner product space. Show that $v \perp w$ if, and only if, Pythagoras' theorem $\|v + w\|^2 = \|v\|^2 + \|w\|^2$ holds.

21

Hilbert space ℌ

Let $(V, \langle \cdot, \cdot \rangle)$ be an inner product space. As we have seen in Chapter 20, $(V, \|\cdot\| := \langle \cdot, \cdot \rangle^{1/2})$ is a normed space and the norm resembles in many aspects the Euclidean, resp. unitary norm in $\mathbb{R}^n$ and $\mathbb{C}^n$. In particular, we have a notion of *convergence*:[1] a sequence $(v_j)_{j \in \mathbb{N}} \subset V$ *converges* to an element $v \in V$ if $(\|v - v_j\|)_{j \in \mathbb{N}}$ converges to 0 in $\mathbb{R}^+$,

$$\lim_{j \to \infty} v_j = v \iff \lim_{j \to \infty} \|v - v_j\| = 0.$$

But it is *completeness* and the study of *Cauchy sequences* in V,

$$(v_j)_{j \in \mathbb{N}} \subset V \text{ Cauchy sequence} \iff \lim_{j,k \to \infty} \|v_j - v_k\| = 0,$$

that gets analysis really going. This leads to the very natural

21.1 Definition A *Hilbert space* ℌ is a complete inner product space, i.e. an inner product space where every Cauchy sequence converges. We will usually write ℌ for a Hilbert space.

21.2 Example The spaces $\mathbb{R}^n$, $\mathbb{C}^n$, $\ell_{\mathbb{K}}^2$ and $L_{\mathbb{K}}^2(\mu)$ over any measure space $(X, \mathcal{A}, \mu)$ are Hilbert spaces and, indeed, the 'typical' ones. This follows from Example 20.5 and the Riesz – Fischer theorem 12.7.

Since every Hilbert space is an inner product space, we have the notion of orthogonality of $g, h \in$ ℌ, see Definition 20.7:

$$g \perp h \iff \langle g, h \rangle = 0.$$

21.3 Definition Let $\mathfrak{H}$ be a Hilbert space. The *orthogonal complement $M^\perp$* of a subset $M \subset \mathfrak{H}$ is by definition

$$M^\perp := \{h \in \mathfrak{H} : h \perp m \quad \forall m \in M\}$$
$$= \{h \in \mathfrak{H} : \langle h, m \rangle = 0 \quad \forall m \in M\}. \tag{21.1}$$

21.4 Lemma *Let $\mathfrak{H}$ be a Hilbert space and $M \subset \mathfrak{H}$ be any subset. The orthogonal complement $M^\perp$ is a closed linear subspace of $\mathfrak{H}$ and $M \subset (M^\perp)^\perp$.*

Proof If $g, h \in M^\perp$ we find for all $\alpha, \beta \in \mathbb{C}$ that

$$\langle \alpha g + \beta h, m \rangle = \alpha \langle g, m \rangle + \beta \langle h, m \rangle = 0 \qquad \forall m \in M,$$

i.e. $\alpha g + \beta h \in M^\perp$ and $M^\perp$ is a linear subspace of $\mathfrak{H}$. To see the closedness we take a sequence $(h_k)_{k \in \mathbb{N}} \subset M^\perp$ such that $\lim_{k \to \infty} h_k = h$. Then, for all $m \in M$,

$$|\langle h, m \rangle| = |\langle h, m \rangle - \underbrace{\langle h_k, m \rangle}_{=0}| = |\langle h - h_k, m \rangle| \overset{20.3}{\leqslant} \|h - h_k\| \cdot \|m\| \xrightarrow{k \to \infty} 0;$$

this shows that $M^\perp$ is closed since $h \in M^\perp$. Finally, if $m \in M$ we get

$$0 = \langle h, m \rangle = \overline{\langle m, h \rangle} \quad \forall h \in M^\perp \implies m \in (M^\perp)^\perp. \qquad \blacksquare$$

The next theorem is central for the study of (the geometry of) Hilbert spaces. Recall that a set $C \subset \mathfrak{H}$ is *convex* if

$$u, w \in C \implies tu + (1-t)w \in C \quad \forall t \in (0, 1).$$

21.5 Theorem (Projection theorem) *Let $C \neq \emptyset$ be a closed convex subset of the Hilbert space $\mathfrak{H}$. For every $h \in \mathfrak{H}$ there is a unique minimizer $u \in C$ such that*

$$\|h - u\| = \inf_{w \in C} \|h - w\| =: d(h, C). \tag{21.2}$$

This element $u = P_C h$ is called (orthogonal) projection *of h onto C and is equally characterized by the property*

$$P_C h \in C \quad \text{and} \quad \operatorname{Re} \langle h - P_C h, w - P_C h \rangle \leqslant 0 \quad \forall w \in C. \tag{21.3}$$

Proof *Existence*: Let $d := \inf_{w \in C} \|h - w\|$. By the very definition of the infimum, there is a sequence $(w_k)_{k \in \mathbb{N}} \subset C$ such that

$$\lim_{k \to \infty} \|h - w_k\| = d.$$

If we can show that $(w_k)_{k \in \mathbb{N}}$ is a Cauchy sequence, we know that the limit $u := \lim_{k \to \infty} w_k$ exists because of the completeness of $\mathfrak{H}$ and is in C since C is closed. Applying the parallelogram law (20.10) with $v = h - w_k$ and $w = h - w_\ell$ gives

$$\left\| h - \frac{w_k + w_\ell}{2} \right\|^2 + \left\| \frac{w_k - w_\ell}{2} \right\|^2 = \frac{1}{2} \left(\|h - w_k\|^2 + \|h - w_\ell\|^2 \right).$$

Since C is convex, $\frac{1}{2} w_k + \frac{1}{2} w_\ell \in C$, thus $d \leqslant \left\| h - \frac{1}{2}(w_k + w_\ell) \right\|$ and

$$d^2 + \frac{1}{4} \|w_k - w_\ell\|^2 \leqslant \frac{1}{2} \left(\|h - w_k\|^2 + \|h - w_\ell\|^2 \right) \xrightarrow{k, \ell \to \infty} d^2.$$

This proves that $(w_k)_{k \in \mathbb{N}}$ is a Cauchy sequence.

Uniqueness: Assume that $u, \tilde{u} \in C$ satisfy both (21.2), i.e.

$$\|u - h\| = d = \|\tilde{u} - h\|.$$

Since by convexity $\frac{1}{2} u + \frac{1}{2} \tilde{u} \in C$, the parallelogram law (20.10) gives

$$d^2 \leqslant \underbrace{\left\| h - \left(\tfrac{1}{2} u + \tfrac{1}{2} \tilde{u} \right) \right\|^2}_{\geqslant d^2} + \left\| \tfrac{1}{2}(u - \tilde{u}) \right\|^2 = \frac{1}{2} \left(\|h - u\|^2 + \|h - \tilde{u}\|^2 \right) = d^2,$$

and we conclude that $\|u - \tilde{u}\|^2 = 0$ or $u = \tilde{u}$.

Equivalence of (21.2),(21.3): Assume that $u \in C$ satisfies (21.2) and let $w \in C$. By convexity, $(1 - t)u + tw \in C$ for all $t \in (0, 1)$ and by (21.2)

$$\|h - u\|^2 \leqslant \|h - (1 - t)u - tw\|^2$$
$$= \|(h - u) - t(w - u)\|^2$$
$$= \|h - u\|^2 - 2t \operatorname{Re} \langle h - u, w - u \rangle + t^2 \|w - u\|^2.$$

Hence, $2 \operatorname{Re} \langle h - u, w - u \rangle \leqslant t \|w - u\|^2$ and (21.3) follows as $t \to 0$.

Conversely, if (21.3) holds, we have for $u = P_C h \in C$

$$\|h - u\|^2 - \|h - w\|^2 = 2 \operatorname{Re} \langle h - u, w - u \rangle - \|u - w\|^2 \leqslant 0 \qquad \forall w \in C,$$

which implies (21.2). ∎

We will now study the properties of the projection operator P_C. If $V, W \subset \mathfrak{H}$ are two subspaces with $V \cap W = \{0\}$, we call $V + W = \{v + w : v \in V, w \in W\}$ the *direct sum* and write $V \oplus W$.

21.6 Corollary (i) *Let* $\emptyset \neq C \subset \mathfrak{H}$ *be a closed convex subset. The projection* $P_C : \mathfrak{H} \to C$ *is a contraction, i.e.*

$$\|P_C g - P_C h\| \leqslant \|g - h\| \qquad \forall g, h \in \mathfrak{H}. \tag{21.4}$$

(ii) *If* $\emptyset \neq C = F$ *is a closed linear subspace of* $\mathfrak{H}$, P_F *is a linear operator and* $f = P_F h$ *is the unique element with*

$$f \in F \qquad and \qquad h - f \in F^{\perp}. \tag{21.5}$$

In particular, $\mathfrak{H} = F \oplus F^{\perp}$.

(iii) *If* F *is not closed, then* $\mathfrak{H} = \bar{F} \oplus F^{\perp}$ *or, equivalently,* $\bar{F} = (F^{\perp})^{\perp}$.

Proof **(i)** follows from the inequality

$$\|P_C g - P_C h\|^2 = \mathrm{Re}\left(\langle P_C g, P_C g - P_C h\rangle - \langle P_C h, P_C g - P_C h\rangle\right)$$
$$= \mathrm{Re}\left(\langle P_C g - g, P_C g - P_C h\rangle + \langle P_C h - h, P_C h - P_C g\rangle\right.$$
$$\left. + \langle g - h, P_C g - P_C h\rangle\right)$$
$$\overset{(21.3)}{\leqslant} \mathrm{Re}\langle g - h, P_C g - P_C h\rangle$$
$$\leqslant \|g - h\| \cdot \|P_C g - P_C h\|,$$

where we used the Cauchy – Schwarz inequality L20.3 for the last estimate.

(ii) Since F is a linear subspace, $v \in F \implies \zeta v \in F$ for all $\zeta \in \mathbb{C}$ and (21.3) reads in this case

$$\mathrm{Re}\langle h - P_F h, \zeta v - P_F h\rangle \leqslant 0 \qquad \forall \zeta \in \mathbb{C}, \ v \in F,$$

or, equivalently,

$$\mathrm{Re}\left(\zeta \langle h - P_F h, v\rangle\right) \leqslant \mathrm{Re}\langle h - P_F h, P_F h\rangle \qquad \forall \zeta \in \mathbb{C}, \ v \in F,$$

which is only possible if $\langle h - P_F h, v\rangle = 0$ for all $v \in F$ and for $v = P_F h$, in particular, $\langle h - P_F h, P_F h\rangle = 0$; this shows (21.5).

If, on the other hand, (21.5) is true, we get for all $v \in F$

$$0 = \mathrm{Re}\langle h - f, v\rangle - \mathrm{Re}\langle h - f, f\rangle = \mathrm{Re}\langle h - f, v - f\rangle,$$

and $f = P_F h$ follows by the uniqueness of the projection.

The decomposition $\mathfrak{H} = F \oplus F^{\perp}$ follows immediately as $h = P_F h + (h - P_F h)$ and $h \in F \cap F^{\perp} \iff \langle h, h\rangle = 0 \iff h = 0$. The decomposition also proves the linearity of P_F since for all $g, h \in \mathfrak{H}$ and $\alpha, \beta \in \mathbb{K}$

$$\Big\langle \underbrace{(\alpha g - \alpha P_F g)}_{\in F^{\perp}} + \underbrace{(\beta h - \beta P_F h)}_{\in F^{\perp}}, \underbrace{\alpha g + \beta h}_{\in F} \Big\rangle = 0$$

as well as

$$\langle (\alpha g + \beta h) - P_F(\alpha g + \beta h), \alpha g + \beta h\rangle = 0$$

which implies, again by uniqueness of the projection, that $P_F(\alpha g + \beta h) = \alpha P_F g + \beta P_F h$.

(iii) We know from Lemma 21.4 that $F \subset (F^\perp)^\perp$ and that $(F^\perp)^\perp$ is closed; therefore, $\bar{F} \subset (F^\perp)^\perp$. Moreover, $F \subset \bar{F}$ implies $\bar{F}^\perp \subset F^\perp$,[✓] showing that

$$\mathfrak{H} \stackrel{21.6(\mathrm{ii})}{=} \bar{F} \oplus (\bar{F})^\perp \subset \bar{F} + F^\perp \subset (F^\perp)^\perp \oplus F^\perp \stackrel{21.6(\mathrm{ii})}{=} \mathfrak{H},$$

and $\mathfrak{H} = \bar{F} \oplus F^\perp$ or $\bar{F} = (F^\perp)^\perp$ follows. ∎

21.7 Remarks (i) It is easy to show that the projection P_F onto a subspace $F \subset \mathfrak{H}$ is *symmetric*, i.e. that

$$\langle P_F g, h \rangle = \langle g, P_F h \rangle \qquad \forall g, h \in \mathfrak{H}, \tag{21.6}$$

and that $P_F^2 = P_F$, i.e.

$$\langle P_F^2 g, h \rangle = \langle P_F g, P_F h \rangle = \langle P_F g, h \rangle \qquad \forall g, h \in \mathfrak{H}. \tag{21.7}$$

In fact, (21.7) implies (21.6). Since $P_F g \in F$, $P_F(P_F g) = P_F g$ by the uniqueness of the projection and $\langle P_F^2 g, h \rangle = \langle P_F g, h \rangle$ follows. Finally,

$$\langle P_F g, h \rangle = \langle P_F g, P_F h \rangle + \underbrace{\langle P_F g, h - P_F h \rangle}_{=0} = \langle P_F g, P_F h \rangle.$$

(ii) *Pythagoras' theorem* has a particularly nice form for projections:

$$\|h\|^2 = \|P_F h\|^2 + \|h - P_F h\|^2 \qquad \forall h \in \mathfrak{H}. \tag{21.8}$$

(iii) A very useful interpretation of C21.6(iii) is the following: a linear subspace $F \subset \mathfrak{H}$ is dense in $\mathfrak{H}$ if, and only if, $F^\perp = \{0\}$. In other words,

$$F \subset \mathfrak{H} \quad \text{is dense} \iff \langle f, h \rangle = 0 \quad \forall f \in F \quad \text{entails} \quad h = 0.$$

Let us briefly discuss two important consequences of the projection theorem 21.5: F. Riesz' representation theorem on the structure of continuous linear functionals on $\mathfrak{H}$ and the problem of finding a basis in $\mathfrak{H}$.

21.8 Definition A *continuous linear functional* on $\mathfrak{H}$ is a map $\Lambda : \mathfrak{H} \to \mathbb{K}$, $h \mapsto \Lambda(h)$ which is linear,

$$\Lambda(\alpha g + \beta h) = \alpha \Lambda(g) + \beta \Lambda(h) \qquad \forall \alpha, \beta \in \mathbb{K}, \forall g, h \in \mathfrak{H}$$

and satisfies

$$|\Lambda(g - h)| \leqslant c(\Lambda) \|g - h\| \qquad \forall g, h \in \mathfrak{H}$$

with a constant $c(\Lambda) \geqslant 0$ independent of $g, h \in \mathfrak{H}$.

It is easy to find examples of continuous linear functionals on $\mathfrak{H}$. Just fix some $g \in \mathfrak{H}$ and set

$$\Lambda_g(h) := \langle h, g \rangle, \qquad h \in \mathfrak{H}. \tag{21.9}$$

Linearity is clear and the Cauchy–Schwarz inequality L20.3 shows

$$|\Lambda_g(h - \hbar)| = |\langle h - \hbar, g \rangle| \leqslant \underbrace{\|g\|}_{= c(\Lambda)} \cdot \|h - \hbar\|.$$

That, in fact, *all* continuous linear functionals of $\mathfrak{H}$ arise in this way is the content of the next theorem, due to F. Riesz.

21.9 Theorem (Riesz representation theorem) *Each continuous linear functional Λ on the Hilbert space $\mathfrak{H}$ is of the form* (21.9), *i.e. there exists a unique $g \in \mathfrak{H}$ such that*

$$\Lambda(h) = \Lambda_g(h) = \langle h, g \rangle \qquad \forall\, h \in \mathfrak{H}.$$

Proof Set $F := \Lambda^{-1}(\{0\})$ which is, due to the continuity and linearity of Λ, a closed linear subspace of $\mathfrak{H}$.[✓] If $F = \mathfrak{H}$, $\Lambda \equiv 0$ and $g = 0 \in \mathfrak{H}$ does the job. Otherwise we can pick some $g_0 \in \mathfrak{H} \setminus F$ and set

$$g := \frac{g_0 - P_F g_0}{\|g_0 - P_F g_0\|} \overset{(21.5)}{\in} F^\perp \implies \Lambda(g) \neq 0.$$

Since $\mathfrak{H} = F \oplus F^\perp$, we can write every $h \in \mathfrak{H}$ in the form

$$h = \frac{\Lambda(h)}{\Lambda(g)} g + \left(h - \frac{\Lambda(h)}{\Lambda(g)} g \right) \in F^\perp \oplus F,$$

hence

$$\left\langle h - \frac{\Lambda(h)}{\Lambda(g)} g, \frac{\Lambda(h)}{\Lambda(g)} g \right\rangle = 0 \iff \langle h, g \rangle = \frac{\Lambda(h)}{\Lambda(g)} \underbrace{\langle g, g \rangle}_{= 1}$$

$$\iff \Lambda(h) = \langle h, \overline{\Lambda(g)}\, g \rangle,$$

and the proof is finished. ∎

We will finally see how to represent elements of a Hilbert space using an orthonormal base (ONB, for short). We begin with a definition.

21.10 Definition Let $\mathfrak{H}$ be a Hilbert space. **(i)** The *(linear) span* of a family $\{e_k : k = 1, 2, \ldots, N\} \subset \mathfrak{H}$, $N \in \mathbb{N} \cup \{\infty\}$, is the set of all *finite* linear combinations of the e_k, i.e.

$$\operatorname{span}\{e_1, e_2, \ldots, e_N\} = \left\{ \sum_{k=1}^{n} \alpha_k \, e_k \, : \, \alpha_1, \ldots, \alpha_n \in \mathbb{R}, \, n \in \mathbb{N}, \, n \leqslant N \right\}.$$

(ii) A sequence $(e_k)_{k \in \mathbb{N}} \subset \mathfrak{H}$ is called a *(countable) orthonormal system* (ONS, for short) if

$$\langle e_k, e_\ell \rangle = \begin{cases} 0, & \text{if } k \neq \ell, \\ 1, & \text{if } k = \ell, \end{cases}$$

that is, $\|e_k\| = 1$ and $e_k \perp e_\ell$ whenever $k \neq \ell$.

21.11 Theorem *Let $(e_k)_{k \in \mathbb{N}}$ be an ONS in the Hilbert space $\mathfrak{H}$ and denote by $E = E(N) = \operatorname{span}\{e_1, \ldots, e_N\}$ the linear span of $e_1, \ldots, e_N$, $N \in \mathbb{N}$.*

(i) *$E = E(N)$ is a closed linear subspace, $P_E g = \sum_{k=1}^{N} \langle g, e_k \rangle \, e_k$ and*

$$\left\| g - \sum_{k=1}^{N} \langle g, e_k \rangle \, e_k \right\| < \|g - f\| \qquad \forall f \in E, \, f \neq P_E g,$$

and also

$$\|P_E g\|^2 = \sum_{k=1}^{N} |\langle g, e_k \rangle|^2.$$

(ii) **(Pythagoras' theorem)** *For $g \in \mathfrak{H}$*

$$\|g\|^2 = \|g - P_E g\|^2 + \|P_E g\|^2 = \left\| g - \sum_{k=1}^{N} \langle g, e_k \rangle \, e_k \right\|^2 + \sum_{k=1}^{N} |\langle g, e_k \rangle|^2.$$

(iii) **(Bessel's inequality)** *For $g \in \mathfrak{H}$*

$$\sum_{k=1}^{\infty} |\langle g, e_k \rangle|^2 \leqslant \|g\|^2.$$

(iv) **(Parseval's identity)** *The sequence $(\sum_{k=1}^{m} c_k \, e_k)_{m \in \mathbb{N}}$, $c_k \in \mathbb{K}$, converges to an element $g \in \mathfrak{H}$ if, and only if, $\sum_{k=1}^{\infty} |c_k|^2 < \infty$. In this case, Parseval's identity holds:*

$$\sum_{k=1}^{\infty} |c_k|^2 = \sum_{k=1}^{\infty} |\langle g, e_k \rangle|^2 = \|g\|^2.$$

Proof **(i)** That $E(N)$ is a linear subspace is due to the very definition of 'span'. The closedness follows from the fact that $E(N)$ is generated by finitely many e_k: if $f \in E(N)$ is of the form $f = \sum_{j=1}^{N} c_j e_j, c_j \in \mathbb{K}$, then

$$\langle f, e_k \rangle = \Big\langle \sum_{j=1}^{N} c_j e_j, e_k \Big\rangle = \sum_{j=1}^{N} c_j \langle e_j, e_k \rangle = c_k.$$

Let $(f^{(n)})_{n \in \mathbb{N}} \subset E(N)$ be a sequence with $f^{(n)} \xrightarrow{n \to \infty} f \in \mathfrak{H}$. Then

$$\Big\| f^{(n)} - \sum_{j=1}^{N} \langle f, e_j \rangle e_j \Big\| = \Big\| \sum_{j=1}^{N} \langle f^{(n)} - f, e_j \rangle e_j \Big\|$$

$$\leqslant \sum_{j=1}^{N} |\langle f^{(n)} - f, e_j \rangle| \cdot \|e_j\|$$

$$\leqslant \sum_{j=1}^{N} \|f^{(n)} - f\| \qquad \qquad (\text{L20.3}, \|e_j\| = 1)$$

$$= N \|f^{(n)} - f\| \xrightarrow{n \to \infty} 0,$$

which shows that $\lim_{n \to \infty} f^{(n)} = \sum_{j=1}^{N} \langle f, e_j \rangle e_j \in E(N)$.

If $g \in \mathfrak{H}$, we observe that $g - \sum_{j=1}^{N} \langle g, e_j \rangle e_j \perp e_k$ for all $k = 1, 2, \ldots, N$, since for these k

$$\Big\langle g - \sum_{j=1}^{N} \langle g, e_j \rangle e_j, e_k \Big\rangle = \langle g, e_k \rangle - \sum_{j=1}^{N} \langle g, e_j \rangle \langle e_j, e_k \rangle$$

$$= \langle g, e_k \rangle - \langle g, e_k \rangle = 0.$$

Since $\mathfrak{H} = E(N) \oplus E(N)^{\perp}$, we get $P_{E(N)}g = \sum_{j=1}^{N} \langle g, e_j \rangle e_j$, while (21.2) implies $\|g - \sum_{j=1}^{N} \langle g, e_j \rangle e_j\| \leqslant \|g - f\|$ for $f \in E(N)$, with equality holding only if $f = P_{E(N)}g$ because of uniqueness of $P_{E(N)}g$. Finally,

$$\|P_{E(N)}g\|^2 = \langle P_{E(N)}g, P_{E(N)}g \rangle = \Big\langle \sum_{j=1}^{N} \langle g, e_j \rangle e_j, \sum_{k=1}^{N} \langle g, e_k \rangle e_k \Big\rangle$$

$$= \sum_{j,k=1}^{N} \langle g, e_j \rangle \overline{\langle g, e_k \rangle} \langle e_j, e_k \rangle = \sum_{j=1}^{N} |\langle g, e_j \rangle|^2,$$

where we used that e_j is an ONS.

(ii) follows from (21.8) and (i).

(iii) From (ii) we get for all $N \in \mathbb{N}$

$$\sum_{j=1}^{N} |\langle g, e_j \rangle|^2 = \|g\|^2 - \|g - P_E g\| \leqslant \|g\|^2.$$

Since the right-hand side is independent of $N \in \mathbb{N}$, we can let $N \to \infty$ and the claim follows.

(iv) Since $\mathfrak{H}$ is complete, it is enough to show that $\left(\sum_{k=1}^{m} c_k e_k \right)_{m \in \mathbb{N}}$ is a Cauchy sequence. Because of the orthogonality of the e_k we see (as in (i))

$$\left\| \sum_{k=m-1}^{n} c_k e_k \right\|^2 = \sum_{k=m-1}^{n} |c_k|^2 \|e_k\|^2 = \sum_{k=m-1}^{n} |c_k|^2,$$

which means that $\left(\sum_{k=1}^{m} c_k e_k \right)_{m \in \mathbb{N}}$ is a Cauchy sequence in $\mathfrak{H}$ if, and only if, $\sum_{k=1}^{\infty} |c_k|^2$ converges. In the latter case, Parseval's identity follows from (iii): for $g = \sum_{k=1}^{\infty} c_k e_k$ we have $P_{E(N)} g = \sum_{k=1}^{N} c_k e_k$ and $c_k = \langle g, e_k \rangle$ by (i). Thus by (ii),

$$\|g\|^2 = \|g - P_{E(N)} g\|^2 + \sum_{j=1}^{N} |\langle g, e_j \rangle|^2 \xrightarrow{N \to \infty} \sum_{j=1}^{\infty} |\langle g, e_j \rangle|^2 = \sum_{j=1}^{\infty} |c_j|^2. \quad \blacksquare$$

Two questions remain: can we always find a countable ONS? If so, can we use it to represent all elements of $\mathfrak{H}$? The answer to the first question is 'yes', while the second question has to be answered by 'no', unless we are looking at *separable* Hilbert spaces, see Definition 21.14 below. Here we will restrict ourselves to the latter situation but we will point towards references where the general case is treated.

21.12 Definition An ONS $(e_k)_{k \in \mathbb{N}}$ in the Hilbert space $\mathfrak{H}$ is said to be *maximal* (also *complete, total, an orthonormal basis*) if for every $g \in \mathfrak{H}$

$$\langle g, e_k \rangle = 0 \quad \forall k \in \mathbb{N} \implies g = 0.$$

The idea behind maximality is that we can obtain $\mathfrak{H}$ as limit of finite-dimensional projections, '$\mathfrak{H} = \lim_N P_{\text{span}\{e_1, \ldots, e_N\}} \mathfrak{H}$' or '$\mathfrak{H} = \bigoplus_{k \in \mathbb{N}} \{e_k\}$', if the limits and summations are understood in the right way. Here we see that the countability of the ONS entails that $\mathfrak{H}$ can be represented as closure of the span of countably many elements – and that this is indeed a restriction should be obvious. Let us make all this more precise.

21.13 Theorem *Let* $(e_k)_{k \in \mathbb{N}}$ *be an ONS in the Hilbert space* $\mathfrak{H}$. *Then the following assertions are equivalent.*

(i) $(e_k)_{k \in \mathbb{N}}$ *is maximal;*

(ii) $\bigcup\limits_{N \in \mathbb{N}} \mathrm{span}\{e_1, \ldots, e_N\}$ *is dense in* $\mathfrak{H}$;

(iii) $g = \sum\limits_{j=1}^{\infty} \langle g, e_j \rangle \, e_j \qquad \forall g \in \mathfrak{H}$;

(iv) $\sum\limits_{j=1}^{\infty} |\langle g, e_j \rangle|^2 = \|g\|^2 \qquad \forall g \in \mathfrak{H}$;

(v) $\sum\limits_{j=1}^{\infty} \langle g, e_j \rangle \, \overline{\langle h, e_j \rangle} = \langle g, h \rangle \qquad \forall g, h \in \mathfrak{H}$.

Proof **(i)**$\Rightarrow$**(ii):** Since $F := \bigcup_{N \in \mathbb{N}} \mathrm{span}\{e_1, \ldots, e_N\} = \mathrm{span}\{e_j : j \in \mathbb{N}\}$ is a linear subspace of $\mathfrak{H}$, the assertion follows from the definition of maximality and Remark 21.7(iii).

(ii)$\Rightarrow$**(iii)** is obvious since

$$\sum_{j=1}^{\infty} \langle g, e_j \rangle \, e_j = \lim_{N \to \infty} \sum_{j=1}^{N} \langle g, e_j \rangle \, e_j = \lim_{N \to \infty} P_{E(N)} g.$$

(iii)$\Rightarrow$**(iv)** follows from Theorem 21.11(iv).
(iv)$\Rightarrow$**(v)** follows from the polarization identity (20.13).
(v)$\Rightarrow$**(i):** If $\langle u, e_k \rangle = 0$ for some $u \in \mathfrak{H}$ and all $k \in \mathbb{N}$, we get from (v) with $g = h = u$ that

$$0 = \sum_{j=1}^{\infty} \langle u, e_j \rangle \, \overline{\langle u, e_j \rangle} = \langle u, u \rangle = \|u\|^2,$$

and therefore $u = 0$. ∎

Theorem 21.13 solves the representation issue. To find an ONS, we recall first what we do in a finite-dimensional vector space V to get a basis. If $V =$

span$\{v_1, \dots, v_N\}$, we remove recursively all $v'_1, \dots, v'_k$ such that still $V = \text{span}$ $(\{v_1, \dots, v_N\} \setminus \{v'_1, \dots, v'_k\})$. This procedure gives us in at *most N steps* a minimal system $\{w_1, \dots, w_n\} \subset \{v_1, \dots, v_N\}$, $N = n + k$, with the property that $V = \text{span}\{w_1, \dots, w_n\}$. Note that this is, at the same time, a maximally independent system of vectors in V. We can now rebuild $\{w_1, \dots, w_n\}$ into an ONS by the

Gram–Schmidt orthonormalization procedure:

$$\left.\begin{array}{l} e_1 := \dfrac{w_1}{\|w_1\|}, \quad \text{and recursively} \\[2mm] \tilde{e}_{j+1} := w_{j+1} - P_{\text{span}\{e_1, \dots, e_j\}} w_{j+1} \\[2mm] \qquad = w_{j+1} - \displaystyle\sum_{\ell=1}^{j} \langle w_{j+1}, e_\ell \rangle \, e_\ell, \\[2mm] e_{j+1} := \dfrac{\tilde{e}_{j+1}}{\|\tilde{e}_{j+1}\|}. \end{array}\right\} \tag{21.10}$$

Another interpretation of (21.10) is this: If we had unleashed the Gram–Schmidt procedure on the set $\{v_1, \dots, v_N\}$, we would have obtained again n orthonormal vectors[✓], say, $f_1, \dots, f_n$ (which are, in general, different from $e_1, \dots, e_n$ constructed from $w_1, \dots, w_n$). A close inspection of (21.10) shows that at each step $V = \text{span}\{f_1, \dots, f_j, v_{j+1}, \dots, v_N\}$, so that (21.10) extends an partially existing basis $\{f_1, \dots, f_j\}$ to a full ONB $\{f_1, \dots, f_n\}$. This means that (21.10) is also a 'basis extension procedure'.

To get (21.10) to work in infinite dimensions we must make sure that $\mathfrak{H}$ is the closure of the span of countably many vectors. This motivates the following convenient (but somewhat restrictive)

21.14 Definition A Hilbert space $\mathfrak{H}$ is said to be *separable* if $\mathfrak{H}$ contains a countable dense subset $G \subset \mathfrak{H}$.

21.15 Theorem *Every separable Hilbert space $\mathfrak{H}$ has a maximal ONS.*

Proof Let $G = \{g_j\}_{j \in \mathbb{N}}$ be an enumeration of some countable dense subset of $\mathfrak{H}$ and consider the subspaces $F_k = \text{span}\{g_1, \dots, g_k\}$. Note that $F_k \subseteq F_{k+1}$, $\dim F_k \leqslant k$ and that $\bigcup_{k \in \mathbb{N}} F_k$ is dense in $\mathfrak{H}$. Now construct an ONB in the finite-dimensional space F_k and extend this ONB using (21.10) to an ONB in F_{k+1}, etc. This produces a sequence $(e_j)_{j \in \mathbb{N}}$ of orthonormal elements such that $\text{span}\{e_j : j \in \mathbb{N}\} = \bigcup_{k \in \mathbb{N}} F_k = G$ is dense in $\mathfrak{H}$ and T21.13 completes the proof. ∎

21.16 Remarks **(i)** Assume that $\mathfrak{H}$ is *separable*. Then we have the following 'algebraic' interpretation of the results in 21.11–21.15. Consider the maps

coordinate projection	(re-)construction map
$\prod : \mathfrak{H} \to \ell^2_{\mathbb{K}}(\mathbb{N})$	$\coprod : \ell^2_{\mathbb{K}}(\mathbb{N}) \to \mathfrak{H}$
$g \mapsto (\langle g, e_j \rangle)_{j \in \mathbb{N}}$	$(c_j)_{j \in \mathbb{N}} \mapsto \displaystyle\sum_{j=1}^{\infty} c_j \, e_j.$

Because of Theorem 21.11(iv), both $\prod$ and $\coprod$ are well-defined maps, and Theorem 21.11 shows that Diagram 1 (below, left) commutes, i.e. $\prod \circ \coprod = \mathrm{id}_{\ell^2_{\mathbb{K}}(\mathbb{N})}$.

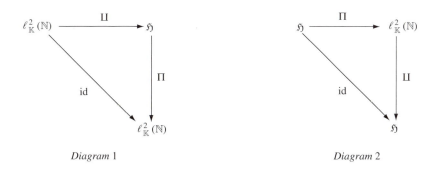

Diagram 1   *Diagram* 2

This means that, if we start with a square-summable sequence, associate an element from $\mathfrak{H}$ with it and project to the coordinates, we get the original sequence back.

The converse operation, if we start with some $h \in \mathfrak{H}$, project h down to its coordinates, and then try to reconstruct h from the (square integrable) coordinate sequence, is much more difficult, as we have seen in Theorems 21.13 and 21.15. Nevertheless, it can be done in every *separable* Hilbert space, and Diagram 2 (above, right) becomes commutative, i.e. $\coprod \circ \prod = \mathrm{id}_{\mathfrak{H}}$.

*This shows that every **separable** Hilbert space $\mathfrak{H}$ can be isometrically mapped onto $\ell^2_{\mathbb{K}}(\mathbb{N})$. The isometry is given by Parseval's identity* 21.11(iv):

$$\sum_{j \in \mathbb{N}} |\langle h, e_j \rangle|^2 = \left\| \prod h \right\|^2_{\ell^2} = \|h\|^2_{\mathfrak{H}}.$$

(ii) If $\mathfrak{H}$ is **not** separable, we can still construct an ONB but now we need transfinite induction or Zorn's lemma. A reasonably short account is given in Rudin's book [40, pp. 83–88]. The results 21.11–21.13 carry over to this case if one makes some technical (what is an uncountable sum? etc.) modifications.

Problems

21.1. Show that every convergent sequence in $\mathfrak{H}$ is a Cauchy sequence.

21.2. Show that $g \mapsto \langle g, h \rangle$, $h \in \mathfrak{H}$, is continuous.

21.3. Show that $\|(g, h)\| := \left(\|g\|^p + \|h\|^p\right)^{1/p}$ is for every $p \geqslant 1$ a norm on $\mathfrak{H} \times \mathfrak{H}$. For which values of p does $\mathfrak{H} \times \mathfrak{H}$ become a Hilbert space?

21.4. Show that $(g, h) \mapsto \langle g, h \rangle$ and $(t, h) \mapsto t\,h$ are continuous on $\mathfrak{H} \times \mathfrak{H}$ resp. $\mathbb{R} \times \mathfrak{H}$.

21.5. Show that a Hilbert space $\mathfrak{H}$ is separable if, and only if, $\mathfrak{H}$ contains a countable maximal orthonormal system.

21.6. Show that for $\mathfrak{H} = L^2(X, \mathcal{A}, \mu)$ and $w \in L^2$ the set $M_w^\perp := \{u \in L^2 : \int u\,w\,d\mu = 0\}^\perp$ is either $\{0\}$ or a one-dimensional subspace of $\mathfrak{H}$.

21.7. Let $(e_j)_{j \in \mathbb{N}} \subset \mathfrak{H}$ be an orthonormal system.

   (i) Show that no subsequence of $(e_j)_{j \in \mathbb{N}}$ converges. However, $\lim_{j \to \infty} \langle e_j, h \rangle = 0$ for every $h \in \mathfrak{H}$.

   [Hint: show that it can't be a Cauchy sequence. Use Bessel's inequality.]

   (ii) The *Hilbert cube* $Q := \left\{ h \in \mathfrak{H} : h = \sum_{j=1}^\infty c_j e_j, \ |c_j| \leqslant \frac{1}{j}, \ j \in \mathbb{N} \right\}$ is closed, bounded and compact (i.e. every sequence has a convergent subsequence).

   (iii) The set $R := \bigcup_{j=1}^\infty \overline{B_{1/j}(e_j)}$ is closed, bounded but not compact (cf. (ii)).

   (iv) The set $S := \left\{ h \in \mathfrak{H} : h = \sum_{j=1}^\infty c_j e_j, \ |c_j| \leqslant \delta_j, \ j \in \mathbb{N} \right\}$ is closed, bounded and compact (cf. (ii)) if, and only if, $\sum_{j=1}^\infty \delta_j^2 < \infty$.

21.8. Let $\mathfrak{H}$ be a real Hilbert space.

   (i) Show that $\|h\| = \sup_{g \neq 0} \dfrac{|\langle g, h \rangle|}{\|g\|} = \sup_{\|g\| \leqslant 1} |\langle g, h \rangle| = \sup_{\|g\| = 1} |\langle g, h \rangle|.$

   (ii) Can we replace in (i) $|\langle \cdot, \cdot \rangle|$ by $\langle \cdot, \cdot \rangle$?

   (iii) Is it enough to take g in (i) from a dense subset rather than from $\mathfrak{H}$ (resp. $\overline{B_1(0)}$ or $\{k \in \mathfrak{H} : \|k\| = 1\}$)?

21.9. Show that the linear span of a sequence $(e_k)_{k \in \mathbb{N}} \subset \mathfrak{H}$, $\mathrm{span}\{e_k : e_k \in \mathfrak{H}, \ k \in \mathbb{N}\}$, is a linear subspace of $\mathfrak{H}$.

21.10. A weak form of the **uniform boundedness principle.** Consider the real Hilbert space $\ell^2 = \ell_{\mathbb{R}}^2(\mathbb{N})$ and let $a = (a_j)_{j \in \mathbb{N}}$ and $b = (b_j)_{j \in \mathbb{N}}$ be two sequences of real numbers.

   (i) Assume that $\sum_{j=1}^\infty a_j^2 = \infty$. Construct a sequence $(j_k)_{k \in \mathbb{N}}$ such that $j_1 = 0$ and $\sum_{j_k < j \leqslant j_{k+1}} a_j^2 > 1$ for all $k \in \mathbb{N}$.

   (ii) Define $b_j := \gamma_k\,a_j$ for all $j_k < j \leqslant j_{k+1}$, $k \in \mathbb{N}$ and show that one can determine the γ_k in such a way that $\sum_{j=1}^\infty a_j b_j = \infty$ while $\sum_{j=1}^\infty b_j^2 < \infty$.

   (iii) Conclude that if $\langle a, b \rangle < \infty$ for all $a \in \ell^2$, we have necessarily $b \in \ell^2$.

   (iv) State and prove the analogue of (iii) for all separable Hilbert spaces.

Remark. The general *uniform boundedness principle* states that in every Hilbert space $\mathfrak{H}$ and for any $H \subset \mathfrak{H}$ one has

$$\sup_{h \in H} |\langle h, g \rangle| < \infty \quad \forall g \in \mathfrak{H} \quad \Longrightarrow \quad \sup_{h \in H} \|h\| < \infty.$$

Interpreting $\Lambda_h : g \mapsto \langle g, h \rangle$ as linear map, this says that the boundedness of the orbits $\Lambda_h(\mathfrak{H})$ for all $h \in H$ implies that the set H is bounded. This formulation perseveres even in Banach spaces. The proof is normally based on Baire's category theorem, cf. Rudin [40].

21.11. Let $F, G \subset \mathfrak{H}$ be linear subspaces. An operator P defined on G is called ($\mathbb{K}$-) *linear*, if $P(\alpha f + \beta g) = \alpha P f + \beta P g$ holds for all $\alpha, \beta \in \mathbb{K}$ and $f, g \in G$.

(i) Assume that F is closed and that $P : \mathfrak{H} \to F$ is the orthogonal projection. Then

$$P^2 = P \quad \text{and} \quad \langle Pg, h \rangle = \langle g, Ph \rangle \quad \forall g, h \in \mathfrak{H}. \tag{21.11}$$

(ii) If $P : \mathfrak{H} \to \mathfrak{H}$ is a map satisfying (21.11), then P is linear and P is the orthogonal projection onto the closed subspace $P(\mathfrak{H})$.

(iii) If $P : \mathfrak{H} \to \mathfrak{H}$ is a linear map satisfying

$$P^2 = P \quad \text{and} \quad \|Ph\| \leqslant \|h\| \quad \forall h \in \mathfrak{H},$$

then P is the orthogonal projection onto the closed subspace $P(\mathfrak{H})$.

21.12. Let $(X, \mathcal{A}, \mu)$ be a measure space and $(A_j)_{j \in \mathbb{N}} \subset \mathcal{A}$ be mutually disjoint sets such that $X = \biguplus_{j \in \mathbb{N}} A_j$. Set

$$Y_j := \left\{ u \in L^2(\mu) : \int_{A_j^c} |u|^2 \, d\mu = 0 \right\}, \qquad j \in \mathbb{N}.$$

(i) Show that $Y_j \perp Y_k$ if $j \neq k$.

(ii) Show that span $\left(\biguplus_{j \in \mathbb{N}} Y_j \right)$, (i.e. the set of all linear combinations of *finitely* many elements from $\biguplus_{j \in \mathbb{N}} Y_j$) is dense in $L^2(\mu)$.

(iii) Find the projection $P_j : L^2(\mu) \to Y_j$.

21.13. Let $(X, \mathcal{A}, \mu)$ be a measure space and assume that $(A_j)_{j \in \mathbb{N}} \subset \mathcal{A}$ is a sequence of pairwise disjoint sets such that $\biguplus_{j \in \mathbb{N}} A_j = X$ and $0 < \mu(A_j) < \infty$. Denote by $\mathcal{A}_n := \sigma(A_1, A_2, \ldots, A_n)$ and by $\mathcal{A}_\infty := \sigma(A_j : j \in \mathbb{N})$.

(i) Show that $L^2(\mathcal{A}_n) \subset L^2(\mathcal{A})$ and that $L^2(\mathcal{A}_n)$ is a closed subspace.

(ii) Find an explicit formula for $E^{\mathcal{A}_n} u$ where $E^{\mathcal{A}_n}$ is the orthogonal projection $E^{\mathcal{A}_n} : L^2(\mathcal{A}) \to L^2(\mathcal{A}_n)$.

(iii) Determine the orthogonal complement of $L^2(\mathcal{A}_n)$.

(iv) Show that $\left(E^{\mathcal{A}_n} u \right)_{n \in \mathbb{N} \cup \{\infty\}}$ is a martingale.

(v) Show that $E^{\mathcal{A}_n} u \xrightarrow{n \to \infty} E^{\mathcal{A}_\infty} u$ a.e. and in L^2.

(vi) Conclude that $L^2(\mathcal{A}_\infty)$ is separable.

22

Conditional expectations in L^2

Throughout this chapter $(X, \mathcal{A}, \mu)$ will be some measure space.

We have seen in Chapter 20 that $L^2_\mathbb{C}(\mathcal{A}) = L^2_\mathbb{R}(\mathcal{A}) \oplus i L^2_\mathbb{R}(\mathcal{A})$. By considering real and imaginary parts separately, we can reduce many assertions concerning $L^2_\mathbb{C}(\mathcal{A})$ to $L^2_\mathbb{R}(\mathcal{A})$. From Chapter 21 we know that $L^2_\mathbb{R}(\mathcal{A})$ is a Hilbert space with inner product, resp. norm

$$\langle u, v \rangle = (u, v)_2 = \int u v \, d\mu, \qquad \text{resp.} \qquad \|u\| = \|u\|_2 = \left(\int u^2 \, d\mu \right)^{1/2}.$$

Since a function[1] $u \in L^2_\mathbb{R}(\mathcal{A})$ is only almost everywhere defined and since (square-) integrable functions with values in $\overline{\mathbb{R}}$ are almost everywhere finite, hence $\mathbb{R}$-valued, cf. Remark 12.5, we can identify $L^2_\mathbb{R}(\mathcal{A})$ and $L^2_{\overline{\mathbb{R}}}(\mathcal{A})$. We will do so and simply write $L^2(\mathcal{A})$.

Caution: Note that for functions $u, v \in L^2(\mathcal{A})$ expressions of the type $u = v$, $u \leqslant v \ldots$ *always mean* $u(x) = v(x)$, $u(x) \leqslant v(x) \ldots$ *for all x outside some μ-null set.*

In this chapter we are mainly interested in linear subspaces of $L^2(\mathcal{A})$ and projections onto them. One particularly important class arises in the following way: if $\mathcal{G} \subset \mathcal{A}$ is a sub-σ-algebra of $\mathcal{A}$, then any $\mathcal{G}$-measurable function is certainly $\mathcal{A}$-measurable. Since $(X, \mathcal{G}, \mu|_\mathcal{G})$ is also a measure space, it seems natural to interpret $L^2(\mathcal{G}, \mu|_\mathcal{G})$ (with norm $\| \cdot \|_{L^2(\mathcal{G})}$) as a subspace of $L^2(\mathcal{A}, \mu)$ (with norm $\| \cdot \|_{L^2(\mathcal{A})}$). This can indeed be done.

22.1 Lemma *Let $\mathcal{G} \subset \mathcal{A}$ be a sub-σ-algebra of $\mathcal{A}$. Then*

$$\iota : \mathcal{L}^2(\mathcal{G}, \mu) \to \mathcal{L}^2(\mathcal{A}, \mu) \qquad \text{and} \qquad \jmath : L^2(\mathcal{G}, \mu) \to L^2(\mathcal{A}, \mu)$$

---

[1] Strictly speaking we should call it an equivalence class of functions, cf. Remark 12.5.

are isometric imbeddings, i.e. linear maps satisfying $\|\iota(u)\|_{\mathcal{L}^2(\mathcal{A})} = \|u\|_{\mathcal{L}^2(\mathcal{G})}$ and $\|j(w)\|_{L^2(\mathcal{A})} = \|w\|_{L^2(\mathcal{G})}$ for all $u \in \mathcal{L}^2(\mathcal{G}, \mu)$, resp. $w \in L^2(\mathcal{G}, \mu)$. In particular $\mathcal{L}^2(\mathcal{G}) [L^2(\mathcal{G})]$ is a closed linear subspace of $\mathcal{L}^2(\mathcal{A}) [L^2(\mathcal{A})]$.

Proof Since $\mathcal{G} \subset \mathcal{A}$ and since $\mu|_\mathcal{G}$ and μ coincide on $\mathcal{G}$, we have $\mathcal{E}(\mathcal{G}) \subset \mathcal{E}(\mathcal{A})$ for the simple functions. The map

$$\iota_\circ : \mathcal{E}(\mathcal{G}) \to \mathcal{E}(\mathcal{A}), \quad f \mapsto \iota_\circ(f) := f = \sum_{j=0}^{N} \alpha_j \mathbf{1}_{G_j},$$

where the latter is a standard representation of f with $\alpha_j \in \mathbb{R}$ and $G_j \in \mathcal{G}$, clearly satisfies

$$\|f\|_{\mathcal{L}^2(\mathcal{A})}^2 = \sum_{j=0}^{N} \alpha_j^2 \mu(G_j) = \sum_{j=0}^{N} \alpha_j^2 (\mu|_\mathcal{G})(G_j) = \|f\|_{\mathcal{L}^2(\mathcal{G})}^2 \qquad \forall f \in \mathcal{E}(\mathcal{G}).$$

According to Corollary 12.11 we can find for every $u \in \mathcal{L}^2(\mathcal{G})$ a sequence $(f_k)_{k \in \mathbb{N}} \subset \mathcal{E}(\mathcal{G}) \cap \mathcal{L}^2(\mathcal{G})$ such that $\lim_{k \to \infty} \|u - f_k\|_{\mathcal{L}^2(\mathcal{G})} = 0$. Therefore,

$$\|\iota_\circ(f_k) - \iota_\circ(f_\ell)\|_{\mathcal{L}^2(\mathcal{A})} = \|\iota_\circ(f_k) - \iota_\circ(f_\ell)\|_{\mathcal{L}^2(\mathcal{G})} \xrightarrow{k,\ell \to \infty} 0,$$

which shows because of the completeness of $\mathcal{L}^2(\mathcal{A})$ (cf. Theorem 12.7) that $\iota(u) := \mathcal{L}^2(\mathcal{A})\text{-}\lim_{k \to \infty} \iota_\circ(f_k)$ is a linear isometry from $\mathcal{L}^2(\mathcal{G})$ to $\mathcal{L}^2(\mathcal{A})$. Since $\mathcal{L}^2(\mathcal{G})$ is complete, $\iota(\mathcal{L}^2(\mathcal{G}))$ is a closed linear subspace of $\mathcal{L}^2(\mathcal{A})$.

Denote by $[u] \in L^2$ the equivalence class containing the function $u \in \mathcal{L}^2$. Since for any two $u, w \in \mathcal{L}^2(\mathcal{G})$ with $u = w$ a.e. we also have $\iota(u) = \iota(w)$ a.e., the map $j([u]) := [\iota(u)]$ is independent of the chosen representative for $[u]$ and defines a linear isometry $j : L^2(\mathcal{G}) \to L^2(\mathcal{A})$. As before, $j(L^2(\mathcal{G}))$ is a closed linear subspace of $L^2(\mathcal{A})$. ∎

It is customary to identify $u \in L^2(\mathcal{G})$ with $j(u) \in L^2(\mathcal{A})$ and we will do so in the sequel. Unless we want to stress the σ-algebra, we will write μ instead of $\mu|_\mathcal{G}$ and $\|\cdot\|_2$ for the norm in $L^2(\mathcal{G}, \mu|_\mathcal{G})$ and $L^2(\mathcal{A}, \mu)$.

A key observation is that the choice of $\mathcal{G} \subset \mathcal{A}$ determines our knowledge about a function u.

22.2 Example Consider a finite measure space $(X, \mathcal{A}, \mu)$ and the sub-σ-algebra $\mathcal{G} := \{\emptyset, G, G^c, X\}$ where $G \in \mathcal{A}$ and $\mu(G) > 0$, $\mu(G^c) > 0$. Let $f \in \mathcal{E}(\mathcal{A})$ be a simple function in standard representation:

$$f = \sum_{j=0}^{m} y_j \mathbf{1}_{A_j}, \qquad y_j \in \mathbb{R}, \ A_j \in \mathcal{A}.$$

Then

$$\int_G f \, d\mu = \sum_{j=0}^{m} y_j \int_G \mathbf{1}_{A_j} \, d\mu = \underbrace{\sum_{j=0}^{m} y_j \frac{\mu(A_j \cap G)}{\mu(G)}}_{=: \, \eta_1} \mu(G) = \eta_1 \int \mathbf{1}_G \, d\mu, \quad (22.1)$$

and similarly,

$$\int_{G^c} f \, d\mu = \underbrace{\sum_{j=0}^{m} y_j \frac{\mu(A_j \cap G^c)}{\mu(G^c)}}_{=: \, \eta_2} \mu(G^c) = \eta_2 \int \mathbf{1}_{G^c} \, d\mu. \quad (22.2)$$

This indicates that we could have obtained the same results in the integrations (22.4) and (22.5) if we had not used $f \in \mathcal{E}(\mathcal{A})$ but the $\mathcal{G}$-simple function

$$g := \eta_1 \mathbf{1}_G + \eta_2 \mathbf{1}_{G^c} \in \mathcal{E}(\mathcal{G}) \quad (22.3)$$

with η_1, η_2 from above. In other words, f and g are indistinguishable, if we evaluate (i.e. integrate) both of them on sets of the σ-algebra $\mathcal{G}$. Note that g is much simpler than f, but we have lost nearly all information of what f looks like on sets from $\mathcal{A}$ save $\mathcal{G}$: if we take a set from the standard representation of f, say, $A_{j_0} \subsetneq G, A_{j_0} \in \mathcal{A}$, then

$$\int_{A_{j_0}} f \, d\mu = y_{j_0} \mu(A_{j_0}) \neq \int_{A_{j_0}} g \, d\mu = \eta_1 \mu(A_{j_0} \cap G)$$

$$= \underbrace{\left(\sum_{j=0}^{m} y_j \frac{\mu(A_j \cap G)}{\mu(G)} \right)}_{= \, \eta_1} \cdot \mu(A_{j_0}),$$

i.e. we would get a weighted *average* of the y_j rather than precisely y_{j_0}.

Let us extend the process sketched in Example 22.2 to σ-finite measures and general square-integrable functions. Our starting point is the observation that, with the notation of 22.2,

$$\int f g \, d\mu = \eta_1 \int_G f \, d\mu + \eta_2 \int_{G^c} f \, d\mu = \eta_1^2 \mu(G) + \eta_2^2 \mu(G^c) = \int g^2 \, d\mu,$$

that is, $\langle f - g, g \rangle = 0$ or $(f - g) \perp g$ in the space $L^2(\mathcal{A})$.

22.3 Definition Let $(X, \mathcal{A}, \mu)$ be a measure space and $\mathcal{G} \subset \mathcal{A}$ be a sub-σ-algebra. The *conditional expectation* of $u \in L^2(\mathcal{A})$ relative to $\mathcal{G}$ is the orthogonal projection onto the closed subspace $L^2(\mathcal{G})$

$$\mathbf{E}^{\mathcal{G}} : L^2(\mathcal{A}) \to L^2(\mathcal{G}), \quad u \mapsto \mathbf{E}^{\mathcal{G}} u.$$

Sometimes one writes $\mathbf{E}(u \mid \mathcal{G})$ instead of $\mathbf{E}^{\mathcal{G}} u$.

The terminology 'conditional expectation' comes from probability theory where this notion is widely used and where $(X, \mathcal{A}, \mu)$ is usually a probability space. In slight abuse of language we continue to call $\mathbf{E}^{\mathcal{G}}$ conditional *expectation* even if μ is not a probability measure. Let us collect some properties of $\mathbf{E}^{\mathcal{G}}$.

22.4 Theorem *Let $(X, \mathcal{A}, \mu)$ be a measure space and $\mathcal{G} \subset \mathcal{A}$ a sub-σ-algebra. The conditional expectation $\mathbf{E}^{\mathcal{G}}$ has the following properties ($u, w \in L^2(\mathcal{A})$):*

  (i) $\mathbf{E}^{\mathcal{G}} u \in L^2(\mathcal{G})$;

 (ii) $\|\mathbf{E}^{\mathcal{G}} u\|_{L^2(\mathcal{G})} \leqslant \|u\|_{L^2(\mathcal{A})}$;

(iii) $\langle \mathbf{E}^{\mathcal{G}} u, w \rangle = \langle u, \mathbf{E}^{\mathcal{G}} w \rangle = \langle \mathbf{E}^{\mathcal{G}} u, \mathbf{E}^{\mathcal{G}} w \rangle$;

 (iv) $\mathbf{E}^{\mathcal{G}} u$ *is the unique minimizer in $L^2(\mathcal{G})$ such that*

$$\|u - \mathbf{E}^{\mathcal{G}} u\|_{L^2(\mathcal{A})} = \inf_{g \in L^2(\mathcal{G})} \|u - g\|_{L^2(\mathcal{A})};$$

  (v) $u = w \implies \mathbf{E}^{\mathcal{G}} u = \mathbf{E}^{\mathcal{G}} w$;

 (vi) $\mathbf{E}^{\mathcal{G}} (\alpha u + \beta w) = \alpha \mathbf{E}^{\mathcal{G}} u + \beta \mathbf{E}^{\mathcal{G}} w$ *for all $\alpha, \beta \in \mathbb{R}$*;

(vii) *If $\mathcal{H} \subset \mathcal{G}$ is a further sub-σ-algebra, then $\mathbf{E}^{\mathcal{H}} \mathbf{E}^{\mathcal{G}} u = \mathbf{E}^{\mathcal{H}} u$*;

(viii) $\mathbf{E}^{\mathcal{G}} (g u) = g \mathbf{E}^{\mathcal{G}} u$ *for all $g \in L^{\infty}(\mathcal{G})$*;

 (ix) $\mathbf{E}^{\mathcal{G}} g = g$ *for all $g \in L^2(\mathcal{G})$*;

  (x) $0 \leqslant u \leqslant 1 \implies 0 \leqslant \mathbf{E}^{\mathcal{G}} u \leqslant 1$;

 (xi) $u \leqslant w \implies \mathbf{E}^{\mathcal{G}} u \leqslant \mathbf{E}^{\mathcal{G}} w$;

(xii) $|\mathbf{E}^{\mathcal{G}} u| \leqslant \mathbf{E}^{\mathcal{G}} |u|$;

(xiii) $\mathbf{E}^{\{\emptyset, X\}} u = \dfrac{1}{\mu(X)} \displaystyle\int u \, d\mu$ *for all $u \in L^1(\mathcal{A}) \cap L^2(\mathcal{A})$* $\qquad (\frac{1}{\infty} := 0).$

Before we turn to the proof of the above properties let us stress again that all (in-)equalities in (i)–(xiii) are between L^2-functions, i.e. they hold only μ-almost everywhere. In particular, $\mathbf{E}^{\mathcal{G}} u$ is itself only determined up to a μ-null set $N \in \mathcal{G}$.

Proof (of Theorem 22.4) Properties **(i)**–**(vi)** and **(ix)** follow directly from Theorem 21.5, Corollary 21.6 and Remark 21.7.

(vii): For all $u, w \in L^2(\mathcal{A})$ we find because of (iii)

$$\langle \mathbf{E}^{\mathcal{H}} \mathbf{E}^{\mathcal{G}} u, w \rangle \overset{(iii)}{=} \langle \mathbf{E}^{\mathcal{G}} u, \mathbf{E}^{\mathcal{H}} w \rangle = \langle u, \mathbf{E}^{\mathcal{G}} \underbrace{\mathbf{E}^{\mathcal{H}} w}_{\in L^2(\mathcal{G})} \rangle$$

$$\overset{(ix)}{=} \langle u, \mathbf{E}^{\mathcal{H}} w \rangle$$

$$= \langle \mathbf{E}^{\mathcal{H}} u, w \rangle$$

as $\mathbf{E}^{\mathcal{H}} w \in L^2(\mathcal{H}) \subset L^2(\mathcal{G})$. Since $w \in L^2(\mathcal{A})$ was arbitrary, we conclude that $\mathbf{E}^{\mathcal{H}} \mathbf{E}^{\mathcal{G}} u = \mathbf{E}^{\mathcal{H}} u$.

(viii): Writing $u = f + f^\perp \in L^2(\mathcal{G}) \oplus L^2(\mathcal{G})^\perp$, we get $g\,u = g\,f + g\,f^\perp$. Moreover, we have for any $\phi \in L^2(\mathcal{G})$ and $g \in L^\infty(\mathcal{G})$ that $g\,\phi \in L^2(\mathcal{G})$, thus

$$\langle g\,f^\perp, \phi \rangle = \langle f^\perp, g\,\phi \rangle = 0,$$

and from the uniqueness of the orthogonal decomposition we infer that

$$(g\,f)^\perp = g\,f^\perp \qquad \text{or} \qquad \mathbf{E}^{\mathcal{G}}(g\,u) = g\,f = g\,\mathbf{E}^{\mathcal{G}} u.$$

(x): Let $0 \leqslant u \leqslant 1$. Since $\mathbf{E}^{\mathcal{G}} u \in L^2(\mathcal{G})$, the Markov inequality P10.12 implies

$$\mu\big(\{|\mathbf{E}^{\mathcal{G}} u| > \tfrac{1}{n}\}\big) \leqslant n^2 \,\|\mathbf{E}^{\mathcal{G}} u\|^2_{L^2(\mathcal{G})} \leqslant n^2 \,\|u\|^2_{L^2(A)} < \infty, \qquad (22.4)$$

and so $\phi_n := \mathbf{1}_{\{|\mathbf{E}^{\mathcal{G}} u| > 1/n\}} \in L^2(\mathcal{G})$. Therefore,

$$\int \mathbf{E}^{\mathcal{G}} u\, \mathbf{1}_{\{\mathbf{E}^{\mathcal{G}} u < 0\}} \phi_n \, d\mu \overset{22.4\text{(iii)}}{=} \int u\, \mathbf{E}^{\mathcal{G}}\big(\mathbf{1}_{\{\mathbf{E}^{\mathcal{G}} u < 0\}} \phi_n\big) \, d\mu$$

$$\overset{22.4\text{(ix)}}{=} \int u\, \mathbf{1}_{\{\mathbf{E}^{\mathcal{G}} u < 0\}} \phi_n \, d\mu \;\geqslant\; 0,$$

which is only possible if $\mu\big(\{\mathbf{E}^{\mathcal{G}} u < 0\} \cap \{|\mathbf{E}^{\mathcal{G}} u| > 1/n\}\big) = 0$, that is, if

$$\mu\big(\{\mathbf{E}^{\mathcal{G}} u < 0\}\big) = \mu\Big(\bigcup_{n \in \mathbb{N}} \{\mathbf{E}^{\mathcal{G}} u < 0\} \cap \{|\mathbf{E}^{\mathcal{G}} u| > 1/n\}\Big)$$

$$= \sup_{n \in \mathbb{N}} \mu\big(\{\mathbf{E}^{\mathcal{G}} u < 0\} \cap \{|\mathbf{E}^{\mathcal{G}} u| > 1/n\}\big) = 0,$$

hence $\mathbf{E}^{\mathcal{G}} u \geqslant 0$.

With very similar arguments we see that $\mathbf{1}_{\{\mathbf{E}^{\mathcal{G}} u > 1\}} \in L^2(\mathcal{G})$, and since $u \leqslant 1$ we have

$$\int \mathbf{E}^{\mathcal{G}} u\, \mathbf{1}_{\{\mathbf{E}^{\mathcal{G}} u > 1\}} \, d\mu \overset{22.4\text{(iii),(ix)}}{=} \int u\, \mathbf{1}_{\{\mathbf{E}^{\mathcal{G}} u > 1\}} \, d\mu \leqslant \mu\big(\{\mathbf{E}^{\mathcal{G}} u > 1\}\big),$$

which entails $\mu(\{\mathbf{E}^{\mathcal{G}} u > 1\}) = 0$ or $\mathbf{E}^{\mathcal{G}} u \leqslant 1$.

(xi): Using that $w - u \geqslant 0$, the first part of the proof of (x) shows $\mathbf{E}^{\mathcal{G}}(w - u) \geqslant 0$, so that by linearity $\mathbf{E}^{\mathcal{G}} w \geqslant \mathbf{E}^{\mathcal{G}} u$.

(xii): Again by the proof of (x) we find for $|u| \pm u \geqslant 0$ that $\mathbf{E}^{\mathcal{G}}(|u| \pm u) \geqslant 0$, and by linearity $\pm \mathbf{E}^{\mathcal{G}} u \leqslant \mathbf{E}^{\mathcal{G}} |u|$. This proves $|\mathbf{E}^{\mathcal{G}} u| \leqslant \mathbf{E}^{\mathcal{G}} |u|$.

(xiii): If $\mu(X) = \infty$, we have $L^2(\{\emptyset, X\}) = \{0\},^{[\checkmark]}$ thus $\mathbf{E}^{\{\emptyset, X\}} u = 0$ and the formula clearly holds.

If $\mu(X) < \infty$, we have $L^2(\{\emptyset, X\}) \simeq \mathbb{R}$, and $\mathbf{E}^{\{\emptyset, X\}} u = c$ is a constant. By (iv), $c = \mathbf{E}^{\{\emptyset, X\}} u$ minimizes $\|u - c\|_{L^2(\mathcal{A})}$, and

$$\int |u - c|^2 \, d\mu = \int u^2 \, d\mu - 2c \int u \, d\mu + c^2 \int d\mu$$

$$= \int u^2 \, d\mu - \frac{1}{\mu(X)} \left(\int u \, d\mu \right)^2 + \frac{1}{\mu(X)} \left(\int u \, d\mu - c \mu(X) \right)^2$$

shows that $c = \frac{1}{\mu(X)} \int u \, d\mu$ is the unique minimizer. ∎

In Chapter 23 we will extend the operator $\mathbf{E}^{\mathcal{G}}$ to $\bigcup_{p \geqslant 1} L^p(\mathcal{A})$ and add a few further properties.

On[2] the structure of subspaces of L^2

In the rest of this chapter we want to address a different question. As we have seen, $\mathbf{E}^{\mathcal{G}} : L^2(\mathcal{A}) \to L^2(\mathcal{G})$ is a symmetric orthogonal projection onto the closed subspace $L^2(\mathcal{G})$ of the Hilbert space $L^2(\mathcal{A})$. It is natural to ask whether *every* orthogonal projection $\pi : L^2(\mathcal{A}) \to \mathfrak{H}$ onto a closed subspace $\mathfrak{H} \subset L^2(\mathcal{A})$ is a conditional expectation. Equivalently we could ask under which conditions a closed subspace $\mathfrak{H}$ of $L^2(\mathcal{A})$ is of the form $L^2(\mathcal{G}) = \mathfrak{H}$ for a suitable sub-σ-algebra $\mathcal{G} \subset \mathcal{A}$.

22.5 Theorem *Let $(X, \mathcal{A}, \mu)$ be a σ-finite measure space. For a closed linear subspace $\mathfrak{H} \subset L^2(\mathcal{A})$ and its orthogonal projection $\pi = P_{\mathfrak{H}} : L^2(\mathcal{A}) \to \mathfrak{H}$, the following assertions are equivalent.*

(i) *$\mathfrak{H} = L^2(\mathcal{G})$ and $\pi = \mathbf{E}^{\mathcal{G}}$ for some sub-σ-algebra $\mathcal{G} \subset \mathcal{A}$ containing an exhausting sequence $(G_j)_{j \in \mathbb{N}} \subset \mathcal{G}$ with $G_j \uparrow X$ and $\mu(G_j) < \infty$.*

(ii) *π is a sub-Markovian operator, i.e. $0 \leqslant u \leqslant 1 \implies 0 \leqslant \pi(u) \leqslant 1$, $u \in L^2(\mathcal{A})$, and for some $u_0 \in L^2(\mathcal{A})$ with $u_0 > 0$ we have $\pi(u_0) > 0$.*

(iii) *$\mathfrak{H} \cap L^\infty(\mathcal{A})$ is an algebra – i.e. it is closed under pointwise products: $f, h \in \mathfrak{H} \cap L^\infty(\mathcal{A}) \implies f h \in \mathfrak{H} \cap L^\infty(\mathcal{A})$ – which is L^2-dense in $\mathfrak{H}$ and contains an (everywhere) strictly positive function $h_0 > 0$.*

(iv) *$\mathfrak{H}$ is a lattice – i.e. $f, h \in \mathfrak{H} \implies f \wedge h \in \mathfrak{H}$ – containing an (everywhere) strictly positive function $h_0 > 0$, and for all $h \in \mathfrak{H}$ also $h \wedge 1 \in \mathfrak{H}$.*

Proof We show that (i)$\Rightarrow$(ii)$\Rightarrow$(iv)$\Rightarrow$(i)$\Rightarrow$(iii)$\Rightarrow$(iv).

(i)$\Rightarrow$(ii) The sub-Markov property of $\pi = \mathbf{E}^{\mathcal{G}}$ follows from Theorem 22.4(x), while

$$u_0 := \sum_{j=1}^{\infty} \frac{2^{-j}}{\sqrt{\mu(G_j) + 1}} \mathbb{1}_{G_j} \in \mathcal{M}^+(\mathcal{G})$$

---

[2] This section can be left out at first reading.

clearly satisfies $0 < u_0 \leqslant 1$,

$$\|u_0\|_2 = \left\| \sum_{j=1}^{\infty} \frac{2^{-j}}{\sqrt{\mu(G_j)+1}} \mathbf{1}_{G_j} \right\|_2 \leqslant \sum_{j=1}^{\infty} \frac{2^{-j}}{\sqrt{\mu(G_j)+1}} \|\mathbf{1}_{G_j}\|_2$$

$$= \sum_{j=1}^{\infty} 2^{-j} \sqrt{\frac{\mu(G_j)}{\mu(G_j)+1}} \leqslant 1,$$

so that $u_0 \in L^2(\mathcal{G})$ and, therefore, $0 < \pi(u_0) = u_0 \leqslant 1$.

(ii)⇒(iv) Since π preserves positivity, we find for all $u \in L^2(\mathcal{A})$ that $\pi(u^+) \geqslant 0$ and $\pi(u) = \pi(u^+) - \pi(u^-) \leqslant \pi(u^+)$, thus $\pi(u) \vee 0 \leqslant \pi(u^+)$.

On the other hand, $\mathfrak{H} = \{h \in L^2(\mathcal{A}) : \pi(h) = h\}$ and the above calculation shows for $h \in \mathfrak{H}$

$$h^+ = (\pi(h))^+ = \pi(h) \vee 0 \leqslant \pi(h^+). \tag{22.5}$$

Since π is a contraction, see (21.4), we find also

$$\|h^+\|_2 \overset{(22.5)}{\leqslant} \|\pi(h^+)\|_2 \leqslant \|h^+\|_2,$$

which implies $\langle \pi(h^+), \pi(h^+) \rangle = \langle \pi(h^+), h^+ \rangle = \langle h^+, h^+ \rangle$. Because of (22.5) we get $\underbrace{\langle \pi(h^+) - h^+, h^+ \rangle}_{\geqslant 0} = 0$ or $\pi(h^+) = h^+$ on the set $\{h^+ > 0\}$. But then

$$\|\pi(h^+)\|_2 = \|h^+\|_2 \implies \int_{\{h^+=0\}} \pi(h^+)^2 \, d\mu = \int_{\{h^+=0\}} (h^+)^2 \, d\mu = 0,$$

which shows that $\pi(h^+) = 0$ on $\{h^+ = 0\}$ or $\mu(\{h^+ = 0\}) = 0$. In either case, $\pi(h^+) = h^+$ (almost everywhere) and $h^+ \in \mathfrak{H}$.

Consequently, $f \wedge h = f - (f-h)^+ \in \mathfrak{H}$. Similarly, $h \wedge 1 = h - (h-1)^+$ and, if $h \in \mathfrak{H}$, we see $\pi(h \wedge 1) \leqslant \pi(h) \wedge 1 = h \wedge 1$. Further,

$$\pi((h-1)^+) = \pi(h) - \pi(h \wedge 1) \geqslant h - h \wedge 1 = (h-1)^+,$$

and since π is a contraction, the same argument which we used to get $\pi(h^+) = h^+$ yields $\pi((h-1)^+) = (h-1)^+$, hence $(h-1)^+, h \wedge 1 \in \mathfrak{H}$. Finally, $h_0 := \pi(u_0)$, u_0 as in (ii), satisfies $h_0 \in \mathfrak{H}$ and $h_0 > 0$.

(iv)⇒(i) We set

$$\mathcal{G} := \{G \in \mathcal{A} : h \wedge \mathbf{1}_G \in \mathfrak{H} \quad \forall h \in \mathfrak{H}\}.$$

Let us first show that $\mathcal{G}$ is a σ-algebra. Clearly, $\emptyset, X \in \mathcal{G}$. If $G \in \mathcal{G}$, then

$$h \wedge \mathbf{1}_{G^c} + \underbrace{h \wedge \mathbf{1}_G}_{\in \mathfrak{H}} = h \wedge 1 + h \wedge 0 \in \mathfrak{H} \qquad \forall h \in \mathfrak{H},$$

which means that $h \wedge \mathbf{1}_{G^c} \in \mathfrak{H}$ and $G^c \in \mathcal{G}$. For any two sets $G, H \in \mathcal{G}$ we see[3]

$$h \wedge \mathbf{1}_{G \cup H} = h \wedge (\mathbf{1}_G \vee \mathbf{1}_H) = (h \wedge \mathbf{1}_G) \vee (h \wedge \mathbf{1}_H) \in \mathfrak{H} \qquad \forall h \in \mathfrak{H},$$

so that $G \cup H \in \mathcal{G}$. Finally, let $(G_j)_{j \in \mathbb{N}} \subset \mathcal{G}$; since $\mathcal{G}$ is $\cup$-stable, we may assume that $G_j \uparrow G := \bigcup_{j \in \mathbb{N}} G_j$. Then

$$(h \wedge \mathbf{1}_{G_j})_{j \in \mathbb{N}} \subset \mathfrak{H} \quad \text{and} \quad \lim_{j \to \infty} (h \wedge \mathbf{1}_{G_j}) = h \wedge \mathbf{1}_G \in L^2(\mathcal{A}) \qquad \forall h \in \mathfrak{H}.$$

Since $|h \wedge \mathbf{1}_{G_j}| \leqslant h \wedge \mathbf{1}_G \in L^2$, an application of the dominated convergence theorem 12.9 shows that in $L^2(\mathcal{A})$ $\lim_{j \to \infty} h \wedge \mathbf{1}_{G_j} = h \wedge \mathbf{1}_G$ for all $h \in \mathfrak{H}$. Since $\mathfrak{H}$ is a closed subspace, we conclude that $h \wedge \mathbf{1}_G \in \mathfrak{H}$ for all $h \in \mathfrak{H}$, thus $G \in \mathcal{G}$.

We will now show that $L^2(\mathcal{G}) = \mathfrak{H}$. If $f \in \mathfrak{H}$ we know from our assumptions that $(\pm f) \wedge 0 \in \mathfrak{H}$, so that $f^+ = -((-f) \wedge 0)$, $f^- = -(f \wedge 0) \in \mathfrak{H}$. Thus $\mathfrak{H} = \mathfrak{H}^+ - \mathfrak{H}^+$, and since also $L^2(\mathcal{G}) = L^2_+(\mathcal{G}) - L^2_+(\mathcal{G})$ it is clearly enough to show that $\mathfrak{H}^+ = L^2_+(\mathcal{G})$.

Assume that $f \in \mathfrak{H}^+$. Then for $a > 0$, $f \wedge a = a(\frac{f}{a} \wedge 1) \in \mathfrak{H}$, and by monotone convergence T11.1 and the closedness of $\mathfrak{H}$,

$$h \wedge \mathbf{1}_{\{f > a\}} = h \wedge \sup_{n \in \mathbb{N}} \left\{ [nf - n(f \wedge a)] \wedge 1 \right\} \in \mathfrak{H} \qquad \forall h \in \mathfrak{H},$$

proving that $\{f > a\} \in \mathcal{G}$ for all $a > 0$. Moreover, $\{f > a\} = X$ if $a < 0$ and $\{f > 0\} = \bigcup_{k \in \mathbb{N}} \{f > 1/k\}$, which shows that $\{f > a\} \in \mathcal{G}$ for all $a \in \mathbb{R}$ and, consequently, $\mathfrak{H}^+ \subset L^2_+(\mathcal{G})$.

Conversely, if $g \in L^2_+(\mathcal{G})$, we can write g as limit of simple functions, $g_n = \sum_{j=1}^{N(n)} y_j^{(n)} \mathbf{1}_{G_j^{(n)}}$ with disjoint sets $G_1^{(n)}, \ldots, G_{N(n)}^{(n)} \in \mathcal{G}$ and $y_j^{(n)} > 0$. For all $h \in \mathfrak{H}$ we find

$$g_n \wedge h = \sum_{j=1}^{N(n)} \left(y_j^{(n)} \mathbf{1}_{G_j^{(n)}} \right) \wedge h = \sum_{j=1}^{N(n)} y_j^{(n)} \left(\mathbf{1}_{G_j^{(n)}} \wedge \frac{h}{y_j^{(n)}} \right) \in \mathfrak{H},$$

and dominated convergence T12.9 and the closedness of $\mathfrak{H}$ imply that $g \wedge h \in \mathfrak{H}$. Choosing, in particular, $h = n h_0$ for some a.e. strictly positive function h_0 and letting $n \to \infty$ gives

$$g = L^2\text{-} \lim_{n \to \infty} (n h_0) \wedge g \in \mathfrak{H}^+,$$

where we again used monotone convergence T11.1 and the closedness of $\mathfrak{H}$. This proves $L^2_+(\mathcal{G}) \subset \mathfrak{H}^+$.

Finally, the sets $G_j := \{h_0 > 1/j\} \in \mathcal{G}$ satisfy $G_j \uparrow X$ and, because of the Markov inequality P10.12, $\mu(G_j) = \mu(\{h_0 > 1/j\}) \leqslant j^2 \int h_0^2 \, d\mu < \infty$.

---

[3] Use that $\mathfrak{H}$ is a vector space and $a \vee b = -((-a) \wedge (-b))$.

(i)$\Rightarrow$(iii): Note that $L^2(\mathcal{G}) \cap L^\infty(\mathcal{G}) = L^2(\mathcal{G}) \cap L^\infty(\mathcal{A})$. An application of the dominated convergence theorem 12.9 shows that the sequence $f_n := (-n) \vee f \wedge n$, $n \in \mathbb{N}$, $f \in L^2(\mathcal{G})$, converges in $L^2(\mathcal{G})$ to f, i.e. $L^2(\mathcal{G}) \cap L^\infty(\mathcal{G})$ is a dense subset of $L^2(\mathcal{G})$. The element $h_0 > 0$ is now constructed as in the proof of (i)$\Rightarrow$(ii). That $L^2(\mathcal{G}) \cap L^\infty(\mathcal{G})$ is an algebra is trivial.

(iii)$\Rightarrow$(iv): Let us show, first of all, that $\mathfrak{H} \cap L^\infty(\mathcal{A})$ is stable under minima. To this end we define recursively a sequence of polynomials in $\mathbb{R}$,

$$p_0(x) := 0, \qquad p_{n+1}(x) := p_n(x) + \tfrac{1}{2}(x^2 - p_n^2(x)), \quad n \in \mathbb{N}_0.$$

By induction it is easy to see that $p_n(0) = 0$ for all $n \in \mathbb{N}_0$ and that

$$0 \leqslant p_n(x) \leqslant p_{n+1}(x) \leqslant |x| \qquad \forall x \in [-1, 1].$$

For $n = 0$ there is nothing to show. Otherwise we can use the induction assumption $p_n(x) \leqslant p_{n+1}(x) \leqslant |x|$ to get

$$0 \leqslant p_{n+1}(x) \leqslant \overbrace{p_{n+1}(x) + \underbrace{\tfrac{1}{2}\left(x^2 - p_{n+1}^2(x)\right)}_{\geqslant 0}}^{\overset{\text{def}}{=}\, p_{n+2}(x)}$$

$$= |x| - \underbrace{(|x| - p_{n+1}(x))}_{\geqslant 0} \cdot \underbrace{\left(1 - \tfrac{1}{2}(|x| + p_{n+1}(x))\right)}_{\geqslant 0 \text{ for } x \in [-1,1]}$$

$$\leqslant |x|.$$

Therefore, $\lim_{n \to \infty} p_n(x) = \sup_{n \in \mathbb{N}} p_n(x) = \ell(x)$ exists for all $|x| \leqslant 1$ and, according to the recursion relation, $\ell(x) = |x|$.

Since $\mathfrak{H}$ is a linear subspace which is stable under products, we get for every $h \in \mathfrak{H} \cap L^\infty(\mathcal{A})$ that $p_n(h/\|h\|_\infty) \in \mathfrak{H}$, and monotone convergence T11.1 and the closedness of $\mathfrak{H}$ show

$$\sup_{n \in \mathbb{N}} p_n\left(\frac{h}{\|h\|_\infty}\right) = \frac{|h|}{\|h\|_\infty} \in \mathfrak{H} \implies |h| \in \mathfrak{H}.$$

As $\mathfrak{H} \cap L^\infty(\mathcal{A})$ is dense in $\mathfrak{H}$, we find for $h \in \mathfrak{H}$ a sequence $(h_k)_{k \in \mathbb{N}} \subset \mathfrak{H} \cap L^\infty(\mathcal{A})$ such that $L^2\text{-}\lim_{k \to \infty} h_k = h$. From above we know, however, that $|h_k| \in \mathfrak{H}$ and $|h_k| \xrightarrow{k \to \infty} |h|$ in $L^2(\mathcal{A})$, thus $|h| \in \mathfrak{H}$. This shows, in particular, that

$$f \wedge h = \tfrac{1}{2}(f + h - |f - h|) \in \mathfrak{H} \qquad \forall f, h \in \mathfrak{H}.$$

Since $0 < h_0 \leqslant \|h_0\|_\infty$, we get for all $n \geqslant \|h_0\|_\infty$ that

$$\frac{n}{n - h_0} = \sum_{j=0}^{\infty}\left(\frac{h_0}{n}\right)^j = \sup_{N \in \mathbb{N}} \sum_{j=0}^{N}\left(\frac{h_0}{n}\right)^j.$$

and for $h \in \mathfrak{H}$,

$$h \wedge \frac{n}{n - h_0} = \lim_{N \to \infty} \underbrace{\sum_{j=0}^{N} \left(\frac{h_0}{n} \right)^j \wedge h}_{\in \mathfrak{H}}.$$

By monotone convergence T11.1 we conclude that $h \wedge \frac{n}{n-h_0} \in \mathfrak{H}$. Finally, as $\frac{n}{n-h_0} \downarrow 1$ and $h^2 \wedge \left(\frac{n}{n-h_0} \right)^2 \leqslant h^2$, we can use the dominated convergence theorem 12.9 and the closedness of $\mathfrak{H}$ to see that $h \wedge 1 \in \mathfrak{H}$. ∎

Problems

22.1. Let $(X, \mathcal{A}, \mu)$ be a measure space and $\mathcal{H} \subset \mathcal{G}$ be two sub-σ-algebras. Show that
$$\mathbf{E}^{\mathcal{H}} \mathbf{E}^{\mathcal{G}} u = \mathbf{E}^{\mathcal{G}} \mathbf{E}^{\mathcal{H}} u = \mathbf{E}^{\mathcal{H}} u \qquad \forall u \in L^2(\mathcal{A}).$$

22.2. Let $(X, \mathcal{A}, \mu)$ be a measure space, $\mathcal{G} \subset \mathcal{A}$ be a sub-σ-algebra and let $\nu := f \mu$ where $f \in \mathcal{M}^+(\mathcal{A})$ is a density $f > 0$.

(i) Denote by $\mathbf{E}_\nu^{\mathcal{G}}$ resp. $\mathbf{E}_\mu^{\mathcal{G}}$ the projections in the spaces $L^2(\mathcal{A}, \mu)$, resp. $L^2(\mathcal{A}, \nu)$. Express $\mathbf{E}_\nu^{\mathcal{G}}$ in terms of $\mathbf{E}_\mu^{\mathcal{G}}$.
[Hint: $\mathbf{E}_\nu^{\mathcal{G}} u = \mathbf{E}_\mu^{\mathcal{G}}(fu)/\mathbf{E}_\mu^{\mathcal{G}} f \, \mathbf{1}_{\{\mathbf{E}_\mu^{\mathcal{G}} f > 0\}}$]

(ii) Under which condition do we have $\mathbf{E}_\mu^{\mathcal{G}} u = \mathbf{E}_\nu^{\mathcal{G}} u$ for all $u \in L^2(\mathcal{A}, \mu) \cap L^2(\mathcal{A}, \nu)$?

Remark. The above result allows us to study conditional expectations for finite measures μ only and to *define* for more general measures a conditional expectation by

$$\mathbf{E}_\nu^{\mathcal{G}} u := \frac{\mathbf{E}_\mu^{\mathcal{G}}(fu)}{\mathbf{E}_\mu^{\mathcal{G}} f} \, \mathbf{1}_{\{\mathbf{E}_\mu^{\mathcal{G}} f > 0\}}.$$

22.3. Let $(X, \mathcal{A}, \mu)$ be a finite measure space, $G_1, \ldots, G_n \in \mathcal{A}$ such that $\bigcup_{j=1}^n G_j = X$ and $\mu(G_j) > 0$ for all $j = 1, 2, \ldots, n$. Then

$$\mathbf{E}^{\mathcal{G}} u = \sum_{j=1}^n \left[\int_{G_j} u(x) \frac{\mu(dx)}{\mu(G_j)} \right] \mathbf{1}_{G_j}.$$

Remark. The measure $\mathbf{1}_{G_j} \mu / \mu(G_j) = \mu(\cdot \cap G_j)/\mu(G_j)$ is often called the *conditional probability* given G_j.

23

Conditional expectations in L^p

Throughout this chapter $(X, \mathcal{A}, \mu)$ will be some measure space.

Our aim is to extend the operator $\mathbf{E}^{\mathcal{G}}$ from $L^2(\mathcal{A})$ to a wider class of functions including the spaces $L^p(\mathcal{A})$, $1 \leqslant p \leqslant \infty$. We will use the same technique that allowed us in Chapters 9 and 10 to extend the integral from the simple functions $\mathcal{E}(\mathcal{A})$ to the positive measurable functions $\mathcal{M}^+(\mathcal{A})$ and integrable functions $\mathcal{L}^1(\mathcal{A})$.

Since we are now considering the spaces L^p of (equivalence classes of) pth power integrable functions, it is convenient to have an analogous notion for measurable functions.

23.1 Definition Let $(X, \mathcal{A}, \mu)$ be a measure space. Two functions $u, v \in \mathcal{M}(\mathcal{A})$ are called *equivalent*, $u \sim v$, if $\{u \neq v\} \in \mathcal{N}_\mu$ is a μ-null set. We write $M(\mathcal{A}) := \mathcal{M}(\mathcal{A})/_\sim$ for the set of all equivalence classes of measurable functions $u \in \mathcal{M}(\mathcal{A})$.

As with L^p-functions, all (in-)equalities between elements from $M(\mathcal{A})$ hold *pointwise almost everywhere*.

23.2 Lemma Let $(X, \mathcal{A}, \mu)$ be a σ-finite measure space. Then $u \in M^+(\mathcal{A})$ if, and only if, there exists an increasing sequence $(u_j)_{j \in \mathbb{N}} \subset L^2_+(\mathcal{A})$ such that $u = \sup_{j \in \mathbb{N}} u_j$.

Proof The 'only if' part '$\Leftarrow$' is trivial since suprema of countably many measurable functions are again measurable (C8.9). For '$\Rightarrow$' let $(A_j)_{j \in \mathbb{N}} \subset \mathcal{A}$ be an exhausting sequence such that $A_j \uparrow X$ and $\mu(A_j) < \infty$. If $u \in M^+(\mathcal{A})$, then $u_k := (u \wedge k) \mathbf{1}_{A_k} \in L^2_+(\mathcal{A})$ and $\sup_{k \in \mathbb{N}} u_k = u$. ∎

23.3 Remark Lemma 23.2 is no longer true if $(X, \mathcal{A}, \mu)$ is not σ-finite. In fact, if $1 \in M^+(\mathcal{A})$ can be approximated by an increasing sequence $(u_k)_{k \in \mathbb{N}} \subset L^2_+(\mathcal{A})$, $1 = \sup_{k \in \mathbb{N}} u_k$, the sets $A_k := \{u_k > 1/k\}$ would form an increasing sequence

$A_k \uparrow X$ with $\mu(A_k) = \mu(\{u_k > 1/k\}) \leqslant k^2 \int u_k^2 \, d\mu < \infty$ by the Markov inequality P10.12.

The key technical point is the following result.

23.4 Lemma *Let $(X, \mathcal{A}, \mu)$ be a measure space, $\mathcal{G} \subset \mathcal{A}$ be a sub-σ-algebra, and $(u_j)_{j \in \mathbb{N}}, (w_j)_{j \in \mathbb{N}} \subset L^2(\mathcal{A})$ be two increasing sequences. Then*

$$\sup_{j \in \mathbb{N}} u_j = \sup_{j \in \mathbb{N}} w_j \implies \sup_{j \in \mathbb{N}} \mathbf{E}^{\mathcal{G}} u_j = \sup_{j \in \mathbb{N}} \mathbf{E}^{\mathcal{G}} w_j. \tag{23.1}$$

If $u := \sup_{j \in \mathbb{N}} u_j$ is in $L^2(\mathcal{A})$, the following conditional monotone convergence *property holds:*

$$\sup_{j \in \mathbb{N}} \mathbf{E}^{\mathcal{G}} u_j = \mathbf{E}^{\mathcal{G}} \left(\sup_{j \in \mathbb{N}} u_j \right) = \mathbf{E}^{\mathcal{G}} u, \tag{23.2}$$

in $L^2(\mathcal{G})$ and almost everywhere.

Proof Let us first of all assume that $u_j \uparrow u$ and $u \in L^2(\mathcal{A})$. Monotone convergence 11.1 and Theorem 12.7 show that $u_j \xrightarrow{j \to \infty} u$ also in L^2-sense. By Theorem 22.4(xi), $(\mathbf{E}^{\mathcal{G}} u_j)_{j \in \mathbb{N}}$ is again an increasing sequence and $\mathbf{E}^{\mathcal{G}} u_j \leqslant \mathbf{E}^{\mathcal{G}} u$. From 22.4(ii), (vi) we get

$$\left\| \mathbf{E}^{\mathcal{G}} u - \mathbf{E}^{\mathcal{G}} u_j \right\|_2 = \left\| \mathbf{E}^{\mathcal{G}} (u - u_j) \right\|_2 \leqslant \| u - u_j \|_2,$$

i.e. $L^2\text{-}\lim_{j \to \infty} \mathbf{E}^{\mathcal{G}} u_j = \mathbf{E}^{\mathcal{G}} u$. For a subsequence $(u_{j(k)})_{k \in \mathbb{N}} \subset (u_j)_{j \in \mathbb{N}}$ we even have $\lim_{k \to \infty} \mathbf{E}^{\mathcal{G}} u_{j(k)} = \mathbf{E}^{\mathcal{G}} u$ a.e., cf. Corollary 12.8. Because of the monotonicity of the sequence $(\mathbf{E}^{\mathcal{G}} u_j)_{j \in \mathbb{N}}$, we get for all $j > j(k)$

$$\left| \mathbf{E}^{\mathcal{G}} u - \mathbf{E}^{\mathcal{G}} u_j \right| = \mathbf{E}^{\mathcal{G}} u - \mathbf{E}^{\mathcal{G}} u_j \leqslant \mathbf{E}^{\mathcal{G}} u - \mathbf{E}^{\mathcal{G}} u_{j(k)},$$

and letting first $j \to \infty$ and then $k \to \infty$ gives $\mathbf{E}^{\mathcal{G}} u_j \uparrow \mathbf{E}^{\mathcal{G}} u$ a.e. This finishes the proof of (23.2).

If $(u_j)_{j \in \mathbb{N}}, (w_j)_{j \in \mathbb{N}} \subset L^2(\mathcal{A})$ are any two increasing sequences[1] such that $\sup_{j \in \mathbb{N}} u_j = \sup_{j \in \mathbb{N}} w_j$, we can apply (23.2) to the increasing sequences $u_j \wedge w_k \uparrow u_j$ (as $k \to \infty$ and for fixed j) and $u_j \wedge w_k \uparrow w_k$ (as $j \to \infty$ and for fixed k). This shows

$$\sup_{j \in \mathbb{N}} \mathbf{E}^{\mathcal{G}} u_j \overset{(23.2)}{=} \sup_{j \in \mathbb{N}} \sup_{k \in \mathbb{N}} \mathbf{E}^{\mathcal{G}} (u_j \wedge w_k) = \sup_{k \in \mathbb{N}} \sup_{j \in \mathbb{N}} \mathbf{E}^{\mathcal{G}} (u_j \wedge w_k) \overset{(23.2)}{=} \sup_{k \in \mathbb{N}} \mathbf{E}^{\mathcal{G}} w_k. \qquad \blacksquare$$

A combination of Lemmata 23.2 and 23.4 allows us to define conditional expectations for positive measurable functions in a σ-finite measure space.

---

[1] We do not assume that $\sup_{j \in \mathbb{N}} u_j, \sup_{j \in \mathbb{N}} w_j \in L^2(\mathcal{A})$.

23.5 Definition Let $(X, \mathcal{A}, \mu)$ be a σ-finite measure space and $\mathcal{G} \subset \mathcal{A}$ be a sub-σ-algebra. Let $u \in M^+(\mathcal{A})$ and let $(u_j)_{j \in \mathbb{N}} \subset L^2_+(\mathcal{A})$ be an increasing sequence such that $u = \sup_{j \in \mathbb{N}} u_j$. Then

$$E^{\mathcal{G}} u := \sup_{j \in \mathbb{N}} \mathbf{E}^{\mathcal{G}} u_j \tag{23.3}$$

is called the *conditional expectation* of u with respect to $\mathcal{G}$. If $u \in M(\mathcal{A})$ and $E^{\mathcal{G}} u^{\pm} \in \mathbb{R}$ almost everywhere, we define (almost everywhere)

$$E^{\mathcal{G}} u := E^{\mathcal{G}} u^+ - E^{\mathcal{G}} u^- = \lim_{j \to \infty} \left(\mathbf{E}^{\mathcal{G}} u_j^+ - \mathbf{E}^{\mathcal{G}} u_j^- \right), \tag{23.4}$$

where $u_j^{\pm} \uparrow u^{\pm}$ are suitable approximating sequences from $L^2_+(\mathcal{A})$.

We write $L^{\mathcal{G}}(\mathcal{A})$ for the set of all functions $u \in M(\mathcal{A})$ such that (almost everywhere) $E^{\mathcal{G}} u$ exists and is finite.

23.6 Theorem *Let $(X, \mathcal{A}, \mu)$ be a σ-finite measure space. The conditional expectation $E^{\mathcal{G}}$ extends $\mathbf{E}^{\mathcal{G}}$, i.e. $L^2(\mathcal{A}) \subset L^{\mathcal{G}}(\mathcal{A})$ and $E^{\mathcal{G}} u = \mathbf{E}^{\mathcal{G}} u$ for all $u \in L^2(\mathcal{A})$.*

Proof Applying (23.2) to u^+ and u^- shows $E^{\mathcal{G}} u^{\pm} = \mathbf{E}^{\mathcal{G}} u^{\pm}$ and, in particular, $E^{\mathcal{G}} u^{\pm} \in L^2(\mathcal{G})$. As such, $E^{\mathcal{G}} u^{\pm}$ is a.e. real-valued, so that (23.4) is always defined in $M(\mathcal{A})$, resp. $M(\mathcal{G})$. ∎

23.7 Theorem *Let $(X, \mathcal{A}, \mu)$ be a σ-finite measure space. Then $L^{\mathcal{G}}(\mathcal{A})$ is a vector space and $L^p(\mathcal{A}) \subset L^{\mathcal{G}}(\mathcal{A})$ for all $1 \leqslant p \leqslant \infty$.*

Proof Let $1 < p < \infty$ and take $u \in L^p(\mathcal{A}) \cap L^2(\mathcal{A})$. Since $\mathbf{E}^{\mathcal{G}} u \in L^2(\mathcal{G})$, the Markov inequality P10.12 shows that $\{|\mathbf{E}^{\mathcal{G}} u| = \infty\} \in \mathcal{G}$ is a μ-null set and that the sets $G_n := \{n > |\mathbf{E}^{\mathcal{G}} u| > 1/n\} \in \mathcal{G}$ have finite μ-measure,

$$\mu(G_n) \leqslant \mu\left(\{|\mathbf{E}^{\mathcal{G}} u| > 1/n\}\right) \leqslant n^2 \int (\mathbf{E}^{\mathcal{G}} u)^2 \, d\mu < \infty.$$

Moreover,

$$\begin{aligned} \left\| \mathbf{E}^{\mathcal{G}} u \, \mathbf{1}_{G_n} \right\|_p^p &= \left\langle \mathbf{E}^{\mathcal{G}} u, |\mathbf{E}^{\mathcal{G}} u|^{p-1} \operatorname{sgn}(\mathbf{E}^{\mathcal{G}} u) \, \mathbf{1}_{G_n} \right\rangle \\ &= \left\langle u, |\mathbf{E}^{\mathcal{G}} u|^{p-1} \operatorname{sgn}(\mathbf{E}^{\mathcal{G}} u) \, \mathbf{1}_{G_n} \right\rangle \quad \text{(by T22.4(iii), (ix))} \\ &\leqslant C_q \|u\|_p, \end{aligned}$$

where we used Hölder's inequality T12.2 with $p^{-1} + q^{-1} = 1$ and

$$C_q = \left(\int |\mathbf{E}^{\mathcal{G}} u|^{(p-1)q} \, \mathbf{1}_{G_n} \, d\mu \right)^{1/q} = \left(\int |\mathbf{E}^{\mathcal{G}} u|^p \, \mathbf{1}_{G_n} \, d\mu \right)^{1/q}.$$

Dividing the above inequality by C_q – if $C_q = 0$ there is nothing to show since in this instance $\mathbf{E}^{\mathcal{G}} u = 0^{[\checkmark]}$ – gives

$$\left\| \mathbf{E}^{\mathcal{G}} u \, \mathbf{1}_{G_n} \right\|_p \leqslant \|u\|_p.$$

As we have seen above, $\mu(\{\mathbf{E}^{\mathcal{G}} u = \infty\}) = 0$, so we can use Beppo Levi's theorem 9.6 to find for all $u \in L^2(\mathcal{A}) \cap L^p(\mathcal{A})$

$$\left\| \mathbf{E}^{\mathcal{G}} u \right\|_p = \left\| \mathbf{E}^{\mathcal{G}} u \, \mathbf{1}_{\{0 < |E^{\mathcal{G}} u| < \infty\}} \right\|_p = \sup_{n \in \mathbb{N}} \left\| \mathbf{E}^{\mathcal{G}} u \, \mathbf{1}_{G_n} \right\|_p \leqslant \|u\|_p. \tag{23.5}$$

For general $u \in L^p(\mathcal{A}) \subset M(\mathcal{A})$, $1 < p < \infty$, we use Lemma 23.2 to find sequences $(u_j^{\pm})_{j \in \mathbb{N}} \subset L^2_+(\mathcal{A})$ such that $0 \leqslant u_j^{\pm} \uparrow u^{\pm}$. Since $u_j^+ - u_j^- \leqslant |u_j| \leqslant |u|$, T22.4(x)–(xii) and Fatou's lemma T9.11 show

$$\left\| E^{\mathcal{G}} u \right\|_p \overset{(23.4)}{=} \left\| \lim_{j \to \infty} \mathbf{E}^{\mathcal{G}}(u_j^+ - u_j^-) \right\|_p$$

$$\leqslant \left\| \liminf_{j \to \infty} \mathbf{E}^{\mathcal{G}} |u_j| \right\|_p$$

$$\leqslant \liminf_{j \to \infty} \left\| \mathbf{E}^{\mathcal{G}} |u_j| \right\|_p$$

$$\overset{(23.5)}{\leqslant} \liminf_{j \to \infty} \left\| |u_j| \right\|_p \leqslant \|u\|_p, \tag{23.6}$$

which, in turn, shows that $E^{\mathcal{G}} u$ is a well-defined function in $L^p(\mathcal{G})$.

If $p = 1$ and $u \in L^2(\mathcal{A}) \cap L^1(\mathcal{A})$, the estimate (23.5) follows more easily from

$$\left\| \mathbf{E}^{\mathcal{G}} u \, \mathbf{1}_{G_n} \right\|_1 = \left\langle \mathbf{E}^{\mathcal{G}} u, \mathbf{1}_{G_n} \right\rangle \overset{21.4(\text{iii}),(\text{ix})}{=} \left\langle u, \mathbf{1}_{G_n} \right\rangle \leqslant \|u\|_1$$

for all $u \in L^2(\mathcal{A}) \cap L^1(\mathcal{A})$ (notation as above), and we conclude that $\|E^{\mathcal{G}} u\|_1 \leqslant \|u\|_1$ for $u \in L^1(\mathcal{A})$ and that $E^{\mathcal{G}} u$ is well-defined.

If $p = \infty$ and $u \in L^2(\mathcal{A}) \cap L^\infty(\mathcal{A})$ we get from T22.4 and the observation that $|u|/\|u\|_\infty \leqslant 1$

$$\left| \mathbf{E}^{\mathcal{G}}(u/\|u\|_\infty) \right| \leqslant \mathbf{E}^{\mathcal{G}}(|u|/\|u\|_\infty) \leqslant 1,$$

thus $\|\mathbf{E}^{\mathcal{G}} u\|_\infty \leqslant \|u\|_\infty$ for all $u \in L^2(\mathcal{A}) \cap L^\infty(\mathcal{A})$. This extends to $E^{\mathcal{G}} u$ for general $u \in L^\infty(\mathcal{A})$ as in the cases where $p \in [1, \infty)$.

It remains to show that $L^{\mathcal{G}}(\mathcal{A})$ is a vector space. This follows immediately from the formula $E^{\mathcal{G}}(\alpha u + \beta w) = \alpha E^{\mathcal{G}} u + \beta E^{\mathcal{G}} w$, which is easily proved from the definition of $E^{\mathcal{G}}$ via approximating sequences and the corresponding property (vi) for $\mathbf{E}^{\mathcal{G}}$ of Theorem 22.4. ∎

The properties of $E^{\mathcal{G}}$ resemble those of $\mathbf{E}^{\mathcal{G}}$. The theorem below is the analogue of Theorem 22.4.

23.8 Theorem *Let $(X, \mathcal{A}, \mu)$ be a σ-finite measure space and let $\mathcal{H} \subset \mathcal{G} \subset \mathcal{A}$ be sub-σ-algebras. The conditional expectation $E^{\mathcal{G}}$ has the following properties $(u, w \in L^{\mathcal{G}}(\mathcal{A}))$:*

   (i) $E^{\mathcal{G}}u \in M(\mathcal{G})$;

  (ii) $u \in L^p(\mathcal{A}) \implies E^{\mathcal{G}}u \in L^p(\mathcal{G})$ *and* $\|E^{\mathcal{G}}u\|_p \leqslant \|u\|_p$; $(p \in [1, \infty])$;

 (iii) $\langle E^{\mathcal{G}}u, w\rangle = \langle u, E^{\mathcal{G}}w\rangle = \langle E^{\mathcal{G}}u, E^{\mathcal{G}}w\rangle$

       (for $u, w \in L^{\mathcal{G}}(\mathcal{A})$, $u E^{\mathcal{G}}w \in L^1(\mathcal{A})$, e.g. if $u \in L^p(\mathcal{A})$ and $w \in L^q(\mathcal{A})$ with $p^{-1} + q^{-1} = 1$);

  (iv) $u = w \implies E^{\mathcal{G}}u = E^{\mathcal{G}}w$;

   (v) $E^{\mathcal{G}}(\alpha u + \beta w) = \alpha E^{\mathcal{G}}u + \beta E^{\mathcal{G}}w$ $(\alpha, \beta \in \mathbb{R})$;

  (vi) $\mathcal{G} \supset \mathcal{H} \implies E^{\mathcal{H}}E^{\mathcal{G}}u = E^{\mathcal{H}}u$;

 (vii) $g \in M(\mathcal{G})$, $u \in L^{\mathcal{G}}(\mathcal{A}) \implies g u \in L^{\mathcal{G}}(\mathcal{A})$ *and* $E^{\mathcal{G}}(g u) = g E^{\mathcal{G}}(u)$;

(viii) $M(\mathcal{G}) \subset L^{\mathcal{G}}(\mathcal{A})$ *and* $E^{\mathcal{G}}g = g E^{\mathcal{G}}1$ *for all* $g \in M(\mathcal{G})$;

(viii') *if $\mu|_{\mathcal{G}}$ is σ-finite* $1 = E^{\mathcal{G}}1$ *and* $g = E^{\mathcal{G}}g$ *for all* $g \in M(\mathcal{G})$;

  (ix) $0 \leqslant u \leqslant 1 \implies 0 \leqslant E^{\mathcal{G}}u \leqslant 1$;

   (x) $u \leqslant w \implies E^{\mathcal{G}}u \leqslant E^{\mathcal{G}}w$;

  (xi) $|E^{\mathcal{G}}u| \leqslant E^{\mathcal{G}}|u|$;

 (xii) $E^{\{\emptyset, X\}}u = \dfrac{1}{\mu(X)} \displaystyle\int u \, d\mu$ *for all* $u \in L^1(\mathcal{A})$ $(\frac{1}{\infty} := 0)$.

Proof (i) is clear from the definition of $E^{\mathcal{G}}$, (ii), (v) were already proved in Theorem 23.7. (iii), (iv), (ix), (x) follow by approximation from the corresponding properties of $\mathbf{E}^{\mathcal{G}}$ from Theorem 22.4, and (xi) is derived from (ix) exactly as in the L^2-case.

(vi) Without loss of generality it is enough to consider the case $u \geqslant 0$. Pick a sequence $(u_j)_{j \in \mathbb{N}} \subset L^2(\mathcal{A})$ such that $u_j \uparrow u$. By Lemma 23.4 and the definition of $E^{\mathcal{G}}$ we know that $\mathbf{E}^{\mathcal{G}}u_j \uparrow E^{\mathcal{G}}u$ as well as $\mathbf{E}^{\mathcal{H}}u_j \uparrow E^{\mathcal{H}}u$ and $\mathbf{E}^{\mathcal{H}}(\mathbf{E}^{\mathcal{G}}u_j) \uparrow E^{\mathcal{H}}(E^{\mathcal{G}}u)$. Since $\mathbf{E}^{\mathcal{H}}(\mathbf{E}^{\mathcal{G}}u_j) = \mathbf{E}^{\mathcal{H}}u_j$, by Theorem 22.4(vii), we are done.

(vii) Assume first that $g, u \geqslant 0$. Define $g_j := g \wedge j \in L^\infty_+(\mathcal{G})$ and let $(u_j)_{j \in \mathbb{N}} \subset L^2_+(\mathcal{A})$ be an increasing sequence such that $\sup_{j \in \mathbb{N}} u_j = u$. Then $g_j u_j \in L^2(\mathcal{A})$, $g_j u_j \uparrow g u$ and T22.4 shows that $\mathbf{E}^{\mathcal{G}}(g_j u_j) = g_j \mathbf{E}^{\mathcal{G}}u_j$. Hence,

$$E^{\mathcal{G}}(g u) = \sup_{j \in \mathbb{N}} \mathbf{E}^{\mathcal{G}}(g_j u_j) = \sup_{j \in \mathbb{N}} \left(g_j \mathbf{E}^{\mathcal{G}}u_j\right) = g E^{\mathcal{G}}u.$$

If $g \geqslant 0$ and $u \in L^{\mathcal{G}}(\mathcal{A})$, the conditional expectation $E^{\mathcal{G}}u^+ - E^{\mathcal{G}}u^-$ is well-defined and we find from the previous calculations

$$g E^{\mathcal{G}}u = g\left(E^{\mathcal{G}}u^+ - E^{\mathcal{G}}u^-\right) = E^{\mathcal{G}}(g u^+) - E^{\mathcal{G}}(g u^-) = E^{\mathcal{G}}(g u).$$

Finally, if $g \in M(\mathcal{G})$ we see, using $g^+ g^- = 0$, that

$$(g^+ - g^-) E^{\mathcal{G}}u = E^{\mathcal{G}}(g^+ u) - E^{\mathcal{G}}(g^- u) = E^{\mathcal{G}}(g u).$$

(viii) Since $(X, \mathcal{A}, \mu)$ is σ-finite, $E^{\mathcal{G}}1 = \sup_{j \in \mathbb{N}} \mathbf{E}^{\mathcal{G}}\mathbf{1}_{A_j}$ for some exhausting sequence $(A_j)_{j \in \mathbb{N}} \subset \mathcal{A}$ with $A_j \uparrow X$ and $\mu(A_j) < \infty$. We can now argue as in (vii) to get $E^{\mathcal{G}}g = g\,E^{\mathcal{G}}1$.

(viii$'$) If $\mu|_{\mathcal{G}}$ is σ-finite, we can find an exhausting sequence $(G_j)_{j \in \mathbb{N}} \subset \mathcal{G}$ with $G_j \uparrow X$ and $\mu(G_j) < \infty$. Since $\mathbf{1}_{G_j} \in L^2(\mathcal{G})$, we find from T22.4(ix) that

$$E^{\mathcal{G}}1 = \sup_{j \in \mathbb{N}} \mathbf{E}^{\mathcal{G}}\mathbf{1}_{G_j} = \sup_{j \in \mathbb{N}} \mathbf{1}_{G_j} = 1.$$

For $g \in M(\mathcal{G})$, we use $g_j^{\pm} := (g^{\pm} \wedge j)\mathbf{1}_{G_j} \in L^2(\mathcal{G})$ as approximating sequences and finish the proof as before.

(ix) If $(u_j)_{j \in \mathbb{N}} \subset L^2(\mathcal{A})$ approximates $0 \leqslant u \leqslant 1$ such that $u_j \uparrow u := \sup_{k \in \mathbb{N}} u_k$, we still have $v_j := u_j^+ \in L^2(\mathcal{A})$ and $v_j \uparrow u$. Thus T22.4(x) implies $0 \leqslant E^{\mathcal{G}}u \leqslant 1$.

(xii) Considering positive and negative parts separately we may assume that $u \geqslant 0$. Since $u \in L^{\mathcal{G}}(\mathcal{A})$, there is an approximating sequence $(u_j)_{j \in \mathbb{N}} \subset L^2_+(\mathcal{A})$, $u = \sup_{j \in \mathbb{N}} u_j$, and as $0 \leqslant u_j \leqslant u \in L^1(\mathcal{G})$, we have $u_j \in L^1(\mathcal{A})$. Theorem 22.4(xiii) gives, together with the definition of $E^{\mathcal{G}}$ and Beppo Levi's theorem 9.6,

$$E^{\{\emptyset, X\}}u = \sup_{j \in \mathbb{N}} \mathbf{E}^{\{\emptyset, X\}}u_j = \sup_{j \in \mathbb{N}} \frac{1}{\mu(X)} \int u_j\, d\mu = \frac{1}{\mu(X)} \int u\, d\mu. \qquad \blacksquare$$

Classical conditional expectations

From now on we will no longer distinguish between $\mathbf{E}^{\mathcal{G}}$ and its extension $E^{\mathcal{G}}$ but always write $\mathbf{E}^{\mathcal{G}}$. In particular, we can now show that the operator $\mathbf{E}^{\mathcal{G}}$ coincides with the *traditional* definition of conditional expectation for L^1-functions. The latter turns out to be a rather elegant way to rewrite the martingale property introduced in Definition 17.1.

23.9 Theorem *Let $(X, \mathcal{A}, \mu)$ be a σ-finite measure space and $\mathcal{G} \subset \mathcal{A}$ be a sub-σ-algebra such that $\mu|_{\mathcal{G}}$ is σ-finite. For $u \in L^1(\mathcal{A})$ and $g \in L^1(\mathcal{G})$ the following conditions are equivalent:*

(i) $\mathbf{E}^{\mathcal{G}}u = g$;

(ii) $\displaystyle\int_G u\, d\mu = \int_G g\, d\mu \qquad \forall\, G \in \mathcal{G}$;

(iii) $\displaystyle\int_G u\, d\mu = \int_G g\, d\mu \qquad \forall G \in \mathcal{G},\ \mu(G) < \infty$.

If $\mathcal{G}$ is generated by a $\cap$-stable family $\mathcal{F} \subset \mathcal{P}(X)$ containing an exhausting sequence $(\mathrm{F}_j)_{j \in \mathbb{N}}, \mathrm{F}_j \uparrow X$ then (i)–(iii) are also equivalent to

(iv) $\displaystyle\int_F u\, d\mu = \int_F g\, d\mu \qquad \forall F \in \mathcal{F}$.

Proof We begin with the general remark that by Theorem 23.8(iii), (viii′) we have for all $G \in \mathcal{G}$ and $u \in L^1(\mathcal{A})$

$$\int_G \mathbf{E}^{\mathcal{G}} u \, d\mu = \langle \mathbf{E}^{\mathcal{G}} u, \mathbf{1}_G \rangle = \langle u, \mathbf{E}^{\mathcal{G}} \mathbf{1}_G \rangle = \langle u, \mathbf{1}_G \rangle = \int_G u \, d\mu. \qquad (23.7)$$

(i)⇒(ii): Because of (23.7) we get for all $G \in \mathcal{G}$ and $k \in \mathbb{N}$

$$\int_G \mathbf{E}^{\mathcal{G}} u \, d\mu \overset{(23.7)}{=} \int_G u \, d\mu \overset{23.9(i)}{=} \int_G g \, d\mu.$$

(ii)⇒(iii) is obvious.

(iii)⇒(i): Take an exhausting sequence $(G_k)_{k \in \mathbb{N}} \subset \mathcal{G}$ with $G_k \uparrow X$ and $\mu(G_k) < \infty$. Then we have for all $G \in \mathcal{G}$

$$\int_{G \cap G_k} \mathbf{E}^{\mathcal{G}} u \, d\mu \overset{(23.7)}{=} \int_{G \cap G_k} u \, d\mu \overset{23.9(iii)}{=} \int_{G \cap G_k} g \, d\mu.$$

Since $|\mathbf{1}_{G \cap G_k} \mathbf{E}^{\mathcal{G}} u| \leqslant |\mathbf{E}^{\mathcal{G}} u| \in L^1$ and $|\mathbf{1}_{G \cap G_k} g| \leqslant |g| \in L^1$, we can use dominated convergence T11.2 to let $k \to \infty$ and get

$$\int_G \mathbf{E}^{\mathcal{G}} u \, d\mu = \int_G g \, d\mu \qquad \forall G \in \mathcal{G},$$

from which we conclude that $\mathbf{E}^{\mathcal{G}} u = g$ a.e. by Corollary 10.14(i).

Assume now, in addition, that $\mathcal{G} = \sigma(\mathcal{F})$. In this case, **(ii)⇒(iv)** is obvious, while **(iv)⇒(ii)** follows with the technique used in Remark 17.2(i): because of (iv) the measures

$$\rho(G) := \int_G u^+ + g^- \, d\mu \quad \text{and} \quad \nu(G) := \int_G u^- + g^+ \, d\mu$$

coincide on $\mathcal{F}$, and by the uniqueness theorem for measures 5.7, on $\mathcal{G}$. ∎

If we combine Theorem 23.9 with the Beppo Levi theorem 9.6 or other convergence theorems we can derive all sorts of 'conditional' versions of these theorems.

23.10 Corollary (Conditional Beppo Levi theorem) *Let $(X, \mathcal{A}, \mu)$ be a σ-finite measure space and $\mathcal{G} \subset \mathcal{A}$ be a sub-σ-algebra such that $\mu|_{\mathcal{G}}$ is σ-finite. For every increasing sequence $(u_j)_{j \in \mathbb{N}} \subset L^1_+(\mathcal{A})$ of positive functions the limit $u := \sup_{j \in \mathbb{N}} u_j$ admits a conditional expectation with values in $[0, \infty]$ and*

$$\sup_{j \in \mathbb{N}} \mathbf{E}^{\mathcal{G}} u_j = \mathbf{E}^{\mathcal{G}} \left(\sup_{j \in \mathbb{N}} u_j \right) = \mathbf{E}^{\mathcal{G}} u. \qquad (23.8)$$

Proof Let $(A_j)_{j \in \mathbb{N}} \subset \mathcal{A}$ be an exhausting sequence of sets, i.e. $A_j \uparrow X$ and $\mu(A_j) < \infty$. Then the functions $w_j := (u_j \wedge j) \mathbf{1}_{A_j} \in L^\infty_+(\mathcal{A}) \cap L^1_+(\mathcal{A})$ and, in particular, $w_j \in L_+^2(\mathcal{A})$.[✓] Moreover, the sequence w_j increases towards u. From Definition 23.5 we get that

$$\mathbf{E}^{\mathcal{G}} u \overset{\text{def}}{=} \sup_{j \in \mathbb{N}} \mathbf{E}^{\mathcal{G}} w_j,$$

which is a numerical function with values in $[0, \infty]$. On the other hand, we know from Theorem 23.9 that for all $G \in \mathcal{G}$

$$\int_G \mathbf{E}^{\mathcal{G}} w_j \, d\mu = \int_G w_j \, d\mu \qquad \text{and} \qquad \int_G \mathbf{E}^{\mathcal{G}} u_j \, d\mu = \int_G u_j \, d\mu$$

holds. Since $\sup_{j \in \mathbb{N}} w_j = u = \sup_{j \in \mathbb{N}} u_j$ and since the sequences $\left(\mathbf{E}^{\mathcal{G}} u_j \right)_{j \in \mathbb{N}}$ and $\left(\mathbf{E}^{\mathcal{G}} w_j \right)_{j \in \mathbb{N}}$ are positive and increasing, cf. Theorem 22.4(xi) and 23.8(x), we conclude from Beppo Levi's theorem 9.6 that

$$\int_G \sup_{j \in \mathbb{N}} \mathbf{E}^{\mathcal{G}} u_j \, d\mu = \sup_{j \in \mathbb{N}} \int_G \mathbf{E}^{\mathcal{G}} u_j \, d\mu = \sup_{j \in \mathbb{N}} \int_G u_j \, d\mu = \int_G \sup_{j \in \mathbb{N}} u_j \, d\mu = \int_G u \, d\mu.$$

With a similar calculation we find

$$\int_G \sup_{j \in \mathbb{N}} \mathbf{E}^{\mathcal{G}} w_j \, d\mu = \int_G u \, d\mu,$$

and, consequently,

$$\int_G \sup_{j \in \mathbb{N}} \mathbf{E}^{\mathcal{G}} u_j \, d\mu = \int_G \sup_{j \in \mathbb{N}} \mathbf{E}^{\mathcal{G}} w_j \, d\mu \qquad \forall G \in \mathcal{G}.$$

By Corollary 10.14 we conclude that $\sup_{j \in \mathbb{N}} \mathbf{E}^{\mathcal{G}} u_j = \sup_{j \in \mathbb{N}} \mathbf{E}^{\mathcal{G}} w_j \overset{\text{def}}{=} \mathbf{E}^{\mathcal{G}} u$ almost everywhere. ∎

In the same way as we deduced Fatou's lemma T9.11 and Lebesgue's dominated convergence theorem 11.2 from the monotonicity property of the integral and Beppo Levi's theorem 9.6, we can get their conditional versions from T22.4(xi), (xii) and C23.10. We leave the simple proofs to the reader.

23.11 Corollary (Conditional Fatou's lemma) *Let* $(X, \mathcal{A}, \mu)$ *be a* σ*-finite measure space,* $\mathcal{G} \subset \mathcal{A}$ *be a sub-*σ*-algebra such that* $\mu|_{\mathcal{G}}$ *is* σ*-finite, and* $(u_j)_{j \in \mathbb{N}} \subset L_+^1(\mathcal{A})$. *Then*

$$\mathbf{E}^{\mathcal{G}} \left(\liminf_{j \to \infty} u_j \right) \leqslant \liminf_{j \to \infty} \mathbf{E}^{\mathcal{G}} u_j. \tag{23.9}$$

23.12 Corollary (Conditional dominated convergence theorem) *Let* $(X, \mathcal{A}, \mu)$ *be a* σ*-finite measure space,* $\mathcal{G} \subset \mathcal{A}$ *be a sub-*σ*-algebra such that* $\mu|_{\mathcal{G}}$ *is* σ*-finite,* and $(u_j)_{j\in\mathbb{N}} \subset L^1(\mathcal{A})$ such that $|u_j| \leqslant w$ for some $w \in L^1_+(\mathcal{A})$. Then

$$\mathbf{E}^{\mathcal{G}}\left(\lim_{j\to\infty} u_j\right) = \lim_{j\to\infty} \mathbf{E}^{\mathcal{G}} u_j. \tag{23.10}$$

23.13 Corollary (Conditional Jensen inequality) *Let $(X, \mathcal{A}, \mu)$ be a σ-finite measure space and $\mathcal{G} \subset \mathcal{A}$ be a sub-σ-algebra such that $\mu|_{\mathcal{G}}$ is σ-finite. Assume that $V : \mathbb{R} \to \mathbb{R}$ is a convex function with $V(0) \leqslant 0$ and $\Lambda : \mathbb{R} \to \mathbb{R}$ a concave function with $\Lambda(0) \geqslant 0$. Then*

$$\mathbf{E}^{\mathcal{G}}\Lambda(u) \leqslant \Lambda(\mathbf{E}^{\mathcal{G}} u) \qquad \forall u \in L^{\mathcal{G}}(\mathcal{A}), \tag{23.11}$$

and, in particular, $\Lambda(u) \in L^{\mathcal{G}}(\mathcal{A})$. Moreover,

$$V(\mathbf{E}^{\mathcal{G}} u) \leqslant \mathbf{E}^{\mathcal{G}} V(u) \qquad \forall u \in L^{\mathcal{G}}(\mathcal{A}) \text{ s.t. } V(u) \in L^{\mathcal{G}}(\mathcal{A}). \tag{23.12}$$

Proof The argument is very similar to the proof of Jensen's inequality T12.14. Note, however, that we do not have to require the finiteness of the reference measure – which was $w\mu$ in T12.14. Let us, for example, prove (23.12). Using Lemma 12.13 and denoting by $\sup_{\ell}$ the supremum over all linear functions ℓ such that $\ell(x) = ax + b \leqslant V(x)$ for all $x \in \mathbb{R}$, we get using T22.4(v), (x), (ix)

$$V(\mathbf{E}^{\mathcal{G}} u) = \sup_{\ell}(a\,\mathbf{E}^{\mathcal{G}} u + b) \leqslant \sup_{\ell}\mathbf{E}^{\mathcal{G}}(au + b) \leqslant \mathbf{E}^{\mathcal{G}} V(u)$$

since $b \leqslant \mathbf{E}^{\mathcal{G}} b$ where we observed that $b \leqslant V(0) \leqslant 0$.

The inequality (23.11) is proved in the same way. ∎

Because of Theorem 23.9 it is now very easy and convenient to express the martingale property D17.1 in terms of conditional expectations. In fact,

23.14 Corollary *Let $(X, \mathcal{A}, \mathcal{A}_j, \mu)$ be a σ-finite filtered measure space. A sequence $(u_j)_{j\in\mathbb{N}} \subset L^1(\mathcal{A})$ such that $u_j \in L^1(\mathcal{A}_j)$ is a martingale (resp. sub- or supermartingale) if, and only if, for all $j \in \mathbb{N}$*

$$\mathbf{E}^{\mathcal{A}_j} u_{j+1} = u_j \qquad \left(resp. \quad \mathbf{E}^{\mathcal{A}_j} u_{j+1} \geqslant u_j \quad or \quad \mathbf{E}^{\mathcal{A}_j} u_{j+1} \leqslant u_j\right).$$

A great advantage of this way of putting things is that we can now formulate the convergence theorem for uniformly integrable martingales T18.6 in a very striking way:

23.15 Theorem (Closability of martingales) *Let $(X, \mathcal{A}, \mathcal{A}_j, \mu)$ be a σ-finite filtered measure space and $\mathcal{A}_\infty = \sigma\left(\bigcup_{j\in\mathbb{N}} \mathcal{A}_j\right)$.*

(i) *For every $u \in L^1(\mathcal{A})$ the sequence $(\mathbf{E}^{\mathcal{A}_j} u)_{j\in\mathbb{N}}$ is a uniformly integrable martingale. In particular, $\mathbf{E}^{\mathcal{A}_j} u \xrightarrow{j\to\infty} \mathbf{E}^{\mathcal{A}_\infty} u$ in L^1 and a.e.*

(ii) *Conversely, if* $(u_j)_{j \in \mathbb{N}}$ *is a uniformly integrable martingale, there exists a function* $u_\infty \in L^1(\mathcal{A}_\infty)$ *such that* $u_j \xrightarrow{j \to \infty} u_\infty$ *in* L^1 *and a.e. such that* $(u_j)_{j \in \mathbb{N} \cup \{\infty\}}$ *is a martingale. In particular,* $u_j = \mathbf{E}^{\mathcal{A}_j} u_\infty$. *In this sense,* u_∞ *closes the martingale* $(u_j)_{j \in \mathbb{N}}$.

Proof **(i)** That $\left(\mathbf{E}^{\mathcal{A}_j} u\right)_{j \in \mathbb{N}}$ is a martingale follows at once from Theorem 23.8(vi). By assumption there exists an exhausting sequence $(A_k)_{k \in \mathbb{N}} \subset \mathcal{A}_0$ with $A_k \uparrow X$ and $\mu(A_k) < \infty$. Therefore, the function

$$w := \sum_{k=1}^{\infty} \frac{2^{-k}}{1 + \mu(A_k)} \mathbf{1}_{A_k}$$

is strictly positive and integrable. Since $u \in L^1(\mathcal{A})$ and $\{|u| \geqslant N w\} \downarrow \emptyset$ as $N \to \infty$, we find by dominated convergence T11.2 that

$$\lim_{N \to \infty} \int_{\{|u| \geqslant N w\}} |u| \, d\mu = 0.$$

This shows that, for all $\epsilon > 0$, large enough $N = N(\epsilon)$ and any $A \in \mathcal{A}$

$$\int_A |u| \, d\mu = \int_{A \cap \{u < N w\}} |u| \, d\mu + \int_{A \cap \{u \geqslant N w\}} |u| \, d\mu$$

$$\leqslant N \int_A w \, d\mu + \int_{\{|u| \geqslant N w\}} |u| \, d\mu \leqslant N \int_A w \, d\mu + \frac{\epsilon}{2}.$$

We can rephrase this as: for all $\epsilon > 0$ there exists $\eta = \eta(\epsilon) > 0$ such that

$$\int_A w \, d\mu < \eta \implies \int_A |u| \, d\mu < \epsilon \qquad \forall A \in \mathcal{A}. \tag{23.13}$$

Since for all $j \in \mathbb{N}$ and $c > 0$

$$\int_{\left\{|\mathbf{E}^{\mathcal{A}_j} u| > c w\right\}} c w \, d\mu \leqslant \int |\mathbf{E}^{\mathcal{A}_j} u| \, d\mu \leqslant \int \mathbf{E}^{\mathcal{A}_j} |u| \, d\mu = \int |u| \, d\mu,$$

we may choose $c = c_0 := \eta^{-1} \int |u| \, d\mu$, which implies that for a given $\epsilon > 0$

$$\int_{\left\{|\mathbf{E}^{\mathcal{A}_j} u| > c_0 w\right\}} w \, d\mu < \eta \overset{(23.13)}{\implies} \int_{\left\{|\mathbf{E}^{\mathcal{A}_j} u| > c_0 w\right\}} |u| \, d\mu < \epsilon.$$

Since $\{|\mathbf{E}^{\mathcal{A}_j} u| > c_0 w\} \in \mathcal{A}_j$, the martingale property implies

$$\int_{\left\{|\mathbf{E}^{\mathcal{A}_j} u| > c_0 w\right\}} |\mathbf{E}^{\mathcal{A}_j} u| \, d\mu \overset{23.8(\mathrm{xi})}{\leqslant} \int_{\left\{|\mathbf{E}^{\mathcal{A}_j} u| > c_0 w\right\}} \mathbf{E}^{\mathcal{A}_j} |u| \, d\mu$$

$$\overset{23.9}{\leqslant} \int_{\left\{|\mathbf{E}^{\mathcal{A}_j} u| > c_0 w\right\}} |u| \, d\mu \leqslant \epsilon \qquad \forall j \in \mathbb{N},$$

which is but uniform integrability of the family $\left(\mathbf{E}^{\mathcal{A}_j} u\right)_{j \in \mathbb{N}}$.

The convergence assertions follow now from the convergence theorem for UI submartingales T18.6.

(ii) follows directly from Theorem 18.6.   ∎

Since the conditional Jensen inequality needs fewer assumptions than the classical Jensen inequality we can improve Example 17.3(v), (vi).

23.16 Corollary *Let* $(X, \mathcal{A}, \mathcal{A}_j, \mu)$ *be a* σ*-finite filtered measure space and* $(u_j)_{j \in \mathbb{N}}$ *be a family of measurable functions* $u_j \in L^{\mathcal{A}_j}(\mathcal{A})$ *which satisfies the* [*sub-*]*martingale property*[2]

$$u_j = \mathbf{E}^{\mathcal{A}_j} u_{j+1} \qquad \left[resp. \quad u_j \leqslant \mathbf{E}^{\mathcal{A}_j} u_{j+1} \right].$$

If $V \colon \mathbb{R} \to \mathbb{R}$ *is a* [*monotone increasing*] *convex function such that* $V(u_j) \in L^1(\mathcal{A}_j)$, *then* $(V(u_j))_{j \in \mathbb{N}}$ *is a submartingale.*

Proof Since $(u_j)_{j \in \mathbb{N}}$ satisfies the [sub-]martingale property, we find from Jensen's inequality C23.13 [and the monotonicity of V] that

$$V(u_j) \leqslant V\left(\mathbf{E}^{\mathcal{A}_j} u_{j+1}\right) \leqslant \mathbf{E}^{\mathcal{A}_j} V(u_{j+1}).$$ ∎

23.17 Example In Example 17.3(ix) we introduced a *dyadic filtration* on the measure space $\left([0, \infty)^n, \mathcal{B}([0, \infty)^n), \lambda = \lambda^n|_{[0,\infty)^n}\right)$ given by

$$\mathcal{A}_j^\Delta = \sigma\left(z + [0, 2^{-j})^n : z \in 2^{-j}\mathbb{N}_0^n\right), \quad j \in \mathbb{N}_0.$$

For $u \in L^1([0, \infty)^n, \lambda)$ and all $j \in \mathbb{N}_0$ we can now rewrite (17.4) as

$$\mathbf{E}^{\mathcal{A}_j^\Delta} u(x) = \sum_{z \in 2^{-j}\mathbb{N}_0^n} \left\{ \int u \frac{\mathbf{1}_{z+[0,2^{-j})^n}}{\lambda\left(z + [0, 2^{-j})^n\right)} \, d\lambda \right\} \mathbf{1}_{z+[0,2^{-j})^n}(x).$$

23.18 Remark In Theorem 22.5 we found necessary and sufficient conditions that a projection in L^2 is a conditional expectation. This result has a counterpart in the spaces L^p, $p \neq 2$, which we want to mention here without proof. Details can be found in the monograph by Neveu [31, pp. 12–16].
Let $(X, \mathcal{A}, \mu)$ *be a finite measure space. Then*

(i) *Let* $p \in [1, \infty)$. *Every bounded linear operator* $T \colon L^p(\mathcal{A}) \to L^p(\mathcal{A})$ *such that* $\int Tf \, d\mu = \int f \, d\mu$, $f \in L^p(\mathcal{A})$, *and* $T(f \, Tg) = (Tf)(Tg)$, $f \in L^\infty(\mathcal{A})$, $g \in L^1(\mathcal{A})$, *is a conditional expectation w.r.t. some sub-*σ*-algebra* $\mathcal{G} \subset \mathcal{A}$.

(ii) *Let* $p \in [1, \infty)$, $p \neq 2$. *Every linear contraction* $T \colon L^p(\mathcal{A}) \to L^p(\mathcal{A})$ *such that* $T^2 = T$ *and* $T1 = 1$ *is a conditional expectation w.r.t. some sub-*σ*-algebra* $\mathcal{G} \subset \mathcal{A}$.

---

[2] This is slightly more general than assuming that $(u_j)_{j \in \mathbb{N}}$ is a [sub-]martingale since [sub-]martingales are, by definition, integrable.

Separability criteria for the spaces $L^p(X, \mathcal{A}, \mu)$

Let $(X, \mathcal{A}, \mu)$ be a measure space. Recall that $L^p(\mathcal{A})$ is *separable* if it contains a countable dense subset $(d_j)_{j \in \mathbb{N}} \subset L^p(\mathcal{A})$. We have seen in Chapter 21 that the Hilbert space $L^2(\mathcal{A})$ is separable if we can find a countable *complete* ONS $(e_j)_{j \in \mathbb{N}} \subset L^2(\mathcal{A})$ since the system $\{q_1 e_1 + \cdots + q_N e_N : N \in \mathbb{N}, q_j \in \mathbb{Q}\}$ is both countable and dense. Conversely, using any countable dense subset $(d_j)_{j \in \mathbb{N}}$ as input for the Gram–Schmidt orthonormalization procedure (21.10), produces a complete countable ONS.

Here is a simple sufficient criterion for the separability of L^p.

23.19 Lemma *Let* $(X, \mathcal{A}, \mu)$ *be a σ-finite measure space and assume that the σ-algebra $\mathcal{A}$ is countably generated, i.e.* $\mathcal{A} = \sigma(A_j : j \in \mathbb{N})$, $A_j \subset X$. *Then* $L^p(X, \mathcal{A}, \mu)$, $1 \leqslant p < \infty$, *is separable.*

Proof *Step 1:* Let us first assume that μ is a *finite measure*. Consider the σ-algebras $\mathcal{A}_n := \sigma(A_1, \ldots, A_n)$; then

$$\mathcal{A}_1 \subset \mathcal{A}_2 \subset \ldots \subset \mathcal{A}_\infty = \sigma(\mathcal{A}_j : j \in \mathbb{N})$$

is a filtration, $\mu|_{\mathcal{A}_j}$ is trivially σ-finite for every $j \in \mathbb{N}$ and $\mathcal{A} = \mathcal{A}_\infty$.

Set $u_j := \mathbf{E}^{\mathcal{A}_j} u$ for $u \in L^1(\mathcal{A})$. By Theorem 23.15 $(u_j, \mathcal{A}_j)_{j \in \mathbb{N}}$ is a uniformly integrable martingale, hence $u_j \xrightarrow{j \to \infty} u$ in L^1 and a.e.

If $v \in L^p(\mathcal{A})$, we set $v_j := |\mathbf{E}^{\mathcal{A}_j} v|^p \leqslant \mathbf{E}^{\mathcal{A}_j}(|v|^p)$ (by Corollary 23.13) and observe that $(v_j, \mathcal{A}_j)_{j \in \mathbb{N}}$ is a submartingale, cf. Theorem 23.15, which is uniformly integrable. The latter follows easily from

$$\int_{\{v_j > w\}} v_j \, d\mu \leqslant \int_{\{v_j > w\}} \mathbf{E}^{\mathcal{A}_j}(|v|^p) \, d\mu \leqslant \int_{\left\{\mathbf{E}^{\mathcal{A}_j}(|v|^p) > w\right\}} \mathbf{E}^{\mathcal{A}_j}(|v|^p) \, d\mu$$

and the uniform integrability of the family $\left(\mathbf{E}^{\mathcal{A}_j}(|v|^p)\right)_{j \in \mathbb{N}}$, see Theorem 23.15. From the (sub-)martingale convergence Theorem 18.6 we conclude that $v_j \xrightarrow{j \to \infty} |\mathbf{E}^{\mathcal{A}_\infty} v|^p = |v|^p$ in L^1 and a.e., and Riesz' theorem 12.10 shows $|v_j|^{1/p} = |\mathbf{E}^{\mathcal{A}_j} v| \xrightarrow{j \to \infty} |v|$ in L^p. Consequently, $\mathbf{E}^{\mathcal{A}_j} v \xrightarrow{j \to \infty} v$ in L^p.

Since the σ-algebra $\mathcal{A}_j$ is generated by finitely many sets, $\mathbf{E}^{\mathcal{A}_j} u$, resp., $\mathbf{E}^{\mathcal{A}_j} v$ are simple functions with canonical representations of the form

$$s = \sum_{k=1}^N y_k \mathbf{1}_{B_k}, \qquad y_k \neq 0, \; B_1, \ldots, B_N \in \mathcal{A}_j \text{ disjoint};$$

as $\mathcal{A}_j$ is kept fixed, we suppress the dependence of y_k, B_k, N on j. If $y_k \notin \mathbb{Q}$, we find for every $\epsilon > 0$ numbers $y_k^\epsilon \in \mathbb{Q}$ such that

$$\left| y_k - y_k^\epsilon \right| \leqslant \frac{\epsilon}{N \, \mu(X)^{1/p}}\,.$$

The triangle inequality now shows

$$\left\| s - \sum_{k=1}^{N} y_k^\epsilon \mathbf{1}_{B_k} \right\|_p \leqslant \sum_{k=1}^{N} \left| y_k - y_k^\epsilon \right| \mu(B_k)^{1/p} \leqslant \epsilon,$$

which proves that the system

$$D := \left\{ \sum_{k=1}^{N} q_k \, \mathbf{1}_{B_k} \; : \; N \in \mathbb{N}, \; q_k \in \mathbb{Q}, \; B_k \in \bigcup_{j \in \mathbb{N}} \mathcal{A}_j \right\}$$

is a countable dense subset of the space $L^p(X, \mathcal{A}, \mu)$, $1 \leqslant p < \infty$.

Step 2: If μ is σ-finite but not finite, we choose an exhausting sequence $(C_j)_{j \in \mathbb{N}} \subset \mathcal{A}$ such that $C_j \uparrow X$ and $\mu(C_j) < \infty$ and consider the finite measures $\mu_j := \mu(\,\cdot\, \cap C_j)$, $j \in \mathbb{N}$, on $C_j \cap \mathcal{A}$. Since every $u \in L^p(\mu_j) = L^p(C_j, C_j \cap \mathcal{A}, \mu_j)$ can be extended by 0 on the set $X \setminus C_j$ and becomes an element of $L^p(\mu_{j+1})$, we can interpret the sets $L^p(\mu_j)$ as a chain of increasing subspaces of each other and of $L^p(\mu)$:

$$L^p(C_j, C_j \cap \mathcal{A}, \mu_j) \subset L^p(C_{j+1}, C_{j+1} \cap \mathcal{A}, \mu_{j+1}) \subset \ldots \subset L^p(X, \mathcal{A}, \mu).$$

Applying the construction from step 1 to each of the sets $L^p(\mu_j)$ furnishes countable dense subsets D_j. Obviously, $D := \bigcup_{j \in \mathbb{N}} D_j$ is a countable set but it is also dense in $L^p(\mu)$. To see this, fix $\epsilon > 0$ and $u \in L^p(\mu)$. Since $X \setminus C_j \downarrow \emptyset$, we find by Lebesgue's dominated convergence theorem 12.9 some $N \in \mathbb{N}$ such that $\int_{X \setminus C_j} |u|^p \, d\mu < \epsilon^p$ for all $j \geqslant N$. Since D_j is dense in $L^p(\mu_j)$ and since $u \mathbf{1}_{C_j} \in L^p(\mu_j)$, there is some $d_j \in D_j \subset D$ with $\| u \mathbf{1}_{C_j} - d_j \|_{L^p(\mu_j)} \leqslant \epsilon$, and altogether we get for large $j \geqslant N$

$$\| u - d_j \|_p \leqslant \| u \mathbf{1}_{C_j} - d_j \|_{L^p(\mu_j)} + \| u \mathbf{1}_{X \setminus C_j} \|_{L^p(\mu)} \leqslant 2\epsilon. \qquad \blacksquare$$

If the underlying set X is a *separable metric space* (cf. Appendix B), the criterion of Lemma 23.19 becomes particularly simple.

23.20 Corollary *Let (X, ρ) be a separable metric space equipped with its Borel σ-algebra $\mathcal{B} = \mathcal{B}(X)$. Then $L^p(X, \mathcal{B}, \mu)$, $1 \leqslant p < \infty$ is separable for every σ-finite measure μ on $(X, \mathcal{B})$. If μ is not σ-finite, $L^p(X, \mathcal{B}, \mu)$ need not be separable.*

Proof Denote by $D \subset X$ a countable dense subset and consider the countable system of open balls $B_r(d) := \{x \in X : \rho(x, d) < r\}$

$$\mathcal{F} := \big\{B_r(d) : d \in D, \ r \in \mathbb{Q}^+\big\} \subset \mathcal{O}(X).$$

Since every open set $U \in \mathcal{O}(X)$ can be written as

$$U = \bigcup_{\substack{B_r(d) \subset U \\ B_r(d) \in \mathcal{F}}} B_r(d) \ ^3$$

which shows that $\mathcal{O}(X) \subset \sigma(\mathcal{F}) \subset \sigma(\mathcal{O}(X)) = \mathcal{B}(X)$. Thus the Borel sets $\mathcal{B}(X) = \sigma(\mathcal{F})$ are countably generated, and the assertion follows from Lemma 23.19.

If μ is not σ-finite, we have the following counterexample: take $X = [0, 1]$ with its natural Euclidean metric $\rho(x, y) = |x - y|$ and let μ be the counting measure on $([0, 1], \mathcal{B}[0, 1])$, i.e. $\mu(B) := \#B$. Obviously, μ is not σ-finite. The p th power μ-integrable *simple* functions are of the form

$$\mathcal{E}(\mathcal{B}) \cap L^p(\mu) = \left\{ \sum_{j=1}^{N} y_j \mathbf{1}_{A_j} : N \in \mathbb{N}, \ y_j \in \mathbb{R}, \ A_j \in \mathcal{B}, \ \#A_j < \infty \right\},$$

so that

$$L^p(\mu) = \Big\{ u : [0, 1] \to \mathbb{R} : \exists\, (x_j)_{j \in \mathbb{N}} \subset [0, 1], \ u(x) = 0 \quad \forall x \neq x_j$$
$$\text{and} \quad \sum_{j=1}^{\infty} |u(x_j)|^p < \infty \Big\}.$$

Obviously, $(\mathbf{1}_{\{x\}})_{x \in [0,1]} \subset L^p(\mu)$, but no single countable system can approximate this family since

$$\|\mathbf{1}_{\{x\}} - \mathbf{1}_{\{y\}}\|_p^p = 0 \quad \text{or} \quad 2$$

according to whether $x = y$ or $x \neq y$. ∎

With somewhat more effort we can show that the conditions of Lemma 23.19 are even necessary.

23.21 Theorem *Let $(X, \mathcal{A}, \mu)$ be a σ-finite measure space. Then the following assertions are equivalent.*

(i) *$\mathcal{A}$ is (almost) separable, i.e. there exists a countable family $\mathcal{F} \subset \mathcal{A}$ such that $\mu(F) < \infty$ for all $F \in \mathcal{F}$ and $\sigma(\mathcal{F}) \approx \mathcal{A}$ in the sense that every set in $\mathcal{A}$ has, up to a null set, a version in $\sigma(\mathcal{F})$.*

---

[3] The inclusion '$\subset$' is obvious, for '$\supset$' fix $x \in U$. Then there exists some $r \in \mathbb{Q}^+$ with $B_r(x) \subset U$. Since D is dense, $x \in B_{r/2}(d)$ for some $d \in D$ with $\rho(d, x) < r/4$, so that $x \in B_{r/2}(d) \subset U$.

(ii) μ is separable,[4] i.e. there exists a countable family $\mathcal{F} \subset \mathcal{A}$ such that $\mu(F) < \infty$ and for every $A \in \mathcal{A}$ with $\mu(A) < \infty$ we have

$$\forall \epsilon > 0, \; \exists F_\epsilon \in \mathcal{F} \; : \; \mu(A \setminus F_\epsilon) + \mu(F_\epsilon \setminus A) \leqslant \epsilon.$$

(iii) $L^p(X, \mathcal{A}, \mu)$ is separable, $1 \leqslant p < \infty$.

Proof (i)$\Rightarrow$(iii): The proof of Lemma 23.19 shows that $L^p(X, \sigma(\mathcal{F}), \mu)$ is separable. Since for each $A \in \mathcal{A}$ there is an $A^* \in \sigma(\mathcal{F})$ with

$$\mu\big((A \setminus A^*) \cup (A^* \setminus A)\big) = 0 \iff \int |\mathbf{1}_A - \mathbf{1}_{A^*}| \, d\mu = 0,$$

every simple function $\phi \in \mathcal{E}(\mathcal{A})$ has a version $\phi^* \in \mathcal{E}(\sigma(\mathcal{F}))$ such that $\int |\phi - \phi^*| \, d\mu = 0$. This proves that $L^p(X, \sigma(\mathcal{F}), \mu) \supset L^p(X, \mathcal{A}, \mu)$ (we have, in fact, equality since $\sigma(\mathcal{F}) \subset \mathcal{A}$), and we see that $L^p(X, \mathcal{A}, \mu)$ is separable.

(iii)$\Rightarrow$(ii): Denote by $(d_j)_{j \in \mathbb{N}}$ a countable dense subset of $L^p(\mathcal{A})$. Since $\mathcal{E}(\mathcal{A}) \cap L^p(\mathcal{A})$ is dense in $L^p(\mathcal{A})$, cf. Lemma 12.11, we find for each d_j a sequence $(f_{jk})_{k \in \mathbb{N}} \subset \mathcal{E}(\mathcal{A}) \cap L^p(\mathcal{A})$ such that $L^p\text{-}\lim_{k \to \infty} f_{jk} = d_j$. Thus $(f_{jk})_{j,k \in \mathbb{N}}$ is also dense in L^p, and the system of subsets

$$\mathcal{F} := \left\{ \bigcup_{\ell=1}^{N} \{f_{jk} = r_\ell\} : N \in \mathbb{N}, \; j, k \in \mathbb{N}, \; r_\ell \in \mathbb{R} \right\}$$

is countable since each f_{jk} attains only finitely many values.

For every $A \in \mathcal{A}$, $\mu(A) < \infty$, we have $\mathbf{1}_A \in L^p(\mathcal{A})$, and we find a subsequence $(f_\ell^A)_{\ell \in \mathbb{N}} \subset (f_{jk})_{j,k \in \mathbb{N}}$ with $\lim_{\ell \to \infty} \|f_\ell^A - \mathbf{1}_A\|_p^p = 0$.

Set $F_\ell := \{|f_\ell^A - \mathbf{1}_A| \leqslant 1/2\} \cap \{|f_\ell^A| > 1/2\}$. Obviously $F_\ell \in \mathcal{F}$, and $F_\ell \subset A$ since

$$A^c \cap F_\ell = A^c \cap \{|f_\ell^A - \mathbf{1}_A| \leqslant 1/2\} \cap \{|f_\ell^A| > 1/2\}$$

$$= A^c \cap \{|f_\ell^A| \leqslant 1/2\} \cap \{|f_\ell^A| > 1/2\}$$

$$= \emptyset.$$

Thus $\mu(F_\ell \setminus A) = 0$, while

$$\mu(A \setminus F_\ell) \leqslant \mu(A \cap \{|f_\ell^A - \mathbf{1}_A| > 1/2\}) + \mu(A \cap \{|f_\ell^A| \leqslant 1/2\}).$$

Using the triangle inequality, we infer

$$A \cap \{|f_\ell^A| \leqslant 1/2\} \subset A \cap \{|f_\ell^A - 1| \geqslant 1/2\} = A \cap \{|f_\ell^A - \mathbf{1}_A| \geqslant 1/2\},$$

---

[4] This notion derives from the fact that $(\mathcal{A}, \rho_\mu)$, $\rho_\mu(A, B) := \mu(A \setminus B) + \mu(B \setminus A)$, $A, B \in \mathcal{A}$ becomes a separable pseudo-metric space in the usual sense, cf. Appendix B.

and with the above calculation and an application of Markov's inequality we conclude that

$$\mu(A \setminus F_\ell) + \mu(F_\ell \setminus A) \;\leqslant\; 2\,\mu(A \cap \{|f_\ell^A - \mathbf{1}_A| \geqslant 1/2\})$$

$$\overset{10.12}{\leqslant}\; 2^{p+1}\, \|f_\ell^A - \mathbf{1}_A\|_p^p.$$

The right-hand side of the above inequality tends to 0 as $\ell \to \infty$, and (ii) follows.

(ii)$\Rightarrow$(i): Fix $A \in \mathcal{A}$ with $\mu(A) < \infty$. Then we find, by assumption, sets $F_n \in \mathcal{F}$ with $\mu(A \setminus F_n) + \mu(F_n \setminus A) \leqslant 2^{-n}$. Consider the sets

$$F^* := \bigcap_{k=1}^{\infty} \bigcup_{n=k}^{\infty} F_n \qquad \text{and} \qquad F_* := \bigcup_{k=1}^{\infty} \bigcap_{n=k}^{\infty} F_n.$$

Then using the continuity of measures T4.4 and σ-subadditivity C4.6,

$$
\mu(F^* \setminus A) + \mu(A \setminus F_*) = \mu\left(\left[\bigcap_{k=1}^{\infty} \bigcup_{n=k}^{\infty} F_n \right] \setminus A \right) + \mu\left(A \cap \left[\bigcup_{k=1}^{\infty} \bigcap_{n=k}^{\infty} F_n \right]^c \right)
$$

$$
= \mu\left(\bigcap_{k=1}^{\infty} \bigcup_{n=k}^{\infty} (F_n \setminus A) \right) + \mu\left(A \cap \bigcap_{k=1}^{\infty} \bigcup_{n=k}^{\infty} F_n^c \right)
$$

$$
= \lim_{k \to \infty} \left[\mu\left(\bigcup_{n=k}^{\infty} (F_n \setminus A) \right) + \mu\left(\bigcup_{n=k}^{\infty} (A \setminus F_n) \right) \right]
$$

$$
\leqslant \lim_{k \to \infty} \sum_{n=k}^{\infty} \left[\mu(F_n \setminus A) + \mu(A \setminus F_n) \right]
$$

$$
\leqslant \lim_{k \to \infty} \sum_{n=k}^{\infty} 2^{-k} = 0.
$$

This shows that for all $A \in \mathcal{A}$ with $\mu(A) < \infty$

$$\exists F_*, F^* \in \sigma(\mathcal{F}) : F_* \subset F^* \quad \text{and} \quad \mu(F^* \setminus A) + \mu(A \setminus F_*) = 0,$$

implying that $\mu(F^* \setminus A) + \mu(A \setminus F^*) = 0$ also.

If $\mu(A) = \infty$ we pick some exhausting sequence $(A_k)_{k \in \mathbb{N}} \subset \mathcal{A}$ with $A_k \uparrow X$ and $\mu(A_k) < \infty$. Then the sets $A \cap A_k$ have finite μ-measure and we can construct, as before, sets F_k^* and $F_{*,k}$. Setting $F^* := \bigcup_{k \in \mathbb{N}} F_k^*$ we find

$$\left(\bigcup_{k \in \mathbb{N}} F_k^* \right) \setminus \bigcup_{j \in \mathbb{N}} (A \cap A_j) = \bigcup_{k \in \mathbb{N}} \left(F_k^* \setminus \bigcup_{j \in \mathbb{N}} (A \cap A_j) \right) \subset \bigcup_{k \in \mathbb{N}} (F_k^* \setminus (A \cap A_k)),$$

and so

$$\mu(F^* \setminus A) \leqslant \mu\left(\bigcup_{k\in\mathbb{N}} (F_k^* \setminus (A \cap A_k)) \right) \leqslant \sum_{k=1}^{\infty} \underbrace{\mu(F_k^* \setminus (A \cap A_k))}_{=0} = 0.$$

The expression $\mu(A \setminus F^*)$ is handled analogously.

This shows that sets from $\mathcal{A}$ and $\sigma(\mathcal{F})$ differ by at most a null set. ∎

Problems

23.1. Complete the proof of Theorem 23.8.

23.2. Show that $E^{\mathcal{G}}1 = 1$ if, and only if $\mu|_{\mathcal{G}}$ is σ-finite. Find a counterexample showing that $E^{\mathcal{G}}1 \leqslant 1$ is, in general, best possible.
[Hint: use $p = 2$ and $E^{\mathcal{G}} = \mathbf{E}^{\mathcal{G}}$.]

23.3. Let $\mathcal{G}$ be a sub-σ-algebra of $\mathcal{A}$. Show that $\mathbf{E}^{\mathcal{G}}g = g$ for all $g \in L^p(\mathcal{G})$.
[Hint: observe that, a.e., $g = g\,\mathbf{1}_{\bigcup_j\{|g|>1/j\}}$ and $\mu(\{|g| > 1/j\}) < \infty$. This emulates σ-finiteness.]

23.4. Let $\mathcal{H} \subset \mathcal{G}$ be two sub-σ-algebras of $\mathcal{A}$. Show that

$$\mathbf{E}^{\mathcal{G}}\mathbf{E}^{\mathcal{H}}u = \mathbf{E}^{\mathcal{H}}\mathbf{E}^{\mathcal{G}}u = \mathbf{E}^{\mathcal{H}}u$$

for all $u \in L^p(\mathcal{A})$ resp. for all $u \in M(\mathcal{A})$ provided $\mu|_{\mathcal{H}}$ is σ-finite.
[Hint: if $\mu|_{\mathcal{H}}$ is not σ-finite, the set $L^p(\mathcal{H})$ can be *very* small ....]

23.5. Consider on the measure space $([0,\infty), \mathcal{B}[0,\infty), \lambda := \lambda^1|_{[0,\infty)})$ the filtration $\mathcal{A}_n := \sigma([0,1), [0,2), \ldots, [n-1,n), [n,\infty))$. Find $E^{\mathcal{A}_n}u$ for $u \in L^p$.

23.6. Let $(X, \mathcal{A}, \mu)$ be a measure space and $\mathcal{G} \subset \mathcal{A}$ be a sub-σ-algebra. Show that, in general,

$$\int E^{\mathcal{G}}u\,d\mu \leqslant \int u\,d\mu, \qquad u \in L^1(\mathcal{A}),$$

with equality holding only if $\mu|_{\mathcal{G}}$ is σ-finite.

23.7. Prove Corollaries 23.11 and 23.12.

23.8. Let $(X, \mathcal{A}, \mathcal{A}_j, \mu)$ be a σ-finite filtered measure space and denote by $\langle u, \phi \rangle$ the canonical dual pairing between $u \in L^p$ and $\phi \in L^q$, $p^{-1} + q^{-1} = 1$, i.e. $\langle u, \phi \rangle := \int u\phi\,d\mu$. A sequence $(u_j)_{j\in\mathbb{N}} \subset L^p$ is *weakly relatively compact* if there exists a subsequence $(u_{j_k})_{k\in\mathbb{N}}$ such that

$$\langle u_{j_k} - u, \phi \rangle \xrightarrow{k\to\infty} 0$$

holds for all $\phi \in L^q$ and some $u \in L^p$. Show that for a martingale $(u_j)_{j\in\mathbb{N}}$ and every $p \in (1,\infty)$ the following assertions are equivalent:

   (i)   there exists some $u \in L^p(\mathcal{A})$ such that $\lim_{j\to\infty} \|u_j - u\|_p = 0$;
   (ii) there exists some $u_\infty \in L^p(\mathcal{A}_\infty)$ such that $u_j = E^{\mathcal{A}_j}u_\infty$;
   (iii) the sequence $(u_j)_{j\in\mathbb{N}}$ is weakly relatively compact.

23.9. Let $(X, \mathcal{A}, \mu)$ be a measure space and $(u_j)_{j\in\mathbb{N}} \subset L^1(\mathcal{A})$. Show that

$$m_1 := u_1, \qquad m_{j+1} - m_j := u_{j+1} - E^{\mathcal{A}_j} u_{j+1}$$

is a martingale under the filtration $\mathcal{A}_j := \sigma(u_1, \ldots, u_j)$.

23.10. (Continuation of Problem 23.9). If

$$\int u_1 \, d\mu = 0 \qquad \text{and} \qquad E^{\mathcal{A}_j} u_{j+1} = 0,$$

then $(u_j)_{j\in\mathbb{N}}$ is called a *martingale difference sequence*. Assume that $u_j \in L^2(\mathcal{A})$ and denote by $s_k := u_1 + \cdots + u_k$ the partial sums. Show that $(s_j^2, \mathcal{A}_j)_{j\in\mathbb{N}}$ is a submartingale satisfying

$$\int s_k^2 \, d\mu = \sum_{j=1}^{k} \int u_j^2 \, d\mu.$$

23.11. **Doob decomposition.** Let $(X, \mathcal{A}, \mathcal{A}_j, \mu)$ be a σ-finite filtered measure space and let $(s_j, \mathcal{A}_j)_{j\in\mathbb{N}}$ be a submartingale. Show that there exists an a.e. unique martingale $(m_j, \mathcal{A}_j)_{j\in\mathbb{N}}$ and an increasing sequence of functions $(a_j)_{j\in\mathbb{N}}$ such that $a_j \in L^1(\mathcal{A}_{j-1})$ for all $j \geqslant 2$ and

$$s_j = m_j + a_j, \qquad j \in \mathbb{N}.$$

[Hint: set $m_0 := u_0$, $m_{j+1} - m_j := u_{j+1} - E^{\mathcal{A}_j} u_{j+1}$ and $a_0 := 0$, $a_{j+1} - a_j := E^{\mathcal{A}_j} u_{j+1} - u_j$. For uniqueness assume $\tilde{m}_j + \tilde{a}_j$ is a further Doob decomposition and study the measurability properties of the martingale $M_j := m_j - \tilde{m}_j = \tilde{a}_j - a_j$.]

23.12. Let $(\Omega, \mathcal{A}, P)$ be a probability space and let $(X_j)_{j\in\mathbb{N}}$ be a sequence of independent identically distributed random variables such that $P(X_j = 0) = P(X_j = 2) = \frac{1}{2}$. Set $M_k := \prod_{j=1}^{k} X_j$. Show that there does not exist any filtration $(\mathcal{A}_j)_{j\in\mathbb{N}}$ and no random variable M such that $M_k = E^{\mathcal{A}_k} M$.

[Hint: compare with Example 17.3(xi).]

Remark. This example shows that not all martingales can be obtained as conditional expectations of a single function.

24

Orthonormal systems and their convergence behaviour

In Chapter 21 we discussed the importance of orthonormal systems (ONSs) in Hilbert spaces. In particular, countable complete ONSs turned out to be bases of separable Hilbert spaces. We have also seen that a countable ONS gives rise to a family of finite-dimensional subspaces and a sequence of orthogonal projections onto these spaces. In the present chapter we are concerned with the following topics:

- to give concrete examples of (complete) ONSs;
- to see when the associated canonical projections are conditional expectations;
- to understand the L^p $(p \neq 2)$ and a.e. convergence behaviour of series expansions with respect to certain ONSs.

The latter is, in general, not a trivial matter. Here we will see how we can use the powerful martingale machinery of Chapters 17 and 18 to get L^p $(1 \leqslant p < \infty)$ and a.e. convergence.

Throughout this chapter we will consider the Hilbert space $L^2(I, \mathcal{B}(I), \rho \lambda)$ where $I \subset \mathbb{R}$ is a finite or infinite interval of the real line, $\mathcal{B}(I) = I \cap \mathcal{B}(\mathbb{R})$ are the Borel sets in I, $\lambda = \lambda^1|_I$ is Lebesgue measure on I and $\rho(x)$ is a density function. We will usually write $\rho(x)\,dx$ and $\int \ldots dx$ instead of $\rho \lambda$ and $\int \ldots d\lambda$.

One of the most important techniques to construct ONSs is the Gram–Schmidt orthonormalization procedure (21.10), which we can use to turn any countable family $(f_k)_{k \in \mathbb{N}}$ into an orthonormal sequence $(e_k)_{k \in \mathbb{N}}$. Something of a problem, however, is to find a reasonable sequence $(f_k)_{k \in \mathbb{N}}$ which can be used as input to the orthonormalization procedure.

Orthogonal polynomials

For many practical applications, such as interpolation, approximation or numerical integration, a natural set of f_k to begin with is given by the polynomials on I.

Usually one applies (21.10) to the sequence of *monomials*

$$(1, t, t^2, t^3, \ldots) = (t^j)_{j \in \mathbb{N}_0}$$

to construct an ONS consisting of polynomials. Of course, this depends heavily on the underlying measure space where polynomials should be square integrable. With some (partly pretty tedious) calculations[1] one can get the following important classes of orthogonal polynomials in $L^2(I, \mathcal{B}(I), \rho(x)\,dx)$.

24.1 Jacobi polynomials $\left(J_k^{(\alpha,\beta)}\right)_{k \in \mathbb{N}_0}$, $\alpha, \beta > -1$ We choose

$$I = [-1, 1], \qquad \rho(x)\,dx = (1-x)^\alpha (1+x)^\beta\,dx, \qquad \alpha, \beta > -1,$$

and we get

$$J_k^{(\alpha,\beta)}(x) = \frac{(-1)^k}{k!\,2^k} \frac{\dfrac{d^k}{dx^k}\left((1-x)^{\alpha+k}(1+x)^{\beta+k}\right)}{(1-x)^\alpha(1+x)^\beta}$$

$$= \frac{1}{2^k} \sum_{j=0}^{k} \binom{k+\alpha}{j}\binom{k+\beta}{k-j}(x-1)^{k-j}(x+1)^j$$

$$\left\|J_k^{(\alpha,\beta)}\right\|_2^2 = \frac{2^{\alpha+\beta+1}}{2k+\alpha+\beta+1} \frac{\Gamma(k+\alpha+1)\,\Gamma(k+\beta+1)}{\Gamma(k+1)\,\Gamma(k+\alpha+\beta+1)}.$$

Choosing in 24.1 particular values for α and β yields other important families.

24.2 Chebyshev polynomials (of the first kind) $(T_k)_{k \in \mathbb{N}_0}$ We choose

$$I = [-1, 1], \qquad \rho(x)\,dx = (1-x^2)^{-1/2}\,dx,$$

and we get

$$T_k(x) = J_k^{(-1/2,-1/2)}(x) = \begin{cases} \sqrt{\dfrac{2}{\pi}}\cos(k\arccos x) & \text{if } k \in \mathbb{N}, \\ \dfrac{1}{\sqrt{\pi}} & \text{if } k = 0, \end{cases}$$

$$\|T_k\|_2^2 = \frac{1}{2}\left(\frac{\Gamma\left(k+\frac{1}{2}\right)}{\Gamma(k+1)}\right)^2.$$

The first few Chebyshev polynomials are

$$1, \quad x, \quad 2x^2 - 1, \quad 4x^3 - 3x, \quad 8x^4 - 8x^2 + 1, \quad 16x^5 - 20x^3 + 5x, \ldots$$

---

[1] The material in Sections 24.1–24.5 below is taken from Alexits [1, pp. 30–37], Gradshteyn-Ryzhik [17, §8.9] and Kaczmarz-Steinhaus [22, §§IV.1–2, 8–9]. Another classic is the book by Szegö [51, §§1–5], and a good *modern* reference is the monograph by Andrews *et al.* [2, §§5.1, 6.1–6.3].

and the following recursion formula holds:

$$T_{k+1}(x) = 2x\, T_k(x) - T_{k-1}(x), \qquad k \in \mathbb{N}.$$

24.3 Legendre polynomials $(P_k)_{k \in \mathbb{N}_0}$ We choose

$$I = [-1, 1], \qquad \rho(x)\, dx = dx,$$

and we get

$$P_k(x) = J_k^{(0,0)}(x) = \frac{(-1)^k}{k!\, 2^k} \frac{d^k}{dx^k}\big((1-x^2)^k\big), \qquad \|P_k\|_2^2 = \frac{2}{2k+1}.$$

The first few Legendre polynomials are

$$1, \quad x, \quad \tfrac{1}{2}(3x^2-1), \quad \tfrac{1}{2}(5x^3-3x), \quad \tfrac{1}{8}(35x^4-30x^2+3), \quad \tfrac{1}{8}(63x^5-70x^3+15x),$$

and the following recursion formula holds:

$$(k+1)\, P_{k+1}(x) = (2k+1)\, x\, P_k(x) - k\, P_{k-1}(x), \qquad k \in \mathbb{N}.$$

24.4 Laguerre polynomials $(L_k)_{k \in \mathbb{N}_0}$ We choose

$$I = [0, \infty), \qquad \rho(x)\, dx = e^{-x}\, dx,$$

and we get

$$L_k(x) = e^x \frac{d^k}{dx^k}\big(e^{-x} x^k\big) = k! \sum_{j=0}^{k} (-1)^j \binom{k}{j} \frac{x^j}{j!}, \qquad \|L_k\|_2^2 = (k!)^2.$$

The first few Laguerre Polynomials are

$$1, \quad 1-x, \quad x^2-4x+2, \quad -x^3+9x^2-18x+6, \quad x^4-16x^3+72x^2-96x+24, \ldots$$

and the following recursion formula holds:

$$L_{k+1}(x) = (2k+1-x)\, L_k(x) - k^2\, L_{k-1}(x), \qquad k \in \mathbb{N}.$$

24.5 Hermite polynomials $(H_k)_{k \in \mathbb{N}_0}$ We choose

$$I = (-\infty, \infty), \qquad \rho(x)\, dx = e^{-x^2}\, dx,$$

and we get

$$H_k(x) = (-1)^k e^{x^2} \frac{d^k}{dx^k}\big(e^{-x^2}\big), \qquad \|H_k\|_2^2 = 2^k\, k!\, \sqrt{\pi}.$$

The first few Hermite polynomials are

$$1, \quad 2x, \quad 4x^2 - 2, \quad 8x^3 - 12x, \quad 16x^4 - 48x^2 + 12, \dots$$

and the following recursion formula holds:

$$H_{k+1}(x) = 2x\, H_k(x) - 2k\, H_{k-1}(x), \qquad k \in \mathbb{N}.$$

In order to decide if a family of polynomials $(p_k)_{k\in\mathbb{N}} \subset L^2(I, \rho(x)\, dx)$ is a complete ONS we have to show that

$$\int u(x)\, p_k(x)\, \rho(x)\, dx = 0 \quad \forall k \in \mathbb{N}_0 \quad \implies \quad u = 0 \text{ a.e.}$$

The key technical result is the *Weierstraß approximation theorem.*

24.6 Theorem (Weierstraß) *Polynomials are dense in $C[0, 1]$ w.r.t. uniform convergence.*

Proof (S.N. Bernstein) Take a sequence $(X_j)_{j\in\mathbb{N}}$ of independent[2] measurable functions on $([0, 1], \mathcal{B}[0, 1], dx)$ which are all Bernoulli $(p, 1-p)$-distributed, $0 < p < 1$, i.e.

$$\lambda(\{X_j = 1\}) = p \qquad \text{and} \qquad \lambda(\{X_j = 0\}) = 1 - p \qquad \forall j \in \mathbb{N},$$

cf. 17.4 for the construction of such a sequence. Write $S_n := X_1 + \cdots + X_n$ for the partial sum and observe that, due to independence,

$$\lambda(\{S_n = k\}) = \lambda\left(\bigcup_{1 \leqslant j_1 \leqslant \dots \leqslant j_k \leqslant n} \left(\{X_{j_1} = 1\} \cap \dots \cap \{X_{j_k} = 1\} \cap \right.\right.$$

$$\left.\left. \cap \{X_{j_{k+1}} = 0\} \cap \dots \cap \{X_{j_n} = 0\} \right) \right)$$

$$= \binom{n}{k} p^k (1 - p)^{n-k},$$

which shows that

$$\int u\left(\tfrac{S_n(x)}{n}\right) dx = \sum_{k=0}^{n} u\left(\tfrac{k}{n}\right) \binom{n}{k} p^k (1 - p)^{n-k} =: B_n(u; p),$$

---

[2] In the sense of Example 17.3(x).

where $B_n(u; p)$ stands for the nth *Bernstein polynomial*.[3] From 17.4 we also know that

$$\int \left| \frac{S_n(x)}{n} - p \right|^2 dx = \frac{p(1-p)}{n} \leq \frac{1}{4n}, \tag{24.1}$$

since the function $p \mapsto p(1-p)$ attains its maximum at $p = 1/2$. As $u \in C[0,1]$ is uniformly continuous, $|u(x) - u(y)| < \epsilon$ whenever $|x - y| < \delta$ is sufficiently small. Thus

$$|B_n(u; p) - u(p)|$$

$$\leq \int \left| u\left(\tfrac{S_n}{n}\right) - u(p) \right| d\lambda$$

$$= \int_{\left\{ \left| \frac{S_n}{n} - p \right| < \delta \right\}} \left| u\left(\tfrac{S_n}{n}\right) - u(p) \right| d\lambda + \int_{\left\{ \left| \frac{S_n}{n} - p \right| \geq \delta \right\}} \left| u\left(\tfrac{S_n}{n}\right) - u(p) \right| d\lambda$$

$$\leq \epsilon \lambda \left(\left\{ \left| \tfrac{S_n}{n} - p \right| < \delta \right\} \right) + 2 \|u\|_\infty \lambda \left(\left\{ \left| \tfrac{S_n}{n} - p \right| \geq \delta \right\} \right)$$

$$\leq \epsilon + 2 \|u\|_\infty \frac{1}{\delta^2} \int \left| \tfrac{S_n}{n} - p \right|^2 d\lambda$$

$$\overset{(24.1)}{\leq} \epsilon + \frac{\|u\|_\infty}{2 n \delta^2},$$

by Markov's inequality P10.12 (in the penultimate step) and (24.1). The above inequality is independent of $p \in [0,1]$, and the assertion follows by letting first $n \to \infty$ and then $\epsilon \to 0$. ∎

24.7 Remark The key ingredient in the above proof is (24.1) which shows that the variance of the random variable S_n vanishes uniformly (in p) as $n \to \infty$. A short calculation confirms that this is equivalent to saying that

$$B_n\big((\bullet - p)^2; p\big) \xrightarrow{n \to \infty} 0 \quad \text{uniformly for } p \in [0,1].$$

With this information, the proof then yields that $B_n(u; p) \xrightarrow{n \to \infty} u$ uniformly in p for all continuous u. This is, in fact, a special case of

Korovkin's theorem: *A sequence of positive linear operators from $C[0,1]$ to $C[0,1]$ converges uniformly for every $u \in C[0,1]$ if, and only if, it converges uniformly for each of the following three test functions:* $1, x, x^2$.

---

[3] In view of the strong law of large numbers, Example 18.8, we observe that $n^{-1} S_n \xrightarrow{n \to \infty} p$ a.e., so that by dominated convergence $B_n(u, p) \xrightarrow{n \to \infty} u(p)$ for each $p \in (0,1)$. Since our argument includes this result as a particular case, we leave it as a side-remark.

(In the present case, the operators are $u \mapsto B_n(u, p)$.) More on this topic can be found in Korovkin's monograph [24, pp. 1–30] or the expository paper [4] by Bauer.

24.8 Corollary *The monomials $(t^j)_{j \in \mathbb{N}_0}$ are complete in $L^1 = L^1([0, 1], dt)$, that is $\int_{[0,1]} u(t) t^j \, dt = 0$ for all $j \in \mathbb{N}_0$ implies that $u = 0$ a.e.*

Proof Assume first that $u \in C[0, 1]$ satisfies $\int_{[0,1]} u(t) t^j \, dt = 0$ for all $j \in \mathbb{N}_0$. This implies, in particular, that

$$\int_{[0,1]} u(t) p(t) \, dt = 0 \quad \text{for all polynomials } p(t).$$

Using Weierstraß' approximation theorem 24.6 we find a sequence of polynomials $(p_k)_{k \in \mathbb{N}}$ which approximate u uniformly on $[0, 1]$. Since $C[0, 1] \subset L^1[0, 1] \subset L^2[0, 1]$, we see

$$\int_{[0,1]} u^2 \, dt = \int_{[0,1]} u \cdot (u - p_k) \, dt \leqslant \|u\|_2 \cdot \|u - p_k\|_2$$

$$\leqslant \|u\|_2 \cdot \|u - p_k\|_\infty \xrightarrow{k \to \infty} 0,$$

and conclude that $u = 0$ a.e. (even everywhere since u is continuous).

Assume now that $u \in L^1([0, 1], dt) \setminus C[0, 1]$ such that $\int_{[0,1]} u(t) t^j \, dt = 0$ for all $j \in \mathbb{N}_0$. The primitive

$$U(x) := \int_{(x,1]} u(t) \, dt$$

is a continuous function, cf. Problem 11.7, and by Fubini's theorem 13.9 we see for all $j \in \mathbb{N}$

$$\int_{[0,1]} U(x) x^{j-1} \, dx = \int_{[0,1]} \int_{[0,1]} \mathbf{1}_{(x,1]}(t) \, u(t) \, x^{j-1} \, dt \, dx$$

$$= \int_{[0,1]} \left(\int_{[0,1]} \mathbf{1}_{[0,t)}(x) \, x^{j-1} \, dx \right) u(t) \, dt$$

$$= \int_{[0,1]} \frac{t^j}{j!} u(t) \, dt = 0.$$

This means that $\int_{[0,1]} U(x) x^k \, dx = 0$ for all $k \in \mathbb{N}_0$ and, by the first part of the proof, that $U \equiv 0$. Lebesgue's differentiation theorem 19.20 finally shows that $u(x) = U'(x) = 0$ a.e. ∎

It is not hard to see that Theorem 24.6 and Corollary 24.8 also hold for the interval $[-1, 1]$ and even for general compact intervals $[a, b]$ (cf. Problem 24.3). This we can use to show that the Jacobi (hence, Legendre and Chebyshev) polynomials are dense in $L^2(I, \rho(x)\,dx)$ and form a complete ONS. Note that

$$\int_{[-1,1]} |u\rho|\,dx \leqslant \left(\int_{[-1,1]} \left(u\sqrt{\rho}\right)^2 dx\right)^{1/2} \cdot \left(\int_{[-1,1]} \left(\sqrt{\rho}\right)^2 dx\right)^{1/2}$$

$$= \left(\int_{[-1,1]} u^2 \rho\,dx\right)^{1/2} \cdot \left(\int_{[-1,1]} \rho\,dx\right)^{1/2} < \infty$$

implies that $u\rho \in L^1([-1, 1], dx)$, and from Corollary 24.8 and the fact that $\rho > 0$ we get

$$\int_{[-1,1]} u(x)\rho(x)\, x^j\, dx = 0 \quad \Longrightarrow \quad u\rho = 0 \text{ a.e.} \quad \Longrightarrow \quad u = 0 \text{ a.e.}$$

This does not quite work for the Hermite and Laguerre polynomials, which are defined on infinite intervals. For the latter we take $u \in L^2([0, \infty), e^{-x}\,dx)$, and find for all $s \geqslant 1$

$$\int_{[0,\infty)} u(x)\, e^{-sx}\, dx = \int_{[0,\infty)} u(x)\, e^{(1-s)x}\, e^{-x}\, dx$$

$$= \sum_{k=0}^{\infty} \frac{(1-s)^k}{k!} \underbrace{\int_{[0,\infty)} u(x)\, x^k\, e^{-x}\, dx}_{=0} = 0$$

(note that the integral and the sum can be interchanged by dominated convergence). Using Jacobi's formula C15.8 to change coordinates according to $t = e^{-x}$, $dt/dx = -e^{-x}$, we get

$$0 = \int_{[0,\infty)} u(x)\, e^{-sx}\, dx = \int_{[0,1)} u(-\ln t)\, t^{s-1}\, dt, \qquad s \geqslant 1,$$

and for $s \in \mathbb{N}$ the above equality reduces to the case covered by Corollary 24.8. A very similar calculation can be used for the Hermite polynomials since

$$\int_{\mathbb{R}} u(x)\, e^{-sx^2}\, dx = \int_{[0,\infty)} (u(x) + u(-x))\, e^{-sx^2}\, dx$$

$$= \int_{[0,\infty)} \left(u(\sqrt{t}) + u(-\sqrt{t})\right) e^{-st} \frac{dt}{2\sqrt{t}},$$

where we used the obvious substitution $x = \sqrt{t}$.

The trigonometric system and Fourier series

We consider now $L^2 = L^2((-\pi, \pi), \mathcal{B}(-\pi, \pi), \lambda = \lambda^1|_{(-\pi,\pi)})$. As before we use dx as a shorthand for $\lambda(dx)$. The *trigonometric system* consists of the functions

$$\frac{1}{\sqrt{2\pi}}, \quad \frac{\cos x}{\sqrt{\pi}}, \quad \frac{\sin x}{\sqrt{\pi}}, \quad \frac{\cos 2x}{\sqrt{\pi}}, \quad \frac{\sin 2x}{\sqrt{\pi}}, \quad \dots, \quad \frac{\cos kx}{\sqrt{\pi}}, \quad \frac{\sin kx}{\sqrt{\pi}}, \dots \quad (24.2)$$

or, equivalently,

$$\frac{1}{\sqrt{2\pi}} e^{ikx}, \qquad k \in \mathbb{Z}, \ i = \sqrt{-1}. \quad (24.3)$$

Since $e^{ix} = \cos x + i \sin x$, we can see that (24.2) and (24.3) are equivalent, and from now on we will only consider (24.2). Orthogonality of the functions in (24.2) follows easily from the classical result that

$$\int_{(-\pi,\pi)} \cos kx \sin \ell x \, dx = \begin{cases} 0, & \text{if } k \neq \ell, \\ \pi, & \text{if } k = \ell \geqslant 1, \\ 2\pi, & \text{if } k = \ell = 0, \end{cases} \quad (24.4)$$

which we leave as an exercise for the reader, see Problem 24.4.

24.9 Definition A *trigonometric polynomial (of order n)* is an expression of the form

$$T(x) = \alpha_0 + \sum_{j=1}^{n} (\alpha_j \cos jx + \beta_j \sin jx), \quad (24.5)$$

where $n \in \mathbb{N}_0$, $\alpha_j, \beta_j \in \mathbb{R}$ and $\alpha_n^2 + \beta_n^2 > 0$.

It is not hard to see that the representation (24.5) of $T(x)$ is equivalent to

$$T(x) = \sum_{j,k=0}^{n} \gamma_{j,k} \cos^j x \sin^k x$$

with coefficients $\gamma_{j,k} \in \mathbb{R}$, cf. Problem 24.5. It is this way of writing $T(x)$ that justifies the name trigonometric *polynomial*.

24.10 Theorem *The trigonometric system (24.2) is a complete ONS in $L^2 = L^2((-\pi, \pi), dx)$.*

Proof We have to show that

$$\left. \begin{array}{l} \int_{(-\pi,\pi)} u(x) \cos kx \, dx = 0 \quad \forall k \in \mathbb{N}_0 \\[2mm] \int_{(-\pi,\pi)} u(x) \sin \ell x \, dx = 0 \quad \forall \ell \in \mathbb{N} \end{array} \right\} \quad \Longrightarrow \quad u = 0 \quad \text{a.e.} \quad (24.6)$$

Assume first that u is continuous and that, contrary to (24.6), $u(x_0) = c \neq 0$ for some $x_0 \in (-\pi, \pi)$. Without loss of generality we may assume that $c > 0$. Since the trigonometric functions are 2π-periodic, we can extend u periodically onto the whole real line. Then $w(x) := c^{-1} u(x + x_0)$ is continuous around $x = 0$, orthogonal on $(-\pi, \pi)$ to any of the functions in $(2)^{[\checkmark]}$, and satisfies $w(0) = 1$. As w is continuous, there is some $0 < \delta < \pi$ such that

$$w(x) > \tfrac{1}{2} \qquad \forall x \in (-\delta, \delta).$$

Consider the trigonometric polynomial

$$t(x) = 1 - \cos\delta + \cos x.$$

Obviously, $t(x)$ and all powers $t^N(x)$ are polynomials in $\cos x$. From de Moivre's formula

$$e^{ikx} = (\cos x + i \sin x)^k$$

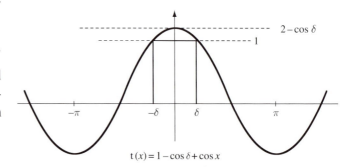

$$t(x) = 1 - \cos\delta + \cos x$$

it is easy to see that $\cos^k x = \sum_{j=0}^{k} c_j \cos jx,^{[\checkmark]}$ see also Gradshteyn and Ryzhik [17, 1.32]. We can thus write $t^N(x)$ as linear combination of $\cos kx$, $k = 0, 1, \ldots, N$. By assumption, w is orthogonal to all of them, and so

$$0 = \int_{(-\pi,\pi)} w(x)\, t^N(x)\, dx = \left(\int_{(-\pi,-\delta]} + \int_{(-\delta,\delta)} + \int_{[\delta,\pi)} \right) w(x)\, t^N(x)\, dx. \qquad (24.7)$$

On $(-\delta, \delta)$ we have $w(x) > \tfrac{1}{2}$ as well as $t(x) > 1$, hence

$$\int_{(-\delta,\delta)} w(x)\, t^N(x)\, dx \geqslant \frac{1}{2} \int_{(-\delta,\delta)} t^N(x)\, dx \xrightarrow{N \to \infty} \infty$$

by monotone convergence T9.6. On the other hand, $|t(x)| \leqslant 1$ for $x \in (-\pi, -\delta] \cup [\delta, \pi)$ and

$$\left| \int_{[\delta,\pi)} w(x)\, t^N(x)\, dx \right| \leqslant (\pi - \delta)\, \|w\|_\infty < \infty \qquad \forall N \in \mathbb{N},$$

which means that (24.7) is impossible, i.e. $w \equiv 0$ and $u \equiv 0$.

An arbitrary function $u \in L^2((-\pi, \pi), dx)$ is, due to the finiteness of the measure, integrable$^{[\checkmark]}$, and we may consider the primitive

$$U(x) := \int_{(-\pi,x)} u(t)\, dt,$$

which is a continuous function, cf. Problem 11.7. Moreover,

$$U(-\pi) = 0 = \int_{(-\pi,\pi)} u(t)\, dt = U(\pi),$$

because of the assumption that u is orthogonal to *every* function from (24.2) and, in particular, to $t \mapsto 1/\sqrt{2\pi}$. By Fubini's theorem 13.9 we get

$$\int_{(-\pi,\pi)} U(x) \cos kx\, dx = \int_{(-\pi,\pi)} \int_{(-\pi,\pi)} \mathbf{1}_{(-\pi,x)}(t)\, u(t) \cos kx\, dt\, dx$$

$$= \int_{(-\pi,\pi)} \left(\int_{(-\pi,\pi)} \mathbf{1}_{(t,\pi)}(x) \cos kx\, dx \right) u(t)\, dt$$

$$= - \int_{(-\pi,\pi)} u(t) \frac{\sin kt}{k}\, dt = 0,$$

and we conclude from the first part of the proof that $U \equiv 0$. Lebesgue's differentiation theorem 19.20 finally shows that $u(x) = U'(x) = 0$ a.e. ∎

Since the trigonometric system is one of the most important ONSs, we provide a further proof of the completeness theorem which gives some more insight into Fourier series and yields even an independent proof of Weierstraß' approximation theorem 24.6 for trigonometric polynomials, cf. Corollary 24.12 below.

We begin with an elementary but fundamental consideration which goes back to Féjer. If $u \in L^2((-\pi, \pi), dt)$, we write

$$a_j := \frac{1}{\pi} \int_{(-\pi,\pi)} u(t) \cos jt\, dt, \qquad b_k := \frac{1}{\pi} \int_{(-\pi,\pi)} u(t) \sin kt\, dt, \qquad (24.8)$$

$(j \in \mathbb{N}_0, k \in \mathbb{N})$ for the *Fourier cosine and sine coefficients* of u and set

$$s_N(u; x) := \sum_{j=1}^{N} \left(a_j \cos jx + b_j \sin jx \right) + \frac{a_0}{2}$$

$$= \frac{1}{\pi} \int_{(-\pi,\pi)} \left(\sum_{j=1}^{N} \left(\cos jt \cos jx + \sin jt \sin jx \right) + \frac{1}{2} \right) u(t)\, dt \qquad (24.9)$$

$$= \frac{1}{\pi} \int_{(-\pi,\pi)} u(t) \underbrace{\left(\frac{1}{2} + \sum_{j=1}^{N} \cos j(t - x) \right)}_{=: D_N(t-x)} dt,$$

where we used the trigonometric formula

$$\cos a \cos b + \sin a \sin b = \cos(a - b). \qquad (24.10)$$

The function $D_N(\bullet)$ is called the *Dirichlet kernel*. In Problem 24.6 we will see that $D_N(\bullet)$ has the following closed-form expression:

$$D_N(x) = \frac{\sin\left(n+\frac{1}{2}\right)x}{2\sin\frac{x}{2}}, \tag{24.11}$$

but we do not need this formula in the sequel.

Now we introduce the *Cesàro C-1 mean*

$$\sigma_N(u;x) := \frac{1}{N+1}\left(s_0(u;x) + s_1(u;x) + \cdots + s_N(u;x)\right), \tag{24.12}$$

and in view of (24.9) we want to compute what is known as the *Féjer kernel*

$$K_N(x) := \frac{1}{N+1}\left(D_0(x) + D_1(x) + \cdots + D_N(x)\right).$$

Using again (24.10) and observing that the cosine is even, we find for every $k = 0, 1, \ldots, N$

$$(1 - \cos x)\, D_k(x) = \frac{1}{2}(1 - \cos x)\sum_{j=-k}^{k}\cos jx$$

$$\overset{(24.10)}{=} \frac{1}{2}\sum_{j=-k}^{k}\left(\cos jx - \cos(j-1)x + \sin x\,\sin jx\right)$$

$$= \frac{1}{2}\left(\cos kx - \cos(k+1)x\right),$$

since $\sin jx = -\sin(-jx)$ is an odd function which cancels if we sum over $-k \leqslant j \leqslant k$. Summing over all values of $k = 0, 1, \ldots, N$ shows

$$K_N(x) = \frac{1}{N+1}\left(D_0(x) + D_1(x) + \cdots + D_N(x)\right)$$

$$= \frac{1}{2(N+1)}\frac{1 - \cos(N+1)x}{1 - \cos x}. \tag{24.13}$$

24.11 Lemma (Féjer) *If $u \in C[-\pi, \pi]$, then $\lim_{N\to\infty}\|\sigma_N(u) - u\|_p = 0$ for all $1 \leqslant p \leqslant \infty$.*

Proof From (24.9), (24.12) and (24.13) we get after a change of variables in the integrals

$$\sigma_N(u;x) = \frac{1}{\pi}\int_{[-\pi,\pi]} u(t)\, K_N(x-t)\, dt$$

$$= \frac{1}{2(N+1)\pi}\int_{[-\pi,\pi]} u(x-t)\,\frac{1 - \cos(N+1)t}{1 - \cos t}\, dt.$$

Since $\frac{1}{\pi} \int_{[-\pi,\pi]} K_N(t) \, dt = 1^{[\checkmark]}$, we see for all $\epsilon > 0$ and sufficiently small $\delta > 0$

$$\|\sigma_N(u) - u\|_p = \left\| \frac{1}{2(N+1)\pi} \int_{[-\pi,\pi]} \frac{1 - \cos(N+1)t}{1 - \cos t} \left(u(\bullet - t) - u \right) dt \right\|_p$$

$$\overset{12.14}{\leqslant} \frac{1}{2(N+1)\pi} \int_{[-\pi,\pi]} \frac{1 - \cos(N+1)t}{1 - \cos t} \left\| u(\bullet - t) - u \right\|_p \, dt$$

$$\leqslant \frac{1}{2(N+1)\pi} \int_{(-\delta,\delta)} \frac{1 - \cos(N+1)t}{1 - \cos t} \left\| u(\bullet - t) - u \right\|_p \, dt$$

$$+ \frac{\|u\|_p}{(N+1)\pi} \int_{[-\pi,-\delta]\cup[\delta,\pi]} \frac{1 - \cos(N+1)t}{1 - \cos t} \, dt$$

$$\leqslant \epsilon + \frac{\|u\|_p}{(N+1)\pi} \frac{4\pi}{1 - \cos \delta},$$

where we used Jensen's inequality and the fact that $\lim_{t \to 0} \|u(\bullet - t) - u\|_p = 0$ by dominated convergence $(p < \infty)$, resp. uniform continuity $(p = \infty)$. Letting first $N \to \infty$ and then $\epsilon \to 0$ finishes the proof. ∎

24.12 Corollary (Weierstraß) *The trigonometric polynomials are dense in* $C[-\pi, \pi]$ *under* $\|\bullet\|_\infty$ *and dense in* $L^p([-\pi, \pi], dt)$, *w.r.t.* $\|\bullet\|_p$, $1 \leqslant p < \infty$.

Proof From (24.9), (24.12) it is obvious that $\sigma_N(u, \bullet)$ is a trigonometric polynomial. The density of the trigonometric polynomials in $C[-\pi, \pi]$ is just Lemma 24.11. Since $C[-\pi, \pi]$ is dense in $L^p([-\pi, \pi], dt)$, cf. Theorem 15.17, we can find for every $\epsilon > 0$ and $u \in L^p[-\pi, \pi]$ some $g_\epsilon \in C[-\pi, \pi]$ with $\|u - g_\epsilon\|_p \leqslant \epsilon$ and a trigonometric polynomial t_ϵ such that $\|g_\epsilon - t_\epsilon\|_\infty \leqslant (2\pi)^{-1/p} \epsilon$. This shows

$$\|u - t_\epsilon\|_p \leqslant \|u - g_\epsilon\|_p + \|g_\epsilon - t_\epsilon\|_p \leqslant \epsilon + (2\pi)^{1/p} \|g_\epsilon - t_\epsilon\|_\infty \leqslant 2\epsilon.$$

For the last estimate we also used that $\|w\|_p \leqslant (2\pi)^{1/p} \|w\|_\infty$. ∎

24.13 Corollary *The trigonometric system (24.2) is a complete ONS in* $L^2 = L^2([-\pi, \pi], dt)$

Proof (of C24.13 and, again, of T24.10) Let $u \in L^2[-\pi, \pi]$ and pick a trigonometric polynomial t_ϵ such that $\|u - t_\epsilon\|_2 \leqslant \epsilon$, cf. Corollary 24.12. Let $n = \text{degree}(t_\epsilon)$. As in the proof of Theorem 24.10 we use de Moivre's formula to see that $\cos^k x$ and $\sin^k x$ can be represented as linear combinations of $1, \cos x, \ldots,$ $\cos kx$ and $\sin x, \ldots, \sin kx.^{[\checkmark]}$

Recall that the partial sum $s_n(u; x) = a_0/2 + \sum_{j=1}^{n}(a_j \cos jx + b_j \sin jx)$ is the projection of u onto span$\{1, \cos x, \sin x, \ldots, \cos nx, \sin nx\}$. Therefore, Theorem 21.11(i) applies and

$$\|u - s_n(u)\|_2 \leqslant \|u - t_\epsilon\|_2 \leqslant \epsilon$$

proves completeness. ∎

The above proof of the completeness of the trigonometric system has a further advantage as it allows a glimpse into other modes of convergence of Fourier series. We have

24.14 Corollary (M. Riesz' theorem) *Let $u \in L^p([-\pi, \pi], dt)$ and $1 \leqslant p < \infty$.*

$$\lim_{n \to \infty} \|u - s_n(u)\|_p = 0 \iff \|s_n(u)\|_p \leqslant C_p \|u\|_p \quad \forall n \in \mathbb{N} \qquad (24.14)$$

with an absolute constant C_p not depending on u or $n \in \mathbb{N}$.

Proof The 'only if' part is a consequence of the uniform boundedness principle (Banach–Steinhaus theorem) from functional analysis, see e.g. Rudin [40, §5.8] or Problem 21.10.

The 'if' part follows from the observation that every trigonometric polynomial T of degree $\leqslant n$ satisfies $s_n(T) = T$.[✓] Choosing for $u \in L^p$ the polynomial $T = t_\epsilon$ with $\|u - t_\epsilon\|_p \leqslant \epsilon$, cf. Corollary 24.12, we infer that for sufficiently large $n >$ degree(t_ϵ)

$$\|u - s_n(u)\|_p \leqslant \|u - t_\epsilon\|_p + \underbrace{\|t_\epsilon - s_n(t_\epsilon)\|_p}_{=0} + \|s_n(t_\epsilon) - s_n(u)\|_p$$

$$\leqslant (1 + C_p)\|u - t_\epsilon\|_p. \qquad ∎$$

Establishing the estimate $\|s_n(u)\|_p \leqslant C_p \|u\|_p$ is an altogether different matter and so is the whole L^p- and pointwise convergence theory for Fourier series. Here we want to mention only a few facts:

- L^p-convergence $(1 \leqslant p < \infty)$ of the Cesàro means $\sigma_n(u)$ follows immediately from Lemma 24.11.

This is in stark contrast to...

- L^p-convergence $(1 \leqslant p < \infty, p \neq 2)$ of the partial sums $s_n(u)$ requires the estimate (24.14); see Corollary 24.14 and, for more details, Wheeden and Zygmund [53, §12.88].

- Pointwise a.e. convergence of the partial sums $s_n(u) \xrightarrow{n \to \infty} u$ when $u \in L^2$ or $u \in L^p$, $1 < p < \infty$, which had been an open problem until 1966. A.N. Kolmogorov constructed in 1922/23 a function $u \in L^1$ whose Fourier series diverges a.e. In his famous 1966 paper L. Carleson proved that a.e. convergence holds for $u \in L^2$, and R.A. Hunt extended this result in 1968 to $u \in L^p$, $1 < p < \infty$.

All these deep results depend on estimates of the type (24.14) and, more importantly, on estimates for $\max_{0 \leqslant j \leqslant n} s_j(u)$ which resemble the maximal martingale estimates which we have encountered in Chapter 19, e.g. T19.12. But there is a catch.

24.15 Lemma *The subspace* $\Sigma_n = \mathrm{span}\{1, \cos x, \sin x, \ldots, \cos nx, \sin nx\}$ *of* $L^2([-\pi, \pi], dx)$ *is not of the form* $L^2(\mathcal{G}_n)$ *where* $\mathcal{G}_n$ *is a sub-σ-algebra of the Borel sets* $\mathcal{B}[-\pi, \pi]$.

Proof The space $L^2(\mathcal{G}_n)$ is a lattice, i.e. if $f \in L^2(\mathcal{G}_n)$, then $|f| \in L^2(\mathcal{G}_n)$. Take $f(x) = \sin x$. Unfortunately,

$$|\sin x| = \frac{2}{\pi} - \frac{4}{\pi}\left(\frac{\cos 2x}{1 \cdot 3} + \frac{\cos 4x}{3 \cdot 5} + \frac{\cos 6x}{5 \cdot 7} + \cdots\right),$$

so that $\sin(\cdot) \in \Sigma_n$ but $|\sin(\cdot)| \notin \Sigma_n$. (You might also want to have a look at Theorem 22.5 for a more systematic treatment.) ∎

This means that martingale methods are not (immediately) applicable to Fourier series.

The Haar system

In contrast to Fourier series, the Haar system allows a complete martingale treatment. Throughout this section we consider $L^2 = L^2([0, 1), \mathcal{B}[0, 1), \lambda)$, $\lambda = \lambda^1|_{[0,1)}$.

24.16 Definition The *Haar system* consists of the functions

$$\left.\begin{aligned}
\chi_{0,0}(x) &:= \mathbf{1}_{[0,1)}(x), \\
\chi_{j,k}(x) &:= 2^{k/2}\left(\mathbf{1}_{\left[\frac{2j-2}{2^{k+1}}, \frac{2j-1}{2^{k+1}}\right)}(x) - \mathbf{1}_{\left[\frac{2j-1}{2^{k+1}}, \frac{2j}{2^{k+1}}\right)}(x)\right), \\
1 &\leqslant j \leqslant 2^k, \ k \in \mathbb{N}_0.
\end{aligned}\right\} \tag{24.15}$$

Obviously, each Haar function is normalized to give $\|\chi_{j,k}\|_2 = 1$. The first few Haar functions are

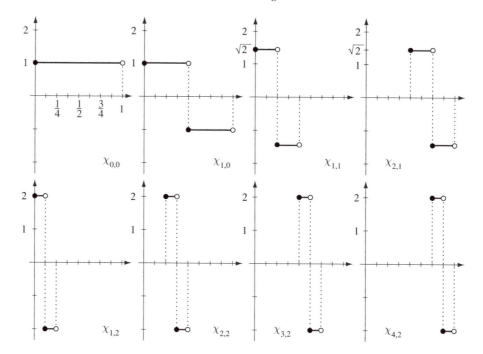

$$\chi_{0,0} \quad \chi_{1,0} \quad \chi_{1,1} \quad \chi_{2,1}$$

$$\chi_{1,2} \quad \chi_{2,2} \quad \chi_{3,2} \quad \chi_{4,2}$$

It is often more convenient to arrange the double sequence (24.15) in lexicographical order: $\chi_{0,0}$; $\chi_{1,0}$; $\chi_{1,1}, \chi_{2,1}$; $\chi_{1,2}, \chi_{2,2}, \chi_{3,2}, \chi_{4,2}$; ...and to relabel them in the following way

$$H_0 := \chi_{0,0}; \quad H_n = H_{2^k+\ell} := \chi_{\ell+1,k}, \quad 0 \leqslant \ell \leqslant 2^k - 1, \tag{24.16}$$

(note that the representation $n = 2^k + \ell$, $0 \leqslant \ell \leqslant 2^k - 1$ is unique). We can now associate with the sequence $(H_n)_{n \in \mathbb{N}}$ a canonical filtration

$$\mathcal{A}_n^H := \sigma(H_0, H_1, \ldots, H_n), \quad n \in \mathbb{N}_0,$$

which is the smallest σ-algebra that makes all functions $H_0, \ldots, H_n$ measurable, cf. Definition 7.5.

24.17 Theorem *The Haar functions are a complete ONS in $L^2([0,1), dx)$. Moreover,*

$$M_N := \sum_{n=0}^{N} a_n H_n, \quad N \in \mathbb{N}_0, \ a_n \in \mathbb{R},$$

is a martingale w.r.t. the filtration $(\mathcal{A}_N^H)_{N \in \mathbb{N}_0}$, and for every $u \in L^p([0,1), dx)$, $1 \leqslant p < \infty$, the Haar–Fourier series

$$s_N(u; x) := \sum_{n=0}^{N} \langle u, H_n \rangle H_n$$

converges to u in L^p and almost everywhere, and the maximal inequality

$$\left\| \sup_{n\in\mathbb{N}} s_n(u) \right\|_p \leqslant \frac{p}{p-1} \|u\|_p$$

holds for all $u \in L^p$ and $1 < p < \infty$.

Proof *Step 1. Orthonormality:* That $\|\chi_{j,k}\|_2 = 1$ is obvious. If the functions $\chi_{j,k} \neq \chi_{\ell,m}$ satisfy $\{\chi_{j,k} \neq 0\} \cap \{\chi_{\ell,m} \neq 0\} = \emptyset$, it is clear that $\int \chi_{j,k} \chi_{\ell,m} \, d\lambda = 0$. Otherwise, we can assume that $k < m$, so that

either $\{\chi_{\ell,m} \neq 0\} \subset \{\chi_{j,k} = +1\}$ or $\{\chi_{\ell,m} \neq 0\} \subset \{\chi_{j,k} = -1\}$

obtains. In either case,

$$\int \chi_{\ell,m} \chi_{j,k} \, d\lambda = \pm \int \chi_{\ell,m} \, d\lambda = 0.$$

Step 2. Martingale property: Let $n = 2^k + \ell$. Then, for all $n \in \mathbb{N}$,

$$\mathcal{A}_n^H = \sigma\left(\chi_{0,0}, \chi_{1,0}, \chi_{1,1}, \ldots, \chi_{2^{k-1},k-1}, \chi_{1,k}, \chi_{2,k}, \ldots, \chi_{\ell+1,k}\right)$$

$$= \sigma\left(\underbrace{\left[0, \frac{1}{2^{k+1}}\right), \ldots, \left[\frac{2\ell+1}{2^{k+1}}, \frac{2\ell+2}{2^{k+1}}\right)}_{=:\mathcal{E}_n}, \underbrace{\left[\frac{\ell+1}{2^k}, \frac{\ell+2}{2^k}\right), \ldots, \left[\frac{2^k-1}{2^k}, 1\right)}_{=:\mathcal{F}_n}\right),$$

where we used that the dyadic intervals are nested and refine. Assume, for simplicity, that $\ell < 2^k - 1$. Then $\{H_{n+1} \neq 0\} \in \mathcal{F}_n$, and so

$$\int_J H_{n+1}(x) \, dx = 0 \qquad \forall J \in \mathcal{E}_n \text{ or } J \in \mathcal{F}_n.$$

(If $\ell = 2^k - 1$ we get an analogous conclusion with a rollover as $\mathcal{A}_n^H$ is just the dyadic σ-algebra generated by all disjoint half-open intervals of length 2^{-k-1} in $[0, 1)$.) By Theorem 23.5 we have $\mathbf{E}^{\mathcal{A}_n^H} H_{n+1} = 0$, and by Theorem 23.8

$$\mathbf{E}^{\mathcal{A}_N^H} M_{N+1} = \mathbf{E}^{\mathcal{A}_N^H}(M_N + a_{N+1} H_{N+1}) = M_N + a_{N+1} \mathbf{E}^{\mathcal{A}_N^H} H_{N+1} = M_N.$$

This shows that $\left(M_N, \mathcal{A}_N^H\right)_{N\in\mathbb{N}}$ is indeed a martingale, cf. Corollary 23.14.

Step 3. Convergence in L^1 and a.e. if $u \in L^1 \cap L^\infty$: Set $a_k := \langle u, H_k \rangle$, so that $M_N = s_N(u)$ becomes the Haar–Fourier partial sum. Using Bessel's inequality (Theorem 21.11) we see

$$\|s_N(u)\|_2^2 = \sum_{k=0}^N |\langle u, H_k \rangle|^2 \leqslant \|u\|_2^2, \tag{24.17}$$

where the right-hand side is finite since $L^1 \cap L^\infty \subset L^2$,[✓] and from the Cauchy–Schwarz C12.3 and Markov P10.12 inequalities we get for all $R > 0$

$$\int_{\{|s_N(u)|>R\}} |s_N(u)| \, d\lambda \leq \|s_N(u)\|_2 \, \lambda(\{|s_N(u)| > R\})^{1/2}$$

$$\leq \frac{1}{R} \|s_N(u)\|_2^2 \leq \frac{1}{R} \|u\|_2^2.$$

Since the constant function R is in $L^2([0, 1), dx)$, the martingale $(s_N(u))_{N \in \mathbb{N}}$ is uniformly integrable in the sense of Definition 16.1, and we conclude from Theorems 18.6 and 23.15 that

$$s_N(u) \xrightarrow{N \to \infty} u_\infty \qquad \text{in } L^1 \text{ and almost everywhere.}$$

Since $(\mathcal{A}_n^H)_{n \in \mathbb{N}}$ contains the sequence $(\mathcal{A}_k^\Delta)_{k \in \mathbb{N}}$ of dyadic σ-algebras – we have indeed $\mathcal{A}_n^\Delta = \mathcal{A}_{2^n-1}^H$ – we know that $\mathcal{A}_\infty^H := \sigma(\mathcal{A}_n^H : n \in \mathbb{N}) = \mathcal{B}[0, 1)$. Just as in Example 23.17 we see that

$$\mathbf{E}^{\mathcal{A}_{2^n-1}^H} u = \mathbf{E}^{\mathcal{A}_n^\Delta} u = s_{2^n-1}(u),$$

and in view of Theorem 23.15 we conclude that $u = u_\infty$ a.e.

Step 4. Convergence in L^p if $u \in L^1 \cap L^\infty$: Observe that $L^1 \cap L^\infty \subset L^p$ for all $1 < p < \infty$.[✓] Applying the inequality

$$|a|^p - |b|^p \leq ||a|^p - |b|^p| = \left| p \int_{|a|}^{|b|} t^{p-1} \, dt \right|$$

$$\leq p \, ||a| - |b|| \max\{|a|^{p-1}, |b|^{p-1}\}$$

$$\leq p \, |a - b| \max\{|a|^{p-1}, |b|^{p-1}\},$$

$a, b \in \mathbb{R}$, $1 < p < \infty$, to the martingale $\mathbf{E}^{\mathcal{A}_N^H} u = s_N(u)$, we get after integrating over $[0, 1)$

$$\pm \int \left(|s_N(u)|^p - |u|^p \right) d\lambda \leq p \, \|s_N(u) - u\|_1 \, \|u\|_\infty^{p-1}, \qquad p > 1,$$

where we also used that $|s_N(u)| = |\mathbf{E}^{\mathcal{A}_N^H} u| \leq \|u\|_\infty$ as $|u| \leq \|u\|_\infty < \infty$, cf. Theorem 23.8(ix). From Riesz' convergence theorem T12.10 we conclude that $s_N(u) \xrightarrow{N \to \infty} u$ in L^p for all $1 < p < \infty$ and all $u \in L^1 \cap L^\infty$.

Step 5. Convergence in L^p if $u \in L^p$: If $u \in L^p$, $1 \leq p < \infty$, is not bounded, we set $u_k := (-k) \vee u \wedge k$. Since we have a finite measure space, $u_k \in L^p \cap L^\infty \subset$

$L^1 \cap L^\infty$, and we see from the triangle inequality and Theorem 23.8(v),(ii)

$$\|s_N(u) - u\|_p \leqslant \|s_N(u) - s_N(u_k)\|_p + \|s_N(u_k) - u_k\|_p + \|u_k - u\|_p$$
$$\leqslant \|s_N(u_k) - u_k\|_p + 2\|u_k - u\|_p.$$

The claim follows as $N \to \infty$ and then $k \to \infty$.

Step 6. A.e. convergence if $u \in L^p$: Since $s_N(u^\pm) = \mathbf{E}^{\mathcal{A}_N^H}(u^\pm) \geqslant 0$, we know from Corollary 23.16 that $\big((s_N(u^\pm))^p\big)_{N \in \mathbb{N}}$ are submartingales which satisfy, by Theorem 23.8(ii),

$$\int (s_N(u^\pm))^p \, d\mu \leqslant \int \big(\mathbf{E}^{\mathcal{A}_N^H}(u^\pm)\big)^p \, d\mu = \big\|\mathbf{E}^{\mathcal{A}_N^H}(u^\pm)\big\|_p^p \leqslant \|u^\pm\|_p^p.$$

Therefore, the submartingale convergence theorem 18.2 applies and shows that $\lim_{N\to\infty} |s_N(u^\pm; x)|^p$ exists a.e., hence, $\lim_{N\to\infty} s_N(u; x)$ exists a.e. Since step 5 and Corollary 12.8 already imply $\lim_{j\to\infty} s_{N_j}(u; x) = u(x)$ a.e. for some subsequence, we can identify the limit and get $\lim_{N\to\infty} s_N(u; x) = u(x)$ a.e.

Step 7. Completeness follows from $\lim_{N\to\infty} \|s_N(u) - u\|_2 = 0$ and T21.13.

Step 8. The maximal inequality is just Doob's maximal L^p-inequality for martingales T19.12 since $(s_n(u))_{n \in \mathbb{N}}$ is a uniformly integrable martingale which is, by step 5 and Theorem 23.15, closed by $s_\infty(u) = u$. ∎

24.18 Remark As a matter of fact, ordering the Haar functions in a sequence like $(H_n)_{n \in \mathbb{N}_0}$ does play a rôle. If $p = 1$, we can find (after some elementary but very tedious calculations) that

$$\left\| \chi_{0,0} + \chi_{1,0} + \sum_{k=1}^{2n} 2^{k/2} \chi_{1,k} \right\|_1 \leqslant \sqrt{2},$$

while the lacunary series satisfies

$$\left\| \chi_{1,0} + \sum_{k=1}^{n} 2^k \chi_{1,2k} \right\|_1 \geqslant cn$$

for some absolute constant $c > 0$. Therefore, we can rearrange $\sum_{n=0}^\infty a_n H_n$ in such a way that it becomes a divergent series $\sum_{n=0}^\infty a_{\sigma(n)} H_{\sigma(n)}$ for some necessarily infinite permutation $\sigma : \mathbb{N}_0 \to \mathbb{N}_0$.

This phenomenon does not happen if $1 < p < \infty$. In fact, $(H_n)_{n \in \mathbb{N}_0}$ is what one calls an *unconditional basis* of L^p, $1 < p < \infty$, which means that every rearrangement of the series $\sum_{n=0}^\infty a_n H_n$ converges in L^p and leads to the same limit. The Haar system is even the litmus test for the existence of unconditional bases: *every Banach space B where $(H_n)_{n \in \mathbb{N}_0}$ is a basis has an unconditional* *basis if, and only if, the basis* $(H_n)_{n\in\mathbb{N}_0}$ *is unconditional,* cf. Olevskiĭ [32, p. 73, Corollary] or Lindenstrauss and Tzafriri [27, vol. II, p. 161, Corollary 2.c.11].

Since the unconditionality of $(H_n)_{n\in\mathbb{N}_0}$ rests on a martingale argument, we include a sketch of its proof. First we need the following *Burkholder–Davis–Gundy inequalities* for a martingale $(u_j)_{j\in\mathbb{N}_0}$ on a *probability space* $(X, \mathcal{A}, \mu)$:

$$\kappa_p \left\| \sup_{0\leqslant j\leqslant N} |u_j| \right\|_p \leqslant \left\| \sqrt{[u_\bullet, u_\bullet]_N} \right\|_p \leqslant K_p \left\| \sup_{0\leqslant j\leqslant N} |u_j| \right\|_p \qquad \text{(BDG)}$$

for all $N \in \mathbb{N}_0$, all $0 < p < \infty$ and some absolute constants $K_p \geqslant \kappa_p > 0$. The expression $[u_\bullet, u_\bullet]_N$ stands for the *quadratic variation* of the martingale

$$[u_\bullet, u_\bullet]_N := |u_0|^2 + \sum_{j=0}^{N-1} |u_{j+1} - u_j|^2.$$

A proof of (BDG) can be found in Rogers and Williams [38, vol. 2, pp. 94–6]. If we combine (BDG) with Doob's maximal L^p-inequality 19.12 we get

$$\kappa_p \|u_N\|_p \leqslant \left\| \sqrt{[u_\bullet, u_\bullet]_N} \right\|_p \leqslant \frac{p K_p}{p-1} \|u_N\|_p \qquad \text{(BDG′)}$$

for all $N \in \mathbb{N}_0$ and $1 < p < \infty$ – mind the different range for p in (BDG′) compared to (BDG). Obviously,

$$u_N := \sum_{k=0}^{N} \langle u, H_k \rangle H_k \qquad \text{and} \qquad w_N := \sum_{k=0}^{N} \epsilon_k \langle u, H_k \rangle H_k,$$

$\epsilon_k \in \{-1, +1\}$, are uniformly integrable martingales (use the argument of the proof of Theorem 24.17) and their quadratic variations $[u_\bullet, u_\bullet]_N = [w_\bullet, w_\bullet]_N$ coincide. Therefore, (BDG′) shows that the martingales $(u_N - u_n)_{N\geqslant n}$ and $(w_N - w_n)_{N\geqslant n}$ satisfy

$$\|u_N - u_n\|_p \sim \left\| [u_\bullet - u_n, u_\bullet - u_n]_N^{1/2} \right\|_p = \left\| [w_\bullet - w_n, w_\bullet - w_n]_N^{1/2} \right\|_p \sim \|w_N - w_n\|_p,$$

where $a \sim b$ means that $\kappa a \leqslant b \leqslant K a$ for some absolute constants $\kappa, K > 0$, so that either both sequences converge or diverge. Let us assume that $(u_N)_{N\in\mathbb{N}_0}$ converges. Then every lacunary series

$$\sum_{j=1}^{\infty} \langle u, H_{k_j} \rangle H_{k_j} \qquad \text{converges} \qquad (24.18)$$

since we can produce its partial sums by adding and subtracting u_N and w_N with suitable ± 1-sequences $(\epsilon_k)_{k\in\mathbb{N}}$. This entails that for every fixed permutation

$\sigma : \mathbb{N}_0 \to \mathbb{N}_0$

$$\left\| \sum_{k=n}^{N} \langle u, H_{\sigma(k)} \rangle H_{\sigma(k)} \right\|_p \leqslant \epsilon, \qquad N > n \quad \text{sufficiently large.}$$

Otherwise, we could find finite sets $\Sigma_0, \Sigma_1, \Sigma_2, \ldots \subset \mathbb{N}_0$ with $(k_j)_{j \in \mathbb{N}} = \bigcup_{n \in \mathbb{N}} \Sigma_n$ and

$$\left\| \sum_{k \in \Sigma_n} \langle u, H_k \rangle H_k \right\|_p > \epsilon \qquad \forall n \in \mathbb{N},$$

contradicting (24.18). For more on this topic we refer to Lindenstrauss and Tzafriri [27].

The Haar wavelet

Let us now consider a Haar system on the whole real line, i.e. in $L^2 = L^2(\mathbb{R}, \mathcal{B}(\mathbb{R}), dx)$. We begin with the remark that the functions $\chi_{0,0} = \mathbf{1}_{[0,1)}$ and $\chi_{1,0} = \mathbf{1}_{[0,1/2)} - \mathbf{1}_{[1/2,1)}$ are the two basic Haar functions, since we can reconstruct all Haar functions $\chi_{j,k}$ from them by scaling and shifting:

$$\chi_{j,k}(x) = 2^{k/2} \chi_{1,0}(2^k x - j + 1), \qquad k \in \mathbb{N}_0, \ j = 1, 2, \ldots, 2^k. \tag{24.19}$$

The advantage of (24.19) over the definition (24.15) is that (24.19) easily extends to all pairs $(j, k) \in \mathbb{Z}^2$ and, thus to a system of functions on $\mathbb{R}$.

24.19 Definition The *Haar wavelets* are the system $(\psi_{j,k})_{j,k \in \mathbb{Z}}$ where the *mother wavelet* is $\psi(x) := \mathbf{1}_{[0,1/2)}(x) - \mathbf{1}_{[1/2,1)}(x)$ and

$$\psi_{j,k}(x) := 2^{k/2} \psi(2^k x - j) = 2^{k/2} \left(\mathbf{1}_{\left[\frac{2j}{2^{k+1}}, \frac{2j+1}{2^{k+1}} \right)}(x) - \mathbf{1}_{\left[\frac{2j+1}{2^{k+1}}, \frac{2j+2}{2^{k+1}} \right)}(x) \right)$$

for all $j, k \in \mathbb{Z}$.

Note that $\psi = \psi_{1,0} = \chi_{1,0}$, $\psi_{j-1,k} = \chi_{j,k}$ for all $j = 1, 2, \ldots, 2^k$ and $k \in \mathbb{N}_0$ while $\psi_{-1,0}(x) = 2^{-1/2} \chi_{0,0}(x)$ for $0 \leqslant x < 1$.

The Haar wavelets can be treated by martingale methods. To do so, we introduce the two-sided dyadic filtration

$$\mathcal{A}_{n+1}^{\Delta} = \sigma \left(\left[\frac{j}{2^{n+1}}, \frac{j+1}{2^{n+1}} \right) : j \in \mathbb{Z} \right) = \sigma(\psi_{j,n} : j \in \mathbb{Z}), \qquad n \in \mathbb{Z},$$

$$\mathcal{A}_{-\infty}^{\Delta} = \bigcap_{n \in \mathbb{Z}} \mathcal{A}_n^{\Delta} = \{ \emptyset, \mathbb{R} \}, \qquad \mathcal{A}_{\infty}^{\Delta} = \sigma \left(\bigcup_{n \in \mathbb{Z}} \mathcal{A}_n^{\Delta} \right) = \mathcal{B}(\mathbb{R}). \tag{24.20}$$

The last assertion follows from the fact that $D = \{j2^{-n-1} : j \in \mathbb{Z}, n \in \mathbb{Z}\}$ is a dense subset of $\mathbb{R}$ and that $\mathcal{B}(\mathbb{R})$ is generated by all intervals of the form $[a, b)$ where $a, b \in D$ (or, indeed, any other dense subset).[✓]

In what follows we have to consider double summations. To keep notation simple, we write

$$\sum_{k=-\infty}^{\infty} \sum_{j=-\infty}^{\infty} a_{j,k} \quad \text{as a shorthand for} \quad \sum_{k=-\infty}^{\infty} \left[\sum_{j=-\infty}^{\infty} a_{j,k} \right]$$

and call $\sum_{k\geqslant\text{const.}} \sum_{j=-\infty}^{\infty}$ the *right tail* and $\sum_{-\infty<k\leqslant\text{const.}} \sum_{j=-\infty}^{\infty}$ the *left tail* of the double sum.

24.20 Theorem *The Haar wavelets $(\psi_{j,k})_{j,k\in\mathbb{Z}}$ are a complete ONS in $L^2 = L^2(\mathbb{R}, \mathcal{B}(\mathbb{R}), dx)$. Moreover, for all $1 < p < \infty$,*

$$u = \sum_{k=-\infty}^{\infty} \sum_{j=-\infty}^{\infty} \langle u, \psi_{j,k}\rangle \psi_{j,k}, \qquad u \in L^p, \tag{24.21}$$

in L^p and almost everywhere, and

$$\left\| \sup_{M,N\in\mathbb{N}} \sum_{k=-M}^{N} \sum_{j=-\infty}^{\infty} \langle u, \psi_{j,k}\rangle \psi_{j,k} \right\|_p \leqslant \frac{2p-1}{p-1} \|u\|_p \tag{24.22}$$

holds for all $1 < p < \infty$ and $u \in L^p$.

Proof *Step 1. Orthonormality* of the family $(\psi_{j,k})_{j,k\in\mathbb{Z}}$ can be seen with arguments similar to those in step 1 of the proof of Theorem 24.17.

Step 2. L^p $(1 \leqslant p < \infty)$ and a.e. convergence of the right tail of (24.21) *if $u \in L^1 \cap L^\infty$:* Note that the inner sum is pointwise convergent since $\psi_{j,k}\psi_{\ell,k} = 0$ whenever $j \neq \ell$. Consider now $u \in L^1 \cap L^\infty \subset L^p$.[✓] Set

$$u_{N,-M} := \sum_{k=-M}^{N} \sum_{j=-\infty}^{\infty} \langle u, \psi_{j,k}\rangle \psi_{j,k} = \mathbf{E}^{\mathcal{A}_{N+1}^{\Delta}} u - \mathbf{E}^{\mathcal{A}_{-M}^{\Delta}} u. \tag{24.23}$$

The latter equality follows from the fact that $\mathbf{E}^{\mathcal{A}_n^{\Delta}}$ is the orthogonal projection onto $L^2(\mathcal{A}_n^{\Delta})$ – whose basis is $\{\psi_{j,k} : j \in \mathbb{Z}, k \in \mathbb{Z}, k \leqslant n-1\}$ – and by 22.4(vii),

$$\mathbf{E}^{\mathcal{A}_{n+1}^{\Delta}} u - \mathbf{E}^{\mathcal{A}_n^{\Delta}} u = \mathbf{E}^{\mathcal{A}_{n+1}^{\Delta}} u - \mathbf{E}^{\mathcal{A}_{n+1}^{\Delta}}\left(\mathbf{E}^{\mathcal{A}_n^{\Delta}} u\right) = \mathbf{E}^{\mathcal{A}_{n+1}^{\Delta}}\left(u - \mathbf{E}^{\mathcal{A}_n^{\Delta}} u\right),$$

which is the orthogonal projection of $L^2(\mathcal{A}_n^{\Delta})^\perp$ onto $L^2(\mathcal{A}_{n+1}^{\Delta})$. This means that

$$\mathbf{E}^{\mathcal{A}_{n+1}^{\Delta}} u - \mathbf{E}^{\mathcal{A}_n^{\Delta}} u = \sum_{j\in\mathbb{Z}} \langle u, \psi_{j,n}\rangle \psi_{j,n}, \tag{24.24}$$

since the resulting function must be $\mathcal{A}_{n+1}^{\Delta}$-measurable as well as orthogonal to $L^2(\mathcal{A}_n^{\Delta})$: i.e. we must include $(\psi_{j,n})_{j\in\mathbb{Z}}$ and exclude $(\psi_{j,k})_{\substack{j\in\mathbb{Z}\\k<n}}$. Summing (24.24) over $n = -M, \ldots, N$ yields (24.23).

Since $(\mathbf{E}^{\mathcal{A}_N^{\Delta}}u)_{N\in\mathbb{N}}$ is by Theorem 23.15 a uniformly integrable martingale, and since $\mathcal{A}_{\infty}^{\Delta} = \mathcal{B}(\mathbb{R})$, we find that

$$\mathbf{E}^{\mathcal{A}_N^{\Delta}}u \xrightarrow{N\to\infty} u \qquad \text{in } L^1 \text{ and a.e. for all } u \in L^1 \cap L^{\infty}.$$

As in step 4 of the proof of Theorem 24.17 we see that this also holds in L^p.

Step 3. L^p $(1 \leqslant p < \infty)$ *convergence of the right tail of* (24.21) *if* $u \in L^p$: For a general $u \in L^p$ we can use dominated convergence T12.9 to see that the functions $u_k := ((-k) \vee u \wedge k)\mathbf{1}_{[-k,k]} \in L^1 \cap L^{\infty}$ approximate u in L^p-sense. By Theorem 23.8(v),(ii),

$$\left\| \mathbf{E}^{\mathcal{A}_N^{\Delta}}u - u \right\|_p \leqslant \left\| \mathbf{E}^{\mathcal{A}_N^{\Delta}}u - \mathbf{E}^{\mathcal{A}_N^{\Delta}}u_k \right\|_p + \left\| \mathbf{E}^{\mathcal{A}_N^{\Delta}}u_k - u_k \right\|_p + \left\| u_k - u \right\|_p$$

$$\leqslant \left\| \mathbf{E}^{\mathcal{A}_N^{\Delta}}u_k - u_k \right\|_p + 2\left\| u_k - u \right\|_p.$$

In view of the result of the previous step, we can let first $N \to \infty$, then $k \to \infty$, and find that $\mathbf{E}^{\mathcal{A}_N^{\Delta}}u \xrightarrow{N\to\infty} u$ in L^p for every $u \in L^p$.

Step 4. *A.e. convergence of the right tail of* (24.21) *if* $u \in L^p$, $1 < p < \infty$ follows from exactly the same arguments that were used in step 6 of the proof of Theorem 24.17.

Step 5. L^p-*convergence* $(1 < p < \infty)$ *of the left tail of* (24.21) *if* $u \in L^p$: It remains to consider $\left(\mathbf{E}^{\mathcal{A}_{-M}^{\Delta}}\right)_{M\in\mathbb{N}}$. Although this is a backwards martingale, we cannot use Theorem 18.7 as $\lambda|_{\mathcal{A}_{-\infty}^{\Delta}}$ is not σ-finite. Instead, we take $u \in L^p$, $1 < p < \infty$ and set $u_R := u\mathbf{1}_{[-R,R]}$, $R > 0$. For all $M \in \mathbb{N}$ with $2^M > R$ we find

$$\mathbf{E}^{\mathcal{A}_{-M}^{\Delta}}u_R = 2^{-M}\int_{[-R,0]}u(x)\,dx\,\mathbf{1}_{[-2^M,0)} + 2^{-M}\int_{[0,R]}u(x)\,dx\,\mathbf{1}_{[0,2^M)}$$

where we used that $\mathbf{E}^{\mathcal{A}_{-M}^{\Delta}}$ projects onto the intervals $[j2^M, (j+1)2^M)$, and we find from the Hölder inequality T12.2 with $p^{-1}+q^{-1}=1$ that

$$\left| \mathbf{E}^{\mathcal{A}_{-M}^{\Delta}}u_R(x) \right| \leqslant 2^{-M}R^{1/q}\|u\|_p\,\mathbf{1}_{[-2^M,2^M)}(x),$$

which implies

$$\left\| \mathbf{E}^{\mathcal{A}_{-M}^{\Delta}}u_R \right\|_p \leqslant 2^{-M}R^{1/q}\|u\|_p(2\cdot 2^M)^{1/p} = c_R\,2^{-M(1-1/p)}\|u\|_p.$$

Finally, by Theorem 23.8(v),(ii),

$$\left\| \mathbf{E}^{\mathcal{A}^{\Delta}_{-M}} u \right\|_p \leqslant \left\| \mathbf{E}^{\mathcal{A}^{\Delta}_{-M}} (u - u_R) \right\|_p + \left\| \mathbf{E}^{\mathcal{A}^{\Delta}_{-M}} u_R \right\|_p$$

$$\leqslant \left\| u - u_R \right\|_p + c_R \, 2^{-M(1-1/p)} \left\| u \right\|_p,$$

and we get $\lim_{M \to \infty} \left\| \mathbf{E}^{\mathcal{A}^{\Delta}_{-M}} u \right\|_p = 0$ for all $u \in L^p$, $1 < p < \infty$, letting first $M \to \infty$ and then $R \to \infty$.

This shows that $u_{-M,N} \xrightarrow{M,N \to \infty} u$ in L^p, $1 < p < \infty$, and the proof of the convergence of (24.21) in L^p, $1 < p < \infty$, is complete.

Step 6. Completeness of the Haar wavelets in L^2 follows if we apply (24.21) in the case $p = 2$, cf. Theorem 21.13.

Step 7. A.e. convergence of the left tail of (24.21)*:* Observe that

$$A := \left\{ \left| \mathbf{E}^{\mathcal{A}^{\Delta}_{-M}} u \right| > \epsilon : \text{ for infinitely many } M \in \mathbb{N} \right\}$$

$$= \bigcap_{M=1}^{\infty} \underbrace{\bigcup_{j=M}^{\infty} \left\{ \left| \mathbf{E}^{\mathcal{A}^{\Delta}_{-j}} u \right| > \epsilon \right\}}_{\in \, \mathcal{A}^{\Delta}_{-M}} \in \mathcal{A}^{\Delta}_{-\infty}.$$

By the martingale maximal inequality, Lemma 19.11, for the reversed martingale $\left(\mathbf{E}^{\mathcal{A}^{\Delta}_{-j}} u \right)_{j \in \mathbb{N}}$ and Theorem 23.8(ii) we see

$$\lambda(A) \leqslant \lambda \left(\bigcup_{j=M}^{\infty} \left\{ \left| \mathbf{E}^{\mathcal{A}^{\Delta}_{-j}} u \right| > \epsilon \right\} \right)$$

$$\leqslant \lambda \left(\left\{ \sup_{j \in \mathbb{N}} \left| \mathbf{E}^{\mathcal{A}^{\Delta}_{-j}} u \right| > \epsilon \right\} \right)$$

$$\leqslant \frac{1}{\epsilon^p} \left\| \mathbf{E}^{\mathcal{A}^{\Delta}_{-1}} u \right\|_p \leqslant \frac{1}{\epsilon^p} \left\| u \right\|_p.$$

This shows that $\lambda(A) < \infty$. Since $\mathcal{A}^{\Delta}_{-\infty} = \{\emptyset, \mathbb{R}\}$ is the trivial σ-algebra, we conclude that $\lambda(A) = 0$ or $A = \emptyset$. Therefore, $\mathbf{E}^{\mathcal{A}^{\Delta}_{-M}} u \xrightarrow{M \to \infty} 0$ almost everywhere and so $u_{N,-M} \xrightarrow{M,N \to \infty} u$ almost everywhere.

Step 8: The maximal inequality (24.22)*:* From step 2 we know that

$$\left\| \sup_{N,M \in \mathbb{N}} u_{N,-M} \right\|_p = \left\| \sup_{N,M \in \mathbb{N}} \sum_{k=-M}^{N} \sum_{j=-\infty}^{\infty} \langle u, \psi_{j,k} \rangle \, \psi_{j,k} \right\|_p$$

$$= \left\| \sup_{N \in \mathbb{N}} \mathbf{E}^{\mathcal{A}^{\Delta}_{N+1}} u - \inf_{M \in \mathbb{N}} \mathbf{E}^{\mathcal{A}^{\Delta}_{-M}} u \right\|_p$$

$$\leqslant \frac{p}{p-1} \left\| u \right\|_p + \left\| \mathbf{E}^{\mathcal{A}^{\Delta}_{-1}} u \right\|_p.$$

The last estimate follows from a combination of Minkowski's inequality, Doob's maximal L^p-inequality for martingales T19.12 applied to the closed (by u) martingale $\left(\mathbf{E}^{A_N^\Delta}u\right)_{N\in\mathbb{N}\cup\{\infty\}}$, cf. step 3 and Theorem 23.15, and the fact that $\left(\left|\mathbf{E}^{A_{-M}^\Delta}u\right|^p\right)_{M\in\mathbb{N}}$ is a reversed submartingale, cf. Example 17.3(vi) or Corollary 23.16, which entails $\|\mathbf{E}^{A_{-M}^\Delta}u\|_p \leqslant \|\mathbf{E}^{A_{-1}^\Delta}u\|_p$. Since by T23.8(ii) conditional expectations are contractions on L^p, we have $\|\mathbf{E}^{A_{-1}^\Delta}u\|_p \leqslant \|u\|_p$, and the proof is completed. ∎

A nice introduction to the Haar and other wavelets is Pinsky [35].

The Rademacher functions

Let $L^2 = L^2\big([0,1), \mathcal{B}[0,1), \lambda\big)$, $\lambda = \lambda^1|_{[0,1)}$. The *Rademacher functions* $(R_k)_{k\in\mathbb{N}_0}$ are functions on L^2 defined by

$$R_0 := \mathbf{1}_{[0,1)}, \quad R_1 := \mathbf{1}_{[0,\frac{1}{2})} - \mathbf{1}_{[\frac{1}{2},0)}, \quad R_2 := \mathbf{1}_{[0,\frac{1}{4})} - \mathbf{1}_{[\frac{1}{4},\frac{1}{2})} + \mathbf{1}_{[\frac{1}{2},\frac{3}{4})} - \mathbf{1}_{[\frac{3}{4},1)}, \dots$$

The graphs of the first four Rademacher functions are

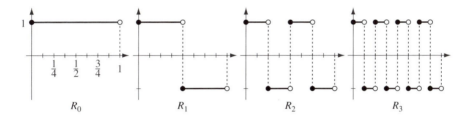

In terms of Haar functions we have

$$R_0 = \chi_{0,0}, \qquad R_{k+1} = \frac{1}{2^{k/2}} \sum_{j=1}^{2^k} \chi_{j,k}, \qquad k \in \mathbb{N}_0. \tag{24.25}$$

Another equivalent definition of the Rademacher system is the following: expand each $x \in [0,1)$ as binary series, $x = \sum_{j=1}^\infty \epsilon_j 2^{-j}$ with $\epsilon_j \in \{0,1\}$ – we exclude expansions terminating with a string of 1s to enforce uniqueness – and set

$$R_0(x) := \mathbf{1}_{[0,1)}(x), \qquad R_k(x) := 2\epsilon_k - 1.$$

Yet another way to think of the functions R_k is as right-continuous versions of sign changes: $R_k(x) \approx \operatorname{sgn} \sin(2^k \pi x)$, $k \in \mathbb{N}_0$.

24.21 Lemma *The system of Rademacher functions $(R_k)_{k\in\mathbb{N}}$ is an ONS of independent[4] functions in $L^2([0,1), dx)$ which is not complete.*

---

[4] In the sense of Example 17.3(x) and Scholium 17.4.

Proof Orthonormality follows since $\int_{\{R_k=\pm1\}} R_\ell \, d\lambda = 0$ for all $k < \ell$, thus $\int R_k R_\ell \, d\lambda = 0$ while $\int R_k^2 \, d\lambda = 1$ is obvious.

In very much the same way we deduce that $\int R_k R_1 R_2 \, d\lambda = 0$ for all $k \in \mathbb{N}_0$ which shows that the system $(R_k)_{k\in\mathbb{N}_0}$ is not complete.

Independence is a special case of Scholium 17.4 with $p = q = 1/2$. ∎

Although $(R_k)_{k\in\mathbb{N}_0}$ is not complete in L^2, it still has good a.e. convergence properties. The reason for this is formula (24.25) and independence.

24.22 Theorem *The Rademacher series $\sum_{k=1}^\infty c_k R_k$, $c_k \in \mathbb{R}$, converges almost everywhere if, and only if, $\sum_{k=0}^\infty c_k^2 < \infty$.*

Proof Assume first that $\sum_{k=0}^\infty c_k^2 < \infty$. In view of (24.25) we set $c_{j,k} := 2^{-k/2} c_k$ and rearrange the absolutely convergent series as

$$\sum_{k=0}^\infty c_k^2 = \sum_{k=0}^\infty \sum_{j=1}^{2^k} c_{j,k}^2 < \infty.$$

We can now interpret the double sequence $(c_{j,k} : 1 \leqslant j \leqslant 2^k, \; k \in \mathbb{N}_0)$ as coefficients of the complete (!) Haar ONS $(\chi_{j,k} : 1 \leqslant j \leqslant 2^k, \; k \in \mathbb{N}_0)$. From Parseval's identity T21.11(iv) we then conclude that the series

$$\sum_{k=0}^\infty c_k R_k = \sum_{k=0}^\infty \sum_{j=1}^{2^k} c_{j,k} \chi_{j,k}$$

converges almost everywhere and in L^2 to some element $u \in L^2$.

Conversely, assume that the series $\sum_{k=0}^\infty c_k R_k$ converges to a finite limit $s(x)$ for all $x \in E \in \mathcal{B}[0,1)$ such that $\lambda(E) > 0$. Writing s_N for the Nth partial sum of this series, we see that

$$A(N) := \bigcup_{j=N}^\infty \left\{ x \in E : |s_j(x) - s(x)| > \tfrac{1}{2} \right\} \quad \text{and} \quad \bigcap_{N\in\mathbb{N}} A(N) = \emptyset.$$

By the continuity of measures T4.4 we find for every $\epsilon > 0$ some $N = N_\epsilon \in \mathbb{N}$ such that

$$\lambda(A(N)) < \epsilon < \tfrac{1}{2}\lambda(E) \qquad \text{and} \qquad \lambda(E \setminus A(N)) > 0.$$

In particular, if $E^* := E \setminus A(N)$,

$$|s_j(x) - s_k(x)| \leqslant |s_j(x) - s(x)| + |s(x) - s_k(x)| \leqslant 1 \qquad \forall j, k > N, \; x \in E^*,$$

and an application of the Cauchy–Schwarz inequality for (double) series, cf. (12.13), shows

$$\lambda(E^*) \geqslant \int_{E^*} \underbrace{\left(\sum_{k=M+1}^{N} c_k R_k \right)^2}_{\leqslant 1} d\lambda$$

$$= \lambda(E^*) \sum_{k=M+1}^{N} c_k^2 + 2 \sum_{M<j<k\leqslant N} c_j c_k \int_{E^*} R_j R_k \, d\lambda$$

$$\geqslant \lambda(E^*) \sum_{k=M+1}^{N} c_k^2 - 2 \left(\sum_{M<j<k\leqslant N} c_j^2 c_k^2 \right)^{1/2} \left(\sum_{M<j<k\leqslant N} \left(\int_{E^*} R_j R_k \, d\lambda \right)^2 \right)^{1/2}$$

$$= \lambda(E^*) \sum_{k=M+1}^{N} c_k^2 - 2 \left(\sum_{M<k\leqslant N} c_k^2 \right) \left(\sum_{M<j<k\leqslant N} \left(\int_{E^*} R_j R_k \, d\lambda \right)^2 \right)^{1/2}. \quad (24.26)$$

Consider now the system $(R_j R_k)_{0 \leqslant j < k < \infty}$. Since for all $m > \ell \geqslant k \geqslant j$ the integral $\int_{\{R_j R_k R_\ell = \pm 1\}} R_m \, d\lambda = 0$, we see that

$$\int (R_j R_k)(R_\ell R_m) \, d\lambda = 0 \qquad \text{if } (j,k) \neq (\ell, m).$$

This shows that $(R_j R_k)_{0 \leqslant j < k < \infty}$ is itself an ONS in L^2, and by Bessel's inequality (Theorem 21.11(iii)) for this ONS and the function $u := \mathbf{1}_{E^*}$ we find

$$\sum_{j<k} \left(\int \mathbf{1}_{E^*} R_j R_k \, d\lambda \right)^2 \leqslant \|\mathbf{1}_{E^*}\|_2^2 = \lambda(E^*) \leqslant 1.$$

For sufficiently large values of $M \in \mathbb{N}$ we can thus achieve that

$$\sum_{M<j<k} \left(\int_{E^*} R_j R_k \, d\lambda \right)^2 \leqslant \left(\frac{\lambda(E^*)}{4} \right)^2,$$

and (24.26) becomes as $N \to \infty$

$$\lambda(E^*) \geqslant \lambda(E^*) \sum_{k>M} c_k^2 - 2 \left(\sum_{k>M} c_k^2 \right) \frac{\lambda(E^*)}{4} = \frac{\lambda(E^*)}{2} \sum_{k>M} c_k^2,$$

which implies $\sum_{k>M} c_k^2 \leqslant 2$, i.e. $\sum_{k=0}^{\infty} c_k^2 < \infty$, and we are done. ∎

It is possible to extend the Rademacher system explicitly to a complete ONS. This can be achieved by the following construction:

$$w_0 := R_0, \qquad w_n := R_{j_1+1} \cdot R_{j_2+1} \cdot \ldots \cdot R_{j_k+1}, \qquad n \in \mathbb{N}, \quad (24.27)$$

where $n = 2^{j_1} + 2^{j_2} + \cdots + 2^{j_k}$ is the unique dyadic representation of $n \in \mathbb{N}$ where $0 \leqslant j_1 < j_2 < \ldots < j_k$. A similar argument to the one used in the second part of the proof of Theorem 24.22 shows that $(w_n)_{n \in \mathbb{N}_0}$ is indeed an ONS. Note that $R_{k+1} = w_{2^k}$, so that $(R_k)_{k \in \mathbb{N}_0} \subset (w_n)_{n \in \mathbb{N}_0}$.

24.23 Definition The system (24.27) is called the *Walsh orthonormal system* (in Paley's ordering).

The Walsh system is a complete ONS, cf. Alexits [1, pp. 61–3] or Schipp *et al.* [41], and it is susceptible to a complete martingale treatment, cf. [41]. Again one considers the filtration of dyadic σ-algebras $(\mathcal{A}_n^\Delta)_{n \in \mathbb{N}_0}$ on $[0, 1)$ and the special partial sums

$$s_{2^n-1}(u) := \sum_{j=1}^{2^n-1} \langle u, w_j \rangle \, w_j.$$

Then $s_n(u) = \mathbf{E}^{\mathcal{A}_n^\Delta} u$ and we have the full martingale toolkit at our disposal. With the methods used so far it is possible to show that $s_{2^n-1}(u) \xrightarrow{n \to \infty} u$ a.e. and in L^p, $1 \leqslant p < \infty$. The case of general partial sums $s_n(u)$ is somewhat harder to handle but it is still doable with some variations of the techniques presented here; see Schipp *et al.* [41, Chapters 4, 6].

Well-behaved orthonormal systems

For the Haar system and the Haar wavelet we could use martingale methods. A close inspection of our proofs reveals that the crucial input for getting martingales is that the ONS $(e_j)_{j \in \mathbb{N}_0}$ satisfies

$$\mathbf{E}^{\mathcal{A}_n} e_{n+1} = 0, \qquad n \in \mathbb{N}_0, \tag{24.28}$$

where $\mathcal{A}_n = \sigma(e_0, e_1, \ldots, e_n)$. This condition implies immediately that the partial sum $\sum_{j=0}^n c_j e_j$ is a martingale w.r.t. the filtration $(\mathcal{A}_n)_{n \in \mathbb{N}_0}$ generated by the ONS $(e_j)_{j \in \mathbb{N}}$.

24.24 Definition Let $(X, \mathcal{A}, \mu)$ be a σ-finite measure space and $1 \leqslant p < \infty$. A family of functions $(e_j)_{j \in \mathbb{N}_0} \subset L^p(X, \mathcal{A}, \mu)$ satisfying (24.28) is called a *system of martingale differences*.

For a system of martingale differences no orthogonality is required. The archetype of martingale differences are sequences of independent[5] functions $(f_k)_{k \in \mathbb{N}_0} \subset$

---

[5] In the sense of Example 17.3(x).

L^2 ($\subset L^1$, since μ is a probability measure) which are normalized such that $\int f_k \, d\mu = 0$ and $\int f_k^2 \, d\mu = 1$. Our methods used in connection with the Haar system and Haar wavelets still apply and yield

24.25 Theorem *Let* $(X, \mathcal{A}, \mu)$ *be a σ-finite measure space and let* $(e_j)_{j \in \mathbb{N}_0}$ *be an ONS of martingale differences in* $L^2(X, \mathcal{A}, \mu)$. *Then*

$$s_n(u; x) := \sum_{j=0}^{n} \langle u, e_j \rangle \, e_j(x), \qquad n \in \mathbb{N}, \ u \in L^1 \cap L^2,$$

is a martingale w.r.t. the filtration $\mathcal{A}_n := \sigma(e_0, e_1, \ldots, e_n)$. *For every* $u \in L^2$ *the sequence* $(s_n(u))_{n \in \mathbb{N}}$ *converges a.e. and satisfies the following maximal inequality:*

$$\left\| \sup_{n \in \mathbb{N}} s_n(u) \right\|_2 \leqslant 2 \, \|u\|_2, \qquad u \in L^2.$$

Proof That the sequence of partial sums satisfies $\mathbf{E}^{\mathcal{A}_n} s_{n+1}(u) = s_n(u)$ for $u \in L^2$ and is, for $u \in L^1 \cap L^2$, a martingale is clear. Therefore, Corollary 23.16 shows that $(|s_n(u^{\pm})|^2)_{n \in \mathbb{N}}$ are submartingales, and from Bessel's inequality, cf. Theorem 21.11,

$$\sup_{n \in \mathbb{N}} \|s_n(u^{\pm})\|_2 \leqslant \|u^{\pm}\|_2, \qquad u \in L^2,$$

we conclude that $(|s_n(u^{\pm})|^2)_{n \in \mathbb{N}}$ satisfy the conditions of the submartingale convergence theorem 18.2. Thus $\lim_{n \to \infty} |s_n(u^{\pm})|^2$ exists a.e. in $[0, \infty)$, and, since $s_n(u^{\pm}) \geqslant 0$, so does $\lim_{n \to \infty} s_n(u)$.

From Doob's maximal inequality T19.12 and Bessel's inequality T21.11 we get

$$\left\| \sup_{n \leqslant N} s_n(u) \right\|_2 \leqslant 2 \, \|s_N(u)\|_2 \leqslant 2 \, \|u\|_2, \qquad N \in \mathbb{N},$$

and the usual monotone convergence argument proves the maximal inequality as $N \to \infty$. ∎

In the situation of Theorem 24.25 we cannot say much more about the limit $\lim_{n \to \infty} s_n(u; x)$ apart from its mere existence. In particular, the partial sums $\lim_{n \to \infty} s_n(u)$ can converge to something completely different from u! Consider, for example, the system of Rademacher functions $(R_n)_{n \in \mathbb{N}_0}$, which is clearly a system of martingale differences. If $u = R_1 R_2$ we get

$$\langle u, R_j \rangle = \langle R_1 R_2, R_j \rangle = \int_{[0,1)} R_1(x) R_2(x) R_j(x) \, dx = 0$$

for all $j \in \mathbb{N}$. Thus $s_n(R_1 R_2) \equiv 0$ is convergent, but $\lim_{n \to \infty} s_n(R_1 R_2) \equiv 0 \neq R_1 R_2$. The reason is that the Rademacher functions are not complete in L^2. This also means that we cannot hope to get L^p-convergence in Theorem 24.25.

24.26 Theorem *Let $(X, \mathcal{A}, \mu)$ be a σ-finite measure space and $(e_j)_{j \in \mathbb{N}_0} \subset L^2(\mathcal{A})$ be an ONS of martingale differences. Denote by $s_n(u)$ the partial sum*

$$s_n(u; x) := \sum_{j=0}^{n} \langle u, e_j \rangle \, e_j(x), \qquad u \in L^2,$$

and by $\mathcal{A}_n := \sigma(e_0, e_1, \ldots, e_n)$ the associated canonical filtration. Then the following assertions are equivalent:

(i) $(e_j)_{j \in \mathbb{N}_0}$ *is a complete ONS.*

(ii) $\int_A s_n(u) \, d\mu = \int_A u \, d\mu$ *for all $A \in \mathcal{A}_n$, $\mu(A) < \infty$, and $u \in L^2(\mu)$.*

(iii) $\mathbf{E}^{\mathcal{A}_n} u = s_n(u)$ *for all $u \in L^2(\mu)$.*

(iv) $\lim_{n \to \infty} \|s_n(u) - u\|_p = 0$ *for all $u \in L^p(\mu)$ and all $1 \leqslant p < \infty$.*

Proof (i)$\Rightarrow$(ii): Since $(e_j)_{j \in \mathbb{N}_0}$ is complete, we know from Theorem 21.13 that $\lim_{n \to \infty} \|s_n(u) - u\|_2 = 0$ for all $u \in L^2$. Using the Cauchy–Schwarz inequality 12.3 we see for every $A \in \mathcal{A}_n$ with $\mu(A) < \infty$

$$\int_A |s_n(u) - u| \, d\mu \leqslant \|s_n(u) - u\|_2 \cdot \|\mathbf{1}_A\|_2 \xrightarrow{n \to \infty} 0.$$

Thus $\lim_{n \to \infty} \int_A s_n(u) \, d\mu = \int_A u \, d\mu$. Since $(e_j)_{j \in \mathbb{N}_0}$ is a system of martingale differences, we know that $\mathbf{E}^{\mathcal{A}_n} e_{n+k} = 0$, $k \in \mathbb{N}$,[✓] and by Theorem 23.9, applied to the function $\mathbf{1}_A \, e_{n+k} \in L^1$ and $A \in \mathcal{A}_n$,

$$\int_A e_{n+k} \, d\mu = \int_A \mathbf{1}_A \, e_{n+k} \, d\mu = 0.$$

Therefore $\int_A u \, d\mu = \lim_{j \to \infty} \int_A s_j(u) \, d\mu = \int_A s_n(u) \, d\mu$ holds for all $n \in \mathbb{N}$ and $A \in \mathcal{A}_n$ with $\mu(A) < \infty$.

(ii)$\Rightarrow$(iii) Since $\mathbf{1}_A u \in L^1$ for all $A \in \mathcal{A}_n$ with $\mu(A) < \infty$ and $u \in L^2$, Theorem 23.9 and 23.8(vii) show

$$\int_A u \, d\mu = \int_A \mathbf{1}_A u \, d\mu = \int_A \mathbf{E}^{\mathcal{A}_n} (\mathbf{1}_A u) \, d\mu$$

$$= \int_A \mathbf{1}_A \, \mathbf{E}^{\mathcal{A}_n}(u) \, d\mu = \int_A \mathbf{E}^{\mathcal{A}_n}(u) \, d\mu.$$

Together with the assumption this gives

$$\int_A \mathbf{E}^{\mathcal{A}_n} u \, d\mu = \int_A s_n(u) \, d\mu \qquad \forall A \in \mathcal{A}_n, \ \mu(A) < \infty.$$

Choose, in particular, for every $k \in \mathbb{N}$ the set $\left\{s_n(u) > \frac{1}{k} + \mathbf{E}^{\mathcal{A}_n}u\right\} \in \mathcal{A}_n$. By Markov's inequality P10.12 we see

$$\mu\left(\left\{s_n(u) - \mathbf{E}^{\mathcal{A}_n}u > \frac{1}{k}\right\}\right) \leqslant k^2 \left\|s_n(u) - \mathbf{E}^{\mathcal{A}_n}u\right\|_2^2 < \infty,$$

so that the above equality becomes

$$0 = \int_{\left\{s_n(u) > \frac{1}{k} + \mathbf{E}^{\mathcal{A}_n}u\right\}} \left(s_n(u) - \mathbf{E}^{\mathcal{A}_n}u\right) d\mu \geqslant \frac{1}{k} \mu\left(\left\{s_n(u) > \frac{1}{k} + \mathbf{E}^{\mathcal{A}_n}u\right\}\right).$$

This is only possible if $\mu\left(\left\{s_n(u) > \frac{1}{k} + \mathbf{E}^{\mathcal{A}_n}u\right\}\right) = 0$. A similar argument for the set $\left\{s_n(u) < \frac{1}{k} + \mathbf{E}^{\mathcal{A}_n}u\right\}$ finally shows

$$\mu\left(\left\{s_n(u) \neq \mathbf{E}^{\mathcal{A}_n}u\right\}\right) = \mu\left(\bigcup_{k\in\mathbb{N}} \left\{|s_n(u) - \mathbf{E}^{\mathcal{A}_n}u| > \frac{1}{k}\right\}\right)$$

$$\leqslant \sum_{k\in\mathbb{N}} \mu\left(\left\{|s_n(u) - \mathbf{E}^{\mathcal{A}_n}u| > \frac{1}{k}\right\}\right) = 0.$$

Therefore, $s_n(u) = \mathbf{E}^{\mathcal{A}_n}u$ a.e.

(iii)$\Rightarrow$(iv) For $u \in L^1 \cap L^\infty$ Theorem 23.15 shows that $\left(\mathbf{E}^{\mathcal{A}_n}u\right)_{n\in\mathbb{N}}$ is a uniformly integrable martingale and that $\mathbf{E}^{\mathcal{A}_n}u \xrightarrow{n\to\infty} u$ in L^1 and a.e. As in step 4 of the proof of Theorem 24.17, we use the inequality

$$|a|^p - |b|^p \leqslant p|a-b| \max\{|a|^{p-1}, |b|^{p-1}\}, \qquad a, b \in \mathbb{R}, \ p > 1,$$

to deduce that

$$\pm \int \left(|\mathbf{E}^{\mathcal{A}_n}u|^p - |u|^p\right) d\mu \leqslant p \left\|\mathbf{E}^{\mathcal{A}_n}u - u\right\|_1 \cdot \|u\|_\infty^{p-1}$$

and, by Riesz' convergence theorem 12.10, that $\mathbf{E}^{\mathcal{A}_n}u \xrightarrow{n\to\infty} u$ in L^p.

If $u \in L^p$ is not bounded, we take an exhausting sequence $(A_k)_{k\in\mathbb{N}} \subset \mathcal{A}$ with $A_k \uparrow X$ and $\mu(A_k) < \infty$ and set $u_k := \left((-k) \vee u \wedge k\right)\mathbf{1}_{A_k}$. Clearly, $u_k \in L^1 \cap L^\infty$, and we see using Theorem 23.8(v),(ii),

$$\left\|\mathbf{E}^{\mathcal{A}_n}u - u\right\|_p \leqslant \left\|\mathbf{E}^{\mathcal{A}_n}u - \mathbf{E}^{\mathcal{A}_n}u_k\right\|_p + \left\|\mathbf{E}^{\mathcal{A}_n}u_k - u_k\right\|_p + \|u_k - u\|_p$$

$$\leqslant \left\|\mathbf{E}^{\mathcal{A}_n}u_k - u_k\right\|_p + 2\|u_k - u\|_p.$$

The claim follows if we let first $n \to \infty$ and then $k \to \infty$.

(iv)$\Rightarrow$(i) is just $p = 2$ combined with Theorem 21.13. ∎

If we know that the elements of the ONS are independent, we obtain the following necessary and sufficient conditions for pointwise convergence which generalize Theorem 24.22.

24.27 Theorem *Let $(X, \mathcal{A}, P)$ be a probability space and $(e_j)_{j \in \mathbb{N}_0} \subset L^2(P)$ be independent random variables such that*

$$\int e_j \, dP = 0 \quad and \quad \int e_j^2 \, dP = 1$$

and let $(c_j)_{j \in \mathbb{N}_0} \subset \mathbb{R}$ be a sequence of real numbers. Then

(i) *The family $(e_j)_{j \in \mathbb{N}_0}$ is an ONS of martingale differences;*
(ii) *If $\sum_{j=0}^{\infty} c_j^2 < \infty$, then $\sum_{j=0}^{\infty} c_j e_j$ converges in $L^2(P)$ and a.e.;*
(iii) *If $\sup_{j \in \mathbb{N}_0} \|e_j\|_\infty \leqslant \kappa < \infty$ and if $\sum_{j=0}^{\infty} c_j e_j$ converges almost everywhere, then $\sum_{j=0}^{\infty} c_j^2 < \infty$.*

Proof (i) We set

$$\mathcal{A}_n := \sigma(e_0, e_1, \dots, e_n), \quad and \quad u_n := \sum_{j=0}^{n} c_j e_j.$$

Since P is a probability measure, $u_j \in L^2(P) \subset L^1(P)$ and under our assumptions it is clear that $(u_j, \mathcal{A}_j)_{j \in \mathbb{N}_0}$ is a martingale.[✓]

By independence we have

$$\int e_j e_k \, dP = \begin{cases} \int e_j \, dP \cdot \int e_k \, dP = 0 & \text{if } j \neq k, \\ \\ \int e_j^2 \, dP = 1 & \text{if } j = k, \end{cases}$$

which entails

$$\int u_n^2 \, dP = \sum_{j,k=0}^{n} c_j c_k \int e_j e_k \, dP = \sum_{j=0}^{n} c_j^2 \tag{24.29}$$

and also

$$\int u_n u_{n+k} \, dP = \int \mathbf{E}^{\mathcal{A}_n} u_n u_{n+k} \, dP = \int u_n \mathbf{E}^{\mathcal{A}_n} u_{n+k} \, dP = \sum_{j=0}^{n} c_j^2. \tag{24.30}$$

(ii) Because of (24.29) we see that

$$\|u_n\|_1^2 \leqslant \|u_n\|_2^2 = \sum_{j=0}^{n} c_j^2 \leqslant \sum_{j=0}^{\infty} c_j^2 < \infty,$$

and the martingale convergence theorem C18.3 shows that $u_n \xrightarrow{n \to \infty} u_\infty$ a.e. Using (24.30) we conclude that

$$\int (u_{n+k} - u_n)^2 \, dP = \int \left(u_{n+k}^2 - 2 u_n u_{n+k} + u_n^2 \right) dP$$

$$= \int u_{n+k}^2 \, dP - \int u_n^2 \, dP = \sum_{j=n+1}^{n+k} c_j^2 \leqslant \sum_{j=n+1}^{\infty} c_j^2.$$

Thus, by Fatou's lemma 9.11,

$$\int (u_\infty - u_n)^2 \, dP \leqslant \liminf_{k\to\infty} \int (u_{n+k} - u_n)^2 \, dP \leqslant \sum_{j=n+1}^{\infty} c_j^2 \xrightarrow{n\to\infty} 0,$$

and $u_n \xrightarrow{n\to\infty} u_\infty$ follows in the L^2-sense.

(iii) Since e_j and $\mathcal{A}_{j-1}$ are independent, we find for all $A \in \mathcal{A}_{j-1}$

$$\int_A (u_j - u_{j-1})^2 \, dP = \int_A c_j^2 \, e_j^2 \, dP \stackrel{(17.6)}{=} c_j^2 \, P(A) = c_j^2 \int_A dP. \tag{24.31}$$

Essentially the same calculation that was used in (24.30) also yields

$$\int_A (u_j - u_{j-1})^2 \, dP = \int (\mathbf{1}_A u_j - \mathbf{1}_A u_{j-1})^2 \, dP$$

$$= \int (\mathbf{1}_A u_j)^2 \, dP - \int (\mathbf{1}_A u_{j-1})^2 \, dP$$

$$= \int_A (u_j^2 - u_{j-1}^2) \, dP,$$

which can be combined with (24.31) to give

$$\int_A \left(u_n^2 - \sum_{j=0}^{n} c_j^2 \right) dP = \int_A \left(u_{n-1}^2 - \sum_{j=0}^{n-1} c_j^2 \right) dP \qquad \forall A \in \mathcal{A}_{n-1}.$$

This means, however, that $w_n := u_n^2 - \sum_{j=0}^{n} c_j^2$ is a martingale.

Consider the stopping time $\tau = \tau_\gamma := \inf\{n \in \mathbb{N}_0 : |u_n| > \gamma\}$, $\inf \emptyset = \infty$. Since the series $\sum_{j=0}^{\infty} c_j \, e_j$ converges a.e., we can choose $\gamma > 0$ in such a way that

$$\kappa^2 P(\{\tau < \infty\}) < \tfrac{1}{2} P(\{\tau = \infty\}).$$

Without loss of generality we may also take $\gamma^2 > |\int w_0 \, dP| + |\int u_0^2 \, dP|$.

The optional sampling theorem 17.8 proves that $(w_{n\wedge\tau})_{n\in\mathbb{N}}$ is again a martingale and, therefore,

$$\int w_0 \, dP = \int w_{n\wedge\tau} \, dP = \int u_{n\wedge\tau}^2 \, dP - \int \sum_{j=0}^{n\wedge\tau} c_j^2 \, dP. \tag{24.32}$$

Taking into account the very definition of τ we find furthermore

$$\int u_{n\wedge\tau}^2 \, dP = \int_{\{\tau > n\}} u_{n\wedge\tau}^2 \, dP + \int_{\{1\leqslant\tau\leqslant n\}} u_{n\wedge\tau}^2 \, dP + \int_{\{\tau=0\}} u_{n\wedge\tau}^2 \, dP$$

$$\leqslant 2\gamma^2 + \int_{\{1\leqslant\tau\leqslant n\}} u_\tau^2 \, dP$$

$$= 2\gamma^2 + \int\limits_{\{1\leqslant\tau\leqslant n\}} (c_\tau\,e_\tau + u_{\tau-1})^2\,dP$$

$$\leqslant 2\gamma^2 + 2\int\limits_{\{1\leqslant\tau\leqslant n\}} \left(c_\tau^2\,e_\tau^2 + u_{\tau-1}^2\right)dP,$$

where we used the elementary inequality $(a+b)^2 \leqslant 2a^2 + 2b^2$ in the last line. Since the e_j are uniformly bounded by κ and since $|u_{\tau-1}| \leqslant \gamma$, we get

$$\int u_{n\wedge\tau}^2\,dP \leqslant 4\gamma^2 + \kappa^2\int\limits_{\{\tau\leqslant n\}} c_\tau^2\,dP$$

$$\leqslant 4\gamma^2 + \kappa^2 P(\{\tau\leqslant n\})\sum_{j=0}^{n} c_j^2 \qquad (24.33)$$

$$\leqslant 4\gamma^2 + \tfrac{1}{2}\,P(\{\tau=\infty\})\sum_{j=0}^{n} c_j^2,$$

since, by construction, $\kappa^2\,P(\{\tau\leqslant n\}) \leqslant \kappa^2\,P(\{\tau<\infty\}) < \tfrac{1}{2}\,P(\{\tau=\infty\})$.

Rearranging (24.32) and combining this with the above estimates we obtain

$$P(\{\tau=\infty\})\sum_{j=0}^{n} c_j^2 = \int\limits_{\{\tau=\infty\}} \sum_{j=0}^{n\wedge\tau} c_j^2\,dP \;\leqslant\; \int \sum_{j=0}^{n\wedge\tau} c_j^2\,dP$$

$$\overset{(24.32)}{=} \int u_{n\wedge\tau}^2\,dP - \int w_0\,dP$$

$$\overset{(24.33)}{\leqslant} 4\gamma^2 + \tfrac{1}{2}\,P(\{\tau=\infty\})\sum_{j=0}^{n} c_j^2 + \gamma^2,$$

uniformly for all $n \in \mathbb{N}$. Since, by assumption, $P(\{\tau=\infty\}) > 0$ for sufficiently large γ, we conclude that $\sum_{j=0}^{\infty} c_j^2 < \infty$. ∎

Theorem 24.27 has an astonishing corollary if we apply the Burkholder–Davis–Gundy inequalities (BDG) from p. 294 to the martingale

$$w_n := u_{n+k} - u_k = \sum_{j=k+1}^{n+k} c_j\,e_j$$

w.r.t. the filtration $\mathcal{F}_n := \mathcal{A}_{n+k} := \sigma(e_0, e_1, \ldots, e_{n+k})$. The part of the inequalities which is important for our purposes reads

$$\kappa_p\,\|w_n\|_p \leqslant \kappa_p\,\Big\|\sup_{0\leqslant j\leqslant n} |w_n|\Big\|_p \leqslant \Big\|\sqrt{[w_\bullet, w_\bullet]_n}\,\Big\|_p, \qquad (24.34)$$

where $n \in \mathbb{N}_0$, $0 < p < \infty$, and the quadratic variation is given by

$$[w_\bullet, w_\bullet]_n = [u_{\bullet+k} - u_k, u_{\bullet+k} - u_k]_n = \sum_{j=0}^{n-1} |u_{j+k+1} - u_{j+k}|^2 = \sum_{j=k+1}^{n+k} c_j^2 \, e_j^2 .$$

If we happen to know that $\sup_{j\in\mathbb{N}} \|e_j\|_\infty \leqslant \kappa < \infty$, we even find

$$\left\| \sqrt{[w_\bullet, w_\bullet]_n} \right\|_\infty \leqslant \kappa \left(\sum_{j=k+1}^{n+k} c_j^2 \right)^{1/2}$$

and we conclude from (24.34) that for all $n, k \in \mathbb{N}$ and $0 < p < \infty$

$$\kappa_p \|u_{n+k} - u_k\|_p \leqslant \left\| [u_{\bullet+k} - u_k, u_{\bullet+k} - u_k]_n^{1/2} \right\|_p$$

$$\leqslant \left\| [u_{\bullet+k} - u_k, u_{\bullet+k} - u_k]_n^{1/2} \right\|_\infty \leqslant \kappa \left(\sum_{j=k+1}^{n+k} c_j^2 \right)^{1/2}$$

holds. This proves immediately the following

24.28 Corollary *Let $(X, \mathcal{A}, P)$ be a probability space and let $(e_j)_{n\in\mathbb{N}_0}$ be a sequence of independent random variables such that*

$$\sup_{j\in\mathbb{N}_0} \|e_j\|_\infty < \infty, \quad \int e_j \, dP = 0 \quad and \quad \int e_j^2 \, dP = 1.$$

Then $u_n := \sum_{j=0}^{n} c_j e_j$ converges in L^2 and a.e. to some $u \in L^2$ if, and only if, $\sum_{j=0}^{\infty} c_j^2 < \infty$.

If the latter is the case, $u \in L^p$ and the convergence takes place in L^p-sense for all $0 < p < \infty$.

Unfortunately, many ONSs of martingale differences are incomplete and seem to behave more often like Rademacher functions than Haar functions. More on this topic can be found in the paper by Gundy [18] and the book by Garsia [16].

24.29 Epilogue The combination of martingale methods and orthogonal expansions opens up a whole new world. Let us illustrate this by a rapid construction of one of the most prominent stochastic process: the *Wiener process* or *Brownian motion*.

Choose in Theorem 24.27 $(X, \mathcal{A}, P) = ([0, 1], \mathcal{B}[0, 1], \lambda)$ where λ is one-dimensional Lebesgue measure on $[0, 1]$; denoting points in $[0, 1]$ by ω, we will often write $d\omega$ instead of $\lambda(d\omega)$. Assume that the independent, identically distributed random variables e_j are all standard normal Gaussian random variables, i.e.

$$P(e_j \in B) = \frac{1}{\sqrt{2\pi}} \int_B e^{-x^2/2} \, dx, \qquad B \in \mathcal{B}[0, 1],$$

and consider the series expansion

$$W_t(\omega) := \sum_{n=0}^{\infty} e_n(\omega)\langle \mathbf{1}_{[0,t]}, H_n \rangle, \qquad \omega \in [0, 1].$$

Here $t \in [0, 1]$ is a parameter, $\langle u, v \rangle = \int_0^1 u(x)v(x) \, dx$, and H_n, $n = 2^k + j$, $0 \leqslant j < 2^k$, denote the lexicographically ordered Haar functions (24.16). A short calculation confirms for $n \geqslant 1$

$$\langle \mathbf{1}_{[0,t]}, H_n \rangle = \int_0^t H_n(x) \, dx = \tfrac{1}{2} 2^{k/2} \int_0^t H_1(2^k x - j) \, dx = \tfrac{1}{2} 2^{-k/2} F_n(t),$$

where $F_1(t) = \int_0^t H_1(x) \, dx \, \mathbf{1}_{[0,1]}(t) = 2t \mathbf{1}_{[0,\frac{1}{2})}(t) - (2t - 2)\mathbf{1}_{[\frac{1}{2},1]}(t)$ is a tent-function and $F_n(t) := F_1(2^k t - j)$. Since $0 \leqslant F_n \leqslant 1$, we see

$$\sum_{n=0}^{\infty} |\langle \mathbf{1}_{[0,t]}, H_n \rangle|^2 \leqslant \frac{1}{4} \sum_{n=0}^{\infty} 2^{-k} = \frac{1}{2},$$

and Theorem 24.27(ii) guarantees that $W_t(\omega)$ exists, for each $t \in [0, 1]$, both in $L^2(d\omega)$-sense and $\lambda(d\omega)$-almost everywhere.

More is true. Since the e_n are independent Gaussian random variables, so are their finite linear combinations (e.g. Bauer [5, §24]) and, in particular, the partial sums

$$S_N(t; \omega) := \sum_{n=0}^{N} e_n(\omega)\langle \mathbf{1}_{[0,t]}, H_n \rangle.$$

Gaussianity is preserved under L^2-limits;[6] we conclude that $W_t(\omega)$ has a Gaussian distribution for each t. The mean is given by

$$\int_0^1 W_t(\omega) \, d\omega = \sum_{n=0}^{\infty} \int_0^1 e_n(\omega) \, d\omega \, \langle \mathbf{1}_{[0,t]}, H_n \rangle = 0$$

(to change integration and summation use that $L^2(d\omega)$-convergence entails $L^1(d\omega)$-convergence on a finite measure space). Since $\int e_n e_m \, d\omega = 0$ or 1

---

[6] (cf. [5, §§23, 24]) if X_n is normal distributed with mean 0 and variance σ_n^2, its Fourier transform is $\int e^{i\xi X_n} \, dP = e^{\sigma_n^2 \xi^2/2}$. If $X_n \xrightarrow{n \to \infty} X$ in L^2-sense, we have $\sigma_n^2 \to \sigma^2$ and, by dominated convergence, $\int e^{i\xi X} \, dP = \lim_n \int e^{i\xi X_n} \, dP = \lim_n e^{\sigma_n^2 \xi^2/2} = e^{\sigma^2 \xi^2/2}$; the claim follows from the uniqueness of the Fourier transform.

according to $n \neq m$ or $n = m$, we can calculate for $0 \leqslant s < t \leqslant 1$ the variance by

$$\int_0^1 (W_t(\omega) - W_s(\omega))^2 \, d\omega$$

$$= \sum_{n,m=0}^{\infty} \int_0^1 e_n(\omega)e_m(\omega) \, d\omega \, \langle \mathbf{1}_{[0,t]} - \mathbf{1}_{[0,s]}, H_n \rangle \langle \mathbf{1}_{[0,t]} - \mathbf{1}_{[0,s]}, H_m \rangle$$

$$= \sum_{n=0}^{\infty} \langle \mathbf{1}_{(s,t]}, H_n \rangle^2 \overset{24.17,21.13}{=} \langle \mathbf{1}_{(s,t]}, \mathbf{1}_{(s,t]} \rangle = t - s.$$

In particular, the increment $W_t - W_s$ has the same probability distribution as W_{t-s}. In the same vein we find for $0 \leqslant s < t \leqslant u < v \leqslant 1$ that

$$\int_0^1 (W_t(\omega) - W_s(\omega))(W_v(\omega) - W_u(\omega)) \, d\omega = \langle \mathbf{1}_{(s,t]}, \mathbf{1}_{(u,v]} \rangle = 0.$$

Since $W_t - W_s$ is Gaussian, this proves already the independence of the two increments $W_t - W_s$ and $W_v - W_u$, cf. [5, §24]. By induction, we conclude that

$$W_{t_n} - W_{t_{n-1}}, \ldots, W_{t_1} - W_{t_0},$$

are independent for all $0 \leqslant t_0 \leqslant \cdots \leqslant t_n \leqslant 1$.

Let us finally turn to the dependence of $W_t(\omega)$ on t. Note that for $M < N$

$$\int_0^1 \sup_{t \in [0,1]} |S_N(t; \omega) - S_M(t; \omega)| \, d\omega = \int_0^1 \sup_{t \in [0,1]} \left| \sum_{n=M+1}^N e_n(\omega) \langle \mathbf{1}_{[0,t]}, H_n \rangle \right| d\omega$$

$$\leqslant \sum_{n=M+1}^N \underbrace{\int_0^1 |e_n(\omega)| \, d\omega}_{= \text{const.}} \sup_{t \in [0,1]} |\langle \mathbf{1}_{[0,t]}, H_n \rangle|$$

$$\leqslant C \sum_{n=M+1}^N \tfrac{1}{2} 2^{-k/2} < \infty,$$

which means that the partial sums $S_N(t; \omega)$ of $W_t(\omega)$ converge in $L^1(d\omega)$ uniformly for all $t \in [0, 1]$. By C12.8 we can extract a subsequence, which converges (uniformly in t) for $\lambda(d\omega)$-almost all ω to $W_t(\omega)$; since for fixed ω the partial sums $t \mapsto S_N(t; \omega)$ are continuous functions of t, this property is inherited by the a.e. limit $W_t(\omega)$.

The above construction is a variation of a theme by Lévy [26, Chap. I.1, pp. 15–20] and Ciesielski [10]. In one or another form it can be found in many probability textbooks, e.g. Bass [3, pp. 11–13] or Steele [45, pp. 35–39]. A related construction of Wiener, see Paley and Wiener [34, Chapter XI], using random Fourier series, is discussed in Kahane [23, §16.1–3].

Problems

24.1. Prove the orthogonality relation for the Jacobi polynomials 24.1.

24.2. Use the Gram–Schmidt orthonormalization procedure to verify the formulae for the first few Chebyshev, Legendre, Laguerre and Hermite polynomials given in 24.1–24.5.

24.3. State and prove Theorem 24.6 and Corollary 24.8 for an arbitrary compact interval $[a, b]$.

24.4. Prove the orthogonality relations (24.4) for the trigonometric system.
[Hint: observe that $\mathrm{Im}\left(e^{i(x+y)} + e^{i(x-y)}\right) = 2\sin x \cos y$.]

24.5. (i) Show that for suitable constants $c_j, s_j \in \mathbb{R}$ and all $k \in \mathbb{N}_0$

$$\cos^k x = \sum_{j=0}^{k} c_j \cos jx \qquad \text{and} \qquad \sin^{k+1} x = \sum_{j=1}^{k+1} s_j \sin jx.$$

(ii) Show that for suitable constants $a_j, b_j \in \mathbb{R}$ and all $k \in \mathbb{N}$

$$\cos kx = \sum_{j=0}^{k} a_j \cos^{k-j} x \sin^j x \quad \text{and} \quad \sin kx = \sum_{j=1}^{k} b_j \cos^{k-j} x \sin^j x.$$

(iii) Deduce that every trigonometric polynomial $T_n(x)$ of order n can be written in the form

$$U_n(x) = \sum_{j,k=0}^{n} \gamma_{j,k} \cos^j x \sin^k x$$

and vice versa.

24.6. Use the formula $\sin a - \sin b = 2\cos \frac{a+b}{2} \sin \frac{a-b}{2}$ to show that $D_N(x)\sin \frac{x}{2} = \frac{1}{2}\sin\left(N+\frac{1}{2}\right)x$. This proves (24.11).

24.7. Find the Fourier series expansion for the function $|\sin x|$.

24.8. Let $u(x) = \mathbf{1}_{[0,1)}(x)$. Show that the Haar–Fourier series for u converges for all $1 \leqslant p < \infty$ in L^p-sense to u. Is this also true for the Haar wavelet expansion?

24.9. Show that the Haar–Fourier series for $u \in C_c(\mathbb{R})$ converges uniformly for every $x \in \mathbb{R}$ to $u(x)$. Show that this remains true for functions $u \in C_\infty(\mathbb{R})$, i.e. the set of continuous functions such that $\lim_{|x|\to\infty} u(x) = 0$.
[Hint: use the fact that $u \in C_c$ is uniformly continuous. For $u \in C_\infty$ observe that $C_\infty = \overline{C_c}^{\|\cdot\|_\infty}$ (closure in sup-norm) and check that $|s_N(u; x)| \leqslant \|u\|_\infty$.]

24.10. Extend Problem 24.9 to the Haar wavelet expansion.
[Hint: use Problem 24.9 and show that $\left\|\mathbf{E}^{A_{-N}^\Delta} u\right\|_\infty \xrightarrow{N\to\infty} 0$ for all $u \in C_c(\mathbb{R})$.]

24.11. Let $u(x) = \mathbf{1}_{[0,1/3)}(x)$. Prove that the Haar–Fourier diverges at $x = \frac{1}{3}$.
[Hint: verify $\liminf_{N\to\infty} s_N(u, \frac{1}{3}) < \limsup_{N\to\infty} s_N(u, \frac{1}{3})$.]

Appendix A

lim inf and lim sup

For a sequence of real numbers $(a_j)_{j \in \mathbb{N}} \subset \mathbb{R}$ the *limes inferior* or *lower limit* is defined as

$$\liminf_{j \to \infty} a_j := \sup_{k \in \mathbb{N}} \inf_{j \geq k} a_j, \tag{A.1}$$

and the *limes superior* or *upper limit* is defined as

$$\limsup_{j \to \infty} a_j := \inf_{k \in \mathbb{N}} \sup_{j \geq k} a_j. \tag{A.2}$$

Lower and upper limits of a sequence are always defined as numbers in $[-\infty, +\infty)$ and $(-\infty, +\infty]$, respectively. This is due to the fact that the sequences $\left(\inf_{j \geq k} a_j\right)_{k \in \mathbb{N}} \subset [-\infty, +\infty)$ and $\left(\sup_{j \geq k} a_j\right)_{k \in \mathbb{N}} \subset (-\infty, +\infty]$ are in- resp. decreasing, so that the $\sup_{k \in \mathbb{N}}$ and $\inf_{k \in \mathbb{N}}$ in (A.1) and (A.2) are actually (improper) limits $\lim_{k \to \infty}$.

Let us collect a few simple properties of lim inf and lim sup.

A.1 Properties (of lim inf and lim sup). Let $(a_j)_{j \in \mathbb{N}}$ and $(b_j)_{j \in \mathbb{N}}$ be sequences of real numbers.

(i) $\displaystyle \liminf_{j \to \infty} a_j = \lim_{k \to \infty} \inf_{j \geq k} a_j$ and $\displaystyle \limsup_{j \to \infty} a_j = \lim_{k \to \infty} \sup_{j \geq k} a_j$.

(ii) $\displaystyle \liminf_{j \to \infty} a_j = -\limsup_{j \to \infty}(-a_j)$.

(iii) $\displaystyle \liminf_{j \to \infty} a_j \leq \limsup_{j \to \infty} a_j$.

(iv) $\displaystyle \liminf_{j \to \infty} a_j$ and $\displaystyle \limsup_{j \to \infty} a_j$ are limits of subsequences of $(a_j)_{j \in \mathbb{N}}$ and all other limits L of subsequences of $(a_j)_{j \in \mathbb{N}}$ satisfy

$$\liminf_{j \to \infty} a_j \leq L \leq \limsup_{j \to \infty} a_j.$$

313

(v) $\lim_{j\to\infty} a_j \in \mathbb{R}$ exists $\iff -\infty < \liminf_{j\to\infty} a_j = \limsup_{j\to\infty} a_j < +\infty.$

In this case $\lim_{j\to\infty} a_j = \liminf_{j\to\infty} a_j = \limsup_{j\to\infty} a_j.$

(vi) $\liminf_{j\to\infty} a_j + \liminf_{j\to\infty} b_j \leqslant \liminf_{j\to\infty} (a_j + b_j),$

$\limsup_{j\to\infty} (a_j + b_j) \leqslant \limsup_{j\to\infty} a_j + \limsup_{j\to\infty} b_j.$

(vii) If $a_j, b_j \geqslant 0$ for all $j \in \mathbb{N}$, then

$$\liminf_{j\to\infty} a_j \, \liminf_{j\to\infty} b_j \leqslant \liminf_{j\to\infty} a_j b_j,$$

$$\limsup_{j\to\infty} a_j b_j \leqslant \limsup_{j\to\infty} a_j \, \limsup_{j\to\infty} b_j.$$

(viii) $\liminf_{j\to\infty} (a_j + b_j) \leqslant \liminf_{j\to\infty} a_j + \limsup_{j\to\infty} b_j \leqslant \limsup_{j\to\infty} (a_j + b_j).$

(ix) If, for all $j \in \mathbb{N}$, $a_j, b_j \geqslant 0$, then

$$\liminf_{j\to\infty} a_j b_j \leqslant \liminf_{j\to\infty} a_j \, \limsup_{j\to\infty} b_j \leqslant \limsup_{j\to\infty} a_j b_j.$$

(x) If the limit $\lim_{j\to\infty} a_j$ exists, then

$$\liminf_{j\to\infty} (a_j + b_j) = \lim_{j\to\infty} a_j + \liminf_{j\to\infty} b_j,$$

$$\limsup_{j\to\infty} (a_j + b_j) = \lim_{j\to\infty} a_j + \limsup_{j\to\infty} b_j.$$

(xi) If $a_j, b_j \geqslant 0$ for all $j \in \mathbb{N}$ and if $\lim_{j\to\infty} a_j$ exists, then

$$\liminf_{j\to\infty} a_j b_j = \lim_{j\to\infty} a_j \, \liminf_{j\to\infty} b_j,$$

$$\limsup_{j\to\infty} a_j b_j = \lim_{j\to\infty} a_j \, \limsup_{j\to\infty} b_j.$$

(xii) $\limsup_{j\to\infty} |a_j| = 0 \implies \lim_{j\to\infty} a_j = 0.$

Proof **(i)** follows from the remark preceding A.1, **(ii)** is clear since

$$\inf_j a_j = -\sup_j (-a_j),$$

and **(iii)** follows from the inequality $\inf_{j\geqslant k} a_j \leqslant \sup_{j\geqslant k} a_j$ where we can pass to the limit $k \to \infty$ on both sides.

Notice that **(ii)** reduces any statement about $\limsup$ to a dual statement for $\liminf$. This means that we need to show **(iv)**–**(xi)** for the lower limit only.

(iv): Let $(a_{n(j)})_{j \in \mathbb{N}} \subset (a_j)_{j \in \mathbb{N}}$ be some subsequence with (improper) limit $L = \lim_{j \to \infty} a_{n(j)}$. Then

$$\inf_{j \geqslant k} a_j \leqslant \inf_{j \geqslant k} a_{n(j)} \leqslant L \implies \liminf_{k \to \infty} \inf_{j \geqslant k} a_j \leqslant L,$$

i.e. $\liminf_{j \to \infty} a_j$ is smaller than any limit of any subsequence. Let us now construct a subsequence which has $L_* := \liminf_{j \to \infty} a_j > -\infty$ as its limit. By the very definition of L_* and the infimum we find for all $\epsilon > 0$ some $N_\epsilon \in \mathbb{N}$ such that

$$\left| L_* - \inf_{j \geqslant k} a_j \right| \leqslant \epsilon \qquad \forall k \geqslant N_\epsilon.$$

Since then $\inf_{j \geqslant k} a_j > -\infty$, we find by the definition of the infimum some $\ell \geqslant k \geqslant N_\epsilon$, $\ell = \ell_{\epsilon,k}$, and a_ℓ with

$$\left| a_\ell - \inf_{j \geqslant k} a_j \right| \leqslant \epsilon.$$

Specializing $\epsilon = \frac{1}{n}$, $n \in \mathbb{N}$, we obtain an infinite family of $a_{\ell(n)}$ from which we can extract a subsequence with limit L_*.

If $L_* = -\infty$, the sequence $(a_j)_{j \in \mathbb{N}}$ is unbounded from below and it is obvious that there must exist a subsequence tending to $-\infty$.

(v): If $\lim_{j \to \infty} a_j$ exists, then all subsequences converge and have the same limit, thus $\liminf_{j \to \infty} a_j = \lim_{j \to \infty} a_j = \limsup_{j \to \infty} a_j$ by (iv).

Conversely, if $L = \liminf_{j \to \infty} a_j = \limsup_{j \to \infty} a_j$, we get for all $k \in \mathbb{N}$

$$0 \leqslant a_k - \inf_{j \geqslant k} a_j \leqslant \sup_{j \geqslant k} a_j - \inf_{j \geqslant k} a_j \xrightarrow{k \to \infty} 0,$$

and $\lim_{k \to \infty} a_k = \lim_{k \to \infty} \inf_{j \geqslant k} a_j = L$ follows from a sandwiching argument.

(vi) follows immediately from

$$\inf_{j \geqslant k} a_j + \inf_{j \geqslant k} b_j \leqslant a_\ell + b_\ell \quad \forall \ell \geqslant k \implies \inf_{j \geqslant k} a_j + \inf_{j \geqslant k} b_j \leqslant \inf_{\ell \geqslant k} (a_\ell + b_\ell)$$

if we pass to the limit $k \to \infty$ on both sides.

(vii): We have $0 \leqslant \inf_{j \geqslant k} b_j \leqslant b_\ell$ for all $\ell \geqslant k$ and multiplying this inequality with $0 \leqslant \inf_{j \geqslant k} a_j \leqslant a_\ell$, $\ell \geqslant k$, gives

$$\inf_{j \geqslant k} a_j \inf_{j \geqslant k} b_j \leqslant a_\ell b_\ell \quad \forall \ell \geqslant k \implies \inf_{j \geqslant k} a_j \inf_{j \geqslant k} b_j \leqslant \inf_{\ell \geqslant k} a_\ell b_\ell.$$

The assertion follows as we go to the limit $k \to \infty$ on both sides.

(viii): We have

$$\inf_{j \geqslant k} (a_j + b_j) \leqslant a_\ell + b_\ell \leqslant a_\ell + \sup_{j \geqslant k} b_j \qquad \forall \ell \geqslant k,$$

so that $\inf_{j \geqslant k}(a_j + b_j) \leqslant \inf_{j \geqslant k} a_j + \sup_{j \geqslant k} b_j$, and the assertion follows as we go to the limit $k \to \infty$ on both sides.

(ix) is similar to (viii) taking into account the precautions set out in (vii).

(x): If $\lim_{j \to \infty} a_j$ exists, we know from (v) that $\lim_{j \to \infty} a_j = \liminf_{j \to \infty} a_j = \limsup_{j \to \infty} a_j$. Thus

$$\lim_{j \to \infty} a_j + \liminf_{j \to \infty} b_j \overset{\text{A.1(v)}}{=} \liminf_{j \to \infty} a_j + \liminf_{j \to \infty} b_j \overset{\text{A.1(vi)}}{\leqslant} \liminf_{j \to \infty}(a_j + b_j)$$

$$\overset{\text{A.1(viii)}}{\leqslant} \limsup_{j \to \infty} a_j + \liminf_{j \to \infty} b_j$$

$$\overset{\text{A.1(v)}}{\leqslant} \lim_{j \to \infty} a_j + \liminf_{j \to \infty} b_j.$$

(xi) is similar to (x) using (v),(vii) and (ix).

(xii): since $|a_j| \geqslant 0$,

$$0 \leqslant \liminf_{j \to \infty} |a_j| \overset{\text{A.1(iii)}}{\leqslant} \limsup_{j \to \infty} |a_j| = 0,$$

and we conclude from (v) that

$$\lim_{j \to \infty} |a_j| = \liminf_{j \to \infty} |a_j| = \limsup_{j \to \infty} |a_j| = 0.$$

Thus $\lim_{j \to \infty} a_j = 0$. ∎

$$* \quad * \quad *$$

Sometimes the following definitions for upper and lower limits of a sequence of sets $(A_j)_{j \in \mathbb{N}}$, $A_j \subset X$, are used:

$$\liminf_{j \to \infty} A_j := \bigcup_{k \in \mathbb{N}} \bigcap_{j \geqslant k} A_j \quad \text{and} \quad \limsup_{j \to \infty} A_j := \bigcap_{k \in \mathbb{N}} \bigcup_{j \geqslant k} A_j. \tag{A.3}$$

The connection between set-theoretic and numerical upper and lower limits is given by

A.2 Lemma *For all $x \in X$ we have*

$$\liminf_{j \to \infty} \mathbf{1}_{A_j}(x) = \mathbf{1}_{\liminf_{j \to \infty} A_j}(x), \tag{A.4}$$

$$\limsup_{j \to \infty} \mathbf{1}_{A_j}(x) = \mathbf{1}_{\limsup_{j \to \infty} A_j}(x). \tag{A.5}$$

Proof Note that

$$\mathbf{1}_{\bigcap_{k\in\mathbb{N}} B_k} = \inf_{k\in\mathbb{N}} \mathbf{1}_{B_k} \qquad \text{and} \qquad \mathbf{1}_{\bigcup_{k\in\mathbb{N}} B_k} = \sup_{k\in\mathbb{N}} \mathbf{1}_{B_k}$$

which follows from

$$\mathbf{1}_{\bigcap_{k\in\mathbb{N}} B_k}(x) = 1 \iff x \in \bigcap_{k\in\mathbb{N}} B_k$$

$$\iff \forall k \in \mathbb{N} : x \in B_k$$

$$\iff \forall k \in \mathbb{N} : \mathbf{1}_{B_k}(x) = 1$$

$$\iff \inf_{k\in\mathbb{N}} \mathbf{1}_{B_k}(x) = 1.$$

A similar argument proves the assertion for $\sup_{k\in\mathbb{N}} \mathbf{1}_{B_k}$. Hence,

$$\mathbf{1}_{\liminf_{j\to\infty} A_j} = \mathbf{1}_{\bigcup_{k\in\mathbb{N}} \bigcap_{j\geqslant k} A_j} = \sup_{k\in\mathbb{N}} \mathbf{1}_{\bigcap_{j\geqslant k} A_j} = \sup_{k\in\mathbb{N}} \inf_{j\geqslant k} \mathbf{1}_{A_j} = \liminf_{j\to\infty} \mathbf{1}_{A_j},$$

and (A.5) follows analogously.

Appendix B

Some facts from point-set topology

The following diagram gives a survey of various types of abstract spaces used in this book. The arrows '⟶' indicate how the spaces are connected. In brackets we mention the key concepts that define the notion of convergence in these spaces.

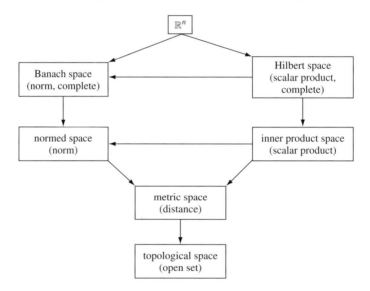

Note that due to the Riesz–Fischer theorem 12.7 the space $(L^2, \langle \cdot, \cdot \rangle)$ is a Hilbert space and all $(L^p, \|\cdot\|_p)$, $1 \leqslant p < \infty$ are Banach spaces.

The material below can be found in many introductory texts on general topology and real analysis. For this compilation we used the books by Willard [54], Steen and Seebach [46] and Rudin [39]. Complete proofs are given in [54] and in the first few chapters of [39].

Topological spaces

Topological spaces are characterized by the notion of *openness* of sets.

B.1 Definition A *topological space* $(X, \mathcal{O})$ consists of a set X and a system $\mathcal{O} = \mathcal{O}(X)$ of subsets of X, called a *topology*, which satisfies the following properties:

$$\emptyset, X \in \mathcal{O}, \tag{$\mathcal{O}_1$}$$

$$U, V \in \mathcal{O} \implies U \cap V \in \mathcal{O}^n, \tag{$\mathcal{O}_2$}$$

$$U_i \in \mathcal{O}, \ i \in I \ (\text{arbitrary}) \implies \bigcup_{i \in I} U_i \in \mathcal{O}. \tag{$\mathcal{O}_3$}$$

A set $U \in \mathcal{O}$ is called an *open set*. A set $F \subset X$ is *closed*, if its complement F^c is open. We write $\mathcal{C} = \mathcal{C}(X)$ for the family of closed sets in X.

From de Morgan's identities (2.2) it is not hard to see that

- X and $\emptyset$ are closed sets,
- unions of finitely many closed sets are again closed,
- intersections of arbitrarily many closed sets are again closed.

B.2 Examples Let X be an arbitrary set.

(i) $\{\emptyset, X\}$ is a topology on X.

(ii) The power set $\mathcal{P}(X)$ is a topology on X.

(iii) Let U be a 'classical' open set in $\mathbb{R}^n$, i.e. for every $x \in U$ one can find some $\epsilon > 0$ such that $B_\epsilon(x) \subset U$. The classical open sets $\mathcal{O}(\mathbb{R}^n)$ are a topology in $\mathbb{R}^n$. *Unless otherwise stated, we will always consider this natural topology on* $\mathbb{R}^n$.

(iv) (*Trace topology*) Let $(X, \mathcal{O}(X))$ be a topological space and $A \subset X$ be any subset. Then the *relatively open* subsets of A,

$$\mathcal{O}(A) := A \cap \mathcal{O}(X) := \{A \cap U : U \in \mathcal{O}(X)\},$$

turn $(A, \mathcal{O}(A))$ into a topological space.

(v) (*Product topology*) Let $(X, \mathcal{O}(X))$ and $(Y, \mathcal{O}(Y))$ be topological spaces. Then $X \times Y$ becomes a topological space under the *product topology* $\mathcal{O}(X \times Y)$: by definition, a set $W \in \mathcal{O}(X \times Y)$ if $W \subset X \times Y$ and if for each $w = (x, y) \in W$ there exist $U \in \mathcal{O}(X)$ and $V \in \mathcal{O}(Y)$ such that

$$w = (x, y) \in U \times V \subset W.$$

This makes $\mathcal{O}(X \times Y)$ the smallest topology containing $\mathcal{O}(X) \times \mathcal{O}(Y)$.

B.3 Definition Let $(X, \mathcal{O})$ be a topological space.

(i) An *open neighbourhood* of a point $x \in X$ is an open set $U = U(x)$ containing x. A *neighbourhood* of x is any set containing an open neighbourhood of x.

(ii) The space X is called *separated* or a *Hausdorff space* if any two different points $x, y \in X$ have disjoint neighbourhoods.

(iii) Let $A \subset X$. The *closure* of A, denoted by $\bar{A}$, is the smallest closed set containing A, i.e. $\bar{A} = \bigcap_{F \in \mathcal{C}, F \supset A} F$.

(iv) Let $A \subset X$. The *(open) interior* of A, denoted by A°, is the largest open set inside A, i.e. $A^\circ = \bigcup_{U \in \mathcal{O}, U \subset A} U$.

(v) A set $A \subset X$ is *dense* in X, if $\bar{A} = X$.

(vi) The space X is *separable* if it contains a countable dense subset.

B.4 Examples **(i)** The space $(\mathbb{R}^n, \mathcal{O}(\mathbb{R}^n))$ is a Hausdorff space.

(ii) The space $(X, \mathcal{P}(X))$ and all spaces mentioned in the diagram at the beginning of the section are Hausdorff spaces.

(iii) The space $(X, \{\emptyset, X\})$ is not separated.

(iv) A set U in a topological space $(X, \mathcal{O})$ is open if, and only if, it is a neighbourhood of each of its points.

(v) The open ball $B_r(x) := \{y \in \mathbb{R}^n : |x - y| < r\}$ in $\mathbb{R}^n$ is an open neighbourhood of x. The closed ball $K_r(x) := \{y \in \mathbb{R}^n : |x - y| \leqslant r\}$ is the closure of $B_r(x)$, thus $\overline{B_r(x)} = K_r(x)$.

(vi) The set of rational numbers $\mathbb{Q}$ is dense in $\mathbb{R}$. Therefore $\mathbb{R}$ is separable. The same is true for $\mathbb{R}^n$ when we consider the countable dense set $\mathbb{Q}^n$.

Density assertions are often expressed through approximation theorems such as Corollary 12.11, Theorem 24.6 or Corollary 24.12.

B.5 Definition A subset K of a Hausdorff space $(X, \mathcal{O})$ is called *compact*, if every cover of K by open sets, $K \subset \bigcup_{i \in I} U_i$, $U_i \in \mathcal{O}$, I is an arbitrary index set, admits a finite sub-cover, i.e. if there are finitely many $U_{i_1}, \ldots, U_{i_n}$ such that $K \subset U_{i_1} \cup \ldots \cup U_{i_n}$. A set L is *relatively compact* if $\overline{L}$ is compact.

B.6 Proposition *Let $(X, \mathcal{O})$ be a Hausdorff space.*

(i) *Every compact set K is closed.*

(ii) *Closed subsets of compact sets are closed.*

(iii) *A family $(K_i)_{i \in I}$ of compact sets (indexed by an arbitrary set I) has non-empty intersection $\bigcap_{i \in I} K_i \neq \emptyset$ if, and only if, every finite subcollection $(K_{i_j})_{j=1}^n$ has non-empty intersection $K_{i_1} \cap K_{i_2} \cap \ldots \cap K_{i_n} \neq \emptyset$.*

B.7 Example A set $K \subset \mathbb{R}^n$ is compact if, and only if, it is closed and bounded. This is also equivalent to saying that every sequence $(x_j)_{j \in \mathbb{N}} \subset K$ has a convergent subsequence. Such a simple characterization of compactness fails in infinite-dimensional spaces, notably in the Hilbert space L^2 or the Banach spaces L^p, $1 \leqslant p < \infty$, see Theorem B.22 and B.27.

Theorem B.6(iii) is an abstract version of the well known interval princi-ple in $\mathbb{R}$: a sequence of *nested* closed intervals $[a_j, b_j] \subset \mathbb{R}$, $j \in \mathbb{N}$, has non-empty intersection $\bigcap_{j \in \mathbb{N}}[a_j, b_j] \neq \emptyset$. If, in addition, $\lim_{j \to \infty}(b_j - a_j) = 0$ then $\bigcap_{j \in \mathbb{N}}[a_j, b_j] = \{L\}$ where $L = \lim_{j \to \infty} a_j = \lim_{j \to \infty} b_j$.

B.8 Definition Let $(X, \mathcal{O}(X))$ and $(Y, \mathcal{O}(Y))$ be two topological spaces. A map $f : X \to Y$ is called *continuous* at $x \in X$, if for every neighbourhood $V = V(f(x))$ we can find a neighbourhood $U = U(x)$ of x such that $f(U) \subset V$. If f is continuous at every $x \in X$, we call f *continuous*.

B.9 Example Definition B.8 coincides on Euclidean spaces with the classical notion of continuity, i.e. a map $f : \mathbb{R}^n \to \mathbb{R}^m$ is continuous at $x \in \mathbb{R}^n$ if, and only if, for every convergent sequence $x_j \xrightarrow{j \to \infty} x$ we have $f(x_j) \xrightarrow{j \to \infty} f(x)$, cf. Theorem B.19.

B.10 Definition Let $(X, \mathcal{O})$ be a topological space. A set $A \subset X$ is called *connected*, if A cannot be written in the form $A = U \cup V$ where $U, V \in \mathcal{O}$ and $U \cap V = \emptyset$.

The set A is called *pathwise connected*, if for any two points $x, y \in A$ there is a continuous curve or *path* $\phi : [0, 1] \to A$ such that $\phi(0) = x$ and $\phi(1) = y$.

B.11 Examples (i) The only connected sets in $\mathbb{R}$ are finite or infinite intervals. The set $(a, b) \cup (c, d)$ where $a < b < c < d$ is not connected.

(ii) Pathwise connected sets are connected; the converse is, in general, wrong: the set $V = \{(x, 0) \in \mathbb{R}^2 : x \leqslant 0\} \cup \{(x, \sin \frac{1}{x}) \in \mathbb{R}^2 : x > 0\}$ is connected, but no path can be found from $(0, 0)$ to any point $(x, \sin \frac{1}{x})$.

B.12 Theorem *Let $f : X \to Y$ be a map between the topological spaces X and Y.*

(i) *The map f is continuous if, and only if, for all open $V \in \mathcal{O}(Y)$ the pre-image $f^{-1}(V) \in \mathcal{O}(X)$ is open.*

(ii) *Let f be continuous. The image $f(K) \subset Y$ of a compact [connected, pathwise connected] set $K \subset X$ is again compact [connected, pathwise connected].*

(iii) *Let $K \subset X$ be a compact set. A continuous map[1] $g : K \to \mathbb{R}$ attains its maximum and minimum.*

(iv) *Let $K \subset X$ be a compact set and $f : K \to \mathbb{R}$ be a injective and continuous map. Then the inverse map $f^{-1} : f(K) \to K$ exists and is continuous.*

Since for our purposes the characterization of continuity by open sets is of central importance, cf. Example 7.3, we include the short

---

[1] We consider here the trace topology $\mathcal{O}(K)$, cf. Example B.2.

Proof (of Theorem B.12(i)) '⇐' Assume first that $f^{-1}(\mathcal{O}(Y)) \subset \mathcal{O}(X)$. Every neighbourhood $\tilde{V}$ of $f(x)$ contains by definition an open set $V \subset \tilde{V}$ with $f(x) \in V$. By assumption, $U := f^{-1}(V)$ is open, and since $x \in U$, U is an (open) neighbourhood of x with $f(U) = f \circ f^{-1}(V) \subset V$.

'⇒' Assume now that f is continuous. Take any open set $B \subset Y$, set $A := f^{-1}(B)$ and fix some $x \in A$. Since B is open, there is some open neighbourhood $V = V(f(x)) \subset B$ and by continuity we find some neighbourhood $U = U(x) \subset X$ of x with $f(U) \subset V$. Thus

$$U \subset f^{-1} \circ f(U) \subset f^{-1}(V) \subset f^{-1}(B) \stackrel{\text{def}}{=} A,$$

which shows that A contains for every of its points a whole neighbourhood. This is to say that A is open.   ∎

B.13 Example Let $g : [a, b] \to \mathbb{R}$ be a continuous function. Since $[a, b]$ is compact, g attains its maximum $M = \sup g([a, b]) = g(x_{\max})$ and minimum $m = \inf g([a, b]) = g(x_{\min})$ at some points $x_{\max}, x_{\min} \in [a, b]$. Since $[a, b]$ is compact and pathwise connected, so is $g([a, b])$, hence it is of the form $[m, M]$. In particular, we have recovered the intermediate value theorem for functions of a real variable.

B.14 Definition Let $(x_j)_{j \in \mathbb{N}} \subset X$, be a sequence in the topological space $(X, \mathcal{O})$. We say that x_j *converges* to $x \in X$ and write $\lim_{j \to \infty} x_j = x$ or $x \xrightarrow{j \to \infty} x$ if for every open neighbourhood $U = U(x)$ there is some $N = N_U \in \mathbb{N}$ such that $x_j \in U$ for all $j \geq N_U$.

This is also the 'usual' convergence in the spaces $\mathbb{R}$ and $\mathbb{R}^n$. Note that limits are only unique if X is a Hausdorff space. Sometimes we can use limits of sequences to give an equivalent description of the topology. This is always the case if every point $x \in X$ has a countable system of open neighbourhoods $(U_n)_{n \in \mathbb{N}}$ with the property that for every neighbourhood $V = V(x)$ of x there is at least one $U_{n_0} \subset V$; this is always true in metric spaces, cf. B.19.

Metric spaces

In metric spaces we have a notion of *distance* between any two points.

B.15 Definition A metric space (X, d) is a set X with a *distance function* or *metric* $d : X \times X \to [0, \infty]$ such that for all $x, y, z \in X$

(definiteness)	$d(x, y) = 0 \iff x = y,$	(d_1)
(symmetry)	$d(x, y) = d(y, x),$	(d_2)
(triangle inequality)	$d(x, y) \leq d(x, z) + d(z, y).$	(d_3)

B.16 Examples (i) Let (X, d) be a metric space. Then $(A, d = d|_{A \times A})$ is again a metric space for all $A \subset X$.

(ii) The real line $\mathbb{R}$ is a metric space with $d(x, y) = |x - y|$. The space $\mathbb{R}^n$ becomes a metric space with each of the following metrics:

$$d_p(x, y) = \begin{cases} \left(\sum_{j=1}^{n} |x_j - y_j|^p \right)^{1/p} & \text{if } 1 \leqslant p < \infty, \\ \max_{1 \leqslant j \leqslant n} |x_j - y_j| & \text{if } p = \infty. \end{cases}$$

(iii) The topological space $(X, \mathcal{P}(X))$ is a metric space with metric

$$d(x, y) = \begin{cases} 1 & \text{if } x \neq y, \\ 0 & \text{if } x = y. \end{cases}$$

(iv) Let (X_j, d_j), $j = 1, 2$, be two metric spaces. Then $X_1 \times X_2$ becomes a metric space for any of the following metrics $(x_j, y_j \in X_j,\ 1 \leqslant p < \infty)$:

$$\rho_p((x_1, x_2), (y_1, y_2)) := \left(d_1^p(x_1, y_1) + d_2^p(x_2, y_2) \right)^{1/p},$$

or $$\rho_\infty((x_1, x_2), (y_1, y_2)) := \max_{j=1,2} d_j(x_j, y_j).$$

B.17 Definition Let (X, d) be a metric space. We call

$$B_r(x) := \{x \in X : d(x, y) < r\} \quad \text{resp.} \quad K_r(x) := \{x \in X : d(x, y) \leqslant r\}$$

an *open* resp. *closed ball* with centre x and radius $r > 0$. An *open set* is a set $U \subset X$ such that for every $x \in U$ there is some $\epsilon > 0$ and $B_\epsilon(x) \subset U$. *Closed sets* arise as complements of open sets.

Using the triangle inequality it is easy to see that open balls in X are also open sets and that closed balls are closed sets. Mind, however, that in general $\overline{B_r(x)} \subsetneq K_r(x)$.

B.18 Lemma *The family of open sets $\mathcal{O}$ of a metric space X is a topology in the sense of Definition B.1. $(X, \mathcal{O})$ is a separated topological space.*

The converse of Lemma B.18 is wrong: the topology $\{\emptyset, \{a\}, X\}$ of the space $X = \{a, b\}$ cannot be generated by any metric.

The topology of metric spaces can be described by sequences.

B.19 Theorem *Let* (X, d), (Y, ρ) *be a metric spaces.*

(i) *A sequence* $(x_j)_{j\in\mathbb{N}} \subset X$ *converges to* x, $x_j \xrightarrow{j\to\infty} x$, *if, and only if,* $d(x_j, x) \xrightarrow{j\to\infty} 0$. *Moreover, the limit* x *is unique.*

(ii) *A set* $F \subset X$ *is closed if, and only if, every convergent sequence* $(x_j)_{j\in\mathbb{N}} \subset F$ *has its limit* $\lim_{j\to\infty} x_j \in F$.

(iii) *A set* $K \subset X$ *is compact if, and only if, every sequence* $(x_j)_{j\in\mathbb{N}} \subset K$ *has a convergent subsequence whose limit is in* K.

(iv) *A set* $A \subset X$ *is dense if, and only if, for every* $x \in X$ *there is a sequence* $(a_j)_{j\in\mathbb{N}}$ *with* $d(a_j, x) \xrightarrow{j\to\infty} 0$.

(v) *A function* $f : X \to Y$ *is continuous at* $x \in X$ *if, and only if, for every sequence* $x_j \xrightarrow{j\to\infty} x$ *we have* $f(x_j) \xrightarrow{j\to\infty} f(x)$.

Since for our purposes the characterization of continuity is of central importance, cf. Example 7.3, we include the short

Proof (of Theorem B.19(v)) We begin with the observation that every neighbourhood $\tilde{U} = \tilde{U}(x)$ of a point $x \in X$ contains some open set $U \subset \tilde{U}$ with $x \in U$. Since U is open, we find by definition some $\delta > 0$ such that $B_\delta(x) \subset U$. This shows that we can restate the definition of continuity B.8 at a point x in the following form:

$$\forall \epsilon > 0 \quad \exists \delta > 0 : f(B_\delta(x)) \subset B_\epsilon(f(x))$$

(mind that the balls are taken in X and Y, respectively).

'$\Rightarrow$': If $x_j \xrightarrow{j\to\infty} x$, we know from the definition of convergence that for every $\delta > 0$ there is some $N = N_\delta$ such that $x_j \in B_\delta(x)$ for all $j \geqslant N_\delta$. Since f is continuous at x, we can choose for every $\epsilon > 0$ some $\delta = \delta_\epsilon > 0$ such that $f(B_\delta(x)) \subset B_\epsilon(f(x))$. Thus

$$f(x_k) \in f(\{x_j : j \geqslant N\}) \subset f(B_\delta(x)) \subset B_\epsilon(f(x)) \quad \forall k \geqslant N,$$

which shows that $f(x_k) \xrightarrow{k\to\infty} f(x)$.

'$\Leftarrow$': Assume that $x_j \xrightarrow{j\to\infty} x$ implies $f(x_j) \xrightarrow{j\to\infty} f(x)$ but that f is not continuous at x. Thus there is some $\epsilon > 0$, such that for all $n \in \mathbb{N}$ the set $f(B_{1/n}(x))$ is not (entirely) contained in $B_\epsilon(f(x))$. Thus we can pick for each $n \in \mathbb{N}$ some $x_n \in B_{1/n}(x)$, such that $f(x_n) \notin B_\epsilon(f(x))$. This means, however, that $x_n \xrightarrow{n\to\infty} x$ while $d(f(x_n), f(x)) \geqslant \epsilon > 0$ for all $n \in \mathbb{N}$, contradicting that $f(x_n)$ converges to $f(x)$. ∎

B.20 Definition Let (X, d) be a metric space. A sequence $(x_j)_{j \in \mathbb{N}}$ is a *Cauchy sequence*, if

$$\forall \epsilon > 0 \quad \exists N = N_\epsilon \in \mathbb{R} \quad \forall j, k \geqslant N_\epsilon : d(x_j, x_k) \leqslant \epsilon.$$

A metric space is *complete* if every Cauchy sequence converges.

An *isometry* is a surjective map $\jmath : X \to Y$ between two metric spaces (X, d) and (Y, ρ) which satisfies $d(x, x') = \rho(\jmath(x), \jmath(x'))$.

B.21 Theorem (Completion) *For every metric space (X, d) there exists a complete metric space $(\widehat{X}, \hat{d})$ such that $\hat{d}|_{X \times X} = d$ and $X \subset \widehat{X}$ is a dense subset. Any two completions of X are, up to isometries, identical.*

By covering a compact set K with the open sets $(B_1(x))_{x \in K}$ and extracting a finite subcover we can easily see that K has finite diameter $\text{diam}(K) := \sup_{x, y \in K} d(x, y)$ and is, therefore, *bounded*. Thus compact sets are closed and bounded. The converse is, in general, not true; however,

B.22 Theorem (Heine–Borel) *A subset of $\mathbb{R}^n$ is compact if, and only if, it is closed and bounded. Moreover, all metrics on $\mathbb{R}^n$ are equivalent in the sense that for any two metrics d and ρ there are absolute constants $c, C > 0$ such that*

$$c\, d(x, y) \leqslant \rho(x, y) \leqslant C\, d(x, y) \qquad \forall x, y \in \mathbb{R}^n.$$

Normed spaces

B.23 Definition A *normed space* $(X, \|\cdot\|)$ is a $\mathbb{K}$-vector space[2] X with a *norm* $\|\cdot\|$, i.e. a map $\|\cdot\| : X \to [0, \infty]$ which satisfies for $x, y \in X$ and $\alpha \in \mathbb{K}$ the following properties:

(definiteness)	$\|x\| > 0 \iff x \neq 0,$	(N_1)
(pos. homogeneity)	$\|\alpha x\| = |\alpha| \cdot \|x\|,$	(N_2)
(triangle inequality)	$\|x + y\| \leqslant \|x\| + \|y\|.$	(N_3)

If we drop the definiteness (N_1), $\|\cdot\|$ is called a *semi-norm* and $(X, \|\cdot\|)$ is a *semi-normed space*.

B.24 Examples **(i)** The spaces $\mathbb{R}$, $\mathbb{C}$, $\mathbb{R}^n$, $\mathbb{C}^n$ equipped with

$$\|x\| := \left(\sum_{j=1}^n |x_j|^p \right)^{1/p} \qquad \text{or} \qquad \|x\| := \max_{1 \leqslant j \leqslant n} |x_j|$$

$(1 \leqslant p < \infty)$ are normed spaces.

---

[2] $\mathbb{K}$ stands for either $\mathbb{C}$ or $\mathbb{R}$.

(ii) Let $(X_j, \|\cdot\|_j)$, $j = 1, 2$, be two normed spaces. Then $X_1 \times X_2$ becomes a normed space under any of the following norms ($x_j \in X_j$, $1 \leqslant p < \infty$):

$$\|(x_1, x_2)\|_{(p)} := \left(\|x_1\|_1^p + \|x_2\|_2^p\right)^{1/p} \quad \text{or} \quad \|(x_1, x_2)\|_{(\infty)} := \max_{j=1,2} \|x_j\|_j.$$

(iii) Every normed space is a metric space with metric given by $d(x, y) = \|x - y\|$. Therefore, all notions and results for metric spaces carry over to normed spaces.

In particular, open and closed balls are given by

$$B_r(x) = \{y \in X : \|x - y\| < r\} \quad \text{and} \quad K_r(x) = \{y \in X : \|x - y\| \leqslant r\}.$$

Since X is a vector space, we have now $\overline{B_r(x)} = K_r(x)$.

However, not every metric space arises from a normed space, e.g. the metric $d(x, y) = 1$ or 0 according to $x \neq y$ or $x = y$ on $\mathbb{R}^n$ cannot be realized by any norm.

B.25 Lemma *Let X be a normed space. Then the following maps are continuous:*

$$X \ni x \mapsto \|x\|, \qquad X \times X \ni (x, y) \mapsto x + y, \qquad \mathbb{K} \times X \ni (\alpha, x) \mapsto \alpha x.$$

B.26 Definition A *Banach space* is a complete normed space.

The following result, due to F. Riesz, says that the Heine–Borel theorem B.22 holds if, and only if, the underlying space is finite-dimensional.

B.27 Theorem (Riesz). *In a normed space V closed and bounded sets are compact if, and only if, V is finite-dimensional.*

Let $\sim$ be an equivalence relation on the normed space X. We write $[x] := \{y \in X : x \sim y\}$ for the equivalence class with representative x. The *quotient space* $X/\sim$ consists of all equivalence classes. It is not hard to see that $X/\sim$ is again a vector space and that

$$[\alpha x + \beta y] = \alpha[x] + \beta[y] \qquad \forall \alpha, \beta \in \mathbb{K}, \ x, y \in X.$$

B.28 Theorem *Let $(X, \|\cdot\|)$ be a (complete) normed space. Then $X/\sim$ is a (complete) normed space under the* quotient norm *given by*

$$\|[x]\|^\sim := \inf\{\|y\| : y \in [x]\}.$$

Essentially the same procedure allows us to turn any semi-normed space $(X, \|\cdot\|)$ into a normed space. We use the following equivalence relation for $x, y \in X$:

$$x \approx y \iff \|x - y\| = 0,$$

and observe that

$$\inf\{\|y\| : y \in [x]\} = \|x\|.$$

B.29 Corollary *Let $(X, \|\cdot\|)$ be a (complete) semi-normed space. Then $X/_{\approx}$ is a (complete) normed space with norm given by $\|[x]\|^{\approx} := \|x\|$.*

B.30 Example Denote by $\mathcal{L}^p(X, \mathcal{A}, \mu)$, $1 \leqslant p < \infty$, the pth power integrable functions of the measure space $(X, \mathcal{A}, \mu)$. Then

$$\|u\|_p := \left(\int |u|^p \, d\mu \right)^{1/p}$$

is a semi-norm on $\mathcal{L}^p(X, \mathcal{A}, \mu)$, and $L^p(X, \mathcal{A}, \mu) := \mathcal{L}^p(X, \mathcal{A}, \mu)/_{\sim}$ is a Banach space if we identify $u, w \in \mathcal{L}^p(X, \mathcal{A}, \mu)$ whenever $\|u - w\|_p = 0$.

Appendix C

The volume of a parallelepiped

In this appendix we give a simple derivation for the volume of the parallelepiped

$$A([0, 1)^n) := \{Ax \in \mathbb{R}^n : x \in [0, 1)^n\}, \quad A \in GL(n, \mathbb{R})$$

for a non-degenerate $n \times n$ matrix $A \in \mathbb{R}^{n \times n}$.

C.1 Theorem $\lambda^n\big[A([0, 1)^n)\big] = |\det A|$ *for all* $A \in GL(n, \mathbb{R})$.

The proof of Theorem C.1 requires two auxiliary results.

C.2 Lemma *If* $D = \text{diag}[\lambda_1, \dots, \lambda_n]$, $\lambda_j > 0$, *is a diagonal* $n \times n$ *matrix, then* $\lambda^n(D(B)) = \det D \, \lambda^n(B)$ *for all Borel sets* $B \in \mathcal{B}(\mathbb{R}^n)$.

Proof Since both D and D^{-1} are continuous maps, $D(B)$ is a Borel set if $B \in \mathcal{B}(\mathbb{R}^n)$, cf. Example 7.3. In view of the uniqueness theorem 5.7 for measures it is enough to prove the lemma for half-open rectangles $[\![a, b)\!)$, $a, b \in \mathbb{R}^n$. Obviously,

$$D[\![a, b)\!) = \underset{j=1}{\overset{n}{\times}} [\lambda_j a_j, \lambda_j b_j),$$

and

$$\lambda^n\big(D[\![a, b)\!)\big) = \prod_{j=1}^n (\lambda_j b_j - \lambda_j a_j) = \lambda_1 \cdot \ldots \cdot \lambda_n \prod_{j=1}^n (b_j - a_j)$$

$$= \det D \, \lambda^n\big([\![a, b)\!)\big). \qquad \blacksquare$$

C.3 Lemma *Every* $A \in GL(n, \mathbb{R})$ *can be written as* $A = SDT$, *where* $S, T \in O(n)$ *are orthogonal* $n \times n$ *matrices and* $D = \text{diag}[\lambda_1, \dots, \lambda_n]$ *is a diagonal matrix with positive entries* $\lambda_j > 0$.

Proof The matrix ${}^tAA$ is symmetric and so we can find some orthogonal matrix $U \in O(n)$ such that

$$ {}^tU({}^tAA)U = \tilde{D} = \mathrm{diag}[\mu_1, \ldots, \mu_n]. $$

Since for $e_j := (\underbrace{0, \ldots, 0, 1, 0 \ldots, 0}_{j})$ and the Euclidean norm $\|\bullet\|$

$$ \mu_j = {}^te_j \tilde{D} e_j = ({}^te_j {}^tU {}^tA)(AU e_j) = \|AU e_j\|^2 > 0, $$

we can define $D := \sqrt{\tilde{D}} = \mathrm{diag}[\lambda_1, \ldots, \lambda_n]$ where $\lambda_j := \sqrt{\mu_j}$. Thus

$$ D^{-1} {}^tU {}^tAAUD^{-1} = \mathrm{id}_n, $$

and this proves that $S := AUD^{-1} \in O(n)$. Since $T := {}^tU \in O(n)$, we easily see that

$$ SDT = (AUD^{-1})D {}^tU = A. \qquad \blacksquare $$

Proof (of Theorem C.1) We have for $A \in GL(n, \mathbb{R})$

$$ \lambda^n\big[A([0,1)^n)\big] \overset{C.3}{=} \lambda^n\big[SDT\,([0,1)^n)\big] $$

$$ \overset{7.9}{=} \lambda^n\big[DT\,([0,1)^n)\big] $$

$$ \overset{C.2}{=} \det D\, \lambda^n\big[T\,([0,1)^n)\big] $$

$$ \overset{C.3}{=} \det D\, \lambda^n\big([0,1)^n\big). $$

Since $S, T \in O(n)$, their determinants are either $+1$ or -1, and we conclude that $|\det A| = |\det(SDT)| = |\det S| \cdot |\det D| \cdot |\det T| = \det D$. $\qquad \blacksquare$

Appendix D

Non-measurable sets

Let $(X, \mathcal{A}, \mu)$ be a measure space and denote by $(X, \mathcal{A}^*, \bar{\mu})$ its completion, cf. Problem 4.13 for the definition and Problems 6.2, 10.11, 10.12, 13.11 and 15.3 for various properties. Here we only need that

$$\mathcal{A}^* = \{A \cup N : A \in \mathcal{A}, \quad N \text{ is a subset of some } \mathcal{A}\text{-measurable } \mu\text{-null set}\}$$

is the completion of $\mathcal{A}$ with respect to the measure μ. It is a natural question to ask how big $\mathcal{A}$ and $\mathcal{A}^*$ are and whether $\mathcal{A} \subset \mathcal{A}^* \subset \mathcal{P}(X)$ are proper inclusions.

Sometimes, see Problems 6.10 or 6.11, these questions are easy to answer. For the Borel σ-algebra $\mathcal{A} = \mathcal{B}(\mathbb{R}^n)$ and Lebesgue measure $\mu = \lambda^n$ this is more difficult. The following definition helps to distinguish between sets in $\mathcal{B}(\mathbb{R}^n)$ and the completion $\mathcal{B}^*(\mathbb{R}^n)$ w.r.t. Lebesgue measure.

D.1 Definition The *Lebesgue σ-algebra* is the completion $\mathcal{B}^*(\mathbb{R}^n)$ of the Borel σ-algebra w.r.t. Lebesgue measure λ^n. A set $B \in \mathcal{B}^*(\mathbb{R}^n)$ is called *Lebesgue measurable*.

The next theorem shows that there are 'as many' Lebesgue measurable sets as there are subsets of $\mathbb{R}^n$.

D.2 Theorem *We have* $\#\mathcal{B}^*(\mathbb{R}^n) = \#\mathcal{P}(\mathbb{R}^n)$ *for all* $n \in \mathbb{N}$.

Proof Since $\mathcal{B}^*(\mathbb{R}^n) \subset \mathcal{P}(\mathbb{R}^n)$ we have that $\#\mathcal{B}^*(\mathbb{R}^n) \leqslant \#\mathcal{P}(\mathbb{R}^n)$. On the other hand, we have seen in Problem 7.10 that the Cantor ternary set C is an uncountable Borel measurable λ^1-null set of cardinality $\#\mathbb{R} = \mathfrak{c}$. Consequently, $\mathbb{R}^{n-1} \times C$ is a λ^n-null set. By definition of the Lebesgue σ-algebra, all sets in $\mathcal{P}(\mathbb{R}^{n-1} \times C)$ are Lebesgue measurable (null) sets, i.e. $\mathcal{P}(\mathbb{R}^{n-1} \times C) \subset \mathcal{B}^*(\mathbb{R}^n)$, and therefore $\#\mathcal{P}(\mathbb{R}^{n-1} \times C) \leqslant \#\mathcal{B}^*(\mathbb{R}^n)$. Using the fact that there is a bijection between C and $\mathbb{R}$ we also get $\#\mathcal{P}(\mathbb{R}^n) \leqslant \#\mathcal{P}(\mathbb{R}^{n-1} \times C) \leqslant \#\mathcal{B}^*(\mathbb{R}^n)$, and the Cantor–Bernstein theorem 2.7 proves that $\#\mathcal{P}(\mathbb{R}^n) = \#\mathcal{B}^*(\mathbb{R}^n)$. ∎

Unfortunately, we cannot use Theorem D.2 to decide whether there are sets which are not Lebesgue measurable. To answer this question we need the axiom of choice.

D.3 Axiom of choice (AC) *Let $\{M_i : i \in I\}$ be a collection of non-empty and mutually disjoint subsets of X. Then there exists a set $L \subset \bigcup_{i \in I} M_i$ which contains exactly one element from each set M_i, $i \in I$.*

Note that AC only asserts the existence of the set L but does not tell us how or if the set L can be constructed at all. (This problem is at the heart of the controversy over whether one should or should not accept AC.)

D.4 Theorem *Assuming the axiom of choice, there exist non-Lebesgue measurable sets in $\mathbb{R}^n$.*

Proof Assume first that $n = 1$. We will construct a non-Lebesgue measurable subset of $\mathbb{I} = [0, 1)$. We call any two $x, y \in \mathbb{I}$ equivalent if

$$x \sim y \quad \Longleftrightarrow \quad x - y \in \mathbb{Q}.$$

The equivalence class containing x is given by $[x] = \{y \in \mathbb{I} : x - y \in \mathbb{Q}\} = (x + \mathbb{Q}) \cap \mathbb{I}$. By construction, $\mathbb{I}$ is partitioned by a family of mutually disjoint equivalence classes $[x_j]$, $j \in J$.

By the axiom of choice[1] there exists a set L which contains exactly one element, say m_j, from each of the classes $[x_j]$, $j \in J$. We will show that L cannot be Lebesgue measurable.

Assume L were Lebesgue measurable. Since for every $x \in \mathbb{I}$ we have $[x] \cap L = \{m_{j_0}\}$, $j_0 = j_0(x) \in J$, we can find some $q \in \mathbb{Q}$ such that $x = m_{j_0} + q$. Obviously, $-1 < q < 1$. Thus

$$\mathbb{I} \subset L + \big(\mathbb{Q} \cap (-1, 1)\big) \subset \mathbb{I} + (-1, 1) = [-1, 2),$$

which we can rewrite as

$$[0, 1) \subset \bigcup_{q \in \mathbb{Q} \cap (-1,1)} (q + L) \subset [-1, 2).$$

Moreover, $(r + L) \cap (q + L) = \emptyset$ for all $r \neq q$, $r, q \in \mathbb{Q}$. Otherwise $r + x = q + y$ for $x, y \in L$, so that $x \sim y$ which is impossible since L contains only one representative

---

[1] We have to use the axiom of choice since J is uncountable. This follows from the observation that the uncountable set $\mathbb{I} = \bigcup_{j \in J}[x_j]$ is the disjoint union of countable sets $[x_j] = (x + \mathbb{Q}) \cap \mathbb{I}$. It is known that all proofs for Theorem D.4 must use the axiom of choice or some equivalent statement, cf. Solovay [44].

of each equivalence class. Therefore we can use the σ-additivity of the measure $\bar{\lambda}^1$ to find

$$1 = \bar{\lambda}^1([0,1)) \leqslant \sum_{q \in \mathbb{Q} \cap (-1,1)} \bar{\lambda}^1(q + L) \leqslant \bar{\lambda}^1([-1,2)) = 3.$$

Since $\bar{\lambda}^1$ is invariant under translations, we get $\bar{\lambda}^1(q + L) = \bar{\lambda}^1(L)$ for all $q \in \mathbb{Q} \cap (-1,1)$. We conclude that

$$1 \leqslant \sum_{q \in \mathbb{Q} \cap (-1,1)} \bar{\lambda}^1(L) \leqslant 3$$

which is not possible. This proves that L cannot be Lebesgue measurable.

If $n > 1$, a similar argument shows that $[0,1)^{n-1} \times L$ is not Lebesgue measurable. ∎

The question whether there are Lebesgue measurable sets which are not Borel measurable can be answered constructively. Since this is quite tedious, we content ourselves with the fact that there are 'fewer' Borel sets than there are Lebesgue measurable sets.

D.5 Theorem *We have* $\#\mathcal{B}(\mathbb{R}^n) = \mathfrak{c}$.

D.6 Corollary *There are Lebesgue measurable sets which are not Borel measurable.*

Proof (of D.6) We know from Theorem D.2 that $\#\mathcal{B}^*(\mathbb{R}^n) = \#\mathcal{P}(\mathbb{R}^n)$ and from Theorem D.5 that $\#\mathcal{B}(\mathbb{R}^n) = \mathfrak{c}$. Since by Theorem 2.9 and Problem 2.17 $\#\mathcal{P}(\mathbb{R}^n) > \#\mathbb{R}^n = \mathfrak{c}$, we conclude that $\mathcal{B}(\mathbb{R}^n) \subsetneq \mathcal{B}^*(\mathbb{R}^n)$. ∎

To prove Theorem D.5 we show that the Borel sets are contained in a family of sets which has cardinality $\mathfrak{c}$. Let $\mathcal{F} := \bigcup_{k=1}^{\infty} \mathbb{N}^k$ be the set of all *finite* sequences of natural numbers and write $\mathcal{C}$ for the family of open balls $B_r(x) \subset \mathbb{R}^n$ with radius $r \in \mathbb{Q}^+$ and centre $x \in \mathbb{Q}^n$. We have seen in Problems 2.19 and 2.9 that

$$\#\mathcal{F} = \#\mathbb{N} \quad \text{and} \quad \#\mathcal{C} = \#(\mathbb{Q}^+ \times \mathbb{Q}^n) = \#\mathbb{N}.$$

Therefore, the collection of all *Souslin schemes*

$$\mathfrak{s} : \mathcal{F} \to \mathcal{C}, \quad (i_1, i_2, \ldots, i_k) \mapsto C_{i_1 i_2 \ldots i_k}$$

has cardinality $\#\mathcal{C}^{\mathcal{F}} = \#\mathbb{N}^{\mathbb{N}} = \mathfrak{c}$, cf. Problem 2.18. With each Souslin scheme $\mathfrak{s}$ we can associate a set $A \subset \mathbb{R}^n$ in the following way: take any sequence $(i_j)_{j \in \mathbb{N}}$ of natural numbers and consider the sequence of finite tuples $(i_1), (i_1, i_2), (i_1, i_2, i_3), \ldots, (i_1, i_2, \ldots, i_k), \ldots$ formed by the first $1, 2, \ldots, k, \ldots$ members of the sequence

$(i_j)_{j\in\mathbb{N}}$. Using the Souslin scheme $\mathfrak{s}$ we pick for each tuple $(i_1, i_2, \ldots, i_k)$ the corresponding set $C_{i_1 i_2 \ldots i_k} \in \mathcal{C}$ to get a sequence of sets $C_{i_1}, C_{i_1 i_2}, C_{i_1 i_2 i_3}, \ldots, C_{i_1 i_2 \ldots i_k}, \ldots$ from $\mathcal{C}$. Finally, we form the intersection of all these sets $C_{i_1} \cap C_{i_1 i_2} \cap C_{i_1 i_2 i_3} \cap \ldots \cap C_{i_1 i_2 \ldots i_k} \cap \ldots$ and consider the union over all possible sequences $(i_j)_{j\in\mathbb{N}}$ of natural numbers:

$$A := A(\mathfrak{s}) := \bigcup_{(i_j : j\in\mathbb{N})\in\mathbb{N}^\mathbb{N}} \bigcap_{k=1}^{\infty} C_{i_1 i_2 \ldots i_k}$$

Note that this union is uncountable, so that A is not necessarily a Borel set.

It is often helpful to visualize this construction as tree:

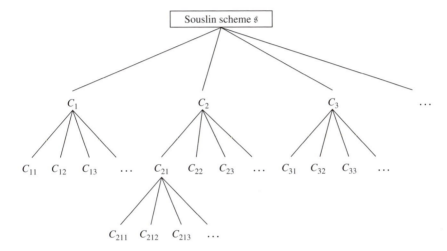

where the $C_{i_1}, C_{i_1 i_2}, C_{i_1 i_2 i_3}, \ldots \in \mathcal{C}$ are the sets of the 1st, 2nd, 3rd, etc. generation. We will also call $C_{i_1 i_2}$ or $C_{i_1 i_2 i_3}$ children or grandchildren of C_{i_1}.

D.7 Definition (Souslin) Let $\mathcal{F}, \mathcal{C}, \mathfrak{s}$ and $A(\mathfrak{s})$ be as above. The sets in $\alpha(\mathcal{C}) := \{A(\mathfrak{s}) : \mathfrak{s} \in \mathcal{C}^{\mathcal{F}}\}$ are called *analytic* or *Souslin sets* (generated by $\mathcal{C}$).

D.8 Lemma *Let $\mathcal{F}, \mathcal{C}$ and $\mathfrak{s}$ be as before.*

(i) *$\alpha(\mathcal{C})$ is stable under countable unions and countable intersections;*
(ii) *$\alpha(\mathcal{C})$ contains all open and all closed subsets of $\mathbb{R}^n$;*
(iii) *$\mathcal{B}(\mathbb{R}^n) = \sigma(\mathcal{C}) \subset \alpha(\mathcal{C})$;*
(iv) *$\#\alpha(\mathcal{C}) \leqslant \mathfrak{c}$.*

Proof (i) Let $A_\ell \in \alpha(\mathcal{C})$, $\ell \in \mathbb{N}$, be a sequence of analytic sets

$$A_\ell = \bigcup_{(i_j : j\in\mathbb{N})\in\mathbb{N}^\mathbb{N}} \bigcap_{k=1}^{\infty} C^\ell_{i_1 i_2 \ldots i_k} .$$

Since

$$A := \bigcup_{\ell \in \mathbb{N}} A_\ell = \bigcup_{\ell \in \mathbb{N}} \bigcup_{(i_j : j \in \mathbb{N}) \in \mathbb{N}^{\mathbb{N}}} \bigcap_{k=1}^{\infty} C^{\ell}_{i_1 i_2 \dots i_k}$$

it is obvious that A can be obtained from a Souslin scheme s which arises by the juxtaposition of the Souslin schemes belonging to the A_ℓ: arrange the double sequence $C^{\ell}_{i_1}$, $(i_1, \ell) \in \mathbb{N} \times \mathbb{N}$, in one sequence – e.g. using the counting scheme of Example 2.5(iv) – to get the first generation of sets while all other generations follow suit in genealogical order. Thus $A \in \alpha(\mathcal{C})$.

For the countable intersection of the A_ℓ we observe first that

$$B := \bigcap_{\ell \in \mathbb{N}} A_\ell = \bigcap_{\ell \in \mathbb{N}} \bigcup_{(i_j : j \in \mathbb{N}) \in \mathbb{N}^{\mathbb{N}}} \bigcap_{k=1}^{\infty} C^{\ell}_{i_1 i_2 \dots i_k} \overset{[\checkmark]}{=} \bigcup_{\substack{(i^m_j : j \in \mathbb{N}) \in \mathbb{N}^{\mathbb{N}} \\ m=1,2,3,\dots}} \bigcap_{\ell=1}^{\infty} \bigcap_{k=1}^{\infty} C^{\ell}_{i^\ell_1 i^\ell_2 \dots i^\ell_k}$$

and then we merge the two infinite intersections indexed by $(\ell, k) \in \mathbb{N} \times \mathbb{N}$ into a single infinite intersection. Once again this can be achieved through the counting scheme of Example 2.5(iv):

$$
\begin{array}{ccccccccccc}
C^1_{i^1_1} & \cap & C^1_{i^1_1 i^1_2} & \cap & C^2_{i^2_1} & \cap & C^1_{i^1_1 i^1_2 i^1_3} & \cap & C^2_{i^2_1 i^2_2} & \cap & C^3_{i^3_1} & \cap \dots \\
\updownarrow & & \updownarrow & & \updownarrow & & \updownarrow & & \updownarrow & & \updownarrow & \\
(1,1) & \to & (1,2) & \to & (2,1) & \to & (1,3) & \to & (2,2) & \to & (3,1) & \to \ \dots
\end{array}
$$

and so

$$B = \bigcup_{\substack{(i^m_j : j \in \mathbb{N}) \in \mathbb{N}^{\mathbb{N}} \\ m=1,2,3,\dots}} \left(C^1_{i^1_1} \cap C^1_{i^1_1 i^1_2} \cap C^2_{i^2_1} \cap C^1_{i^1_1 i^1_2 i^1_3} \cap C^2_{i^2_1 i^2_2} \cap C^3_{i^3_1} \cap \dots \right).$$

We will now construct a Souslin scheme which produces B by arranging the sets $C^j_{k\ell m \dots}$ in a tree:

- The first generation are the sets $C^1_{i^1_1}$, $i^1_1 \in \mathbb{N}$.
- The second generation are the sets $C^1_{i^1_1 i^1_2}$, $i^1_2 \in \mathbb{N}$, such that they are for fixed i^1_1 the children of $C^1_{i^1_1}$.
- Each $C^1_{i^1_1 i^1_2}$ has the same offspring, namely the sets $C^2_{i^2_1}$, $i^2_1 \in \mathbb{N}$, which form jointly the third generation.
- The fourth generation are the sets $C^1_{i^1_1 i^1_2 i^1_3}$, $i^1_3 \in \mathbb{N}$, such that they are for fixed i^1_1, i^1_2 the grandchildren of $C^1_{i^1_1 i^1_2}$.
- The fifth generation are the sets $C^2_{i^2_1 i^2_2}$, $i^2_2 \in \mathbb{N}$, such that they are for fixed i^2_1 the grandchildren of $C^2_{i^2_1}$.

- Each $C^2_{i_1^2 i_2^2}$ has the same offspring, namely the sets $C^3_{i_1^3}$, $i_1^3 \in \mathbb{N}$, which form jointly the sixth generation.

- ...

This shows that $B \in \alpha(\mathcal{C})$.

(ii) Every open set can be written as countable union of $\mathcal{C}$-sets

$$U = \bigcup_{\substack{B_r(x) \subset U \\ B_r(x) \in \mathcal{C}}} B_r(x).$$

Indeed, the inclusion '$\supset$' is obvious, for '$\subset$' fix $x \in U$. Then there exists some $r \in \mathbb{Q}^+$ with $B_r(x) \subset U$. Since $\mathbb{Q}^n$ is dense in $\mathbb{R}^n$, $x \in B_{r/2}(y)$ for some $y \in \mathbb{Q}^n$ with $|x - y| < r/4$, so that $x \in B_{r/2}(y) \subset U$. Since there are only countably many sets in $\mathcal{C}$, the union is *a fortiori* countable. By part (i) we then get that $U \in \alpha(\mathcal{C})$, i.e. $\alpha(\mathcal{C})$ contains all open sets.

For a closed set F we know that

$$F = \bigcap_{j \in \mathbb{N}} U_j \quad \text{where} \quad U_j = F + B_{1/j}(0) = \left\{ y : x \in F, \ |x - y| < \tfrac{1}{j} \right\}$$

is a countable intersection of open$^{[\checkmark]}$ sets U_j. Since open sets are analytic, part (i) implies that $F \in \alpha(\mathcal{C})$.

(iii) Consider the system $\Sigma := \{A \in \alpha(\mathcal{C}) : A^c \in \alpha(\mathcal{C})\}$. We claim that Σ is a σ-algebra. Obviously, Σ satisfies conditions $(\Sigma_1), (\Sigma_2)$ – i.e. contains $\mathbb{R}^n$ and is stable under complementation. To see (Σ_3) we take a sequence $(A_j)_{j \in \mathbb{N}} \subset \Sigma$ and observe that, by part (i),

$$\bigcup_{j \in \mathbb{N}} A_j \in \alpha(\mathcal{C}) \quad \text{and} \quad \left(\bigcup_{j \in \mathbb{N}} A_j \right)^c = \bigcap_{j \in \mathbb{N}} \underbrace{A_j^c}_{\in \alpha(\mathcal{C})} \in \alpha(\mathcal{C}),$$

so that $\bigcup_j A_j \in \Sigma$.

Because of (ii) we have $\mathcal{C} \subset \Sigma \subset \alpha(\mathcal{C})$ and this implies that $\sigma(\mathcal{C}) \subset \alpha(\mathcal{C})$. Since, by (ii), all open sets are countable unions of sets from $\mathcal{C}$, we get $\mathcal{C} \subset \mathcal{O} \subset \sigma(\mathcal{C})$ ($\mathcal{O}$ denotes the family of open sets) or $\sigma(\mathcal{C}) = \sigma(\mathcal{O}) \stackrel{\text{def}}{=} \mathcal{B}(\mathbb{R}^n)$.

(iv) follows immediately from the fact that there are $\#\mathcal{C}^{\mathcal{F}} = \#\mathbb{N}^{\mathbb{N}} = \mathfrak{c}$ Souslin schemes, cf. Definition D.7. ∎

The **Proof of Theorem D.5** is now easy: By Lemma D.8 there are at most $\mathfrak{c}$ analytic sets. Since each singleton $\{x\}$, $x \in \mathbb{R}^n$, is a Borel set, there are at least $\mathfrak{c}$

Borel sets (use Problem 2.17 to see $\#\mathbb{R}^n = \mathfrak{c}$). So,

$$\mathfrak{c} \leqslant \#\mathcal{B}(\mathbb{R}^n) \leqslant \#\alpha(\mathcal{C}) \leqslant \mathfrak{c}$$

and an application of Theorem 2.7 finishes the proof. ∎

D.9 Remark Our approach to analytic sets follows the original construction of Souslin [42], which makes it easy to determine the cardinality of $\alpha(\mathcal{C})$. This, however, comes at a price: if one wants to work with this definition, things become messy, as we have seen in the proof of Lemma D.8(i). Nowadays analytic sets are often introduced by one of the following characterizations. *A set $A \subset \mathbb{R}^n$ is analytic if, and only if, one of the following equivalent conditions holds:*

  (i) $A = f(\mathbb{R})$ *for some left-continuous function* $f : \mathbb{R} \to \mathbb{R}^n$;
 (ii) $A = g(\mathbb{N}^\mathbb{N})$ *for some Borel measurable function* $g : \mathbb{N}^\mathbb{N} \to \mathbb{R}^n$;
(iii) $A = h(B)$ *for some Borel set* $B \in \mathcal{B}(X)$, *some Polish space*[2] *X and some Borel measurable function* $h : B \to \mathbb{R}^n$;
 (iv) $A = \pi_2(B)$ *where* $\pi_2 : Y \times \mathbb{R}^n \to \mathbb{R}^n$ *is the coordinate projection onto* $\mathbb{R}^n$, *Y is a compact Hausdorff space*[3] *and* $B \subset Y \times \mathbb{R}^n$ *is a* $\mathcal{K}_{\sigma\delta}$-*set, i.e. B can be written as countable intersection ('δ') of countable unions ('σ') of compact subsets ('$\mathcal{K}$') of* $Y \times \mathbb{R}^n$.

For a proof we refer to Srivastava [43] which is also our main reference for analytic sets. The *Souslin operation* α can be applied to other systems of sets than $\mathcal{C}$. Without proof we mention the following facts:

$$\alpha(\mathcal{C}) = \alpha(\{\text{open sets}\}) = \alpha(\{\text{closed sets}\}) = \alpha(\{\text{compact sets}\})$$

and also

$$\alpha(\mathcal{C}) = \alpha(\alpha(\mathcal{C})) \qquad \text{and} \qquad \mathcal{B}(\mathbb{R}^n) \subsetneqq \alpha(\mathcal{C}) \subsetneqq \mathcal{B}^*(\mathbb{R}^n).$$

Most constructions of sets which are not Borel but still Lebesgue measurable are actually constructions of non-Borel analytic sets, cf. Dudley [14, §13.2].

---

[2] i.e. a space X which can be endowed with a metric for which X is complete and separable.
[3] cf. Appendix B, Definition B.3.

Appendix E

A summary of the Riemann integral

In this appendix we give a brief outline of the Riemann integral on the real line. The notion of integration was well known for a long time and ever since the creation of differential calculus by Newton and Leibniz, integration was perceived as anti-derivative. Several attempts to make this precise were made, but the problem with these approaches was partly that the notion of integral was implicit – i.e. axiomatically given rather than constructively – partly that the choice of possible integrands was rather limited and partly that some fundamental points were unclear.

Out of the need to overcome these insufficiencies and to have a sound foundation, Bernhard Riemann asked in his Habilitationsschrift *Über die Darstellbarkeit einer Function durch eine trigonometrische Reihe*[1] the question *Also zuerst: Was hat man unter $\int_a^b f(x)\,dx$ zu verstehen?*[2] (p. 239) and proposed a general way to define an integral which is constructive, which is (at least for continuous integrands) the anti-derivative, and which can deal with a wider range of integrands than all its predecessors.

We do not follow Riemann's original approach but use the Darboux technique of upper and lower integrals. Riemann's original definition will be recovered in Theorem E.5(iv).

The (proper) Riemann integral

Riemann integrals are defined only for *bounded* functions on *compact* intervals $[a, b] \subset \mathbb{R}$; this avoids all sorts of complications arising when either the domain or the range of the integrand is infinite. Both cases can be dealt with by various extensions of the Riemann integral, one of which – the so-called *improper* Riemann integral – we will discuss later on.

---

[1] On the representability of a function by a trigonometric series.

[2] First of all: what is the meaning of $\int_a^b f(x)\,dx$?

A *partition* π of the interval $[a, b]$ consists of finitely many points satisfying

$$\pi = \{a = t_0 < t_1 < \ldots < t_{k-1} < t_k = b\}, \qquad k = k(\pi).$$

We call mesh $\pi := \max_{1 \leqslant j \leqslant k(\pi)}(t_j - t_{j-1})$ the *mesh* or *fineness* of the partition. Given a partition π and a bounded function $u : [a, b] \to \mathbb{R}$ we define

$$m_j := \inf_{x \in [t_{j-1}, t_j]} u(x) \quad \text{and} \quad M_j := \sup_{x \in [t_{j-1}, t_j]} u(x),$$

for all $j = 1, 2, \ldots, k(\pi)$, and introduce the *lower*, resp. *upper Darboux sums*

$$S_\pi[u] := \sum_{j=1}^{k(\pi)} m_j(t_j - t_{j-1}) \quad \text{resp.} \quad S^\pi[u] := \sum_{j=1}^{k(\pi)} M_j(t_j - t_{j-1}).$$

Obviously, $S_\pi[\cdot], S^\pi[\cdot]$ are linear, and if $|u(x)| \leqslant M$, they satisfy

$$|S_\pi[u]| \leqslant S_\pi[|u|] \leqslant M(b - a), \qquad |S^\pi[u]| \leqslant S^\pi[|u|] \leqslant M(b - a). \tag{E.1}$$

E.1 Lemma *Let π be a partition of $[a, b]$ and $\pi' \supset \pi$ be a refinement of π. Then*

$$S_\pi[u] \leqslant S_{\pi'}[u] \leqslant S^{\pi'}[u] \leqslant S^\pi[u]$$

holds for all bounded functions $u : [a, b] \to \mathbb{R}$.

Proof Since $S_\pi[u] = -S^\pi[-u]$ and since $S_{\pi'}[u] \leqslant S^{\pi'}[u]$ is trivially fulfilled, it is enough to show $S_\pi[u] \leqslant S_{\pi'}[u]$. The partitions π, π' contain only finitely many points and we may assume that $\pi' = \pi \cup \{\tau\}$ where $t_{j_0-1} < \tau < t_{j_0}$ for some index $1 \leqslant j_0 \leqslant k(\pi)$. The rest follows by iteration. Clearly,

$$S_\pi[u] = \sum_{j \neq j_0} m_j(t_j - t_{j-1}) + m_{j_0}(t_{j_0} - \tau) + m_{j_0}(\tau - t_{j_0-1})$$

$$\leqslant \sum_{j \neq j_0} m_j(t_j - t_{j-1}) + \inf_{x \in [\tau, t_{j_0}]} u(x)(t_{j_0} - \tau)$$

$$+ \inf_{x \in [t_{j_0-1}, \tau]} u(x)(\tau - t_{j_0-1}) = S_{\pi'}[u]. \qquad \blacksquare$$

Lemma E.1 shows that the following definition makes sense.

E.2 Definition Let $u : [a, b] \to \mathbb{R}$ be a bounded function. The *lower* and *upper integrals* of u are given by

$$\int_{*a}^{b} u := \sup_{\pi} S_\pi[u] \qquad \text{and} \qquad \int_{a}^{*b} u := \inf_{\pi} S_\pi[u]$$

where $\sup_\pi$ and $\inf_\pi$ range over all finite partitions of $[a, b]$.

E.3 Lemma $\int_{*a}^{b} u \leqslant \int_{a}^{*b} u$ *and* $\int_{*a}^{b} u = -\int_{a}^{*b}(-u).$

E.4 Definition A bounded function $u : [a, b] \to \mathbb{R}$ is said to be (*Riemann*) *integrable*, if the upper and lower integrals coincide. Their common value is denoted by

$$\int_{a}^{b} u(x)\, dx := \int_{*a}^{b} u = \int_{a}^{*b} u$$

and is called the (*Riemann*) *integral* of u. The collection of all Riemann integrable functions in $[a, b]$ is denoted by $\mathcal{R}[a, b]$.

E.5 Theorem (Characterization of $\mathcal{R}[a, b]$) *Let* $u : [a, b] \to \mathbb{R}$ *be a bounded function. Then the following assertions are equivalent*

(i) *$u \in \mathcal{R}[a, b]$.*
(ii) *For every $\epsilon > 0$ there is some partition π such that $S^{\pi}[u] - S_{\pi}[u] \leqslant \epsilon$.*
(iii) *For every $\epsilon > 0$ there is some $\delta > 0$ such that $S^{\pi}[u] - S_{\pi}[u] \leqslant \epsilon$ for all partitions π with mesh $\pi < \delta$.*
(iv) *The limit $I = \lim\limits_{\text{mesh } \pi \to 0} \sum\limits_{j:t_j \in \pi} u(\xi_j)(t_j - t_{j-1})$ exists for every choice of intermediate values $t_{j-1} \leqslant \xi_j \leqslant t_j$; this means that for all $\epsilon > 0$ there exists a $\delta > 0$ such that for all partitions π with mesh $\pi < \delta$*

$$\left| I - \sum_{j:t_j \in \pi} u(\xi_j)(t_j - t_{j-1}) \right| \leqslant \epsilon$$

independently of the intermediate points.
If the limit exists, $I = \int_{a}^{*b} u = \int_{*a}^{b} u.$

Proof We show the implications $(\mathbf{i}) \Rightarrow (\mathbf{ii}) \Rightarrow (\mathbf{iii}) \Rightarrow (\mathbf{iv}) \Rightarrow (\mathbf{i})$.

$(\mathbf{i}) \Rightarrow (\mathbf{ii})$: By the very definition and the lower and upper integrals in terms of of sup and inf, we find for every $\epsilon > 0$ partitions π' and π'' such that

$$\int_{*a}^{b} u - S_{\pi'}[u] \leqslant \frac{\epsilon}{2} \quad \text{and} \quad S^{\pi''}[u] - \int_{a}^{*b} u \leqslant \frac{\epsilon}{2}.$$

Using the common refinement $\pi = \pi' \cup \pi''$ we get from Lemma E.1 and the integrability of u

$$S^{\pi}[u] - S_{\pi}[u] \leqslant S^{\pi''}[u] - S_{\pi'}[u] = \left(S^{\pi''}[u] - \int_{a}^{*b} u \right) + \left(\int_{*a}^{b} u - S_{\pi'}[u] \right) \leqslant \epsilon.$$

(ii)$\Rightarrow$(iii): This is the most intricate step in the proof. Fix $\epsilon > 0$ and denote by $\pi_\epsilon := \{a = t_0^\epsilon < t_1^\epsilon < \ldots < t_k^\epsilon = b\}$ the partition in (ii). We choose $\delta > 0$ in such a way that

$$\delta < \frac{1}{2} \min_{1 \leqslant j \leqslant k} \left(t_j^\epsilon - t_{j-1}^\epsilon\right) \quad \text{and} \quad \delta < \frac{\epsilon}{4k\|u\|_\infty}.$$

If $\pi := \{a = t_0 < t_1 < \ldots < t_N = b\}$ is any partition with mesh $\pi < \delta$ we find

$$
\begin{aligned}
S^\pi[u] - S_\pi[u] &= \sum_{j \,:\, \pi_\epsilon \cap [t_{j-1}, t_j] \neq \emptyset} (M_j^\pi - m_j^\pi)(t_j - t_{j-1}) \\
&\quad + \sum_{j \,:\, \pi_\epsilon \cap [t_{j-1}, t_j] = \emptyset} (M_j^\pi - m_j^\pi)(t_j - t_{j-1}),
\end{aligned}
\tag{E.2}
$$

where M_j^π, m_j^π indicates that the supremum resp. infimum is taken w.r.t. intervals defined by the partition π. The first sum has at most $2k$ terms since $\pi_\epsilon \cap [a, t_1] = \{a\}$, $\pi_\epsilon \cap [t_{N-1}, b] = \{b\}$ and since all other t_j^ϵ, $1 \leqslant j \leqslant k-1$, appear in exactly one or two intervals defined by π. Thus

$$\sum_{j \,:\, \pi_\epsilon \cap [t_{j-1}, t_j] \neq \emptyset} (M_j^\pi - m_j^\pi)(t_j - t_{j-1}) \leqslant 2k \cdot 2\|u\|_\infty \cdot \delta \leqslant \epsilon. \tag{E.3}$$

The second sum in (E.2) can be written as a double sum

$$
\begin{aligned}
\sum_{j \,:\, \pi_\epsilon \cap [t_{j-1}, t_j] = \emptyset} &(M_j^\pi - m_j^\pi)(t_j - t_{j-1}) \\
&= \sum_{j=1}^{k} \left[\sum_{\ell \,:\, [t_{\ell-1}, t_\ell] \subset [t_{j-1}^\epsilon, t_j^\epsilon]} (M_\ell^\pi - m_\ell^\pi)(t_\ell - t_{\ell-1}) \right] \\
&\leqslant \sum_{j=1}^{k} \left[\sum_{\ell \,:\, [t_{\ell-1}, t_\ell] \subset [t_{j-1}^\epsilon, t_j^\epsilon]} (M_j^{\pi_\epsilon} - m_j^{\pi_\epsilon})(t_\ell - t_{\ell-1}) \right] \\
&\leqslant \sum_{j=1}^{k} (M_j^{\pi_\epsilon} - m_j^{\pi_\epsilon})(t_j^\epsilon - t_{j-1}^\epsilon) \\
&= S^{\pi_\epsilon}[u] - S_{\pi_\epsilon}[u] \leqslant \epsilon.
\end{aligned}
\tag{E.4}
$$

Together (E.2)–(E.4) show $S^\pi[u] - S_\pi[u] \leqslant 2\epsilon$ for any partition π with mesh $\pi < \delta$.

(iii)⇒(iv): Fix $\epsilon > 0$ and choose $\delta > 0$ as in (iii). Then we have for any partition $\pi = \{a = t_0 < \ldots < t_{k(\pi)} = b\}$ with mesh $\pi < \delta$ and any choice of intermediate points $\xi_j \in [t_{j-1}, t_j]$,

$$S^\pi[u] - \epsilon \leqslant S_\pi[u] \leqslant \sum_{j=1}^{k(\pi)} u(\xi_j)(t_j - t_{j-1}) \leqslant S^\pi[u] \leqslant S_\pi[u] + \epsilon.$$

This implies

$$\int_a^b{}^* u - \epsilon \leqslant \sum_{j=1}^{k(\pi)} u(\xi_j)(t_j - t_{j-1}) \leqslant \int_a^b{}^* u$$

and

$$\int_{a*}^b u \leqslant \sum_{j=1}^{k(\pi)} u(\xi_j)(t_j - t_{j-1}) \leqslant \int_{a*}^b u + \epsilon,$$

which means that $\sum_{j=1}^{k(\pi)} u(\xi_j)(t_j - t_{j-1}) \xrightarrow{\text{mesh } \pi \to 0} I = \int_{a*}^b u = \int_a^b{}^* u.$

(vi)⇒(i): Assume that $\sum_{j=1}^{k(\pi)} u(\xi_j)(t_j - t_{j-1}) \xrightarrow{\text{mesh } \pi \to 0} I$ exists for any choice of intermediate values. We have to show that $I = \int_a^b{}^* u = \int_{a*}^b u$. By definition of the limit, there is some $\epsilon > 0$ and some partition π with mesh $\pi < \delta$ such that

$$I - \epsilon \leqslant \sum_{j=1}^{k(\pi)} u(\xi_j)(t_j - t_{j-1}) \leqslant I + \epsilon.$$

Since this must hold uniformly for any choice of intermediate values, we can pass to the infimum and supremum of these values and get

$$I - \epsilon \leqslant \sum_{j=1}^{k(\pi)} \inf_{\xi \in [t_{j-1}, t_j]} u(\xi)\,(t_j - t_{j-1}) \leqslant \sum_{j=1}^{k(\pi)} \sup_{\xi \in [t_{j-1}, t_j]} u(\xi)\,(t_j - t_{j-1}) \leqslant I + \epsilon.$$

Thus $I - \epsilon < S_\pi[u] \leqslant S^\pi[u] \leqslant I + \epsilon$, and

$$I - \epsilon < S_\pi[u] \leqslant \int_{a*}^b u \leqslant \int_a^b{}^* u \leqslant S^\pi[u] \leqslant I + \epsilon. \qquad \blacksquare$$

Once we know that u is Riemann integrable, we can work out the value of the integral by particular Riemann sums:

E.6 Corollary *If* $u : [a, b] \to \mathbb{R}$ *is Riemann integrable, then the integral is the limit of Riemann sums*

$$\lim_{n \to \infty} \sum_{j=1}^{k_n} u(\xi_j^{(n)})(t_j^{(n)} - t_{j-1}^{(n)})$$

where $\pi_n = \{a = t_0^{(n)} < t_1^{(n)} < \ldots < t_{k_n}^{(n)} = b\}$ *is any sequence of partitions with mesh* $\pi_n \xrightarrow{n \to \infty} 0$ *and where* $\xi_j^{(n)} \in [t_{j-1}^{(n)}, t_j^{(n)}]$ *are some intermediate points.*

The existence of the limit of Riemann sums for some particular sequence of partitions does not guarantee integrability.

E.7 Example The Dirichlet jump function $u(x) := \mathbf{1}_{[0,1] \cap \mathbb{Q}}(x)$ on $[0, 1]$ is not Riemann integrable, since for each partition π of $[0, 1]$ we have $M_j = 1$ and $m_j = 0$, so that $\int_{*0}^{1} u = S_\pi[u] = 0$ while $\int_0^{*1} u = S^\pi[u] = 1$.

On the other hand, the equidistant Riemann sum

$$\sum_{j=1}^{k} u(\xi_j)\left(\frac{j}{k} - \frac{j-1}{k}\right) = \frac{1}{k} \sum_{j=1}^{k} u(\xi_j)$$

takes the value $\frac{n}{k}$, $0 \leqslant n \leqslant k$ if we choose $\xi_1, \ldots, \xi_n$ rational and $\xi_{n+1}, \ldots, \xi_k$ irrational. This allows us to construct sequences of Riemann sums which converge to any value in $[0, 1]$.

Let us now find concrete functions which are Riemann integrable. A *step function* on $[a, b]$ is a function $f : [a, b] \to \mathbb{R}$ of the form

$$f(x) = \sum_{j=1}^{N} y_j \mathbf{1}_{I_j}(x)$$

where $N \in \mathbb{N}$, $y_j \in \mathbb{R}$ and I_j are (open, half-open, closed, even degenerate) adjacent intervals such that $I_1 \cup \ldots \cup I_N = [a, b]$ and $I_j \cap I_k$, $j \neq k$, intersect in at most one point. We denote by $\mathcal{T}[a, b]$ the family of all step functions on $[a, b]$.

E.8 Theorem *Continuous functions, monotone functions, and step functions on* $[a, b]$ *are Riemann integrable.*

Proof Notice that the functions from all three classes are bounded on $[a, b]$.

Continuous functions: Let $u : [a, b] \to \mathbb{R}$ be continuous. Since $[a, b]$ is compact, u is uniformly continuous and we find for all $\epsilon > 0$ some $\delta > 0$ such that

$$|u(x) - u(y)| \leqslant \epsilon \quad \forall x, y \in [a, b], \ |x - y| < \delta.$$

If π is a partition of $[a, b]$ with mesh $\pi < \delta$ we find

$$S^{\pi}[u] - S_{\pi}[u] = \sum_{t_j \in \pi} (M_j - m_j)(t_j - t_{j-1}) \leqslant \epsilon \sum_{t_j \in \pi} (t_j - t_{j-1}) = \epsilon (b - a)$$

since, by uniform continuity,

$$M_j - m_j = \sup u([t_{j-1}, t_j]) - \inf u([t_{j-1}, t_j]) = \sup_{\xi, \eta \in [t_{j-1}, t_j]} \big(u(\xi) - u(\eta) \big) \leqslant \epsilon.$$

Thus $u \in \mathcal{R}[a, b]$ by Theorem E.5(iii).

Monotone functions: We can safely assume that $u : [a, b] \to \mathbb{R}$ is monotone increasing, otherwise we would consider $-u$. For the equidistant partition π_k with points $t_j = a + j\frac{b-a}{k}, 0 \leqslant j \leqslant k$, we get

$$S^{\pi_k}[u] - S_{\pi_k}[u] = \sum_{j=1}^{k} (u(t_j) - u(t_{j-1}))(t_j - t_{j-1})$$

$$= \frac{b-a}{k} \sum_{j=1}^{k} (u(t_j) - u(t_{j-1})) = \frac{b-a}{k} (u(b) - u(a)),$$

where we used that $\sup u([t_{j-1}, t_j]) = u(t_j)$ and $\inf u([t_{j-1}, t_j]) = u(t_{j-1})$ because of monotonicity. Since $\frac{b-a}{k} (u(b) - u(a))$ can be made arbitrarily small, $u \in \mathcal{R}[a, b]$ by Theorem E.5(ii).

Step functions: Let u be a step function which has value y_j on the interval I_j, $j = 1, \ldots, k$. The endpoints of the non-degenerate intervals form a partition of $[a, b]$, $\pi = \{a = t_0 < t_1 < \ldots < t_N = b\}$, $N \leqslant k$, and we set for every $\epsilon > 0$

$$\pi_\epsilon := \big\{ a = s_0' < s_1 < s_1' < s_2 < \ldots < s_{N-1} < s_{N-1}' < s_N = b \big\}$$

where $s_j < t_j < s_j'$, $1 \leqslant j \leqslant N-1$, and $s_j' - s_j < \epsilon/(2N\|u\|_\infty)$. Since u is constant with value y_j on each interval $[s_{j-1}', s_j]$, we find

$$S^{\pi_\epsilon}[u] - S_{\pi_\epsilon}[u]$$

$$= \sum_{j=1}^{N} (y_j - y_j)(s_j - s_{j-1}') + \sum_{j=1}^{N-1} \Big[\sup u([s_j, s_j']) - \inf u([s_j, s_j']) \Big](s_j' - s_j)$$

$$\leqslant \sum_{j=1}^{N-1} 2\|u\|_\infty \frac{\epsilon}{2N\|u\|_\infty} \leqslant \epsilon.$$

Therefore Theorem E.5(ii) proves that $u \in \mathcal{R}[a, b]$. ∎

With somewhat more effort one can prove the following general theorem.

E.9 Theorem *Any bounded function* $u : [a, b] \to \mathbb{R}$ *with at most countably many points of discontinuity is Riemann integrable.*

An elementary proof of this based on a compactness argument can be found in Strichartz [48, §6.2.3], but since Theorem 11.8 supersedes this result anyway, we do not include a proof here.

A combination of Theorems E.8 and E.5 yields the following quite useful criterion for integrability.

E.10 Corollary $u \in \mathcal{R}[a, b]$ *if, and only if, for every* $\epsilon > 0$ *there are* $f, g \in \mathcal{T}[a, b]$ *such that* $f \leqslant u \leqslant g$ *and* $\int_a^b (g - f)\, dt \leqslant \epsilon$.

E.11 Theorem *The Riemann integral is a positive linear form on the vector lattice* $\mathcal{R}[a, b]$, *that is, for all* $\alpha, \beta \in \mathbb{R}$ *and* $u, w \in \mathcal{R}[a, b]$ *one has*

(i) $\alpha u + \beta w \in \mathcal{R}[a, b]$ *and* $\displaystyle \int_a^b (\alpha u + \beta w)\, dt = \alpha \int_a^b u\, dt + \beta \int_a^b w\, dt;$

(ii) $u \leqslant w \;\;\Longrightarrow\;\; \displaystyle \int_a^b u\, dt \leqslant \int_a^b w\, dt;$

(iii) $u \vee w,\ u \wedge w,\ u^+,\ u^-,\ |u| \in \mathcal{R}[a, b]$ *and* $\displaystyle \left| \int_a^b u\, dt \right| \leqslant \int_a^b |u|\, dt;$

(iv) $|u|^p,\ u w \in \mathcal{R}[a, b],\ 1 \leqslant p < \infty.$

Proof **(i)** follows immediately from the linearity of the limit criterion in Theorem E.5(iv).

(ii): In view of (i) it is enough to show that $v := w - u \geqslant 0$ entails $\int_a^b v\, dt \geqslant 0$. This, however is clear since $v \in \mathcal{R}[a, b]$ and

$$0 \leqslant \int_{*a}^b v = \int_a^b v\, dt.$$

(iii): Since $u \vee w = -((-u) \wedge (-w))$, $u^+ = u \vee 0$, $u^- = (-u) \vee 0$ and $|u| = u^+ - u^-$, it is enough to prove that $u \wedge w \in \mathcal{R}[a, b]$. By Corollary E.10 there are for every $\epsilon > 0$ step functions $f, g, \phi, \gamma \in \mathcal{T}[a, b]$ such that $f \leqslant u \leqslant g$, $\phi \leqslant w \leqslant \gamma$ and $\int_a^b (g - f)\, dt + \int_a^b (\gamma - \phi)\, dt \leqslant \epsilon$. Obviously, $f \wedge g,\ \phi \wedge \gamma$ are again step functions[✓] with $f \wedge \phi \leqslant u \wedge w \leqslant g \wedge \gamma$ and

$$\int_a^b \left[f \wedge \phi - g \wedge \gamma \right] dt \leqslant \int_a^b \left[(g - f) + (\phi - \gamma) \right] dt \leqslant \epsilon,$$

where we used (ii) and the elementary inequality for $a, b, A, B \in \mathbb{R}$

$$a \wedge A - b \wedge B \leqslant \max\{|a - b|, |A - B|\} \leqslant |a - b| + |A - B|.$$

Finally, since $\pm u \leqslant |u|$ we find by parts (i),(ii) that $\pm \int_a^b u\, dt \leqslant \int_a^b |u|\, dt$ which implies $\left| \int_a^b u\, dt \right| \leqslant \int_a^b |u|\, dt$.

(iv): By (iii), $|u| \in \mathcal{R}[a, b]$ and, by Corollary E.10, we find for each $\epsilon > 0$ step functions $f \leqslant |u| \leqslant g$ such that $\int_a^b (g - f) \, dt \leqslant \epsilon$. Without loss of generality, we may assume that $f \geqslant 0$ and $g \leqslant \|u\|_\infty$ – otherwise we could consider f^+ and $g \wedge \|u\|_\infty$ and note that $f^+, g \wedge \|u\|_\infty \in \mathcal{T}[a, b]$, $f^+ \leqslant |u| \leqslant g \wedge \|u\|_\infty$ and

$$\int_a^b \left(g \wedge \|u\|_\infty - f^+ \right) dt \leqslant \int_a^b (g - f) \, dt \leqslant \epsilon.$$

Thus $f^p \leqslant |u|^p \leqslant g^p$, where $f^p, g^p \in \mathcal{T}[a, b]$. By the mean value theorem of differential calculus we get

$$g^p - f^p \leqslant p \|g\|_\infty^{p-1} (g - f) \leqslant p \|u\|_\infty^{p-1} (g - f).$$

Thus, by (ii),

$$\int_a^b \left(g^p - f^p \right) dt \leqslant p \|u\|_\infty^{p-1} \int_a^b (g - f) \, dt \leqslant p \|u\|_\infty^{p-1} \epsilon.$$

Since $uw = \frac{1}{4}((u + w)^2 - (u - w)^2)$, we conclude that $uw \in \mathcal{R}[a, b]$ from (i) and the fact that $u^2 = |u|^2 \in \mathcal{R}[a, b]$. ∎

Note that Theorem E.11(iii) has no converse: $|u| \in \mathcal{R}[a, b]$ does not imply that $u \in \mathcal{R}[a, b]$ (as is the case for the Lebesgue integral, cf. T10.3). This can be seen by the modified Dirichlet jump function $u := \mathbf{1}_{[0,1] \cap \mathbb{Q}} - \mathbf{1}_{[0,1] \setminus \mathbb{Q}}$ which is not Riemann integrable but whose modulus $|u| = \mathbf{1}_{[0,1]}$ is Riemann integrable.

E.12 Corollary (Mean value theorem for integrals) *Let $u \in \mathcal{R}[a, b]$ be either positive or negative and let $v \in C[a, b]$. Then there exists some $\xi \in (a, b)$ such that*

$$\int_a^b u(t)v(t) \, dt = v(\xi) \int_a^b u(t) \, dt. \tag{E.5}$$

Proof The case $u \leqslant 0$ being similar, we may assume that $u \geqslant 0$. By Theorem E.8 and E.11(iv), uv is integrable and because of E.11(ii) we have

$$\inf v([a, b]) \int_a^b u(t) \, dt \leqslant \int_a^b u(t)v(t) \, dt \leqslant \sup v([a, b]) \int_a^b u(t) \, dt.$$

Since v is continuous on $[a, b]$, the intermediate value theorem guarantees the existence of some $\xi \in (a, b)$ such that (E.5) holds. ∎

E.13 Theorem *Let $[c, d] \subset [a, b]$. Then $\mathcal{R}[a, b] \subset \mathcal{R}[c, d]$ in the sense that $u \in \mathcal{R}[a, b]$ satisfies $u|_{[c,d]} \in \mathcal{R}[c, d]$. Moreover, for any $u \in \mathcal{R}[a, b]$*

$$\int_a^b u \, dt = \int_a^c u \, dt + \int_c^b u \, dt.$$

Proof By Theorem E.8 and E.11 we find that $\mathbf{1}_{[c,d]} u \in \mathcal{R}[a, b]$. Since we can always add the points c and d to any of the partitions appearing in one of the criteria of Theorem E.5, we see that $u|_{[c,d]} = (\mathbf{1}_{[c,d]}\,u)|_{[c,d]} \in \mathcal{R}[c,d]$ and

$$\int_a^b \mathbf{1}_{[c,d]}\,u\,dt = \int_c^d u\,dt.$$

Considering $u = \mathbf{1}_{[a,c]}\,u + \mathbf{1}_{[c,b]}\,u$ proves also the formula in the statement of the theorem.   ∎

The fundamental theorem of integral calculus

Since by Theorem E.13 $\mathcal{R}[a,x] \subset \mathcal{R}[a,b]$, we can treat $\int_a^x u(t)\,dt$, $u \in \mathcal{R}[a,b]$, as a function of its upper limit $x \in [a,b]$.

E.14 Lemma *For every $u \in \mathcal{R}[a,b]$ the function $U(x) := \int_a^x u(t)\,dt$ is continuous for all $x \in [a,b]$.*

Proof Since u is bounded, $M := \sup_{x\in[a,b]} |u(x)| < \infty$. For all $x,y \in [a,b]$, $x < y$, we have by Theorem E.13 and E.11

$$|U(y) - U(x)| = \left| \int_a^y u(t)\,dt - \int_a^x u(t)\,dt \right|$$

$$= \left| \int_x^y u(t)\,dt \right| \leqslant \int_x^y |u(t)|\,dt \leqslant M\,(y-x) \xrightarrow{x-y\to 0} 0,$$

showing even uniform continuity.   ∎

We can now discuss the connection between differentiation and integration. Let us begin with a few examples.

E.15 Example (i) Let $[0,1] \subsetneq [a,b]$. Then $u(x) = \mathbf{1}_{[0,1]}(x)$ is an integrable function and

$$U(x) := \int_a^x u(t)\,dt = \left\{ \begin{array}{ll} 0, & \text{if } x \leqslant 0, \\ x, & \text{if } 0 < x < 1, \\ 1, & \text{if } x \geqslant 1, \end{array} \right\} = x^+ \wedge 1.$$

Note that $U'(x)$ does not exist at $x = 0$ or $x = 1$, so that $u(x)$ cannot be the derivative of any function (at every point).

(ii) Let $[a,b] = [0,1]$ and take an enumeration $(q_j)_{j\in\mathbb{N}}$ of $[0,1]\cap\mathbb{Q}$. Then the function

$$u(x) := \sum_{j:\,q_j \leqslant x} 2^{-j} = \sum_{j=1}^{\infty} 2^{-j}\mathbf{1}_{[q_j,1]}(x), \qquad x \in [0,1],$$

is increasing, satisfies $0 \leqslant u \leqslant 1$ and its discontinuities are jumps at the points q_j of height $u(q_j+) - u(q_j-) = 2^{-j}$ – this is as bad as it can get for a monotone function, cf. Lemma 13.12. By Theorem E.8 u is integrable, and since $(q_j)_{j\in\mathbb{N}}$ is dense, there is no interval $[c, d] \subset [0, 1]$ such that $U'(x) = u(x)$ for all $x \in (c, d)$ for any function $U(x)$.

(iii) Consider on $[-1, 1]$ the function

$$u(x) := \begin{cases} x^2 \sin \frac{1}{x^2}, & \text{if } x \neq 0, \\ 0, & \text{if } x = 0. \end{cases}$$

It is an elementary exercise to show that $u'(x)$ exists on $(-1, 1)$ and

$$u'(x) = \begin{cases} 2x \sin \frac{1}{x^2} - \frac{2}{x} \cos \frac{1}{x^2}, & \text{if } x \neq 0, \\ 0, & \text{if } x = 0. \end{cases}$$

Thus u' exists everywhere, but it is not Riemann integrable in any neighbourhood of $x = 0$ since u' is unbounded.

(iv) Let $(q_n)_{n\in\mathbb{N}}$ be an enumeration of $(0, 1) \cap \mathbb{Q}$. The function

$$u(x) := \begin{cases} 2^{-n}, & \text{if } x = q_n, \ n \in \mathbb{N}, \\ 0, & \text{if } x \in ([0, 1] \setminus \mathbb{Q}) \cup \{0, 1\}, \end{cases}$$

is discontinuous for every $x \in (0, 1) \cap \mathbb{Q}$ and continuous otherwise. Moreover, $u \in \mathcal{R}[0, 1]$ which follows from Theorem E.9 or directly from the following argument: fix $\epsilon > 0$ and $n \in \mathbb{N}$ such that $2^{-n} < \epsilon$. Choose a partition $\pi = \{0 = t_0 < t_1 < \ldots < t_N = 1\}$ with mesh $\pi = \delta < \frac{\epsilon}{n}$ in such a way that each q_k from $Q_n := \{q_1, q_2, \ldots, q_n\}$ is the midpoint of some $[t_{j-1}, t_j]$, $j = 1, 2, \ldots, N$. Therefore, if M_j denotes $\sup u([t_{j-1}, t_j])$,

$$0 \leqslant S_\pi[u] \leqslant S^\pi[u] = \sum_{j=1}^{N} M_j (t_j - t_{j-1})$$

$$= \sum_{j:[t_{j-1},t_j] \cap Q_n \neq \emptyset} M_j (t_j - t_{j-1})$$

$$+ \sum_{j:[t_{j-1},t_j] \cap Q_n = \emptyset} M_j (t_j - t_{j-1})$$

$$\leqslant n \frac{\epsilon}{n} + 2^{-n} \sum_{j:[t_{j-1},t_j] \cap Q_n = \emptyset} (t_j - t_{j-1})$$

$$\leqslant \epsilon + \epsilon \sum_{j=1}^{N} (t_j - t_{j-1}) = 2\epsilon.$$

This proves $u \in \mathcal{R}[0, 1]$ and $0 \leqslant \int_0^x u(t) \, dt \leqslant \int_0^1 u(t) \, dt = 0$. Thus $u'(x) = 0 \neq u(x)$ for all x from a dense subset.

The above examples show that the Riemann integral is not always the antiderivative, nor is the antiderivative an extension of the Riemann integral. The two concepts, however, coincide on a large class of functions.

E.16 Definition Let $u : [a, b] \to \mathbb{R}$ be a bounded function. Every function $U \in C[a, b]$ such that $U'(x) = u(x)$ for all but possibly finitely many $x \in (a, b)$ is called a *primitive* of u.

Obviously, primitives are only unique up to constants: for every constant c, $U + c$ is again a primitive of u. On the other hand, if U, W are two primitives of u, we have $U' - W' = 0$ at all but finitely many points $a = x_0 < x_1 < \ldots < x_n = b$. Thus the mean value theorem of differential calculus shows $U = W + \text{const.}$ (cf. Rudin [39, Thm. 5.11]), first on each interval (x_{j-1}, x_j), $j = 1, 2, \ldots, n$, and then, by continuity, on the whole interval $[a, b]$.

E.17 Proposition *Every $u \in C[a, b]$ has $U(x) := \int_a^x u(t)\, dt$ as a primitive. Moreover,*

$$U(b) - U(a) = \int_a^b u(t)\, dt.$$

Proof Since continuous functions are integrable, $U(x)$ is well-defined by Theorem E.13 and continuous by Lemma E.14. For $a < x < x + h < b$ and sufficiently small h we find

$$\left| U(x + h) - U(x) - h\, u(x) \right| = \left| \int_a^{x+h} u(t)\, dt - \int_a^x u(t)\, dt - \int_x^{x+h} u(x)\, dt \right|$$

$$= \left| \int_x^{x+h} \big(u(t) - u(x)\big)\, dt \right|$$

$$\leqslant \int_x^{x+h} \left| u(t) - u(x) \right| dt$$

$$\leqslant \int_x^{x+h} \epsilon\, dt = \epsilon h,$$

where we used that $u(t)$ is continuous at $t = x$. With a similar calculation we get

$$\left| U(x) - U(x - h) - h\, u(x) \right| \leqslant \epsilon h,$$

and a combination of both inequalities shows that

$$\lim_{y \to x} \frac{U(y) - U(x)}{y - x} = u(x).$$

The formula $U(b) - U(a) = \int_a^b u(t)\, dt$ follows from the fact that $U(a) = 0$. ∎

E.18 Theorem (**Fundamental theorem of calculus**) *Assume that U is a primitive of $u \in \mathcal{R}[a, b]$. Then*

$$U(b) - U(a) = \int_a^b u(t)\, dt.$$

Proof Let C be some finite set such that $U'(x) = u(x)$ if $x \in (a, b) \setminus C$. Fix $\epsilon > 0$. Since u is integrable, we find by E.5(ii) a partition π of $[a, b]$ such that $S^\pi[u] - S_\pi[u] \leqslant \epsilon$. Because of Lemma E.1 this inequality still holds for the partition $\pi' := \pi \cup C$ whose points we denote by $a = t_0 < t_1 < \ldots < t_k = b$. Since

$$U(b) - U(a) = \sum_{j=1}^k (U(t_j) - U(t_{j-1}))$$

and since U is differentiable in each segment (t_{j-1}, t_j) and continuous on $[a, b]$, we can use the mean value theorem of differential calculus to find points $\xi_j \in (t_{j-1}, t_j)$ with

$$U(t_j) - U(t_{j-1}) = U'(\xi_j)(t_j - t_{j-1}) = u(\xi_j)(t_j - t_{j-1}), \qquad 1 \leqslant j \leqslant k.$$

Using $m_j = \inf u([t_{j-1}, t_j]) \leqslant u(\xi_j) \leqslant \sup u([t_{j-1}, t_j]) = M_j$ we can sum the above equality over $j = 1, \ldots, k$ and get

$$S^{\pi'}[u] - \epsilon \leqslant S_{\pi'}[u] \leqslant U(b) - U(a) \leqslant S^{\pi'}[u] \leqslant S_{\pi'}[u] + \epsilon.$$

By integrability, $S_{\pi'}[u] \leqslant \int_a^b u\, dt \leqslant S^{\pi'}[u]$, and this shows

$$\int_a^b u\, dt - \epsilon \leqslant U(b) - U(a) \leqslant \int_a^b u\, dt + \epsilon \qquad \forall \epsilon > 0,$$

which proves our claim. ∎

E.19 Remark There is not much room to improve the fundamental theorem E.18. On one hand, Example E.15(ii) shows that an integrable function need not have a primitive and E.15(iv) gives an example where $\int_a^x u\, dt$ exists, but is not a primitive in any interval; on the other hand, E.15(iii) provides an example of a function u' which has a primitive u but which is itself not Riemann integrable since it is *unbounded*. Volterra even constructed an example of a *bounded* but not Riemann integrable function with a primitive, see Sz.-Nagy [30, pp. 155–7].

To overcome this phenomenon was one of the motivations for Lebesgue when he introduced the *Lebesgue integral*. And, in fact, *every bounded function f on the interval $[a, b]$ with a primitive F is Lebesgue integrable*: indeed, since F is continuous, it is measurable in the sense of Chapter 8 and so is the limit $f(x) = \lim_{n \to \infty}(F(x + \frac{1}{n}) - F(x))/\frac{1}{n}$, cf. Corollary 8.9 – the finitely many points where the limit does not exist are a Lebesgue null set and pose no problem. Since

$|f|$ is dominated by the (Lebesgue) integrable function $M \, \mathbf{1}_{[a,b]}$, $M := \sup f([a, b])$, we conclude that $f \in \mathcal{L}^1[a, b]$.

An immediate consequence of the integral as antiderivative are the following integration formulae which are easily proved by 'integrating up' the corresponding differentiation rules.

E.20 Theorem (Integration by parts) *Let u' and v' be integrable functions on $[a, b]$ with primitives u and v. Then uv is a primitive of $u'v + uv'$ and, in particular,*

$$\int_a^b u'(t)v(t) \, dt = u(b)v(b) - u(a)v(a) - \int_a^b u(t)v'(t) \, dt.$$

E.21 Theorem (Integration by substitution) *Let $u \in \mathcal{R}[a, b]$ and assume that $\phi : [c, d] \to [a, b]$ is a strictly increasing differentiable function such that $\phi(c) = a$ and $\phi(d) = b$. If $u \circ \phi$, $\phi' \in \mathcal{R}[c, d]$ and if u has a primitive U, then $U \circ \phi$ is a primitive of $u \circ \phi \cdot \phi'$ as well as*

$$\int_a^b u(t) \, dt = \int_c^d u(\phi(s))\phi'(s) \, ds = \int_{\phi^{-1}(a)}^{\phi^{-1}(b)} u(\phi(s))\phi'(s) \, ds.$$

E.22 Corollary (Bonnet's mean value theorem[3]) *Let $u, v \in \mathcal{R}[a, b]$ have primitives U and V. If $u \leqslant 0$ [resp. $u \geqslant 0$] and $U \geqslant 0$, then there exists some $\xi \in (a, b)$ such that*

$$\int_a^b U(t)v(t) \, dt = U(a) \int_a^\xi v(t) \, dt. \tag{E.6}$$

$$\left[resp. \ \int_a^b U(t)v(t) \, dt = U(b) \int_\xi^b v(t) \, dt. \right] \tag{E.6$'$}$$

Proof By subtracting a suitable constant from V we may assume that $V(a) = 0$ and, by the fundamental theorem E.18, $V(a) = \int_a^x v(t) \, dt$. Integration by parts now shows

$$\int_a^b U(t)v(t) \, dt = U(b)V(b) - \int_a^b u(t)V(t) \, dt.$$

Since $u \leqslant 0$ we get

$$\int_a^b U(t)v(t) \, dt \leqslant U(b)V(b) - \sup V([a, b]) \int_a^b u(t) \, dt$$

$$= U(b)V(b) - \sup V([a, b]) \, (U(b) - U(a))$$

$$= U(b)\big(V(b) - \sup V([a, b])\big) + \sup V([a, b]) \, U(a)$$

$$\leqslant \sup V([a, b]) \, U(a),$$

---

[3] Also known as the second mean value theorem of integral calculus.

and a similar calculation yields the other inequality below:

$$\inf V([a, b])\, U(a) \leqslant \int_a^b U(t) v(t)\, dt \leqslant \sup V([a, b])\, U(a).$$

Applying the intermediate value theorem to the continuous function V furnishes some $\xi \in (a, b)$ such that (E.6) holds. ∎

Integrals and limits

One of the strengths of Lebesgue integration is the fact that we have fairly general theorems that allow interchanging pointwise limits and Lebesgue integrals.

Similar results for the Riemann integral regularly require uniform convergence. Recall that a sequence of functions $(u_n(\bullet))_{n\in\mathbb{N}}$ on $[a, b]$ *converges uniformly* (in x) to u, if

$$\forall \epsilon > 0 \quad \exists N_\epsilon \in \mathbb{N} : \forall x \in [a, b],\ \forall n \geqslant N_\epsilon : \quad |u_n(x) - u(x)| \leqslant \epsilon.$$

The basic convergence result for the Riemann integral is the following.

E.23 Theorem *Let $(u_n)_{n\in\mathbb{N}} \subset \mathcal{R}[a, b]$ be a sequence which converges uniformly to a function u. Then $u \in \mathcal{R}[a, b]$ and*

$$\lim_{n\to\infty} \int_a^b u_n\, dt = \int_a^b \lim_{n\to\infty} u_n\, dt = \int_a^b u\, dt.$$

Proof Let π be a partition of $[a, b]$ and let $\epsilon > 0$ be given. Since $u_n \xrightarrow{n\to\infty} u$ uniformly, we can find some $N_\epsilon \in \mathbb{N}$ such that $|u(x) - u_n(x)| \leqslant \epsilon/(b - a)$ uniformly in $x \in [a, b]$ for all $n \geqslant N_\epsilon$. Because of (E.1) we find for all $n \geqslant N_\epsilon$

$$S^\pi[u] - S_\pi[u] = S^\pi[u - u_n] + S^\pi[u_n] - S_\pi[u_n] - S_\pi[u - u_n]$$
$$\leqslant 2\epsilon + S^\pi[u_n] - S_\pi[u_n],$$

thus

$$\int_a^{b*} u - \int_{a*}^b u \leqslant 2\epsilon + S^\pi[u_n] - S_\pi[u_n] \qquad \forall n \geqslant N_\epsilon.$$

Fixing some $n_0 \geqslant N_\epsilon$ we can use that u_{n_0} is integrable and choose π in such a way that $S^\pi[u_{n_0}] - S_\pi[u_{n_0}] \leqslant \epsilon$. This shows that $\int_a^{b*} u - \int_{a*}^b u \leqslant 3\epsilon$ and $u \in \mathcal{R}[a, b]$.

Once u is known to be integrable, we get for all $n \geqslant N_\epsilon$

$$\left| \int_a^b (u - u_n)\, dt \right| \leqslant \int_a^b |u - u_n|\, dt \leqslant \epsilon\, (b - a) \xrightarrow{\epsilon\to 0} 0. \qquad \blacksquare$$

We can now consider Riemann integrals which depend on a parameter.

E.24 Theorem (Continuity theorem) *Let* $u : [a, b] \times \mathbb{R} \to \mathbb{R}$ *be a continuous function. Then*

$$w(y) := \int_a^b u(t, y) \, dt$$

is continuous for all $y \in \mathbb{R}$.

Proof Since $u(\cdot, y)$ is continuous, the above Riemann integral exists. Fix $y \in \mathbb{R}$ and consider any sequence $(y_n)_{n \in \mathbb{N}}$ with limit y. Without loss of generality we can assume that $(y_n)_{n \in \mathbb{N}} \subset I := [y - 1, y + 1]$. Since $[a, b] \times I$ is compact, $u|_{[a,b] \times I}$ is uniformly continuous, and we can find for all $\epsilon > 0$ some $\delta > 0$ such that

$$\sqrt{(t - \tau)^2 + (y - \eta)^2} < \delta \quad \Longrightarrow \quad |u(t, y) - u(\tau, \eta)| < \epsilon.$$

As $y_n \xrightarrow{n \to \infty} y$, there is some $N_\epsilon \in \mathbb{N}$ with

$$|u(t, y_n) - u(t, y)| < \epsilon \qquad \forall t \in [a, b], \ \forall n \geqslant N_\epsilon,$$

i.e. $u(y_n, t) \xrightarrow{n \to \infty} u(y, t)$ uniformly in $t \in [a, b]$. Theorem E.23 and the continuity of $u(t, \cdot)$ therefore show

$$\lim_{n \to \infty} w(y_n) = \lim_{n \to \infty} \int_a^b u(t, y_n) \, dt = \int_a^b \lim_{n \to \infty} u(t, y_n) \, dt = \int_a^b u(t, y) \, dt = w(y)$$

which is but the continuity of w at y. ∎

E.25 Theorem (Differentiation theorem) *Let* $u : [a, b] \times \mathbb{R} \to \mathbb{R}$ *be a continuous function with continuous partial derivative* $\frac{\partial}{\partial y} u(t, y)$. *Then*

$$w(y) := \int_a^b u(t, y) \, dt$$

is continuously differentiable and

$$w'(y) = \frac{d}{dy} \int_a^b u(t, y) \, dt = \int_a^b \frac{\partial}{\partial y} u(t, y) \, dt.$$

Proof Since $u(\cdot, y)$ and $\frac{\partial}{\partial y} u(\cdot, y)$ are continuous, the above integrals exist. Fix $y \in \mathbb{R}$ and consider any sequence $(y_n)_{n \in \mathbb{N}}$ with limit y. Without loss of generality we can assume that $(y_n)_{n \in \mathbb{N}} \subset I := [y - 1, y + 1]$.

We introduce the following auxiliary function

$$h(t, z) := u(t, z) - u(t, y) - \frac{\partial}{\partial y} u(t, y) \, (z - y).$$

Clearly, $h(t, y) = 0$ and $\frac{\partial}{\partial z} h(t, z) = \frac{\partial}{\partial z} u(t, z) - \frac{\partial}{\partial y} u(t, y)$ is continuous and uniformly continuous on $[a, b] \times I$, i.e. for all $\epsilon > 0$ there is some $\delta > 0$ such that

$$\sqrt{(t - \tau)^2 + (z - \zeta)^2} < \delta \quad \Longrightarrow \quad \left| \frac{\partial}{\partial z} h(t, z) - \frac{\partial}{\partial \zeta} h(\tau, \zeta) \right| < \epsilon.$$

From the mean value theorem of differential calculus we infer that for some ζ between z and y

$$|h(t, z)| = |h(t, z) - h(t, y)| = \left| \frac{\partial}{\partial \zeta} h(\tau, \zeta) \right| \cdot |z - y|$$

$$= \left| \frac{\partial}{\partial \zeta} h(\tau, \zeta) - \frac{\partial}{\partial y} h(t, y) \right| \cdot |z - y|$$

$$\leqslant \epsilon |z - y|$$

whenever $z, y \in I$ and $|z - y| < \delta$. This shows that for some $N_\epsilon \in \mathbb{N}$

$$\left| u(t, y_n) - u(t, y) - \frac{\partial}{\partial y} u(t, y)(y_n - y) \right| \leqslant \epsilon |y_n - y| \quad \forall t \in [a, b], \ \forall n \geqslant N_\epsilon.$$

Theorem E.23 now shows that

$$w'(y) = \lim_{n \to \infty} \frac{w(y_n) - w(y)}{y_n - y} = \lim_{n \to \infty} \int_a^b \frac{u(t, y_n) - u(t, y)}{y_n - y} \, dt$$

$$= \int_a^b \lim_{n \to \infty} \frac{u(t, y_n) - u(t, y)}{y_n - y} \, dt = \int_a^b \frac{\partial}{\partial y} u(t, y) \, dt. \qquad \blacksquare$$

Improper Riemann integrals

Let us finally have a glance at various extensions of the Riemann integral to unbounded intervals and/or unbounded integrands. The following cases can occur:

A. the interval of integration is $[a, +\infty)$ or $(-\infty, b]$;

B. the interval of integration is $[a, b)$ or $(a, b]$, and the integrand $u(t)$ is unbounded as $t \uparrow b$ resp. $t \downarrow a$;

C. the interval of integration is (a, b) with $-\infty \leqslant a < b \leqslant +\infty$ and the integrand may or may not be unbounded.

A. *Improper Riemann integrals of the type* $\int_a^\infty u\,dt$ *or* $\int_{-\infty}^b u\,dt$

E.26 Definition If $u \in \mathcal{R}[a, b]$ for all $b \in (a, \infty)$ [resp. $a \in (-\infty, b)$] and if the limit

$$\lim_{b\to\infty} \int_a^b u\,dt \qquad \left[\text{resp.} \lim_{a\to-\infty} \int_a^b u\,dt\right]$$

exists and is finite, we call u *improperly Riemann integrable* and write $u \in \mathcal{R}[a, \infty)$ [resp. $u \in \mathcal{R}(-\infty, b]$]. The value of the above limit is called the *(improper Riemann) integral* and denoted by $\int_a^\infty u\,dt$ [resp. $\int_{-\infty}^b u\,dt$].

The *typical examples* of improper integrals of this kind are expressions of the type $\int_1^\infty t^\lambda\,dt$ if $\lambda < 0$. In fact, if $\lambda \neq -1$,

$$\int_1^\infty t^\lambda = \lim_{b\to\infty} \int_1^b t^\lambda\,dt = \lim_{b\to\infty} \frac{1}{\lambda+1}(b^{\lambda+1}-1) = \begin{cases} \frac{-1}{\lambda+1} & \text{if } \lambda < -1, \\ \infty & \text{if } \lambda > -1, \end{cases}$$

and a similar calculation confirms that $\int_1^\infty t^{-1}\,dt = \infty$. Thus $t^\lambda \in \mathcal{R}[1, \infty)$ if, and only if, $\lambda < -1$.

From now on we will only consider integrals of the type $\int_a^\infty u\,dt$, the case of a finite upper and infinite lower limit is very similar. The following Cauchy criterion for improper integrals is quite useful.

E.27 Lemma $u \in \mathcal{R}[a, \infty)$ *if, and only if,* $u \in \mathcal{R}[a, b]$ *for all* $b \in (a, \infty)$ *and* $\lim_{x,y\to\infty} \int_x^y u\,dt = 0$ ($x, y \to \infty$ *simultaneously*).

Proof This is just Cauchy's convergence criterion for $U(z) = \int_a^z u(t)\,dt$ as $z \to \infty$. ∎

It is not hard to see that Lemma E.27 implies, in particular, that

- $\mathcal{R}[a, \infty)$ is a vector space, i.e. for all $\alpha, \beta \in \mathbb{R}$ and $u, w \in \mathcal{R}[a, \infty)$,

$$\int_a^\infty (\alpha u + \beta w)\,dt = \alpha \int_a^\infty u\,dt + \beta \int_a^\infty w\,dt.$$

- $u \in \mathcal{R}[a, \infty)$ if, and only if, $\int_b^\infty u\,dt$ exists for all $b > a$.

E.28 Corollary *Let* $u, w : [a, \infty) \to \mathbb{R}$ *be two functions such that* $|u| \leqslant w$. *If* $w \in \mathcal{R}[a, \infty)$, *and if* $u \in \mathcal{R}[a, b]$ *for all* $b > a$, *then* $u, |u| \in \mathcal{R}[a, \infty)$. *In particular,* $|u| \in \mathcal{R}[a, \infty)$ *implies that* $u \in \mathcal{R}[a, \infty)$.

Proof For all $y > x > a$ we find using Theorem E.11 and Lemma E.27 that

$$\left| \int_x^y u \, dt \right| \leqslant \int_x^y |u| \, dt \leqslant \int_x^y w \, dt \xrightarrow{x, y \to \infty} 0$$

which shows, again by E.27, that $u, |u| \in \mathcal{R}[a, \infty)$. ∎

Note that, unlike Lebesgue integrals, improper Riemann integrals are not *absolute integrals* since improper integrability of u does NOT imply improper integrability of $|u|$, see e.g. Remark 11.11 where $\int_0^\infty \sin t/t \, dt$ is discussed. This means that the following convergence theorems for improper Riemann integrals are not necessarily covered by Lebesgue's theory.

E.29 Theorem *Let* $(u_n)_{n \in \mathbb{N}} \subset \mathcal{R}[a, \infty)$. *If for some* $u : [a, \infty) \to \mathbb{R}$

- $u_n(t) \xrightarrow{n \to \infty} u(t)$ *uniformly in* $t \in [a, b]$ *and for every* $b > a$,
- $\lim\limits_{b \to \infty} \int_a^b u_n \, dt$ *exists uniformly for all* $n \in \mathbb{N}$, *i.e. for every* $\epsilon > 0$ *there is some* $N_\epsilon \in \mathbb{N}$ *such that*

$$\sup_{n \in \mathbb{N}} \left| \int_x^y u_n \, dt \right| < \epsilon \qquad \forall y > x > N_\epsilon,$$

then $u \in \mathcal{R}[a, \infty)$ *and* $\lim\limits_{n \to \infty} \int_a^\infty u_n \, dt = \int_a^\infty \lim\limits_{n \to \infty} u_n \, dt = \int_a^\infty u \, dt.$

Proof That $u \in \mathcal{R}[a, b]$ for all $b > a$ follows from Theorem E.23. Fix $\epsilon > 0$ and choose N_ϵ as in the above statement. For all $y > x > N_\epsilon$

$$\left| \int_x^y u \, dt \right| \leqslant \left| \int_x^y (u - u_n) \, dt \right| + \left| \int_x^y u_n \, dt \right| \leqslant (y - x) \sup_{t \in [x, y]} |u(t) - u_n(t)| + \epsilon,$$

and as $n \to \infty$ we find $\left| \int_x^y u \, dt \right| \leqslant \epsilon$ for all $y > x > N_\epsilon$, hence $u \in \mathcal{R}[a, \infty)$ by Lemma E.27. ∎

In pretty much the same way as we derived Theorems E.24, E.25 from the basic convergence result E.23 we get now from E.29 the following continuity and differentiability theorems for improper integrals.

E.30 Theorem *Let* $I \subset \mathbb{R}$ *be an open interval and* $u : [a, \infty) \times I \to \mathbb{R}$ *be continuous such that* $u(\cdot, y) \in \mathcal{R}[a, \infty)$ *for all* $y \in I$ *and*

$$\lim_{b \to \infty} \int_a^b u(t, y) \, dt \quad \text{exists uniformly for all} \quad y \in [c, d] \subset I.$$

Then $U(y) := \int_a^\infty u(t, y) \, dt$ *is continuous for all* $y \in (c, d)$.

Proof (sketch) Fix $y \in (c, d)$ and choose any sequence $(y_n)_{n \in \mathbb{N}} \subset (c, d)$ with limit y. By the assumptions $u_n(t) := u(t, y_n) \xrightarrow{n \to \infty} u(t, y)$ uniformly for all $t \in [a, b]$. Now the basic convergence theorem for improper integrals E.29 applies and shows $U(y_n) \xrightarrow{n \to \infty} U(y)$. ∎

E.31 Theorem *Let $I \subset \mathbb{R}$ be an open interval and $u : [a, \infty) \times I \to \mathbb{R}$ be continuous with continuous partial derivative $\frac{\partial}{\partial y} u(t, y)$. If $u(\bullet, y), \frac{\partial}{\partial y} u(t, y) \in \mathcal{R}[a, \infty)$ for all $y \in I$, and if*

$$\lim_{b \to \infty} \int_a^b u(t, y)\, dt \quad \text{and} \quad \lim_{b \to \infty} \int_a^b \frac{\partial}{\partial y} u(t, y)\, dt$$

exist uniformly for all $y \in [c, d] \subset I$, then $W(y) := \int_a^\infty u(t, y)\, dt$ exists and is differentiable on (c, d) with derivative

$$W'(y) = \frac{d}{dy} \int_a^\infty u(t, y)\, dt = \int_a^\infty \frac{\partial}{\partial y} u(t, y)\, dt.$$

Proof (sketch) Set $U(x, y) := \int_a^x u(t, y)\, dt$. By Theorem E.25 $\frac{\partial}{\partial y} U(x, y)$ exists and equals $\int_a^x \frac{\partial}{\partial y} u(t, y)\, dt$. By assumption,

$$U(x, y) \xrightarrow{x \to \infty} \int_a^\infty u(t, y)\, dt \qquad \text{pointwise for all } y \in [c, d],$$

$$\frac{\partial}{\partial y} U(x, y) \xrightarrow{x \to \infty} \int_a^\infty \frac{\partial}{\partial y} u(t, y)\, dt \qquad \text{uniformly for all } y \in [c, d].$$

By a standard theorem on uniform convergence and differentiability, cf. Rudin [39, Theorem 7.17], we now conclude

$$\frac{d}{dy} \int_a^\infty u(t, y)\, dt = \int_a^\infty \frac{\partial}{\partial y} u(t, y)\, dt. \qquad ∎$$

E.32 Theorem *Let $u, w \in \mathcal{R}[a, b]$ for all $b \in (a, \infty)$ and assume that $u, w \geqslant 0$ and that $\lim_{x \to \infty} u(x)/w(x) = A > 0$ exists. Then $u \in \mathcal{R}[a, \infty)$ if, and only if, $w \in \mathcal{R}[a, \infty)$.*

Proof By assumption we find for every $\epsilon > 0$ some $N_\epsilon \in \mathbb{N}$ such that

$$0 < A - \epsilon \leqslant \frac{u(x)}{w(x)} \leqslant A + \epsilon \qquad \forall x \geqslant N_\epsilon \, (> a).$$

Thus $(A - \epsilon)w(x) \leqslant u(x) \leqslant (A + \epsilon)w(x)$ for all $x \geqslant N_\epsilon$. Thus, if $w \in \mathcal{R}[a, \infty)$, we get $(A + \epsilon)w \in \mathcal{R}[a, \infty)$ (cf. the remark following Lemma E.27) and, by Corollary E.28, $u \in \mathcal{R}[a, \infty)$.

Similarly, if $u \in \mathcal{R}[a, \infty)$, we have $u/(A - \epsilon) \in \mathcal{R}[a, \infty)$ and, again by E.28, $w \in \mathcal{R}[a, \infty)$. ∎

We will finally study the interplay of series and improper integrals.

E.33 Theorem *Let $a = b_0 < b_1 < b_2 < \ldots$ be a strictly increasing sequence with $b_k \to \infty$.*

(i) *If $u \in \mathcal{R}[a, \infty)$, then $\sum_{k=1}^{\infty} \int_{b_{k-1}}^{b_k} u \, dt$ converges.*

(ii) *If $u \geqslant 0$ and $u \in \mathcal{R}[b_{k-1}, b_k]$ for all $k \in \mathbb{N}$, then the convergence of $\sum_{k=1}^{\infty} \int_{b_{k-1}}^{b_k} u \, dt$ implies $u \in \mathcal{R}[a, \infty)$.*

Proof (i): Since $u \in \mathcal{R}[a, \infty)$,

$$\int_a^{\infty} u \, dt = \lim_{n \to \infty} \int_a^{b_n} u \, dt = \lim_{n \to \infty} \sum_{k=1}^{n} \int_{b_{k-1}}^{b_k} u \, dt = \sum_{k=1}^{\infty} \int_{b_{k-1}}^{b_k} u \, dt.$$

(ii): Define $S := \sum_{k=1}^{\infty} \int_{b_{k-1}}^{b_k} u \, dt$. Since b_k increases to ∞, we find for all $b > a$ some $N \in \mathbb{N}$ such that $b_N > b$. Consequently,

$$\int_a^{b} u \, dt \leqslant \int_a^{b_N} u \, dt = \sum_{k=1}^{N} \int_{b_{k-1}}^{b_k} u \, dt \leqslant S$$

which shows that the limit $\lim_{b \to \infty} \int_a^b u \, dt = \sup_{b>0} \int_a^b u \, dt \leqslant S$ exists. ∎

E.34 Theorem (Integral test for series) *Let $u \in C[0, \infty)$, $u \geqslant 0$, be a decreasing function. Then*

$$\int_0^{\infty} u \, dt \qquad and \qquad \sum_{k=0}^{\infty} u(k)$$

either both converge or diverge.

Proof Note that by Theorem E.8 $u \in \mathcal{R}[0, b]$ for all $b > 0$, so that the improper integral can be defined. Since u is decreasing,

$$u(k+1) \leqslant \int_k^{k+1} u(t) \, dt \leqslant u(k),$$

cf. Theorem E.11, and summing these inequalities over $k = 0, 1, \ldots, N$ yields

$$\sum_{k=1}^{N+1} u(k) = \sum_{k=0}^{N} u(k+1) \leqslant \int_0^{N+1} u(t) \, dt \leqslant \sum_{k=0}^{N} u(k).$$

Since u is positive and since the series has only positive terms, it is obvious that $\int_0^{\infty} u \, dt$ converges if, and only if, the series $\sum_{k=0}^{\infty} u(k)$ is finite. ∎

B. Improper Riemann integrals with unbounded integrands

E.35 Definition If $u \in \mathcal{R}[a, c]$ [resp. $u \in \mathcal{R}[c, b]$] for all $c \in (a, b)$ and if the limit

$$\lim_{c \uparrow b} \int_a^c u \, dt \qquad \left[\text{resp. } \lim_{c \downarrow a} \int_c^b u \, dt \right]$$

exists and is finite, we call u *improperly Riemann integrable* and write $u \in \mathcal{R}[a, b)$ [resp. $u \in \mathcal{R}(a, b]$]. The value of the limit is called the *(improper Riemann) integral* and denoted by $\int_a^b u \, dt$.

Notice that the function u in E.35 need not be bounded in (a, b). If it is, the improper integral coincides with the ordinary Riemann integral.

E.36 Lemma *If the function $u \in \mathcal{R}[a, b)$ [or $u \in \mathcal{R}(a, b]$] has an extension to $[a, b]$ which is bounded, then the extension is Riemann integrable over $[a, b]$, and proper and improper Riemann integrals coincide.*

Proof We consider only $[a, b)$, since the other case is similar. Denote, for notational simplicity, the extension of u again by u.

Let $M := \sup u([a, b])$, fix $\epsilon > 0$ and pick $c < b$ with $b - c \leqslant \frac{\epsilon}{M}$. Since $u \in \mathcal{R}[a, c]$, we can find a partition π of $[a, c]$ such that $S^\pi[u] - S_\pi[u] \leqslant \epsilon$. For the partition $\pi' := \pi \cup \{b\}$ of $[a, b]$ we get

$$S^{\pi'}[u] - S^\pi[u] = \sup u([c, b]) \frac{\epsilon}{M} \leqslant M \frac{\epsilon}{M} = \epsilon$$

and

$$S_{\pi'}[u] - S_\pi[u] = \inf u([c, b]) \frac{\epsilon}{M} \leqslant M \frac{\epsilon}{M} = \epsilon,$$

which implies that $S^{\pi'}[u] - S_{\pi'}[u] \leqslant 3\epsilon$ and $u \in \mathcal{R}[a, b]$ by Theorem E.5. The claim now follows from Lemma E.14. ∎

Many of the results for improper integrals of the form $\int_a^\infty u \, dt$ resp. $\int_{-\infty}^b u \, dt$ carry over with minor notational changes to the case of half-open bounded intervals. Note, however, that in the convergence theorems some assertions involving uniform convergence are senseless in the presence of unbounded integrands. We leave the details to the reader.

The *typical examples* of improper integrals of this kind are expressions of the type $\int_0^1 t^\lambda \, dt$ if $\lambda < 0$. In fact, if $\lambda \neq -1$,

$$\int_0^1 t^\lambda = \lim_{\epsilon \to 0} \int_\epsilon^1 t^\lambda \, dt = \lim_{\epsilon \to 0} \frac{1}{\lambda + 1}(1 - \epsilon^{\lambda+1}) = \begin{cases} \frac{1}{\lambda+1} & \text{if } \lambda > -1, \\ \infty & \text{if } \lambda < -1, \end{cases}$$

and a similar calculation confirms that $\int_0^1 t^{-1}\,dt = \infty$. Thus $t^\lambda \in \mathcal{R}(0, 1]$ if, and only if, $\lambda > -1$.

C. Improper Riemann integrals where both limits are critical

Assume now that the integration interval is (a, b) and that both endpoints a and b, $-\infty \leqslant a < b \leqslant +\infty$, are critical, i.e. that the integrand is unbounded at one or both endpoints and/or that one or both endpoints are infinite.

Let $u \in \mathcal{R}(a, c] \cap \mathcal{R}[c, b)$ for some point $a < c < b$ and suppose that d satisfies $c < d < b$. By the remark following Lemma E.27 and Theorem E.13 we find

$$\int_a^c u\,dt + \int_c^b u\,dt = \lim_{x \downarrow a} \int_x^c u\,dt + \lim_{y \uparrow b} \int_c^y u\,dt$$

$$= \lim_{x \downarrow a} \int_x^c u\,dt + \int_c^d u\,dt + \lim_{y \uparrow b} \int_d^y u\,dt$$

$$= \lim_{x \downarrow a} \int_x^d u\,dt + \lim_{y \uparrow b} \int_d^y u\,dt$$

$$= \int_a^d u\,dt + \int_d^b u\,dt,$$

which shows that $u \in \mathcal{R}(a, d] \cap \mathcal{R}[d, b)$. Therefore, the following definition makes sense.

E.37 Definition Let $-\infty \leqslant a < b \leqslant +\infty$ and let $(a, b) \subset \mathbb{R}$ be a bounded or unbounded open interval. Then $u : (a, b) \to \mathbb{R}$ is said to be *improperly integrable* if for some (hence, all) $c \in (a, b)$ the function u is improperly integrable both over $(a, c]$ and $[c, b)$, i.e. we define $\mathcal{R}(a, b) := \mathcal{R}(a, c] \cap \mathcal{R}[c, b)$. The *(improper Riemann) integral* is then given by

$$\int_a^b u\,dt := \int_a^c u\,dt + \int_c^b u\,dt = \lim_{x \downarrow a} \int_x^c u\,dt + \lim_{y \uparrow b} \int_c^y u\,dt.$$

The *typical example* of an improper integral of this kind is Euler's Gamma function

$$\Gamma(x) := \int_0^\infty t^{x-1} e^{-t}\,dt, \qquad x > 0,$$

which is treated in Example 10.14 in the framework of Lebesgue theory, but the arguments are essentially similar. The Gamma function is only for $0 < x < 1$ a two-sided improper integral, since for $x \geqslant 1$ it can be interpreted as a one-sided improper integral over $[0, \infty)$, cf. Lemma E.36.

Further reading

Measure theory is used in many mathematical disciplines. A few of them we have touched in this book and the purpose of this section is to point towards literature which treats these subjects in depth. The choice of books and topics is certainly not comprehensive. On the contrary, it is very personal, limited by my knowledge of the literature and, of course, my own mathematical taste. I decided to include only books in English and which I thought are accessible to readers of the present text.

Real analysis (in particular measure and integration theory for analysts)

Bass, R. F., *Probabilistic Techniques in Analysis*, New York: Springer 1995.

Dudley, R. M., *Real Analysis and Probability* (2nd edn), Cambridge: Cambridge University Press, Studies in Adv. Math. vol. **74**, 2002.

Hewitt, E. and K. R. Stromberg, *Real and Abstract Analysis*, New York: Springer, Grad. Texts in Math. vol. **25**, 1975.

Kolmogorov, A. N. and F. V. Fomin, *Introductory Real Analysis*, Mineola (NY): Dover, 1975.

Lieb, E. H. and M. Loss, *Analysis* (2nd edn), Am. Mathematical Society, Grad. Studies in Math. vol. **14**, Providence (RI) 2001.

Rudin, W., *Real and Complex Analysis* (3rd edn), McGraw-Hill, New York 1987.

Saks, S., *Theory of the Integral* (2nd revised edn), Hafner, Mongrafie Matematyczne Tom **VII**, New York 1937. [Reprinted by Dover, 1964. Free online edition in the *Wirtualna Biblioteka Nauki:* http://matwbn.icm.edu.pl/kstresc.php?tom=7&wyd=10]

Stroock, D., *A Concise Introduction to the Theory of Integration* (3rd edn), Birkhäuser, Boston 1999.

Sz.-Nagy, B., *Introduction to Real Functions and Orthogonal Expansions*, Oxford University Press, Univ. Texts in the Math. Sci., New York 1965.

Wheeden, R. L. and A. Zygmund, *Measure and Integral. An Introduction to Real Analysis*, Marcel Dekker, Pure Appl. Math. vol. **43**, New York 1977.

Functional analysis

Bollobas, B., *Linear Analysis. An Introductory Course* (2nd edn), Cambridge University Press, Cambridge 1999.

Hirsch, F. and G. Lacombe, *Elements of Functional Analysis*, Springer, Grad. Texts in Math. vol. **192**, New York 1999.

Kolmogorov, A. N. and F. V. Fomin, *Introductory Real Analysis*, Mineola (NY): Dover, 1975.

Yosida, K., *Functional Analysis* (6th edn), Springer, Grundlehren math. Wiss. Bd. **123**, Berlin 1980.

Zaanen, A. C., *Integration* (completely revised edn. of *An Introduction to the Theory of Integration*), North-Holland, Amsterdam 1967.

Fourier series, harmonic analysis, orthonormal systems, wavelets

Alexits, G., *Convergence Problems of Orthogonal Series*, Pergamon, Int. Ser. Monogr. Pure Appl. Math. vol. **20**, Oxford 1961.

Andrews, G. E., Askey, R. and R. Roy, *Special Functions*, Cambridge University Press, Encycl. Math. Appl. vol. **71**, Cambridge 1999.

Garsia, A. M., *Topics in Almost Everywhere Convergence*, Markham, Chicago 1970.

Helson, H., *Harmonic Analysis*, Addison-Wesley, London, 1983.

Kahane, J.-P., *Some Random Series of Functions* (2nd edn), Cambridge University Press, Stud. Adv. Math. vol. **5**, Cambridge 1985.

Krantz, S. G., *A Panorama of Harmonic Analysis*, Mathematical Association of America, Carus Math. Monogr. vol. **27**, Washington 1999.

Pinsky, M. A., *Introduction to Fourier Analysis and Wavelets*, Brooks/Cole, Ser. Adv. Math., Pacific Grove (CA) 2002.

Schipp, F., Wade, W. R. and P. Simon, *Walsh Series. An Introduction to Dyadic Harmonic Analysis*, Adam Hilger, Bristol 1990.

Stein, E. M., *Singular Integrals and Differentiability Properties of Functions*, Princeton University Press, Math. Ser. vol. **30**, Princeton (NJ) 1970.

Stein, E. M. and R. Shakarchi, *Fourier Analysis: An Introduction*, Princeton University Press, Princeton (NJ) 2003.

Sz.-Nagy, B., *Introduction to Real Functions and Orthogonal Expansions*, Oxford University Press, Univ. Texts in the Math. Sci., New York 1965.

Wojtaszczyk, P., *A Mathematical Introduction to Wavelets*, Cambridge University Press, London Math. Society Student Texts vol. **37**, Cambridge 1997.

Zygmund, A., *Trigonometric Series* (2nd edn), Cambridge University Press, Cambridge 1959. [Almost unaltered softcover editions: Cambridge: Cambridge University Press, 1969, 1988 and 2003.]

Geometric measure theory, Hausdorff measure, fine properties of functions

Evans, L. C. and R. F. Gariepy, *Measure Theory and Fine Properties of Functions*, CRC Press, Boca Raton (FL) 1992.

Mattila, P., *Geometry of Sets and Measures in Euclidean Spaces*, Cambridge University Press, Studies in Adv. Math. vol. **44**, Cambridge 1995.

Morgan, F., *Geometric Measure Theory: A Beginner's Guide* (3rd edn), Academic Press, San Diego, 2000.

Rogers, C. A., *Hausdorff Measures*, Cambridge University Press, Cambridge Math. Library, Cambridge 1970.

Ziemer, W. P., *Weakly Differentiable Functions*, Springer, Grad. Texts in Math. vol. **120**, New York 1989.

Topological measure theory, functional analytic aspects of integration and measure

Bauer, H., *Measure and Integration Theory*, de Gruyter, Studies in Math. vol. **26**, Berlin 2001.

Choquet, G., *Lectures on Analysis*. vol. 1: Integration and Topological Vector Spaces, W. A. Benjamin, New York 1969.

Dieudonné, J., *Treatise on Analysis*, vol. II, Academic Press, Pure Appl. Math. vol. **10-II**, New York 1969.

Hewitt, E. and K. A. Ross, *Abstract Harmonic Analysis*, vol. 1, Springer, Grundlehren math. Wiss. Bd. **115**, Berlin 1963.

Malliavin, P., *Integration and Probability*, Springer, Grad. Texts in Math. **157**, New York 1995.

Oxtoby, J. C., *Measure and Category* (2nd edn), Springer, Grad. Texts Math. vol. **2**, New York 1980.

Weir, A. J., *General Integration and Measure*, Cambridge University Press, Cambridge 1974.

Borel and analytic sets

Rogers, C. A. *et al.*, *Analytic Sets*, Academic Press, London 1980.

Srivastava, S. M., *A Course on Borel Sets*, Springer, Grad. Texts Math. vol. **180**, New York 1998.

Probability theory (in particular probabilistic measure theory)

Ash, R. B. and C. A. Doléans-Dade, *Probability and Measure Theory* (2nd edn), Academic Press, San Diego (CA) 2000.

Billingsley, P., *Probability and Measure* (3rd edn), Wiley, Ser. Probab. Math. Stat., New York 1995.

Chow, Y. S. and H. Teicher, *Probability Theory. Independence, Interchangeability, Martingales* (3rd edn), Springer, Texts in Stat., New York 1997.

Durrett, R., *Probability: Theory and Examples* (3rd edn), Thomson Brooks/Cole, Duxbury Adv. Studies, Belmont (CA) 2004.

Kallenberg, O., *Foundations of Modern Probability*, Springer, New York 2001.

Malliavin, P., *Integration and Probability*, Springer, Grad. Texts in Math. **157**, New York 1995.

Neveu, J., *Mathematical Foundations of the Calculus of Probability*, Holden Day, San Francisco (CA) 1965.

Stromberg, K., *Probability for Analysts*, Chapman and Hall, Probab. Ser., New York 1994.

Martingales and their applications

Ash, R. B. and C. A. Doléans-Dade, *Probability and Measure Theory* (2nd edn), Academic Press, San Diego (CA) 2000.

Chow, Y. S. and H. Teicher, *Probability Theory. Independence, Interchangeability, Martingales* (3rd edn), Springer, Texts in Stat., New York 1997.

Dellacherie, C. and P. A. Meyer, *Probabilities and Potential Pt. B: Theory of Martingales*, North Holland, Math. Studies, Amsterdam 1982. [Note that *Probabilities and Potential Pt. A*, Amsterdam 1979, by the same authors is a prerequisite for this text.]

Garsia, A. M., *Topics in Almost Everywhere Convergence*, Markham, Chicago 1970.

Meyer, P. A., *Probabilities and Potentials*, Blaisdell, London 1966.

Neveu, J., *Discrete-parameter Martingales*, North Holland, Math. Libr. vol. **10**, Amsterdam 1975.

Rogers, L. C. G. and D. Williams, *Diffusions, Markov Processes and Martingales* (2 vols., 2nd edn), Cambridge Math. Library, Cambridge 2000.

References

[1] Alexits, G., *Convergence Problems of Orthogonal Series*, Oxford: Pergamon, Int. Ser. Monogr. Pure Appl. Math. vol. **20**, 1961.

[2] Andrews, G. E., Askey, R. and R. Roy, *Special Functions*, Cambridge: Cambridge University Press, Encycl. Math. Appl. vol. **71**, 1999.

[3] Bass, R. F., *Probabilistic Techniques in Analysis*, New York: Springer, 1995.

[4] Bauer, H., Approximation and abstract boundaries, *Am. Math. Monthly* **85** (1978), 632–647. Also in: H. Bauer, *Selecta*, Berlin: de Gruyter, 2003, 436–451.

[5] Bauer, H., *Probability Theory*, Berlin: de Gruyter, Studies in Math. vol. **23**, 1996.

[6] Bauer, H., *Measure and Integration Theory*, Berlin: de Gruyter, Studies in Math. vol. **26**, 2001.

[7] Benyamini, Y. and J. Lindenstrauss, *Geometric Nonlinear Functional Analysis*, vol. 1, Providence (RI): Am. Math. Soc., Coll. Publ. vol. **48**, 2000.

[8] Boas, R. P., *A Primer of Real Functions,* Math. Association of America, Carus Math. Monogr. vol. **13**, 1960.

[9] Carathéodory, C., Über das lineare Maß von Punktmengen – eine Verallgemeinerung des Längenbegriffs, *Nachr. Kgl. Ges. Wiss. Göttingen Math.-Phys. Kl.* (1914), 404–426. Also in: C. Carathéodory, *Gesammelte mathematische Schriften* (5 Bde.), München: C.H. Beck, 1954-57, Bd. 4, 249–275.

[10] Ciesielski, Z., Hölder condition for realizations of Gaussian processes, *Trans. Am. Math. Soc.* **99** (1961), 403–413.

[11] Diestel, J. and J.J. Uhl Jr., *Vector Measures*, Providence (RI): American Mathematical Society, Math. Surveys no. **15**, 1977.

[12] Dieudonné, J., Sur un théorème de Jessen, *Fundam. Math.* **37** (1950), 242–248. Also in: J. Dieudonné, *Choix d'œuvres mathématiques* (2 tomes), Paris: Hermann, 1981, t. 1, 369–275.

[13] Doob, J. L., *Stochastic Processes*, New York: Wiley, Ser. Probab. Math. Stat., 1953.

[14] Dudley, R. M., *Real Analysis and Probability*, Pacific Grove (CA): Wadsworth & Brooks/Cole, Math. Ser., 1989.

[15] Dunford, N. and J. T. Schwartz, *Linear Operators I*, New York: Pure Appl. Math. vol. **7**, Interscience, 1957.

[16] Garsia, A. M., *Topics in Almost Everywhere Convergence*, Chicago: Markham, 1970.

[17] Gradshteyn, I. and I. Ryzhik, *Tables of Integrals, Series, and Products* (4th corrected and enlarged edn), San Diego (CA): Academic Press, 1992.

[18] Gundy, R. F., Martingale theory and pointwise convergence of certain orthogonal series, *Trans. Am. Math. Soc.* **124** (1966), 228–248.

[19] Hausdorff, F., *Grundzüge der Mengenlehre*, Leipzig: Veit & Comp., 1914 (1st edn). Reprint of the original edn, New York: Chelsea, 1949.

[20] Hewitt, E. and K. R. Stromberg, *Real and Abstract Analysis*, New York: Springer, Grad. Texts Math. vol. **25**, 1975.

[21] Hunt, G. A., *Martingales et processus de Markov*, Paris: Dunod, Monogr. Soc. Math. France t. **1**, 1966.

[22] Kaczmarz, S. and H. Steinhaus, *Theorie der Orthogonalreihen* (2nd corr. reprint), New York: Chelsea, 1951. First edition appeared under the same title with PWN, Warsaw: Monogr. Mat. Warszawa vol. **VI**, 1935.

[23] Kahane, J.-P., *Some Random Series of Functions*, (2nd edn) Cambridge: Cambridge University Press, Stud. Adv. Math. vol. **5**, 1985.

[24] Korovkin, P. P., *Linear Operators and Approximation Theory*, Delhi: Hindustan Publ. Corp., 1960.

[25] Krantz, S. G., *A Panorama of Harmonic Analysis*, Washington: Mathematical Association of America, Carus Math. Monogr. vol. **27**, 1999.

[26] Lévy, P., *Processus stochastiques et mouvement Brownien*, Paris: Gauthier-Villars, Monographies des Probabilités Fasc. **VI**, 1948.

[27] Lindenstrauss, J. and Tzafriri, L., *Classical Banach Spaces I, II*, Berlin: Springer, Ergeb. Math. Grenzgeb. 2. Ser. Bde. **92**, **97**, 1977–79.

[28] Marcinkiewicz, J. and A. Zygmund, Sur les fonctions indépendantes, *Fundam. Math.* **29** (1937), 309–335. Also in: J. Marcinkiewicz, *Collected Papers*, Warsaw: PWN, 1964, 233–259.

[29] Métivier, M., *Semimartingales. A Course on Stochastic Processes*, Berlin: de Gruyter, Stud. Math. vol. **2**, 1982.

[30] Sz.-Nagy, B., *Introduction to Real Functions and Orthogonal Expansions*, New York: Oxford University Press, Univ. Texts in the Math. Sci., 1965.

[31] Neveu, J., *Discrete-parameter Martingales*, Amsterdam: North Holland, Math. Libr. vol. **10**, 1975. Slightly updated version of the French original: *Martingales à temps discrèt*, Paris: Masson, 1972.

[32] Olevskiĭ, A. M., *Fourier Series with Respect to General Orthogonal Systems*, Berlin: Springer, Ergeb. Math. Grenzgeb. Bd. **2**. Ser. **86**, 1975.

[33] Oxtoby, J. C., *Measure and Category*, (2nd edn), New York: Springer, Grad. Texts Math. vol. **2**, 1980.

[34] Paley, R. E. A. C. and N. Wiener, Providence (RI): *Fourier Transforms in the Complex Domain*, American Mathematical Society, Coll. Publ. vol. **19**, 1934.

[35] Pinsky, M. A., *Introduction to Fourier Analysis and Wavelets*, Pacific Grove (CA): Brooks/Cole, Ser. Adv. Math., 2002.

[36] Pratt, J. W., On interchanging limits and integrals, *Ann. Math. Stat.* **31** (1960), 74–77. [Acknowledgement of Priority, *Ann. Math. Stat.* **37** (1966), 1407.]

[37] Riemann, B., Über die Darstellbarkeit einer Function durch eine trigonometrische Reihe, *Nachr. Kgl. Ges. Wiss. Göttingen* **13** (1867), 227–271. Also in: Bernhard Riemann, *Collected Papers*, Berlin: Springer, 1990, 259–303.

[38] Rogers, L. C. G. and D. Williams, *Diffusions, Markov Processes and Martingales* (2 vols., 2nd edn), Cambridge: Cambridge Mathematical Library, 2000.

[39] Rudin, W., *Principles of Mathematical Analysis* (3rd edn), New York: McGraw-Hill, 1976.

[40] Rudin, W., *Real and Complex Analysis* (3rd edn), New York: McGraw-Hill, 1987.

[41] Schipp, F., Wade, W. R. and P. Simon, *Walsh Series. An Introduction to Dyadic Harmonic Analysis*, Bristol: Adam Hilger, 1990.

[42] Souslin, M. Y., Sur une définition des ensembles mesurables *B* sans nombres transfinis, *C. R. Acad. Sci. Paris* **164** (1917), 88–91.

[43] Srivastava, S. M., *A Course on Borel Sets*, New York: Springer, Grad. Texts Math. vol. **180**, 1998.

[44] Solovay, R. M., A model of set theory in which every set of reals is Lebesgue measurable, *Ann. Math.* **92** (1970), 1–56.

[45] Steele, J. M., *Stochastic Calculus and Financial Applications*, New York: Springer, Appl. Math. vol. **45**, 2000.

[46] Steen, L. A. and J. A. Seebach, *Counterexamples in Topology*, New York: Dover, 1995.

[47] Stein, E. M., *Singular Integrals and Differentiability Properties of Functions*, Princeton (NJ): Princeton University Press, Math. Ser. vol. **30**, 1970.

[48] Strichartz, R. S., *The Way of Analysis* (rev. edn), Sudbury (MA): Jones and Bartlett, 2000.

[49] Stromberg, K., The Banach–Tarski paradox, *Am. Math. Monthly* **86** (1979), 151–161.

[50] Stroock, D. W., *A Concise Introduction to the Theory of Integration* (3rd edn), Boston: Birkhäuser, 1999.

[51] Szegö, G., *Orthogonal Polynomials*, Providence (RI): Am. Math. Soc., Coll. Publ. vol. **23**, 1939.

[52] Wagon, S., *The Banach–Tarski Paradox*, Cambridge: Cambridge University Press, Encycl. Math. Appl. vol. **24**, 1985.

[53] Wheeden, R. L. and A. Zygmund, *Measure and Integral. An Introduction to Real Analysis*, New York: Marcel Dekker, Pure Appl. Math. vol. **43**, 1977.

[54] Willard, S., *General Topology*, Reading (MA): Addison-Wesley, 1970.

[55] Yosida, K., *Functional Analysis* (6th edn), Berlin: Springer, Grundlehren Math. Wiss. Bd. **123**, 1980.

[56] Young, W. H., On semi-integrals and oscillating successions of functions, *Proc. London Math. Soc. (2)* **9** (1910/11), 286–324.

Notation index

This is intended to aid cross-referencing, so notation that is specific to a single section is generally not listed. Some symbols are used locally, without ambiguity, in senses other than those given below. Numbers following entries are page numbers with the occasional (Pr $m.n$) referring to Problem $m.n$ on the respective page.

Unless otherwise stated, binary operations between functions such as $f \pm g$, $f \cdot g$, $f \wedge g$, $f \vee g$, comparisons $f \leqslant g$, $f < g$ or limiting relations $f_j \xrightarrow{j \to \infty} f$, $\lim_j f_j$, $\liminf_j f_j$, $\limsup_j f_j$, $\sup_i f_i$ or $\inf_i f_i$ are always understood pointwise.

Alternatives are indicated by square brackets, i.e., '*if A [B] ... then P [Q]*' should be read as '*if A ... then P*' and '*if B ... then Q*'.

Abbreviations and shorthand notation

a.a.	almost all, 80	$\cup$-stable	stable under finite unions
a.e.	almost every(where), 80	$\cap$-stable	stable under finite
ONB	orthonormal basis, 239		intersections, 32
ONS	orthonormal system, 239	∎	end of proof, x
UI	uniformly integrable, 163, 194	[✓]	indicates that a small intermediate step is
w.r.t.	with respect to		required, x
negative	always in the sense $\leqslant 0$	⇃	(in the margin) caution, x
positive	always in the sense $\geqslant 0$		

Special labels, defining properties

$(\Delta_1), (\Delta_2), (\Delta_3)$	Dynkin system, 31	$(S_1), (S_2), (S_3)$	semi-ring, 37
$(M_1), (M_2)$	measure, 22	$(\Sigma_1), (\Sigma_2), (\Sigma_3)$	σ-algebra, 15

Mathematical symbols

Sub- and superscripts

		$\perp$	orthogonal complement, 235
$+$	positive part,	b	bounded
	positive elements	c	compact support

Name and subject index

This should be used in conjunction with the Bibliography and the Index of Notation. Numbers following entries are page numbers which, if accomplished by (Pr *n.m*), refer to Problem *n.m* on that page; a number with a trailing 'n' indicates that a footnote is being referenced. Unless otherwise started 'integral', integrability' etc. always refer to the (abstract) Lebesgue integral. Within the index we use 'L-...' and 'R-...' as a shorthand for '(abstract) Lebesgue-...' and 'Riemann-...'